Independent Birth of Organisms

INDEPENDENT BIRTH OF ORGANISMS

A New Theory That Distinct Organisms Arose
Independently From the Primordial Pond,
Showing That Evolutionary Theories
Are Fundamentally Incorrect

PERIANNAN SENAPATHY, Ph.D.

Genome Press

MADISON

Copyright © 1994 by Periannan Senapathy.
All rights reserved.
Printed in the United States of America
on 50% recycled paper.

No part of this book may be reproduced in any form by any electronic or mechanical means (including photocopying, recording, or information storage and retrieval) without permission in writing from the publisher.

Library of Congress Catalog Card Number: 94-096112

ISBN 0-9641304-0-8

Genome Press
579 D'Onofrio Drive
Madison, WI 53719

e-mail: Sena@Genome.com

to my wife Roopa

and my daughters Kavin and Sudar

TABLE OF CONTENTS

CHAPTER 1
Introduction   1

CHAPTER 2
Darwin's Theory:
Natural Selection and Evolution   13

CHAPTER 3
The Genome of Every Distinct Organism is
Closed to Evolutionary Changes   31

CHAPTER 4
Genetic Mutation and Rearrangement
Cannot Cause Organismal Evolution   103

Chapter 5
A Prelude to the New Theory   199

Chapter 6
The Primordial Pond: Universal Sequence Pool and Universal Gene Pool   205

Chapter 7
The Abundant Occurrence of Genes is Inevitable in the Primordial Pond   221

Chapter 8
Independent Birth of Multitudes of Organisms from the Primordial Pond   293

Chapter 9
The Molecular Scenario of Life: Evidence for the New Theory   375

Chapter 10
Most Organisms are Distinct and Evolutionarily Unrelatable   453

Chapter 11
A New Look at the Fossil Record   491

Chapter 12
Conclusion   521

Appendix: Genetics Primer   535

Notes and References   567

Index   629

Preface

I have long been an admirer of Charles Darwin and his theory of evolution, and like many other scientists I can attribute much of my original fascination with biology to the obvious intellectual and even aesthetic appeal of evolution theory. But, also like many of my peers, I have long been troubled by seemingly contradictory evidence and "missing links" in the fossil record. Basic evolution theory has remained on the table for well over a century, but despite its longevity and enormous popularity it has never been proven — nor disproven — and most of the problems that have dogged evolution since Darwin persist to this day.

A generation ago evolution theory easily survived Watson and Crick's extraordinary revelations of DNA structure and function, and subsequent researchers' characterizations of genetic mechanisms at the molecular level. Indeed the mere fact that this new knowledge did not contradict evolution probably enhanced the theory's credibility more than it deserved. These insights, after all, do not prove evolution any more than they debunk it; they simply coexist with the still unproven theory. But evolution theory will not be so lucky in the wake of contemporary genome research.

In 1981, while researching molecular evolution at the National Institutes of Health, my findings led me to ponder an alternative explanation for the diversity of life on earth. All evolution theories hang on the premise that only a single primitive organism emerged from the primordial pond. But that critical premise rests on a mathematical assessment of the chances that any organism could have emerged from the random chemical reactions in the primordial pond. Statistically, say the evolutionists, the random assembly of primordial components into even a single primitive genome would have been virtually impossible, so the origin of the first organism can only be attributed to some fantastically improbable accident. But what if that underlying premise is false? What if the primordial pond actually produced an abundant assortment of genes?

If so, then the genomes for numerous organisms might have been assembled independently, and more or less simultaneously, within the primordial pond. If so, the primordial pond would have spawned multitudes of genetically and morphologically distinct life forms, instead of one or just a few. And if so, then virtually all of the evidence that has troubled evolutionists since Darwin — the gaps in the fossil record, the "Cambrian explosion," the complexity of advanced organ systems, all of it — would be easily accommodated by this new explanation of the births of organisms.

Since 1981, I have focused most of my research on this question, and on the questions that arise from its implications. My investigations have benefited immeasurably from the dramatic advances in computers and DNA research over the past decade, which have armed researchers with powerful new analytical tools to solve genome riddles. Modern computing power and our ever-expanding genomic knowledge have also made possible sophisticated statistical characterizations of competing scenarios for the random assembly of primordial components into genes, and for the assembly of genes into viable genomes.

Virtually all of my research points to the same body of dramatic conclusions: that the fundamental premise of all evolution theories is fundamentally flawed, that only slight variations among essentially similar species can be attributed to the mechanisms of evolution, and that the overwhelming majority of earth's biodiversity must have been born in the primordial pond.

This book is the story of my twelve-year journey toward a new theory of the births of organisms. While it is a science book and does provide technical details, I believe that the story of the journey carries the day, and that most educated readers — not just other scientists — will find it engaging. In this book, as in the universe, the scientific details are of secondary importance to the splendors of their significance.

With all of the many layers of detail in a book like this, it is probably inevitable that a few errors will survive the draft reviews and find their way into print. Certainly I accept responsibility for any such errors, but I also remain confident that they will prove to be incidental, and will pose no significant threat to the book's critical assertions.

Acknowledgments

I am deeply grateful to several people who have helped me to capture this story on paper. Dr. Ganesan Ramalingam, now a scientist at the IBM Watson Research Center, wrote many of the complex computer programs for the studies described in the book while still a graduate student at the University of Wisconsin. My sister, Dr. Narmadha Kuppuswami, read the drafts many times and offered many valuable suggestions for improvements — all while maintaining her own busy obstetrics/gynecology practice. The book's readability owes much to the skill of my editor, Jeremy Kohler. Dr. Polani Seshagiri, a scientist in molecular embryology, helped verify important embryological details of portions of this narrative. I am also grateful to Mr. Carl Curtis for assisting with editorial matters in the early drafts, to Mr. Bob Morse for editorial insights and advice in the later drafts, and to Ms. Ann Teuke for her meticulous prepress production work.

Finally, I owe a special thanks to my wife Roopa, not only for her valued literary advice in the course of several passes through the drafts, but also — and most importantly — for cheerfully enduring the hardships that inevitably accompany such an undertaking, and for being a source of constant encouragement.

1

INTRODUCTION

Where did we come from? It seems such a simple question to carry such a monumental significance, and to have kept us stumped throughout human history. And yet after all these millennia the question remains unanswered — at least definitively. Despite the best efforts of generations of capable scientists, the origin of life on earth remains as great a mystery today as it was at the dawn of human curiosity. Much of our fascination with life on earth surely derives from its diversity: a seemingly endless variety of fantastic creatures, each playing an often mysterious part in a complex global interplay of survival. Life is drama, intrigue, splendor, grace, whimsy, passion — and so much more. Life is beautiful, and it's only natural for us to want to know where it came from.

Many theologians, secular philosophers and poets — and even many scientists — have been content to answer the question abstractly, in terms of a divine spark that somehow transformed inanimate matter into living organisms. But such explanations tell us only "who" — not "how." Science is obliged to characterize the "spark" and the "somehow," whether or not the process can be attributed to divine intervention.

Thanks to this past century's breathtaking advances in science and technology, we seem to have arrived at last at the outskirts of a definitive answer. Today we find ourselves pondering a mountain of scientific information extracted from the living world, and with it we can uncover new clues to the origin of life on earth. Among the most significant of our observations is the fact that while there are many similar species that are essentially the variants of a single organism, there are numerous organisms that are unique and distinct. Many similar species of snails, for example, occur as slight variations of the same basic snail organism, while snails as a group are unique and absolutely distinct from other groups such as crabs. Most organisms exhibit unique body parts — distinct kinds of mouths, eyes, digestive systems, sensory organs, and other appendages — that are well suited to their organisms' particular lives and environments. This complex scenario of wildly different organisms, from dragonflies and crabs to dinosaurs and blue whales, each uniquely suited to its environment, suggests strongly that some unifying principles must have governed their origins.

The search for these principles is among the most intriguing challenges ever to confront the human mind, probably because its denouement will explain who we are and where we came from. Any scientific answer to these questions should invoke plausible and logical mechanisms to explain the whole scenario of all life on earth,[1] in every detail — its origin, certainly, but also its history and future.

Darwin's theory of evolution, published in his book *Origin of Species* in 1859,[2] has been the most accepted theory for the origin and diversity of species on earth for over a century. It has been one of the greatest concepts in biological science and has been the most convincing of all explanations given so far on the origin and diversity of creatures. Intense research in various disciplines has appeared to support this theory. Its basic components, natural selection and adaptation, seemed to finely explain not only the origin of species, but also the perfect fit of organisms to their environments and their relationships to each other.

Despite the fact that Darwin's theory has been one of the most convincing explanations, it should be noted that there have been monumental problems with observed facts that go against the theory. In fact Darwin's theory has never been proven in the past 130 years — although some scientists assume that it is well established based on some genetic and organismal similarities. There are many major scenarios of life on earth that are unexplainable by evolutionary theory. For instance, according to the fossil record, multitudes of unique creatures abruptly originated in a simultaneous burst when multicellular life first originated on earth. This cannot be explained based on evolutionary theory. The second problem is about classifying or grouping organisms based on assumed evolutionary connections among them. Organisms are classified into sets of similar organisms first. These groups are further arranged in a nested manner based on assumed evolutionary connections — many species into one genus, many genera into one family, many families into one order and so on. However, the larger groupings, or the "higher taxa," are found to be unconnectable by evolution. If creatures on earth had all originated by evolution from one original ancestral creature as the evolutionary theory states, then the higher taxa should be connectable by evolution. The third problem concerns the evolution of highly complex organs such as the eye. Many evolutionists themselves agree that it is far too difficult to evolve highly complex organs by evolutionary means.

Even with these crucial problems, Darwin's theory works for some aspects of life's scenario making one marvel at the beauty of the theory. There are scientists who completely believe in Darwin's theory, there are those who are almost completely against the theory, and there are those who are in between. The presence of physical and genetic similarities among sets of organisms has been considered in recent times to be very supportive of Darwin's theory of descent with modification. Thus, even those who very well understand the crucial problems existing for the theory are either ridiculed by those who strongly believe in the theory or ignored, for there is no other alternative scientific theory to the theory of evolution. People believe that even though Darwin's theory of natural selection is able to explain the origin of creatures to some limited extent, somehow all organisms should have evolved from one original creature on earth by some yet undiscovered evolutionary mechanisms. This is essentially the story of the theory of evolution in our society today.

As a molecular biologist doing research on the structure and function of genes in the late 1970s and early 80s, I strongly believed in Darwin's theory of evolution. The principles of natural selection and adaptation were both appealing and convincing to me. I believed that evolution of organisms had occurred from one or a few original ancestral creatures, and I never doubted the validity of Darwin's theory. However, the number of problems unsolved by the theory, such as the "missing links" between supposedly related organisms, were puzzling to me. Furthermore, there are many questions concerning the origin of life itself still unanswered by scientific research.

Darwin's theory states that all organisms evolved ultimately from one or a few original organisms,[3] and gives a mechanism for the change of one organism into another. Darwin did not offer an explanation as to how the first organism could have originated from inanimate matter. He simply stated that life had been "originally breathed into a few forms or into one," and that from so simple a beginning, endless forms have been and are being evolved.[4] So far there has been no convincing explanation concerning the origin of the first one or a few organisms from inanimate matter, except for some nebulous and vague speculations.

This prompted me to become very interested in the question of how life itself had originated. I took a molecular biology approach, studying sequences of DNA and proteins, to find out how the genes of the first creature could have originated. Based on already established facts about chemical evolution and simulation experiments, I was convinced that the genetic (DNA) sequences in the primordial pond on earth (the pond where life is supposed to have originated) must have been random, and that they must have coded for the proteins of the first cells. With this in mind, I intensely studied the properties of random genetic sequences to understand how they might have coded for proteins. While studying this problem, it suddenly occurred to me that genes could have been abundantly available in the primordial pond, and genomes (collections of genes) of different organisms could have been assembled independently from this common pool of genes. It immediately showed that numerous creatures could have been born independently from the primordial pond without evolutionary connection. If so, I realized that this concept could solve the many problems unresolved by Darwin's theory. I began investigating DNA and proteins, using the computer to simulate random sequences, to prove that genes could in fact occur in the primordial pond. Although this took me a number of years, my predictions turned out to be true.

Molecular geneticists and evolutionists strongly believe that genes could not occur in a primordial pond by chance, based on a simple probabilistic approach. Their argument is that even a small specific gene with a particular sequence cannot occur purely by chance on earth, because the total amount of random primordial genetic material required for this chance occurrence will have a mass that is far larger than the mass of the whole universe. In contrast, I discovered that this approach to understanding the origin of genes is incorrect. I took a different, more systematic approach to this question, which led to the demonstration that multitudes of genes could in fact abundantly occur in the primordial pond. Once this crucial principle is established, I could see, based on the biochemical richness and complexity of the primordial pond, that there is no difficulty in proving the assembly of these genes into genomes of multicellular organisms. With this background it was not too difficult to see that the probability of assembling the genomes of multicellular organisms is not too much different compared to that of assembling the genomes of simple, one-celled organisms — because it was not difficult to see from recent research on the structure of genes that there was little difference in complexity between the genomes of the typical single-celled eukaryotic microbes (cells with a nucleus) and the typical multicellular animal. Furthermore, I was able to demonstrate that the probabilities of assembling the genomes for different multicellular organisms, whether anatomically and physically simple or complex, are not much different from one another either — again because there does not seem to be much difference in complexity among their genomes.

This enabled me to propose an entirely new theory on the origin of diverse creatures on earth, without involving organismal evolution at all. The new theory is that each of the numerous unique creatures originated independently in the primordial pond directly from its genome assembled in a "seed cell." The basic principle is that if genes were abundantly available in the primordial pond, they could have randomly assembled to form various genomes, each capable of forming an organism. Although only one out of a large number of genomes could form a viable organism, there could have been myriad permutations and combinations of genes leading to the formation of genomes for millions of viable creatures. The genomes were directly assembled into single *seed cells*, analogous to the fertilized eggs of sexually reproducing organisms, from which the development of the individual starts. Based on the analysis of gene structure, I could see that the first cells on earth must have been unicellular eukaryotes (cells with a nucleus) — not simpler bacteria as has been traditionally thought.

When I evaluated the scenario of life on earth based on the new theory of the independent birth of organisms, I saw that it was capable of explaining the origin of diverse creatures convincingly without any of the problems that Darwin's theory faced for over a century. In fact the new theory provided clear explanations for all details of life at different levels — molecules, organisms, and fossils. It provided a coherent explanation for the physical similarities within a set of similar creatures, as well as the distinct differences among the distinct sets of organisms. It could demonstrate that gaps between unique creatures are real and the assumed missing links between such creatures are really nonexistent. The so far enigmatic sudden explosion of creatures at the very start of multicellular life as shown in the fossil record (the Cambrian explosion) as well as the sudden appearance of new creatures in later geological periods, are all explained by the new theory.

Once I formulated the new theory, which could well explain the scenario of life on earth without involving any organismal evolutionary connections, I realized that Darwin's theory must be fundamentally incorrect in claiming that all creatures on earth have come about by organismal descent with modification from one or a few original creatures. Although Darwin's theory has been deeply rooted in scientific literature for over a century, and more so in recent times, I was convinced that it must be incorrect in its very foundation in its claim that a particular creature could evolve into another distinct or unique creature. However, it took me a great deal of thorough research into the mechanisms of genomic change to find out where exactly he went wrong. A systematic analysis of all the possible mechanisms of genomic change showed that a creature's genome cannot be changed into that of another unique creature even over any length of geological time by any mutational mechanism, as erroneously claimed by evolutionary geneticists.

My research demonstrated that the genome of every unique creature is fixed, and all its mutations lead only to slight variations, which are confined within a closed framework for every independent creature. Mutations in trivial characteristics such as body size and coat color could only lead to normal varieties that are not distinct from the prototype creature. While many organisms have unique genes in their genomes, I could show using probabilistic approaches that not even a single entirely new gene for an entirely new biochemical or biological function could evolve through any kind of mechanism within the genome of organisms. Likewise, I was able to show that the developmental genetic pathway (DG pathway) of every distinct creature is unique and rigid and cannot be changed through organismal evolution into that of another distinct organism. By these criteria, I arrived at the conclusion that the genome of every independently born creature is unique and unchangeable into that of another unique creature, and therefore is essentially immutable. Thus, although each genome allows many different kinds of sequence mutations and rearrangements to take place within it, the genome of every distinct creature has a fixed characteristic framework which is constant and cannot change into that of another unique creature. These mutational changes, however, could lead to slightly changed creatures that are not much differing from the prototype, which explains the origin of similar creatures from a unique creature.

If one creature cannot change into another distinct creature in any length of geological time, it means that Darwin's theory of the origin of organisms on earth should be fundamentally incorrect. Where did Darwin go wrong? His theory is based on the individual variations of a species, from which different breeds of domestic animals can be produced by artificial selection. He extended this ability of artificial selection to that of natural selection in the wild, where many similar species could evolve from one given species over the long course of geological time. That is, the ability of artificial selection to produce different breeds from individual variations of a species, when operated and extended by nature over geological time, would produce different similar species starting from the same individual variations of one species. We could see that up to this point Darwin was correct. However, he extended this process to mean that a species could give rise to an entirely distinct creature even with a new body part over geological time, and that all the multitudes of unique creatures on earth have originated by the process of species change from one original ancestral creature. I could see that this extension was his fundamental mistake.

By systematically analyzing random processes, I demonstrated that it is highly improbable to evolve a new organ or appendage by any mutational mechanism in the genome. Without this capability no mechanism can be an evolutionary mechanism. All genomic mutations and rearrangements occur within the closed framework of every independently born unique creature, and have absolutely nothing to do with the supposed evolutionary change of one unique creature into another distinct creature. Evolutionary theories can explain only the production of very similar organisms. Thus, any evolutionary theory is fundamentally incorrect when it tries to explain the origin of all creatures on earth from one original creature.

My twelve years of research led to several principles that demonstrated that Darwin's theory and its later modifications are fundamentally incorrect, and that proved the new theory of the independent birth of creatures from the primordial pond. This book, the culmination of this extensive research, provides a complete account of several scientific principles that demonstrate that the scenario of organismal variety on earth is fundamentally due to the independent birth of creatures from the primordial pond. Evolutionary theories can only account for the presence of sets of similar creatures. They cannot explain the presence of numerous distinct creatures. Thus, the main principle that explains the origin of diverse creatures on earth is the principle of independent birth of creatures from the primordial pond. Each unique creature that was independently born was then modified within a confined framework of its own to produce various similar species.

I realized that mutations can only change the genes in a genome into their normal variants or introduce defects in them. A gene for one particular biochemical function cannot mutate and change into a distinct unique gene for another specific biochemical function. Therefore the biochemical functions of the set of genes within a given genome are fixed. Thus the first fundamental tenet of evolutionary theories — that a new unique gene can be brought forth by mutational mechanisms over geological time from another gene — is killed here. And since new DG pathways required for new body parts cannot be evolved by any kind of mutation within a genome, the second fundamental tenet of evolutionary theories — that new creatures with unique body parts can be evolved from an ancestral creature lacking these body parts — is also killed here.

It is believed that in the primordial soup, first a few genes evolved by building from some short genetic sequences, which made possible the evolution of the single cell. From this organism's genome came forth new and unique genes, evolving a simple multicellular creature. From this original multicellular creature evolved various complex multicellular creatures through the process of further genetic change. As I was working with the origin of genes from random genetic sequences, I realized that simple-to-complex gene evolution was quite unnecessary to explain the origin of complex genes found in multicellular creatures. I could demonstrate that the complex genes of multicellular creatures simply existed in very long random genetic sequences as complete genes, which could assemble to form multitudes of distinct genomes for a large number of unique creatures directly in the primordial pond. This process fully explains the origin and diversity of creatures on earth. Thus, the process of the evolution of simple to complex organisms by organismal descent with modification — by any evolutionary mechanism — is also unnecessary to explain the complexity and diversity of creatures on earth.

The new theory of the independent birth of creatures is able to explain the origin and diversity of complex creatures without the problems that evolutionary theory faces. The new theory unifies all the biological processes into a single coherent process: the origin of all the biological mechanisms and the origin of the genomes of all the numerous unique creatures in a single primordial pond eons ago. Thus, the currently held principle that all biology is coherently explained by evolutionary connections among all creatures on earth is also shown to be incorrect.

This book has two major objectives: 1) to demonstrate that evolutionary theories in general — Darwin's theory and later modifications to it — are fundamentally incorrect, and 2) to propose the new theory of the independent birth of organisms and demonstrate its validity.

A road map of the chapters of the book might be helpful at this juncture. After this first introductory chapter, Darwin's theory and its basic principles as they stand now are described in the second chapter. In the third chapter we shall see that Darwin's mechanisms are superficially appealing, but when scrutinized at the genomic level, they cannot explain the evolution of new unique genes and unique body parts by organismal descent with modification.

In the fourth chapter we shall demonstrate that all the possible and known types of genetic mutations can only change a particular gene into its normal variants or into a defective gene but not to a new unique gene even over geological time. Mutational changes can only change a given organism into its normal variants, or develop one or more of its normal body parts abnormally, or lead to genetic diseases and cancer. They do not have the capability to evolve a unique body part even over geological time.

The new theory of the independent birth of organisms is described in the next four chapters. Chapter 5 provides an overall perspective of the systematic development of the principles of the new theory in the chapters to come.

Because the theory proposes the independent birth of multitudes of creatures directly from the primordial pond, a detailed description of the primordial pond is required; this forms Chapter 6. The primary aim of this chapter is to systematically show that very long DNA sequences could occur in the primordial pond, and that the total amount of DNA in the primordial pond was truly enormous.

The new theory is based on the premise that genes were abundantly available in the primordial pond. Chapter 7 is devoted to showing this. This has been achieved primarily from my own research findings, and to some extent using already established scientific facts.

The principles that led to the new theory and its major implications are described in Chapter 8. We shall see how the random combinations of genes in a primordial pond could lead to the assembly of numerous genomes. Although only rarely could such combinations give rise to a viable genome that could express itself as a living creature, still numerous combinations would be successful leading to multitudes of distinct creatures. Also we shall see that pieces of the first assembled genomes would soon mix among themselves and with other genes in the open primordial pond, producing new creatures with mixed features. A number of new principles that lead to the new theory and that are predicted by the new theory are discussed in this chapter.

In Chapter 9 we shall see how the independent birth of creatures would account for the molecular and genomic scenario found in the variety of creatures that are living today. We shall show ample molecular and genomic evidence to demonstrate that the distinct creatures on earth should have originated independently, although from a common primordial pond.

The theory predicts that organisms should be unique in having unique body structures that do not occur in other organisms, and that cannot be derived by evolution from organisms lacking them — showing that unique creatures could only have originated by their independent births. The genetic and organismal similarities are due to their independent births from a common primordial pond. Chapter 10 demonstrates this in fair detail.

We shall discuss in Chapter 11 how the details of the fossil record strongly support the new theory. The new theory predicts that multitudes of creatures should have originated simultaneously. In addition, the birth of creatures should have continued over a long geological period leading to the sudden birth of further independent creatures in later geological time, but ending eons ago. These predictions are clearly borne out by the fossil record. The sudden burst of organisms at the very start of the appearance of multicellular life is what we see in the "Cambrian explosion." This has been one of the major unsolved problems for evolutionary theory — and is in fact evidence against evolutionary theory. The prediction of the new theory, that a random mixture of unique body parts be present in different organisms, is also excellently illustrated in the fossil record. The random mixture of organs and appendages in fossil creatures has been another enigma to evolutionary theories.

We shall conclude the book in Chapter 12 with a comprehensive summary of the new theory. The new theory and its scientific principles fundamentally change what life on earth holds for its future. It is commonly believed that organisms have been evolving as per Darwin's theory, and that evolution is an ongoing process that would give rise to many newer creatures in the future. But based on the new theory, this is false. No new creature is being or will ever be evolved on earth. Each existing creature varies only slightly from its prototype born in the primordial pond, and can change only within that prototype's genomic framework. Furthermore, extinctions will constantly reduce the number of distinct creatures on earth. If modern society abuses the environment, it could make the earth barren very fast.

This book is written to be accessible to anyone with a modern high school biology background. A brief appendix explains the fundamentals of molecular biology, to help the reader understand the principles discussed in the book, and a quick reading of this appendix would probably be very helpful to any nonspecialist. While some of the technical details in the chapters may not be adequately explained at the level of a lay reader, in most cases you will find that these details are merely supplemental to the surrounding narrative. That is, most lay readers will miss none of the basic concepts by simply skipping the technical details, notes and references, which are provided here only to support and document the narrative.

Perhaps the greatest appeal of the new theory lies in its unification of biological processes into a single, coherent process at the molecular level, without any evolutionary connection of organisms. This may have significant ramifications in many facets of biological science and research. All of the common biological phenomena — commonality in genes, biochemical materials and biological processes — in all of earth's unique creatures are independently derived from the organisms' independent origins in the common primordial pond. The new theory reveals the immense power of the prebiotic processes in the primordial pond in giving rise to numerous unique creatures directly from it. It demonstrates that life was not an accident; its simultaneous expression in numerous independent creatures was an inevitable consequence of the biochemical richness of earth's primordial ponds.

2

DARWIN'S THEORY: NATURAL SELECTION AND EVOLUTION

The exquisite scenario of life on earth, with its abundant variety of life forms, has been an endless source of wonder and fascination for mankind. Myriad living creatures with unique structures and shapes flourish everywhere on earth, from hot springs to arctic ice, from the bottoms of the seas to the tops of the mountains, and in the air. Life permeates our planet in extremely diverse forms — from microscopic bacteria to elephants. How and when all these innumerable living things appeared on the earth has puzzled mankind for centuries. It is one of the most intriguing questions that excites scientists and the lay public alike.

For centuries naturalists have noted physical similarities among organisms that resemble one another, and have classified them into hierarchical groups of organisms.[1] These similarities alone led to a general belief, even before the times of Jean Baptiste Lamarck[2] and Charles Darwin, that species evolved from one to another. It was Darwin who in 1859 proposed a mechanism for the evolution of one organism into another — the theory of natural selection and adaptation, which asserts that all organisms evolved from earlier ancestors, which were ultimately derived from one or only a few most primitive and simple organisms. Since its proposal, Darwin's theory has traversed a difficult and stormy road. Intense debates raged for years, and between 1900 and 1940 Darwin's theory almost died. It was only after 1940 that the theory was resurrected by what we know as "neo-Darwinism" (described later). Although recent discoveries in molecular genetics appear to support Darwin's theory, there remains an intense debate among scientists regarding the mechanisms of evolution.

Darwin's theory

Although thoughts on evolution had existed for several centuries, it was Charles Darwin who first gave the most convincing and elegant explanation as to how organisms could change from one to another.[3] Sailing around the world for five years as a naturalist on board HMS Beagle, he studied the distribution of inhabitants on the continents, the South American continent in particular. He continued his studies for several years at home in England, which led him to propose his theories on the origin of species. He believed that species must have changed from one to another over geological time.

After observing similar species living on different islands of the Galapagos, he envisioned a mechanism — of natural selection and adaptation — for the change of one species into another. He observed that there were many similar species, the giant turtles for instance, each present on a different island, with slight variations from the others. He was convinced that these species were related, having descended from a common ancestor at an earlier time. Because each species with distinct characteristics existed on different geographically isolated islands, he was led to think that geographical isolation was responsible for the characteristic variation. After years of research trying to find a mechanism for such a change, he eventually concluded that "natural selection" was responsible for the change of one species into another.

Here is Darwin's theory of natural selection in a nutshell:

1. A certain amount of variation always exists among individuals of each species. These variations are heritable.
2. Every species has an inherent tendency to produce a great number of offspring, more than required for simple maintenance of their numbers. This leads to a large number of individuals competing for a limited supply of resources.

3. Competition among individuals of a population for the available resources at a time of overpopulation will result in the elimination of weaker individuals and the survival of those that vary in some way that makes them better able to cope with the harsh conditions. These variations are heritable and are thus "selected" as the surviving population. Due to repetition of the same process in subsequent generations, the variations that fit the harsh environment will tend to prevail and will be maintained until the environment itself is changed. In this process, the type of organism that survives is said to have adapted to the new environment, changed from its original form.

The individuals belonging to a particular species are different from each other (except for identical twins). To Darwin, the variations occurring in the individuals of a species seemed to be heritable, because children resembled their parents. Those individuals that are the best fit for a particular environment are selected by that environment to survive, hence, "survival of the fittest." Thus, it is the existence of heritable individual variations in the population of a species that is fundamentally responsible for the phenomenon of natural selection and evolution.

Darwin saw that similar forces were in operation in man's production of breeds through "artificial" selection. Different breeds of horse, sheep or any other animal could be produced starting from a population of its species by allowing only those individuals with the most desirable attributes, or variations, to reproduce. Darwin showed that these heritable variations are the major source of artificial selection of breeds, and, he reasoned, the same variations when subjected to natural selection by differing physical or ecological environments would lead to the gradual transformation of one species into one or more other species. Darwin saw it this way:[4]

> Can the principle of selection, which we have seen is so potent in the hands of man, apply in nature? I think we shall see that it can act most effectually. ... Can it, then, be thought improbable, seeing that variations useful to man have undoubtedly occurred, that other variations useful in some way to each being in the great and complex battle of life, should sometimes occur in the course of thousands of generations? If such do occur, can we doubt (remembering that many more individuals are born than can possibly survive) that individuals having any advantage, however slight, over others, would have the best chance of surviving and of procreating their kind? On the other hand, we may feel sure that any variation in the least degree injurious would be rigidly destroyed. This preservation of favorable variations and the rejection of injurious variations, I call Natural Selection.

He believed that natural selection based on individual variations would have the same effect in the long run as artificial selection, only at a slower pace. By extending the success of artificial selection, Darwin built his case for the plausibility of natural selection:[5]

> As man can produce and certainly has produced a great result by his methodical and unconscious means of selection, what may not nature effect? Man can act only on external and variable characters. Nature ... can act on every internal organ, on every shade of constitutional difference, on the whole machinery of life. Man selects only for his own good; Nature only for that of the being for which she tends... Under Nature, the slightest difference of structure or constitution may well turn the nicely balanced scale in the struggle for life, and so be preserved. How fleeting are the wishes and efforts of man! How short his time! and consequently how poor his products be, compared with those accumulated by nature during whole geological periods. Can we wonder, then, that nature's productions should be far 'truer' in character than man's productions; that they should be infinitely better adapted to the most complex conditions of life, and should plainly bear the stamp of far higher workmanship? ... Natural selection ... is a power incessantly ready for action, and is as immensely superior to man's efforts, as the work of Nature are to those of Art.

Darwin's theory of evolution was to have profound ramifications and implications at many levels of human thinking for more than a century. Right or wrong, it has become not only the fabric of modern science, but also the fabric of modern culture.

Contemporary objections to Darwin's theory, and his answers

Darwin's theory of evolution has had several fundamental objections against it not only by his contemporaries but also by his successors. However, Darwin foresaw most of these possible objections against the concept of natural selection. He tried to answer them in his *Origin of Species*, devoting four chapters to these objections in his first edition. He added one more chapter in a later edition in response to other objections his critics had raised. Some of these objections are: 1) gaps between species (absence of transitional forms among living species), 2) absence of transitional forms in the fossil record, 3) apparent sudden appearance of numerous distinct creatures when life first started, 4) evolution of highly complex organs such as the eye, and 5) evolution of intricate behavior patterns such as instincts. If natural selection and evolution of all organisms from one original ancestor had taken place, such problems should not truly exist. Thus, to defend his theory it was necessary for Darwin to explain these observations.

1. Absence of Transitional Forms in Living Species — Gaps Between Species

Darwin, in arguing that heritable variations fitting an environment are selected as the environment changes, stated that the conversion of one species to the next involves a gradual series of physical alterations. However, if we turn to nature, these series of transitional forms are generally absent. For instance, supposing the bear had evolved from the wolf by gradual change of wolf to bear (or vice versa), then where are the intermediate forms that help us connect the wolf to the bear? Likewise, if humans had evolved from monkey, then why are the forms between monkey and human not present currently?

Darwin assumed that transitional forms between an ancestor and its progeny were short lived, and the ancestor as well as the transitional forms had succumbed to competition by their more advanced descendants. Therefore, in the evolution of two species from a supposed ancestor, there need not be transitional forms between the two descendent species. Rather, the descendent species may differ considerably from each other. Darwin argued that the two different breeds of pigeon (the fantail and the pouter) were descended not from each other but from the common rock pigeon. Therefore, a transitional form between the fantail and the pouter never existed. Similarly, the monkeys existing now on earth and the human evolved along separate branches from a common ancestor; and not one directly from the other. Modern evolutionists feel that this explanation may be acceptable for the absence of the transitional forms between distantly related organisms, but not for the absence of such forms between species.

2. Absence of Transitional Forms in the Fossil Record

Another obvious difficulty for Darwin's theory is the gaps in the fossil record. If transitional forms between two supposedly related living species are absent because they evolved separately from a common ancestor, and because their common ancestor and the transitional forms had become extinct, these forms should have been preserved in the fossil record. But in almost all cases they are not. In fact, entirely new species appeared quite abruptly in the fossil record throughout the history of life.

On this subject, Darwin raised his own objection that such forms ought to have existed in the fossil record. "This, perhaps, is the most obvious and gravest objection which can be urged against my theory," he wrote. He believed that the fossil record itself was incomplete for several reasons, and, he concluded, the gaps between fossil species in the successive strata are due to such incompleteness rather than true gaps between evolving species. With regard to the "imperfections of the fossil record" Darwin wrote,

> I look at the natural geological record, as a history of the world imperfectly kept, and written in a changing dialect; of this history we possess the last volume alone, relating only to two or three countries. Of this volume, only here and there a short chapter has been preserved; and of each page, only here and there a few lines. ... On this view, the difficulties above discussed are greatly diminished, or even disappear.

Darwin also argued that the apparent sudden appearance of species in the fossil record is perhaps illusory. We have good fossil collections only from a small part of the world, and these groups may have evolved over long periods of time elsewhere, only to be discovered on their subsequent migration to these geographical locations where they are found. But, as we shall see later, the fossil record is now more than complete, and yet the problem of the absence of transitional forms has stubbornly remained. This genuine absence of transitional forms poses an unsurmountable problem to modern evolutionists.

3. Sudden outburst of diverse forms of organisms when life first started on earth — the problem of the Cambrian Explosion

According to Darwin's theory, at the beginning of the origin of life on earth one or a few[6] multicellular organisms had somehow originated (but he did not say how). Gradually one evolved into two or three similar organisms, taking many millions of years, which in turn gave rise to correspondingly more organisms, continuing the process until today. Thus, within the first few million years after the appearance of the first original multicellular organism on earth, there would be only a limited number of species, most of which would be similar to each other (this should be shown in the fossil record). But, according to the fossil record, multitudes of unique creatures, so distinct from each other that they are classifiable into many different major groups (higher taxonomic groupings[7]), appeared simultaneously when animal life appeared on earth for the first time. This outburst of numerous very distinct living forms which appeared in the fossil record covering a short geological time is a great difficulty for Darwin's theory. For the problem of the *Cambrian explosion*, the sudden outburst of many groups of organisms during the Cambrian period, about 570 million years ago, Darwin had no satisfactory answer.

He recognized that if his theory were correct, there should have existed a rich record of precursors for the complex animals that appeared in the Cambrian explosion, just before the Cambrian, i.e., in the Precambrian period:

> If my theory be true, it is indisputable that before the lowest Silurian stratum was deposited, long periods elapsed, as long as, or probably far longer than, the whole interval from the Silurian age to the present day; and that during these vast, yet quite unknown periods of time, the world swarmed with living creatures.

But the fossils of such precursor creatures were found to be absent in the Precambrian too. Here Darwin brings his usual argument of the imperfection of the fossil record to explain the absence of such precursors. But the total absence of such precursors made him realize the gravity of the problem. He admitted toward the end of his life:

> The case at present must remain inexplicable; and may be truly urged as a valid argument against the views here entertained.

A rich fossil record from before the Cambrian explosion has been discovered in the past thirty years. Even in them Darwin's predictions of a gradual increase in complexity towards the Cambrian period have not come true. The problem of the Cambrian explosion has remained as stubborn as ever — generating intense debates, and even leading to rejection and modification of his theory of gradual evolution by some modern evolutionists,[8] which we shall discuss later.

4. The difficulty of evolving highly complex and perfect organs by natural selection

One of the grave difficulties of Darwin's theory concerns the existence of highly complex organs in animals. Natural selection is supposed to account for characteristics of only small selective advantage, but Darwin extended this mechanism to explain the evolution of extraordinarily intricate organs. Darwin's opponents wondered how natural selection could produce highly complex organs such as eyes and wings. In particular, there are many parts to an eye, such as the lens, cornea, and retina. Individually these parts could not confer any selective advantage to the organ as a whole until the entire organ had been perfected. How then could the early stages of these parts evolve if they could not confer a selective advantage? Of what use would a lens be without a retina? By Darwin's mechanisms, only gradual series of steps could account for such remarkably perfect organs, whereas the objections were strongly against this. Darwin himself admitted the seriousness of this matter when he wrote,

> If it could be demonstrated that any complex organ existed, which could not possibly have been formed by numerous, successive, slight modifications, my theory would absolutely break down.

To surmount this difficulty, Darwin pointed out the plausibility of early transitional stages, including the reasons for selection at each stage. Among invertebrates, the eye exhibits a series of gradations, from an optic nerve merely coated with pigment to a highly developed one found in vertebrates. Whatever his explanations, he was still perplexed with the difficulty in evolving complex organs such as the eye through natural selection, and he wrote:

> To suppose that the eye, with all its inimitable contrivances for adjusting the focus to different distances, for admitting different amounts of light, and for the correction of spherical and chromatic aberration, could have been formed by natural selection seems, I freely confess, absurd in the highest possible degree.[9]

5. The evolution of intricate behavior patterns

Another difficulty for Darwin's theory concerns the evolution of complex behavior patterns, such as instinctive behaviors of certain insects. The evolution of the behavior of sterile castes of workers or soldiers in some ants, termites and bees through natural selection is a difficult problem for evolutionary theory. There are clear subdivisions in these insects, such as various castes of sterile workers which are dedicated solely to carrying out certain defined functions throughout their lives, sterile soldiers, again dedicated solely to protecting the community of the insect through certain defined acts, a queen dedicated solely to laying eggs, and a king dedicated solely to copulation. How can such distinct and defined subdivisions as these widely different castes evolve from an organism that entirely lacked these subdivisions?

Darwin believed that instincts are evolved analogously to the artificial selection of domestic animals, wherein animal breeders have successfully selected certain behavior patterns in preference to others. He believed natural selection of intricate behavior patterns occurred in the same way as the evolution of intricate structures and functions in complex organs such as the eye.

6. Questions about adaptation

Darwin's mechanism of adaptation, the by-product of natural selection, has also faced objections from even the evolutionary quarters. For instance, although the noted evolutionist Richard Lewontin[10] is convinced that the fit manifested between organisms and their environment is a major outcome of evolutionary change, he has raised some interesting and important questions with respect to adaptation as a mechanism of evolutionary change. He argues that the modern view of adaptation is that the external world sets certain "problems" that organisms need to "solve." Evolution, by means of natural selection, is the mechanism for solving those problems. Adaptation is the process of evolutionary change by which the organism provides an increasingly better "solution" to the "problem."

A difficulty with regard to adaptation that Lewontin has raised is that if evolution is described as the process of adaptation of organisms to their environments, or niches, then the niches must exist before the species that fill them. However, there is an infinite number of ways the world can be broken up into arbitrary niches. It is easy to describe "niches" that are unoccupied — which should be occupied if evolution has been taking place. For example, we do not see any grass-eating snakes, even though snakes live in the grass. Similarly, there are no birds that eat the mature leaves of trees, even though they inhabit trees. Given any description of ecological niche, occupied by an actual organism, one can create millions of descriptions of unoccupied niches simply by adding another arbitrary specification. In essence, Lewontin argues that natural selection does not inevitably lead to adaptation, and that it is sometimes hard to define an adaptation.

Evolutionary theory since Darwin
and the modern synthesis

Since Darwin the study of evolutionary mechanisms has had a turbulent history. The nature of the hereditary substance, the DNA, was not known to the world until long after his time. Darwin himself had proposed an incorrect theory of inheritance, called pangenesis. It stated that every organ of the body produced minute hereditary particles, called gemmules; for instance, the liver produced liver-gemmules and the eye produced eye-gemmules. The gemmules were carried from every organ through the blood and were collected to form sperm and egg.

There were many other kinds of questions and uncertainties about Darwin's theory after his lifetime. Some people did not believe in the mechanism of natural selection, because they thought individual variations were easily homogenized in a population. Because offspring are generally intermediate in attributes between their parents, there was widespread belief in a concept called "blending inheritance."[11] The attributes from the two parents were thought to be mixed as in the case of mixing red paint with white resulting in a pink blend. This would quickly homogenize a population, and so natural selection would have no effect. New variations that arise would also be lost by this blending. At the same time, there was also a widespread belief that, as in Lamarck's theory,[12] variations acquired during an individual's lifetime could be passed on to offspring. This provided an alternative to natural selection. Darwin himself had given some credence to this belief in his later years.

But it was Gregor Mendel who discovered the true "particulate" nature of inheritance, which would deflate the blending inheritance concept. Variation was determined by rearrangements of genetic particles, he said, which could mutate to give rise to new variants. However, instead of supporting Darwin's natural selection mechanism, Mendel's theory was initially considered to undermine it. Early Mendelians such as Hugo de Vries dismissed continuous variation among individuals as inconsequential and largely nongenetic.[13] To them species were forms that differed discretely in their attributes. They believed therefore that species arose in one or a few steps of discrete mutations. If discrete mutations can give rise to species, they said, natural selection was not required for the origin of species. They thus dismissed Darwin's key principle of natural selection and gradual change. In fact, Darwin's theory was nearly dead in the early twentieth century. It was rejected by many paleontologists, who embraced theories such as that of Lamarck.

THE MODERN SYNTHESIS

Nonetheless, around 1940 Darwin's theory sprang back to life in what is called the Modern Synthesis, brought about by the combined efforts of many scientists working in many different fields of biology. This "neo-Darwinian modern synthesis" appeared to reconcile Darwin's theory with the facts of genetics.

Scientists had demonstrated many principles contributing to the modern synthesis. Notable figures such as Theodosius Dobzhansky, Ernst Mayr, George Gaylord Simpson, Julian Huxley, Sewall Wright, and Ledyard Stebbins contributed to the synthesis that evolved into so-called new Darwinism, neo-Darwinism, or neo-Darwinian theory. It accounted for genetic change and the origin of species.[14]

The population of a species contains genetic variations that arise by random mutation and rearrangement in their genes, as well as variations in the frequency of genes occurring in a population caused by random fluctuations (genetic drift[15]), but mostly influenced by natural selection. Most of the genetic variants which are adaptive in different environments result in only slight physical changes, meaning that there are only slight physical variations in a population suited to one environment compared to another population suited to another environment. Physical change is gradual. The process of "speciation," which is supposed to bring about the splitting of one species into two or more species by evolution, brings about the diversification of organisms. Ordinarily, speciation involves reproductive isolation among populations — such as in the geographical isolation that Darwin encountered in the Galapagos Islands. Continued for a sufficiently long time, these processes result in vastly different organisms that bear little resemblance to one another. In this way, all of the numerous unique invertebrates, as well as the vertebrates such as the fish, amphibians, reptiles, birds and mammals would have evolved from a single original organism.

Recent modifications to Darwin's theory of evolution

Darwin foresaw the problems for his theory of evolution, such as questions raised by the fossil record, and he tried to answer them with a belief that they would be solved by the completion of the fossil record in the future. But rather than solve these problems, intense research over the past century has only exacerbated them. Furthermore, although the accumulating molecular evidence appears to support and even prove Darwin's theory of evolution, none of the problems faced by the theory has really been solved. The shortfalls of Darwinian evolution remain fresh and vigorous, causing endless debates in the scientific community and beyond.

To accommodate some of the discrepancies of Darwin's theory, scientists, especially in recent years, have proposed modifications and additions. As it turns out, each only addresses a specific inconsistency of Darwin's theory and does not deal with all of the difficulties that the theory faces. As we shall see, none of these modifications provide real solutions.

Sudden speciation and Goldschmidt's theory of "hopeful monsters"

Speciation, the splitting of one species into two, is believed to be solely responsible for the enormous diversity of life on earth. Speciation has been the subject of much controversy. Confusion surrounds a process called "phyletic speciation" or "phyletic transformation," in which one species gradually changes and becomes sufficiently different to be recognized as distinct from its ancestors. Eli C. Minkoff writes in his book *Evolutionary Biology*:[16]

> Most of Darwin's discussions on the origin of species focus on this process rather than true speciation. Darwin had sought to explain the "origin of species," but he succeeded in explaining only the process of successional change. By confusing phyletic transformation with true speciation, Darwin ultimately failed to solve the problem he had posed for himself.

True speciation is supposed to involve the splitting of one species into two or more species — gradually, through innovations that spread through populations, or suddenly, through innovations[17] in individuals. Theories of sudden speciation were in vogue in the early twentieth century. Species were seen to be separated by "bridgeless gaps"[18] that could hardly be crossed by any gradual process. Hugo De Vries stated in his mutation theory that new species, separated widely from their forerunners, could arise suddenly through a single drastic mutation, or "macromutation."

Richard Goldschmidt was a noted proponent of sudden speciation. He proposed that a "hopeful monster" could arise by a process of mass mutation and could produce a new species all at once. He argued that new species could arise in a very short time, not as a result of accumulation of many small adaptive "micromutations" in genes, but by one or more "systemic" or macromutations. Many evolutionists were critical of this proposal, including Ernst Mayr,[19] a noted evolutionist of recent times. They emphasized that the probability of a viable, successful organism resulting from a mass mutation of this sort is vanishingly small, and that a "hopeless monster," rather than a hopeful one, would inevitably result.

The current thinking on the subject of sudden speciation is clear from Minkoff's writing:[20]

> The sudden creation of a new species by a single individual is fraught with problems. With whom would this individual mate? If the new individual truly belongs to a new species, any back-crossing to its parent species is by definition precluded. Second, most species are estimated to differ from one another by thousands of genetic differences. A "macromutation" of sufficient magnitude to create a species distinction would have to involve a large amount of change, and large, sudden mutational changes are nearly always very harmful. Finally, to maintain a harmonious genotype despite sudden change, a very large number of genes would have to mutate simultaneously in a well-planned manner — a wholesale "mass mutation."

Even with so much opposition to Goldschmidt's proposal, some current evolutionists searching for alternative mechanisms for sudden speciation have begun to view some of his ideas more sympathetically. For instance, Guy L. Bush has written recently, "Can Goldschmidt's macromutations result in rapid evolutionary change and speciation after all? I think the answer is a qualified yes."[21] Stephen J. Gould says, "I do feel that certain forms of macromutational theory are legitimate, and I have supported them, though not in the context of punctuated equilibrium."[22]

The Gould-Eldredge theory of punctuated equilibrium

As we have seen, the fossil record suggests that an organism originates abruptly, not gradually as Darwin would have it, and stays unchanged for millions of years until it becomes extinct. To account for this great discrepancy, Gould and Eldredge[23] have proposed the theory of "punctuated equilibrium," which states that species stay constant (at equilibrium) for a long period of geological time before evolving abruptly into new species.

According to the theory, speciation takes place in isolated, peripheral populations. The resulting new species are less likely to be preserved in the fossil record under these conditions, and will only enter the record when it successfully replaces the prior species over a wide area. The fossil record will then display a series of species suddenly replacing one another, instead of a gradual transition.

On the other hand, Darwin's Origin of Species, also known as the "theory of gradualism," rejects the apparent abrupt replacement of one species by another in the fossil record, accounting for it as a consequence of the record's imperfection. Punctuated equilibrium, in sharp contrast to gradualism, proposes that during the life of a species, very little morphological change takes place, and only during speciation do most morphological conversions occur.

When we reflect upon this, the principal weakness for the theory of punctuated equilibrium is that it does not offer any genetic explanation as to how species could change rapidly during speciation. Gould states:

> Punctuated equilibrium is a specific claim about speciation and its deployment in geological time; it should not be used as a synonym for any theory of rapid evolutionary change at any scale.[24] ... Punctuated equilibrium is not a theory of macromutation; it is not a theory of any genetic process. It is a theory about larger-scale patterns — the geometry of speciation in geological time.[25]

There has been considerable opposition to the Gould-Eldredge model.[26] Their theoretical explanations as to why speciation would be necessary for evolutionary change are strongly opposed by population geneticists. By studying several relatively well-documented species transitions among early mammals, Gignerich and Schoeninger[27] concluded that they support the gradualist model. Whatever be the case, as Gould himself has stated, the theory of punctuated equilibrium does not offer any genetic mechanism for the rapid evolution of new species.

Kimura's neutral theory of evolution

For the past 2-3 decades, there has been a continuing debate among molecular evolutionists as to whether the variations, at the molecular level, found in individuals of a species are selectively neutral or have been selected due to their adaptive value. According to neutral theory,[28] a great majority of the evolutionary changes at the molecular level are caused not by Darwinian selection but by random fixation of selectively neutral[29] mutants in the species. That is, these neutral variations do not affect an organism's fitness, but are preserved nonetheless. And these selectively neutral changes contribute to much of organismal evolution.

One prediction of the neutral model is that the rate of evolution should be higher in those proteins and DNA sequences that are not subject to strong functional constraints; fewer mutations will be selected against, and more will have a neutral effect, because they are less likely to disrupt the molecule's function. The average rate of evolution of a molecule may be estimated by measuring the difference between analogous molecules from two related organisms, and dividing by the time since they diverged from a common ancestor (judged from the fossil record). But there is much controversy about the neutral theory among evolutionists.

Motoo Kimura, in defending the neutral theory of evolution, writes:[30]

> The neutral theory does not deny the possibility that some changes are adaptive. Thus, it is by no means antagonistic to the Darwinian theory of evolution by natural selection. However, because of its emphasis on mutation and random drift, and also because of its accent on negative selection rather than positive Darwinian selection, the neutral theory clearly differs in its theoretical framework from the traditional neo-Darwinian or "synthetic" theory of evolution. ... The proposal of the neutral theory, followed immediately by strong support from King and Jukes with their provocative title "non-Darwinian evolution" and emphasis and extension of the theory by our group, led

to a great deal of controversy. This is often referred to as "the neutralist-selectionist controversy" and has been documented by many authors, particularly by Crow, Cadler, and Lewontin. Such a controversy is not surprising because evolutionary biology has been dominated for more than half a century by the new-Darwinian view that organisms become progressively adapted to their environments by accumulating beneficial mutants, and evolutionists naturally expected this principle to extend to the molecular level. ... However, the traditional synthetic theory is no longer as firm as it was in the late 1950s and early 1960s.

THE CURRENT STATUS OF THE EVOLUTIONARY THEORY

It will be universally admitted that although the modern synthesis rescued evolutionary theory from near demise, much confusion surrounding the evolutionary theory persists. In spite of the fact that most scientists strongly believe in the evolutionary theory, the mechanisms of evolution are still hotly debated. Evolutionists contradict and criticize each other's mechanisms, with quite valid reasoning. Even evolutionists would agree that there is no mechanism that is unanimously acceptable. Without getting into details I quote Guy Bush, an evolutionist, to illustrate the situation. In the course of describing the large deficiency of knowledge in the genetics of speciation, Guy Bush writes the following:[31]

> One has only to peruse the literature to realize that although much has been written, little concrete information is actually available on the genetics of speciation. For instance, a genetic cornerstone of current speciation theory is "coadaptation." Dobzhansky, Mayr, Carson, and many others have championed the view that natural selection inevitably favors combinations of alleles at nearly all loci that must harmoniously interact. ... Unfortunately, the hard data on which the concept of coadaption is based are not impressive. Specific cases where the number, rate, kind, and mode of action of genetic loci involved in speciation have been established are woefully lacking, and I am unaware of any unequivocal cases demonstrating that genetic revolutions have been directly associated with speciation. In fact, in cases where it has been expected, such as the Hawaiian *Drosophila*, which are likely candidates for speciation by the founder effect, it has not been found. In almost all cases, the observed facts are controversial and, upon close scrutiny, clearly open to conflicting interpretations. As emphasized by Hedrick et al. in a detailed review of the problem, coadaptation has often been invoked in numerous speculative writings to explain the origin and retention of multilocus systems, but rarely has it been tested experimentally or even demonstrated in nature.

Here Bush remarks that little concrete information is actually available on the genetics of speciation. Although he believes in speciation, he feels that "the processes involved are, for the most part, unknown."

When one peruses the scientific literature on evolution, one sees that each evolutionary mechanism proposed by an evolutionist is criticized by many other evolutionists as unacceptable or incorrect. For instance, Mayr and some of his colleagues object to the concept of the "hopeful monster" proposed by Richard Goldschmidt.[32] As we saw before, many evolutionists are at variance with Kimura and his "neutral theory." Gould and Eldredge are criticized for their theoretical explanations, concerning the punctuated equilibrium model, by population geneticists. Guy Bush feels that the whole field of the genetics of speciation is utterly inadequate and has many flaws. As we saw before, Bush criticized the "coadaptation" theme of Dobzhansky, Mayr and Carson.

We may conclude from these several observations that no one mechanism of evolution is satisfactory to the evolutionists. There is no one mechanism that can consistently explain the evolutionary process clearly. Each mechanism seems to be applicable in isolated cases, and cannot explain the larger picture of supposed evolutionary change of organisms in nature.

Conclusion

The origin and diversity of creatures on earth are among the most important and fascinating questions that intrigue the human mind. The scenario of life is complex, and all organisms can be grouped into numerous sets of similar creatures, although there are clear gaps among the various sets. Based on the similarities within each set, science has believed for centuries that entirely new and unique creatures can and do evolve from others. No sound mechanism was articulated to explain these changes until 1859, when Charles Darwin appeared to offer a sound explanation in the form of natural selection. Noting the power of artificial selection to produce various breeds starting from a stock of animals, Darwin proposed that natural selection, over longer geological time, would produce different species starting from only a population of one — even producing all the multitudes of unique and distinct organisms with absolutely unique body plans, organs and body parts. If his theory is correct, natural selection would clearly explain the scenario of life on earth as well as the structural and functional details of all organisms. But the theory fails to explain several aspects in the scenario of living beings, the fossil record, and the details of many organisms' body structures. Darwin tried to answer some of these questions in his book *Origin of Species*, but the controversies remain. To better accommodate all known evidence, some modern evolutionists have proposed modifications to the theory, but no such revisions have won acceptance by all the evolutionists, and to this day no mechanism has been proposed that truly explains the evolution of organisms at all the different levels of life on earth — organismal, molecular, and fossil.

At a molecular level, scientists do not agree on whether the mutations and variations that occur in a population are selective or neutral. Even if the neutral theory is correct, it still fails to explain the evolution of a unique creature with new genes and unique body parts. As we shall see in the next two chapters, explanations such as the neutral theory have not solved any of the problems faced by the evolutionary theory. The same thing can be said for the theory of the punctuated equilibrium, which does not offer any real genetic mechanisms. Overall, nothing much has really changed since Darwin. The problems faced by Darwin's theory of evolution remain as fresh as ever, and are even more tenacious for a persistent lack of real solutions. As the next two chapters will show, these problems persist not because of lesser inconsistencies in the theory, but because the theory itself is fundamentally incorrect.

Throughout the history of evolutionary theory, no well-founded, scientific theory has emerged to coherently explain all our observations without involving organismal evolution. This is another reason for the prolonged controversies and debates over Darwin's theory, as though the problems with it are due only to some smaller inconsistencies, rather than to a fundamental problem with the theory. There has simply been no viable alternative model to the concept of evolution. The new theory of the "Independent Birth of Organisms," detailed in later chapters, offers such a non-evolutionary mechanism.

3

THE GENOME OF EVERY DISTINCT ORGANISM IS CLOSED TO EVOLUTIONARY CHANGES

Almost everyone marvels at the beauty of Darwin's theory, which was innovative upon its introduction more than 130 years ago and remains the accepted theory today. Despite its popularity, however, the theory has been subject to 130 years of strong criticisms that are difficult to refute. The most challenging of these criticisms include an improbable sudden outburst of multitudes of creatures at the very first appearance of life, the absence of transitional forms in the fossil record, wide gaps between organisms, and the complexity of certain advanced organs. As we shall see here, these unsolved problems have persisted for so long not because of minor inconsistencies in the theory, but because the theory is fundamentally incorrect in its fundamental premise: that all organisms on earth evolved from one or only a few original ancestors.

I realized during my research with genetic and protein sequences that evolutionary mechanisms are not necessary to explain the origins of diverse creatures on earth. The scenario of life on earth can be explained far better, and without any of the problems that encumber Darwin's theory, by a new, alternative premise: that organisms could originate independently from the primordial pond. If this new concept is correct, then Darwin's theory must be fundamentally incorrect. I researched Darwin's theory to determine where he could have gone wrong, and found a critical error in his assumption that natural selection was capable of forming entirely new organisms.

Obviously, individual variations within a species do exist, and the creation of very different-looking breeds by artificial selection is indeed a reality. My own observations led me to conclude that, while natural selection can produce many similar varieties (now termed species) of a distinct creature, it cannot produce another distinct creature with new genes or unique body parts. Darwin's extrapolation of the effects of artificial selection to the scale of geologic time was a reasonable path to take, but his expectation that such a natural-selection mechanism could produce new and utterly different organisms was entirely a leap of faith — not at all supported either by logic or by observation. This, however, is precisely what Darwin did, and his fundamental error lies there.

By carefully analyzing the basic mechanism required for the hypothetical evolution of one organism into another — the evolution of new genes and the complex genetic network required for the evolution of a new organ — we shall see in this chapter that such an evolution is highly improbable, even through a supposed series of transitional organisms over a long period of geological time. Without the capability to evolve a new organ or body part, no evolutionary mechanism can explain the origin of diverse creatures on earth.

This chapter will further demonstrate that the set of genes in any organism is essentially constant, fixed and unchangeable. We shall address the real meaning of the individual variations at both the genome and organismal levels, and show why they are confined to an absolutely closed framework in every creature. Finally, we shall see how Darwin's fundamental mistake lay in an invalid extrapolation from the known effects of artificial selection, which produces variant breeds of an organism, and from the ability of natural selection to produce many similar species of a single distinct organism. Lacking our modern understanding of genomes and molecular biology, Darwin assumed falsely that these phenomena, working over geological time, could incrementally produce entirely new and unique creatures, across what we now recognize to be fixed and immutable genomic boundaries.

In this chapter, we are going to distinguish between a distinct organism and the similar species that belong to a distinct organism. At the outset, therefore, we need to have some definitions. Let us define a *distinct organism or creature* to be a group of similar species that is distinguished from other groups in having one or more unique genes and/or body parts. In contrast, the *similar species* of a distinct organism all have essentially the same genes and/or the same body structures and parts. Thus, in the living world, there are distinct organisms, each with its own constituent set of similar species. For example, any species of snail would be a member of the same organism/creature, the snail. In most places in the book, we shall simply use the term *organism* or *creature* to mean a distinct organism or creature. Also, we will use the term *evolution* to mean the organismal evolution of one distinct organism into another distinct organism. The change of one species into a similar species is considered here to be not evolution, but rather the production of natural varieties of a distinct organism.

The genome of an organism is closed and locked with respect to evolution. The variability of a creature is confined to the closed framework of its genome.

The modern explanation of Darwin's theory is that several kinds of random genetic mutations cause a variety of changes in the genetic makeup, or genome, of an organism, which are responsible for many kinds of organismal variations. Natural selection operates nonrandomly on these variations evolving one species into new species. The genetic variations lead to the evolution of many new genes, which define the new organism. In other words, the genome of an organism is so highly variable that it can be easily molded into the genomes of many entirely different organisms through long series of changes.

To the contrary, we shall demonstrate that the genome of a distinct creature is a closed framework with respect to evolutionary change and therefore an organism's fundamental characteristics are immutable. We will show that through random mutational processes 1) no new gene can be evolved within the genome of a species, and 2) no organisms can evolve the complex network of genes needed to create a new organ or appendage. But we must first understand that any mutation occurring in a genome must be random.

Random mutational processes cannot lead to the evolution of new genes and genetic networks needed for new organs and appendages

What is randomness in the context of genetic sequences?

What is meant by randomness in the context of supposed evolutionary change? When we toss a coin the probability of obtaining a head or a tail is 1/2. If we throw a die, the probability of obtaining a given number between one and six — say the number four — is 1/6. The probability that two given numbers occur in consecutive throws of a die — say four at the first throw and three at the second throw — is 1/6 x 1/6 = 1/36. This is the same for two given numbers occurring at two nonconsecutive throws — say four at second throw and two at 7th throw.

In the context of evolution, mutations are supposed to occur at random, at any point, or nucleotide, within the DNA sequences that make up the genome of a species. A typical genome may contain over a billion nucleotides. While not all nucleotide locations in a genome mutate at the same rate, and not all possible mutations are equally likely, essentially they are all equally random throughout the genome.[1] Also, the chance that a specific mutation will occur is not affected by how useful that mutation would be for the organism or for the supposed evolution. Each of the nucleotides, adenine (A), thymidine (T), guanine (G) and cytidine (C), have an essentially equal probability to mutate to the other nucleotides. In addition, mutations occur in a genome with the same probability throughout gene sequences as well as the intergenic, "junk" DNA.

IF ORGANISMAL EVOLUTION IS CORRECT, THEN NEW GENES, WHICH ARE PRESENT AS WE GO UP THE SUPPOSED LADDER OF EVOLUTION, MUST BE EVOLVED IN THE GENOMES OF ORGANISMS

New or unique proteins and genes in many distinct creatures

There are a great many organisms that contain entirely unique genes, which are not present in most other organisms. While most organisms have a lot of genes in common,[2] unique genes are also widely found.

As we go up the supposed evolutionary ladder from "lower" to "higher" organisms, we find unique genes at all levels. For instance, the genes for making silk are found only in insects; blood coagulation proteins only in vertebrates (see also chapter 4); protein hormones such as chorionic gonadotropins only in some mammals; and most eye-specific genes are absent in a variety of invertebrate organisms lacking eyes altogether.

There are numerous proteins with exotic functions.[3] The African plant protein Monellin has a powerfully sweet taste. There are Antarctic fish that carry unique "antifreeze" proteins in their blood. Insect wings have at their hinges the protein resilin, with an almost perfect elastic property. To escape their cocoons, silkworm larvae secrete cocoonase to break apart the silk fibers of the cocoon. These are examples of numerous proteins that are structurally and biochemically unique to one or a few species.

Many biochemicals are also unique to the organisms that use them.[4] Often, unique genes create odor-causing biochemicals that are essential for an organism's survival. Plants frequently have a putrid odor to attract pollinating insects; the odor is similar to that of decaying flesh on which these insects feed. Other plants have sweet odors to attract nectar-feeding insects. Many animals use a chemical scent, or pheromone, to attract mates. Additionally, plants and animals can ward off predators with bitter, unpleasant chemicals, such as the plant products known as alkaloids. Such attractants and repellents are highly specific to the organisms that produce them.

Mosquitoes use an anticoagulant to prevent the victim's blood from clotting during feeding. Venomous animals, such as snakes, scorpions, and jellyfish, produce highly specific toxins designed to act on the nervous systems of their victims. These toxins have the ability to digest protein, for destroying a wide variety of tissues in their victims.

Many simple organisms, such as fungi and lichens, secrete specific chemicals that reduce or prevent bacterial growth. Penicillin, synthesized by penicillium molds, is but one of many such chemicals. Penicillinase, a specific enzyme that degrades penicillin, is synthesized by certain bacteria.

When we look throughout the assumed evolutionary tree, it is very clear that entirely unique or new proteins are present in a random manner throughout the organisms. If organismal evolution did take place starting from the most primitive ancestral organism, these unique genes should all have evolved in a few billion years from common ancestors that lacked these genes. But as we shall see below, it is simply impossible to evolve an entirely new gene, even gradually over trillions of years.

THE EVOLUTION OF EVEN ONE NEW GENE WITHIN A GENOME IS IMPROBABLE

Let us take a DNA sequence, 10 nucleotides long, and let mutations happen in it randomly:

$$A\ T\ G\ A\ C\ G\ T\ C\ C\ T$$
$$1\ \ 2\ \ 3\ \ 4\ \ 5\ \ 6\ \ 7\ \ 8\ \ 9\ \ 10$$

The probability that a mutation occurs at a given position, say at the 4th, is 1/10. The probability that at that position, the A is changed to G is 1/3 (because A can be changed to G, C or T). Therefore the probability that at the 4th position the A is changed into G in the given 10-nucleotide sequence is 1/10 x 1/3 = 1/30. Similarly, the probability that the C at the

9th position is changed to T is 1/30. Consequently, in the whole sequence, the probability that the A at the 4th position is changed to G, and the C at the 9th position is changed to T is 1/30 x 1/30 = 1/900. This means that on an average, about 900 random mutations may be required to produce the sequence with the above specific changes. In fact, it can be computed that 4143 random mutations are required for the probability of producing the sequence with the two specific changes to be 0.99. [5]

Extending this computation, the probability of mutating a 100-nucleotide-long gene at a given position is 1/300 (see Figure 3.1). The probability of mutating this gene at two given positions is $(1/300)^2$. Similarly, the probability of mutating this gene at ten given positions is $(1/300)^{10}$. Therefore, if a gene is 1000 nucleotides long, and if it requires specific nucleotide changes at 100 positions (10% change) to change this gene into a new gene, then the probability to achieve this is $(1/3000)^{100}$ or approximately 10^{-350}. Therefore, for such a change to become likely, the original sequence may have to undergo, on average, 10^{350} random mutations. Since most genes are longer, the number of specific nucleotide changes to convert one gene into a new gene is far greater. But even if we assume that with a given 10% change of sequence a gene can be converted into a new gene, the number of random mutations needed for this is absurdly high.

PROBABILITY

1/3 X 1/100
= 1/300

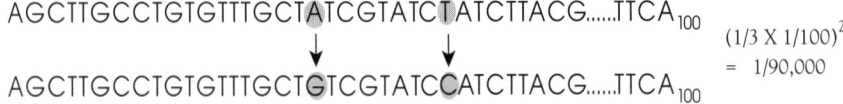

$(1/3 \times 1/100)^2$
= 1/90,000

FIGURE 3.1. EXTREMELY LOW PROBABILITY FOR EVEN A SMALL SPECIFIC CHANGE IN A SEQUENCE MUTATION. Mutations occur randomly at any nucleotide position. In a given gene one hundred nucleotides long, the probability for a mutation to occur at a specific position is 1/100. At this location, the probability for a specific change, of A ⟶ G for instance, is 1/3 x 1/100. The probability of two specific changes is $(1/3 \times 1/100)^2$. Similarly, the probability that specific changes occur at ten given locations is $(1/3 \times 1/100)^{10} \cong (1/10)^{24}$. So even a 10% change has a probability too low to be evolutionarily meaningful. In reality, genes are far longer, and the probabilities far lower.

Mutation rates are supposed to be in the range of 10^{-9} to 10^{-6} per nucleotide per generation in animals.[6] Even assuming a high mutation rate of 10^{-5}, a genome of approximately one billion nucleotides would have a maximum of about 10,000 nucleotide changes per generation. These mutations would be spread throughout the genome randomly, and not directed to one or a few genes. Therefore, if the genome contains 10,000 genes, each gene could be expected to undergo one nucleotide mutation on average per generation. Because we have considered the mutations occurring within each gene at the same time, in our above calculations, the same probability can be applied to another gene. That is, a second gene would undergo a specific 10% change with the same probability. This would be the same probability for any number of genes to change to the same extent. And as we have seen, a typical gene may require 10^{350} such mutations before achieving a specified 10% change. Even if each generation only lasts one year, it would still take 10^{350} years to achieve this. In other words, assuming that a specific 10% change of the genome would convert one creature into another, it would take about 10^{350} years for the transformation of a given organism to the next on the assumed ladder of evolution. Compare this to the age of the earth itself, which is less than 5×10^9 years old, a minute fraction of the time needed. Note that in the above computations, the average length of the gene was taken to be 1000 nucleotides. However, it is usually longer, and the average length is about 10,000 nucleotides, in which ~10% codes for the protein, and the rest is "junk" DNA. To bring about ~10% specific change only in the coding sequence of such a gene, it would require about 10^{450} random mutations (see Chapter 4 for details). Even if it takes only 1% specific change in a genome for a given organism to move to the next organism on the ladder of evolution, it would still take about 10^{45} years.

In the preceding argument, we implied that all mutations should converge within a single genome or the genome of a single individual in a supposed lineage of species. Natural selection will preserve only those mutations that result in a beneficial physical characteristic. Many new characteristics, such as new organs and tissues, cannot appear without a whole host of new genes acting in an integrated system. Appearing individually, the new genes will not be selected and will have no "evolutionary" meaning. Also, unless all the specific changes in a given gene that are required for it to become the new gene occurs, the new gene is not formed, and such a partially-changed gene has no selective value. It is an all-or-none law — all the required changes must be made before any change at all will be preserved.

In essence, for one organism to be transformed into another, it is necessary that several genes should be changed simultaneously within the same individual or all the changes that lead to the new genes should converge in one individual in a supposed evolutionary lineage. Even if all the mutations necessary for changing one organism into another occur scattered in various individuals of the population, they must all come together in one individual at one time through mating within the population. Incompletely-changed genes have no meaning and are not useful, as incompletely-formed organs are not useful. The probability for achieving a useful change decreases drastically as the number of individuals from which the changes must converge increases. It is therefore incorrect to argue that all the mutations occurring within a population of a species must be taken into account in the above computation; the time taken for the change of one organism into another will not decrease significantly, even if we take into account all the mutations occurring in the population of a species.

At the start of the Cambrian period, during the Cambrian explosion, a great number of complex multicellular creatures, classifiable into many different major groups, suddenly appeared without any primitive precursor organisms. Likewise, the numerous unique organisms found in a strikingly well-preserved and complete sample of the fossil record — called the Burgess Shale fauna and classified into 20-30 large distinct groups (called phyla) and smaller distinct groups (called classes and orders) within each phylum — appeared on earth within a period of about 10 million years. Furthermore, based on the fossil record, it is supposed to have taken only about 600 million years for all the organisms (more than one billion species) to evolve from an original organism. On the contrary, as we discussed, it would not be possible even in a trillion trillion trillion (10^{36}) years to achieve the evolution of even one unique creature with new genes from another creature that lacked these genes — with the observed mutation rate. Therefore, the time within which all these organisms are believed to have come about by evolutionary change is totally inconsistent with what is required probabilistically to arrive at even one unique organism through evolutionary change via random mutation.

In the preceding computations, we have considered only point mutations in a genome. We reach similar conclusions even when we consider the combined effects of all the different kinds of mutations, such as transposition, which is supposed to shuffle sequences within a genome (Chapter 4). It is therefore clear that new genes cannot be evolved by mutations within the genomes. This leads to one of the most important concepts — that the set of genes in the genome of an organism is constant, and that mutations cannot change this constancy.[7]

We saw above that there exist many new genes in organisms at successive steps on the ladder of evolution, which are improbable to be evolved. However, a new body part in a new organism on the assumed ladder of evolution may not require new genes, but rather, a new network of genes. Let us call this genetic program, consisting of many different genes working in concert, turning on and off at specific times during the embryonic development of an organism, the developmental-genetic pathway (DG pathway). A complex DG pathway is required to grow each part of the body. In the supposed evolution of an organism, to develop a new body part requires the evolution of a new DG pathway. However, as we shall see below, it is highly improbable to arrive at such a new DG pathway through organismal evolution over any length of geological time, even from the genes already existing in a genome, that is, even without having to evolve new genes.

It is improbable that the DG pathway for a new organ could evolve within the genome of an organism

As we go up the ladder of evolution, starting from the primitive worm, new and unique organs and appendages are present in most organisms that are lacking in the "lower" organisms

According to evolutionary theory, all animals have evolved from one or a few common ancestors, such as a primitive worm. Although a primitive worm itself contains many cell types and systems for digestion, locomotion, and reproduction, it lacks the organs and appendages that most other organisms have. There are innumerable different kinds of organs and appendages in all the animal world used for several purposes. All these body parts are believed to have evolved from the primitive worm by the evolutionary mechanism of natural selection (and other purported evolutionary mechanisms, see Chapter 4). Random genetic mutations are supposed to provide the basis for changes in the genome which would lead to the evolution of these new and unique body parts.

From worms which lack a body cavity between the digestive tract and the body wall (called acoelomate worms), worms that contain a body cavity (coelomate worms) are supposed to have evolved. Today's biology textbooks describe how evolutionists do not have a convincing explanation even for this supposedly first step in the evolution of "higher" organisms.[8] Worms having no legs or any other appendages are presumed to have evolved into a large number of different millipedes and centipedes. These animals then evolved wings and became numerous kinds of insects such as beetles, honeybees, dragonflies and butterflies. The worms are also believed to have evolved into many different kinds of marine and terrestrial invertebrates — such as snails, lobsters, spiders, sea-stars, squids and octopuses, each with antennae, gills, poison glands, compound eyes, and even eyes more advanced than those of vertebrates (in squids). One of the invertebrates then evolved into a bony fish, evolving at the same time vertebrate eyes, bones and fins. Fish then evolved into numerous kinds of amphibians, which led to a large number of reptiles, and then to mammals and birds, all with a number of unique organs and appendages.

Thus, if organismal evolution did take place — whether by natural selection or by any other mechanism — new and unique organs and appendages should have evolved with them. This means that the DG pathways developing these unique organs should have specifically evolved in their genomes. But we shall see that it is virtually impossible to originate even one new organ or appendage starting from the primitive worm, or from any organism to the supposedly next organism on the evolutionary ladder, by any evolutionary mechanism even in a trillion trillion years.

THE DG PATHWAY OF AN ORGANISM IS UNIQUE AND ABSOLUTELY RIGID. THEREFORE IT CANNOT CHANGE.

Even if new genes can be evolved within a genome, would it be enough for the evolution of a new body part? The answer: Certainly not. Whether or not new genes are required, the set of genes that build a new body part must be organized into a specific network, which we call the developmental genetic pathway of the body part. As we shall discuss below (see also Genetics Primer), this is a specific circuit of genetic on-off switches that is far more complex than the electronic circuits used in computers. The probability to attain even a small hypothetical circuit of on-off switches itself is tremendously low. And a typical network of genes forming a new organ is far larger than the hypothetical example that we shall discuss below. We shall see that DG pathways in different organisms with unique body parts are unique and rigid, and that it is improbable for the DG pathway (and the corresponding body parts) of one organism to have changed from that of another, even through an evolutionary series of intermediate organisms.

What is a DG pathway, and how does it lead to the three-dimensional size, shape and function of an organism?

As described in the Genetics Primer, an organism develops from a single cell called the zygote. The zygote is a fertilized egg. It divides to form two cells which divide and form four and so on up to a particular stage. Then cells or groups of cells commit themselves in their timed genetic programs to develop into each tissue and organ. This is done extremely precisely forming the various parts of the body in proper proportions.

The differentiation of the zygote into different body parts is brought about by activating specific sets of genes differentially in the dividing cells, via a network of on-off switches in these genes. In the subsequently dividing cells which eventually build the body parts, diverse subsets of genes will be activated in an exquisitely well-programmed, timed sequence starting from the first cell. This is the DG pathway of an organism, in which genes are differentially activated and repressed to build the different parts of the body through embryonic development until the individual is built.

A DG pathway is built from a complex network of genes: if we consider a series in which each gene is sequentially activated by the previous gene, for instance, gene 1 activates gene 2, which activates gene 3, and so on, just like in a domino effect, many different series such as these are networked together to bring about the development of an organism. Thus, it is analogous to many interconnected rows of dominos. See Figure 3.2

The basic principle in development involves the branching of one cell type into two different cell types. That is, changing a series of on/off genetic switches must change one type of cell (A) into two new series starting with two different types of cell (B and C). For example, an embryonic cell might give rise to two different cells, one destined to form liver, and the other kidney. Similarly, several distinct cell types must be formed, through specific series of on-off switches, to develop a zygote into an organism.

When cell A divides into cells B and C, we find that B and C must be distinct from one another in order for each to embark along a different DG pathway. This difference is most likely biochemically (RNA or protein) mediated. Consider that two proteins are made at the end of the cell cycle of cell A. When cell A divides, one protein winds up only in cell B, and the other only in cell C. Now the two resulting cells are different from each other and from the parent cell — one of these proteins can switch on a new series of on/off genetic switches leading to B-type cells, and the other protein switches another series leading to C-type cells. See Figure 3.3 for this proposed model. Without such differential movement of a protein or some other biochemical, it is difficult to envisage a mechanism whereby the genetic pathway of one type of cell can change into two new pathways leading to two new types of cells.

Although the exact mechanism described above may not happen in organisms, some similar mechanism must occur to achieve cellular differentiation. If this mechanism of cell differentiation is called cellular "track switching," then hundreds or even thousands of track switching events must occur in the development of an organism, all of which are built into precisely timed, extremely complex series as shown in Figure 3.4. It would be the same or identical in two individuals of a species (or two similar

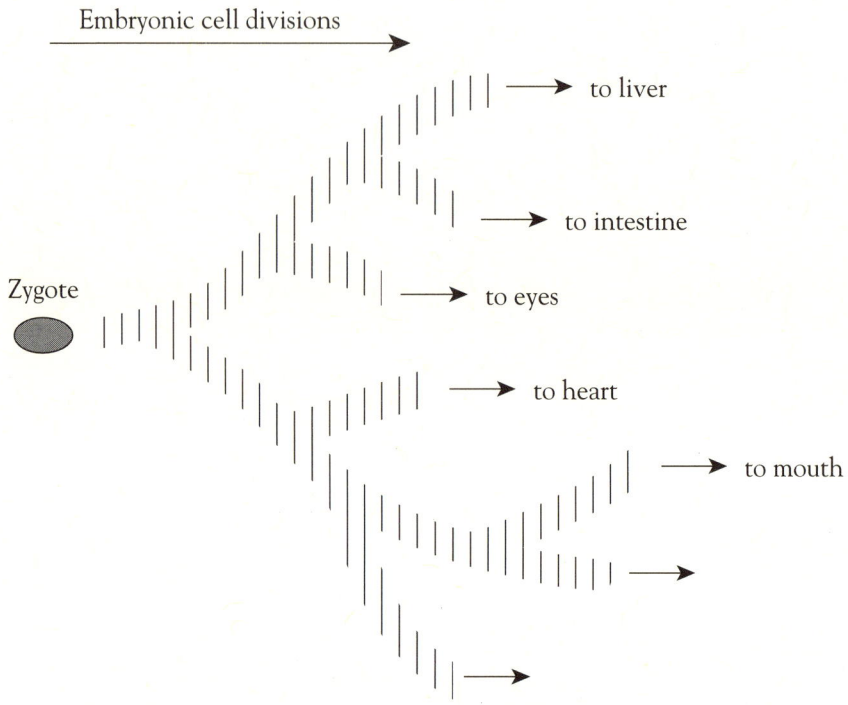

FIGURE 3.2. DIFFERENTIAL ACTIVATION OF DEVELOPMENTALLY REGULATED GENES DURING EMBRYONIC CELL DIVISIONS IS LIKE AN INTERCONNECTED *DOMINO* EFFECT. During embryo growth, the zygote cell divides into many types of cells, such as liver, bone, and muscle. This is brought about by differential switching of genes in subsequently dividing cells until all the cell types are specified and all body parts are constructed. The switching of one particular developmental gene may activate a cascade of subordinate genes that will eventually form a liver. A different developmental gene with different subordinate genes leads to the eye. Like a *domino effect*, one gene activates another, which activates the next and so on, except, in this case, many dominos are interconnected as shown.

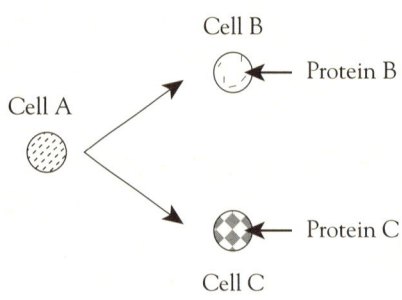

FIGURE 3.3. CHANGING GENETIC NETWORK FROM A STEM CELL TO ITS DAUGHTER CELLS. Consider the early cell divisions in a developing embryo, in which a cell A leads to two cells B and C, which will develop into different body parts. Cell A must trigger two different developmental genetic networks in its two descendant cells. As cell A is dividing, two proteins are synthesized, and one (protein B) is specifically included only in cell B and the other (protein C) only in cell C. The proteins then trigger two entirely distinct genetic networks leading to two different organs.

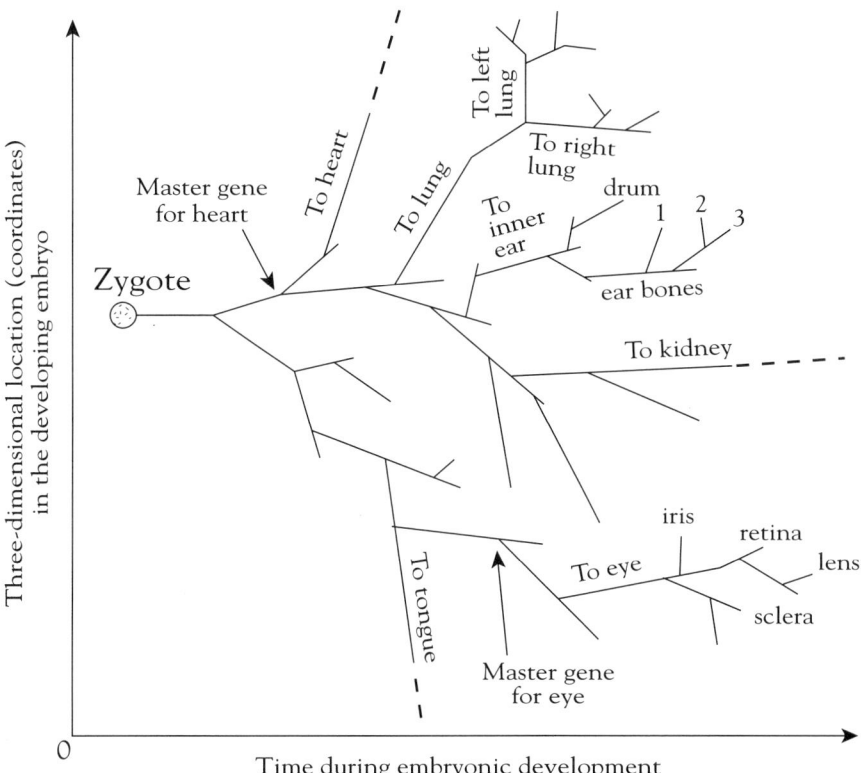

FIGURE 3.4. A DEVELOPMENTAL GENETIC PATHWAY IS A SET OF SPECIFICALLY TIMED GENETIC SWITCHES FOR THE DEVELOPMENT OF DIFFERENT ORGANS AND BODY PARTS, EACH OF WHICH OPERATES AT SPECIFIC THREE-DIMENSIONAL LOCATIONS OF THE DEVELOPING EMBRYO. The hypothetical pattern shown above illustrates the progression of genetic track switching. The differential expression of the genes starts in the zygote. As the zygote starts to divide, each subsequently divided cell (or group of cells) expresses a different set of genes. This process eventually leads to the construction of various body parts. The specification as to which cell develops into which body part is determined during early development of the embryo. This occurs at specific times and in specific cells, located at particular positions in the embryo. The genetic road map traversed by each body part is unique. This pattern of genetic track switching is distinct for different organisms and is rigidly fixed.

species). Small genetic sequence variations, which fall within a closed genomic framework lead to slight variations or fluctuations in the shape and size of the organism. Size and appearance can vary between two different individuals due to variations in the size and shape of individual body parts. However, the overall anatomical proportions of the organs and appendages, and their positions within the individuals of a species will still fall within a constant framework. However, between two distinct creatures — say the snail and the crab — the DG pathways are totally distinct. These distinctions define real differences in the DG pathways of various organisms — that is, a constant DG pathway exists in the various individuals of a species, but two distinct DG pathways in two distinct organisms.[9]

It is extremely improbable that the DG pathway of one organism will change into that of another. Therefore the unique DG pathway of each organism is extremely rigid.

The DG pathways of two distinct organisms are clearly different. Even if we look at two organisms on consecutive steps of an assumed evolutionary ladder — such as the reptile that led to the first mammal and the first mammal — they will have quite distinct DG pathways, at least because of the new and unique body parts. By showing that the DG pathway of a "lower" animal cannot be changed to that of the next "higher" animal, we shall prove that in general an organism with a new body part cannot evolve from another organism that lacks that body part. The developmental genetic network required to construct even a simple organ or body part is itself highly complex. Imagine arriving at this complex network as well as integrating it with the complex genetic network that develops the whole organism. We shall show that it is highly improbable through any kind of genomic change.

Consider a developmental genetic pathway consisting of only ten genes. If we are given one hundred genes in the gene pool, the probability of arriving at this specific genetic network is 100^{-10} or 10^{-20}. This is an extremely low probability. This means that one out of 10^{20} combinations of 100 given genes, taken ten at a time, would be the desired order of genes for a new developmental genetic pathway leading to a new organ. In reality the number of genes in the DG pathway for an organ is over a hundred, tremendously reducing the probability of arriving at a specific DG pathway to 10^{-200}, an insurmountably low level.

Although the total number of genes in the genome of any creature is also far more than our hypothetical example (10,000-100,000), we must take into account the fact that many metabolic cycles are already organized into genetic networks that are common to all cells in the body. The genetic networks that construct the basic cell are also built into the genome of an organism. What we are considering here is only the unique genetic pathways and networks that are different, compared to those already existing in an organism, and, that are needed specifically to build a new organ or limb. For instance, the bone of a vertebrate lacking in an invertebrate, the feather of a bird lacking in a reptile, the placenta of a mammal lacking in a reptile, or even a primitive eye lacking in a worm, is built with unique and complex genetic networks and pathways. Even these series of unique genetic pathways are very long (and networked complexly) with an immense number of elements in them so that the odds against achieving them by any kind of mutations in a given genome is astronomical. Therefore, it is virtually impossible to change one organism into another through Darwin's mechanisms.

The unique DG pathway of every distinct organism is revealed through the "mapping" of embryonic cell lineage. Starting from the zygote, the first cell of the individual, developmental biologists can determine the fate of all the successively dividing cells until the embryo is fully developed. This has been done in several invertebrate organisms such as the C. *elegans*, Aplysia and sea urchin. In Aplysia, for instance, there are only about 1000 cells in its fully developed embryo. The timing and the three-dimensional locations of all the dividing cells, starting from the first until all these cells have been formed in the body of the Aplysia, have been determined. The cell lineage map is absolutely distinct and unique in each organism that has been studied.[10] For our discussions, the cell lineage map can be equated to the DG pathway of the organism, because this map is precisely determined genetically — that is, by the differential expression of the genes in the genome (see also Genetics Primer). Based on the cell lineage maps of different organisms belonging to the various major categories, one of the world's leading developmental and molecular embryologists, Eric H. Davidson from the California Institute of Technology, has stated that they are absolutely distinct (see Chapter 9).[11] One can see that the DG pathway of organisms at successively higher steps on the assumed ladder of evolution are distinctively different because based on comparative anatomy, these organisms are quite distinct and have unique body parts — e.g., fish and amphibians. These considerations plainly demonstrate that the DG pathways of each organism are certainly unique.

From our probabilistic calculations above, it can be easily discerned that the developmental program of one organism cannot be changed either gradually or abruptly into the developmental program of another organism. We can be absolutely convinced that the available knowledge in developmental biology, genetics, and molecular biology is sufficient to indicate clearly that the DG pathway of an organism is unchangeable thereby making every distinct organism immutable. Errors in the DG pathway of one organism can only lead to various defects in the development of that organism. These defects are not the cause of evolutionary change. These principles will become even more clear when we discuss how the unique and rigid DG pathways of the different organisms originated (see Chapter 8).

Here we should note that a distinct organism can give rise to its own varieties which are now called species, but not to another distinct organism with either new genes or a new DG pathway. This is what we mean by the immutability of a distinct organism. For example, a snail can give rise to many different snail varieties, but never to another distinct creature, say a crab or a sea star. The set of genes among the varieties or species of a distinct creature would be essentially the same. Likewise, the DG pathways of the different varieties or species of a distinct creature would be essentially the same (even if their sizes vary considerably). However, each species may not interbreed with another similar species (which is actually another variety of the distinct creature) and therefore will have its own set of individual variations. Thus, whenever we speak about the immutability or unchangeability of an organism into another, we should remember that a distinct creature cannot mutate into another distinct creature, but can change into its own varieties (i.e., similar species).

How the developmental switches of even one organism, such as the original ancestral organism, had evolved first is not explained by neo-Darwinian theory. Recently, evolutionary geneticists have started to speculate that because the "structural-gene" mutations (mutations in the portion of the gene which codes for a protein) do not seem to have any effect in evolution (see Chapter 4), it is the "regulatory-gene" mutations (mutations in the portion of the gene which regulates a gene's activity such as on/off switching, or in genes that control development) that must lead to evolution. In fact, the DG pathway of a distinct creature is so rigid that it cannot be changed even slightly by mutations in regulatory genes or regulatory sequences.[12]

Mutations in a genome can only lead to normal individual variations, or to genetic defects, which are absolutely useless for organismal evolutionary change

Different kinds of mutational and rearrangement mechanisms are constantly operating in the genome of every organism (see Genetics Primer and Chapter 4). However, as we shall see, they can only lead to sequence changes that lead to minor individual variations or to three classes of defects, shown in Figure 3.5. One class is the defects in genes involved in metabolic pathways. The defect in the protein product such as the globin in hemoglobin can lead to sickle cell anemia. The defect in a gene involved in the synthesis of a pig-

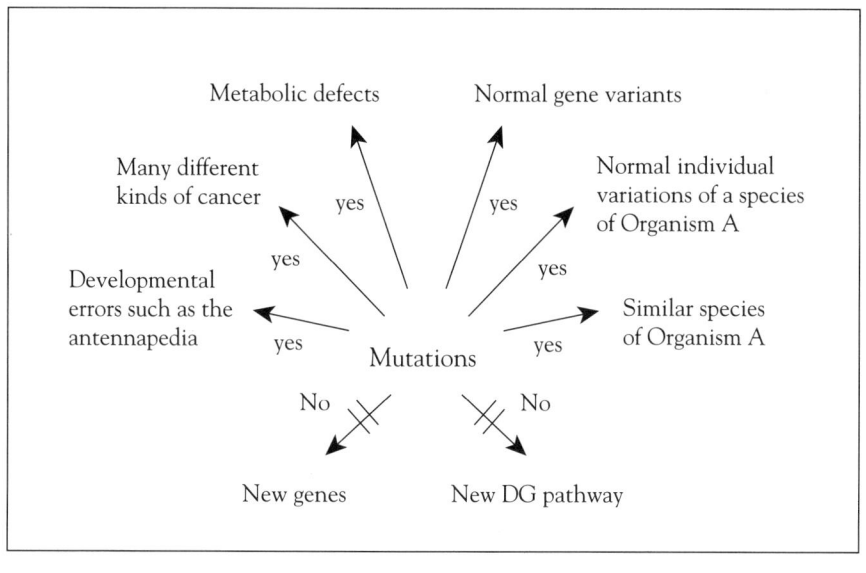

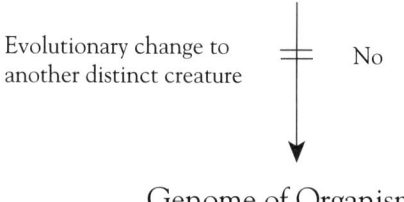

FIGURE 3.5. THE INCAPABILITY OF MUTATIONAL CHANGES IN THE GENOME TO EVOLVE NEW GENES OR A NEW DG PATHWAY FOR A NEW BODY PART LEADS TO THE CONSTANCY OF THE SET OF GENES IN ITS GENOME AND THE FIXITY OF THE DG PATHWAY FOR THE ORGANISM. The several kinds of mutations that occur in the genome of an organism (see Chapter 4) can only lead to the following four types of effects: 1) Normal individual variations and production of similar species; 2) developmental defects (deformities); 3) defects in cell division control genes, resulting in cancer; and 4) metabolic defects (diseases) such as sickle cell anemia and phenylketonuria. Because no new gene or a new DG pathway for a new body part can ever be evolved by any kind of mutational mechanism, the constancy of the set of genes and the rigidity of the DG pathway of a genome of an organism never change.

ment would lead to either no pigment or a differently colored pigment. If the enzyme coded by a gene is involved in the synthesis of a metabolic product, then a defect in such a gene would lead to the deficiency of the product or the accumulation of an unwanted biochemical, such as in the disease phenylketonuria. The second class of mutations are in those genes which principally control the growth and division of each cell. Depending upon the kind of cell in which this mutation occurs, it can lead to many different kinds of cancers, which result from uncontrolled cell division. The third class is the defect in the genes controlling the development of the organism resulting in developmental defects. A mutation in a master gene that specifies the switch for the whole of the subordinate genes responsible for the development of a particular body part, may lead to the development of the animal without the body part. If the mutation occurs in a gene somewhere downstream in the hierarchy of the developmental genetic pathway of a body part, then that body part will be malformed. In some cases, a gene mutation can lead to the duplication of a body part. It can also lead to the misplacement of a body part at an abnormal location in the body (see Figure 3.6). To sum up, mutations can only change a gene into a normal variant of the same gene, or it can become defective with respect to the structure and function of its protein product, leading to one of the organismal diseases as described above. Consequently, the set of genes in a genome is constant as long as the creature exists on earth. Similarly, mutations can never evolve a new DG pathway for a new body part from the DG pathway of an organism lacking that body part. Any mutation in one or more of the genes involved in controlling the development of an organism can only lead to developmental aberrations — abolition, duplication, misplacement, or malformation of a body part already existing in an animal — which do not contribute to evolution. Consequently, the DG pathway of a distinct organism is fixed and rigid.

Cancers are clearly not the monstrous outgrowths that are assumed to aid in evolution. Metabolic defects certainly do not have anything to do with evolutionary change. Likewise, developmental defects also are not the monstrous variations that can aid in evolution. It must be realized that all these developmental defects are on body parts already existing in the body of an animal. Let us remember that no new body part is ever produced by this mechanism. How could these defective aberrations ever contribute to evolution, or be the material basis for natural selection? When absolutely no new body part can be evolved in any organism even over geological time, how can these developmental aberrations, metabolic defects and cancer lead to the evolution of a new organism with a new body part? The conclusions from our analyses based on probabilistic approaches to the evolution of new genes strongly suggest that no mutational change can lead to new genes or a new

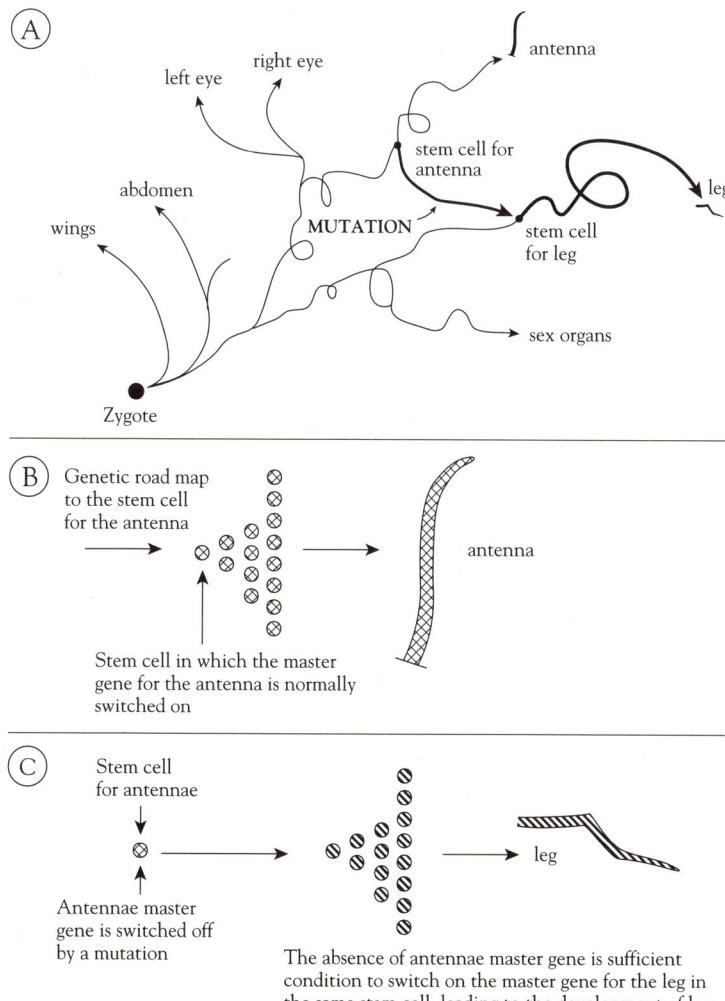

FIGURE 3.6. A POSSIBLE GENETIC MODEL FOR THE REPLACEMENT OF THE ANTENNA BY THE LEG IN THE *DROSOPHILA* BY MUTATION. In the fruit fly with a mutation in a gene called *Antennapedia*, legs develop in the place of the antennae. It is possible that there is a master gene for the development of an antenna and another for the development of a leg. These master genes must be induced in the right cells in the three dimensional space of the embryo for them to be produced at the right places. In all other cells, these master genes must be repressed. It may be that in the genetic road map to the cell that normally switches on the master gene for the antenna, the switching off of the master gene for the antenna can happen by a mutation. This may fortuitously lead to the switching on of a gene (the leg master gene) normally kept repressed in that same cell. This effectively changes the stem cell for the antenna into the stem cell for the leg at the location of the normal antenna stem cell, leading to the development of the leg in place of the antenna. Thus, rare mistakes would misplace a whole body part as well as result in other kinds of developmental aberrations such as missing body parts and malformations.

DG pathway for a new organ even over geological time. We can therefore boldly conclude that all these mutational changes and their organismal effects have absolutely nothing to do with organismal evolution.

The conclusion: The genome of an organism is closed and locked with respect to evolution, because neither a new gene nor a new DG pathway for an organ can be evolved within it

In summary, based on the facts that neither a new gene nor a new DG pathway for a new organ or body part can be evolved within a genome, we can conclude that the genome is a closed framework and that it is locked with respect to evolution (see Figures 3.5 and 3.7). Sequence mutations and rearrangements occur only within the closed framework, which allows many kinds of mutations and rearrangements. All the gene mutations occur without ever mutating the genome of a distinct creature into that of another. No mutational mechanism can change this constancy. This is the most important and crucial principle that establishes that Darwin was incorrect.

The numerous groups of organisms which have new genes and/or unique body parts that we see in the living world cannot be evolved from organisms which lack these; they are evolutionarily unrelated (see Figure 3.8).

Strong corroboration for the rigidity of the DG pathway from the fossil record

In favor of our demonstrated view that the DG pathway of an organism is fixed and can never be changed even over geological time, the fossil record comes to absolute support. What is really meant by a fixed DG pathway of an organism? It means that the organism cannot really change in its anatomical structure over geological time. In fact, one finds in the fossil record that each organism appears suddenly, and remains virtually unchanged until the organism becomes extinct and disappears from the record.[13] If the record of an organism has continued to the present time, and if the organism is found to be living on earth, it is found to be exactly the same organism as when it originated in the fossil record. This is in fact the norm of organisms on earth. Scores of extinct organisms, and living organisms termed "living fossils" are known and serve as examples.[14] The horseshoe crab (Limulus) has changed very little for over 500 million years. Lingula (an inarticulate shellfish) has remained exactly the same since it appeared in the fossil record 600 million years ago. Among vertebrates, the opposum has changed very little in 70 million years; the coelacanth, a fish which first appeared in the record over 100 million years ago, is living now absolutely unchanged. Among mammals, for example, the rat when it appeared in the fossil record for the first time did so as the rat, similarly the bat, rabbit and so on. Indeed, there have been no real transitional forms between any such organisms suddenly appear-

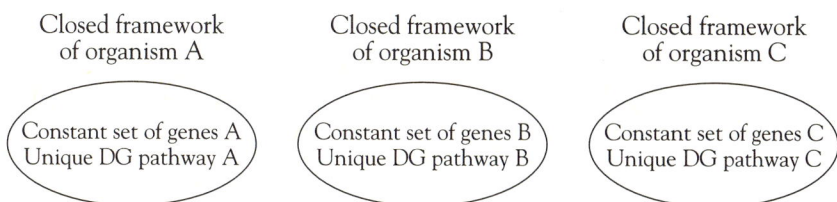

FIGURE 3.7. THE CONSTANCY OF THE GENOME OF AN ORGANISM MAKES IT A CLOSED FRAMEWORK FOR EVOLUTIONARY CHANGE. The genome of an organism is closed evolutionarily because the set of genes in the genome is constant and the DG pathway into which the genes are organized is fixed. All the mutations can occur only within the closed boundary, leading to normal individual variations, similar species, congenital and genetic diseases, developmental aberrations, and cancer as described in Figure 3.5. Because of this fixity of the genome, each organism is immutable.

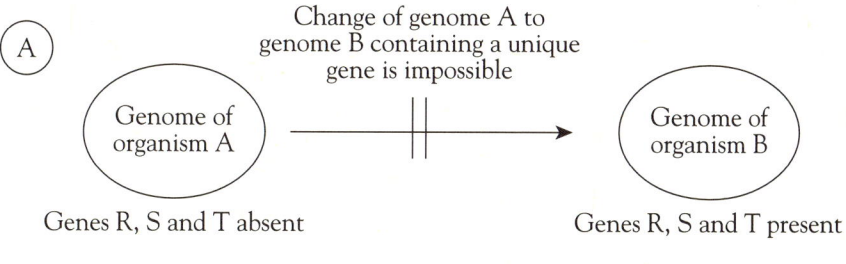

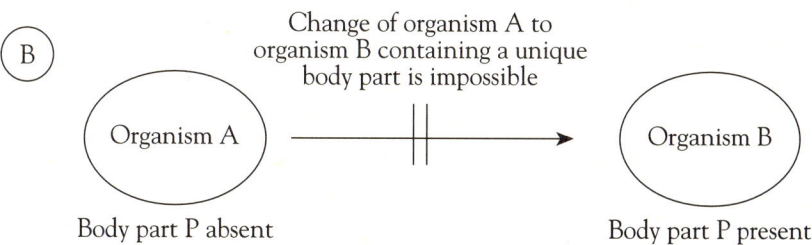

FIGURE 3.8. AN ORGANISM THAT CONTAINS ENTIRELY NEW GENES OR NEW BODY PARTS CAN NEVER BE EVOLVED FROM ANOTHER ORGANISM THAT LACKS THEM. There are numerous organisms on the assumed evolutionary ladder that contain entirely unique genes and/or unique body parts, which are lacking in their ancestral organisms. Because no new genes can be evolved in the genome of an organism even over geological time and even through a supposed series of intermediate organisms, an organism that contains one or more new genes can never be evolved from an organism that lacks them. Likewise, because the new DG pathway for a unique body part cannot be evolved in the genome of an organism, a creature that contains a unique body part, whether it requires new genes or not, can never be evolved from another creature lacking that body part.

ing in the record with any known organisms previously seen in the record. Evolutionists only assume that they must have been evolved from some prior organisms. If at all, what appears to be the change in an organism in the record is just the production of varieties of one organism, which is misconstrued to represent an evolutionary change. The systematic appearance of new organisms in the record, and their remaining virtually unchanged throughout the length of the geological record, is an absolute truth, precisely illustrating and corroborating our principle that the DG pathway of an organism is fixed and can never change in nature.

The individual variations of a species are confined within a closed framework: A new principle

What is individual variation?

Most organisms are distinctly discernible — compare the cat, dog, rabbit, human, gorilla, and deer. Despite the fact that some species resemble each other, the population of one species never genetically mixes with that of another, even a similar species — say between human and guerrilla, or bear and wolf. Thus, when we speak about a species, it is inherently understood to be distinctly different from any other organism.

What is most interesting about a given species is that the individuals in the population of a species are not identical (except for identical twins). Each individual varies slightly from all others in the population — in height, weight, overall appearance, the size and shape of the body parts. However, two individuals, even at extremes of a particular characteristic (say, height in humans), will be easily recognized as belonging to the same species. This is because the overall proportions of body parts are the same, whether one is tall or short. The proportions of the body parts are fairly constant among individuals of a species, and fall within a closed framework. In fact, even the absolute quantities of the different characteristics of the individuals, or of the different limbs, bones and body parts, predictably fall within a given constant framework. The height of mature human beings falls within a range, from about three to nine feet. We do not, however, see any individual 12 feet tall. Similarly the weight of an individual can be 50 pounds to 500 pounds. No individual weighing 1500 pounds is ever found. The ratio of the length of the fingers to body height can be plotted within a fairly constant framework.[15]

Artificial selection and its absolute limitation

Humans have practiced artificial selection for thousands of years. It is a process by which the individuals of a species with desired characteristics are selected and allowed to mate in successive generations in order to arrive at individuals in which these characteristics are magnified in a predicted or desired manner. The size of horns in sheep, for instance, can be increased by artificial selection — by selecting individuals with long horns, in successive generations, to mate and produce offspring. In fact, a characteristic such as the length of the horn can be extended beyond the naturally existing range, up to some absolute limit. This is true with respect to any characteristic, say the size or weight of the sheep, or the length of its wool. Based on this limitation, we can see that the variations among the individuals of a species are confined within a closed framework.

The absolute limitation of artificial selection of a breed in any one particular direction shows that all possible individual variations of a species are confined within a closed, constant framework

The experience of animal breeders is clearly in favor of this view. They have established that artificial selection of a breed cannot be extended endlessly in any one given direction. For instance, no artificial breed of sheep can be produced that can be as tall as six feet. After all, when attempts to produce such artificial breeds go beyond a certain limit, all animals are stubborn and no longer cooperate. This is what happens when sheep breeders select for any trait such as thickness of coat or height. Artificial selection in that direction ceases. One simply cannot get a taller sheep or a thicker coat beyond a certain limit. If attempted, the breed may become sterile or feeble, or its coat may become thinner. If one tries to get too far away from the normal characteristics of sheep by artificial selection, things break down. In the end, the "sheep" characteristic asserts itself. This is true for any animal species and any characteristic that has been so far tried, and, we can be sure, will be true for any untried characteristics as well. For instance, in breeding for miniature dogs, after a certain level of miniaturization, the dogs may develop some defects. The essence is that artificial breeding cannot be endlessly extended in a given direction for a specific set of characteristics. This certainly indicates that the individual variations are confined within a framework that must be characteristic of a given species.

Variations in the population of a species under a given natural environment fall within a fairly well defined framework. Artificial selection can only expand this to a broader, but still limited framework. The framework defines boundaries that are characteristically and inherently fixed for every species in its every trait (see Figure 3.9).

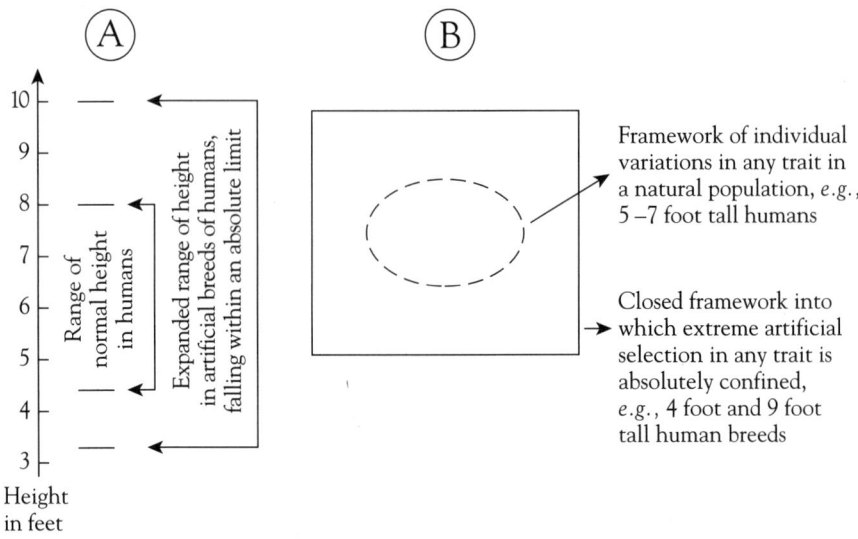

FIGURE 3.9. REALIZATION OF EXTREME TRAITS IN A SPECIES BY ARTIFICIAL SELECTION AND ITS ABSOLUTE LIMITATION IN A CLOSED FRAMEWORK. In a population of a species living under natural conditions, the frequency of individuals with a particular characteristic is distributed in such a manner that they gravitate to a central node; for example in human beings, people about 5-6 feet tall will be most frequent, and on either side of this range, say below 5 feet and above 6 feet, the frequency reduces rapidly. The frequency of individuals who are 4 feet tall, or 8 feet tall, would be nearly zero, and the frequency of 9 foot tall humans will certainly be zero (shown in Figure A). This range, taken for many traits, is what is represented in Figure B as the framework of individual variations under nature. However, by artificial selection, we may be able to select a breed of humans with individuals as tall as 9 or 10 feet, but there is an absolute limit above which this trait cannot be extended, say beyond 11 feet. This is what is represented as the closed framework of artificial selection in Figure B.

Despite the fact that a large number of very different-looking breeds can be produced from a species, no breed can ever be produced that falls outside of the framework of the species.

Also, a new body part, or an anatomically new outgrowth, can never be produced by any extreme of artificial selection. Whatever the individual variations, the set of anatomical structures will be constant in the normal individuals of the species. Consider the case of the individual variations with respect to height in human beings, say the tallest and the shortest person in the world. The set of anatomical structures will be the same in the two individuals. The major difference will be in the sizes of the limbs. No new anatomical structures will be found in either of them. This is true with any type of individual variation (any trait) in any species of organism. However different looking are the various breeds produced by artificial selection, we can boldly say that they contain the same set of anatomical structures. If there are some characteristic differences among the individuals or breeds of a species, such as the case in the number of tail feathers in a bird[16] or the number of ribs in a snake, they still belong to the same framework of the species. For instance, the number of ribs can vary in different individuals of the snake, when eggs from the same litter are incubated at different temperatures.[17,18] Such individual differences therefore, are an inherent property of a species and have nothing to do with evolution.[19]

Natural selection from the population of a species may produce two different species that cannot interbreed — due to, for example, difference in size. Also, mutations in genes for trivial characters such as size may also change one species into another. But the two species are essentially the same except for size in this example. Thus, a species can produce another similar species either by natural selection or by mutation in genes for trivial traits. However, we should note that all such similar species are actually varieties or part of a parent organism that is distinct, and cannot mutate to another distinct creature. For instance, a snail can produce many species (i.e., distinct varieties) of the snail, but the snail cannot change into another creature, such as a scallop or an octopus. See Figure 3.10.

What is the real meaning and source of individual variations?

All variations in the individuals of a species are due to nucleotide sequence differences in the same, constant set of genes in the genome of the species. No new genes or DG pathways are found in any individuals of a species.

The very well established fact that one can mold individual variations to produce breeds with desired characteristics even in one's lifetime influenced Darwin to believe that an organism is highly plastic. This implies, in terms of modern genetics, that a genome is highly plastic, but, as we have shown, the genome of a creature is plastic only within the closed framework char-

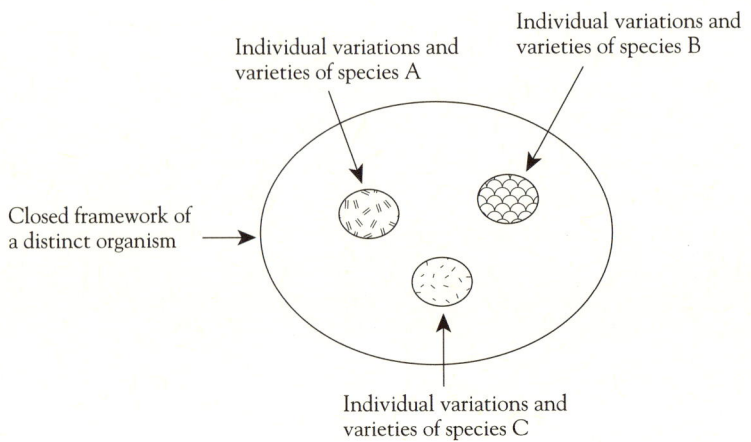

FIGURE 3.10. THE CLOSED FRAMEWORK OF EACH DISTINCT ORGANISM ON EARTH. One species of a distinct organism can give rise to many similar species of the same organism (species A, B, and C in the figure), but never to a new distinct organism. Each species of an organism has its own defined framework of individual variations — which can give rise to artificial breeds, natural varieties and other similar species.

acteristic of that creature. Furthermore, as we have seen, the variations in the genome of a distinct creature can only lead to the similar species of that organism (i.e., natural varieties of that organism) and not to the genome of a distinct unique creature.

Individual variations are due to the variations in the sequences of the same finite set of genes in the genomes of different individuals of the same species. All artificial selections exploit such individual variations. No new genes or new DG pathways will be found in the different breeds or varieties. But if a new distinct creature with a new body part is to be produced by Darwin's principles of descent with modification, these variations should include new genes and/or new DG pathways. This difference between the two phenomena is real and it is important for us to understand.

One might ask, if individual variations are due to sequence differences in the same set of genes, what are these differences, where do they occur, and how do they lead to variations in individuals? The molecular activities of an individual are a set of biochemical reactions, carried out mainly by proteins. The gene sequences carry the messages for these proteins passively (see Genetics Primer). From the start of the fertilization of an egg, complex cycles of biochemical reactions take place. A complex, but well-defined and programmed set of genetic pathways (metabolic and developmental) take the zygote past embryo to the adult, and then through aging and death of the individual.

Let us take two individuals (of the same sex), A and B, from the population of a particular species. The complete set of genes in A and B must be "functionally" the same. In other words, the two individuals must have qualitatively identical genes, that is, the set of enzyme proteins in the two individuals must catalyze functionally identical sets of biochemical reactions. Moreover, the genetic circuits (the set of biochemical reactions and their precise networks) through which the zygotes of the two different individuals traverse in order to reach the full grown state must be exactly the same — that is, the DG pathways of two individuals of the same species are identical.

What is different is the quantity or rate of reaction of the same enzyme in the two different individuals. For example, the enzyme glucokinase breaks down glucose into smaller sugars thereby yielding energy to the animal.[20] But the amino acid sequence of this enzyme in the two individuals can vary, without in any way altering the enzyme's basic function. However, the rate of its activity may vary slightly. Because an enzyme's amino acid sequence is determined by nucleotide sequences in the genes that create it, the nucleotide sequences of the genes coding for the same enzyme may vary in the two individuals. That is, the two are variants of the same gene. There can be many variants of the same gene that do not change the basic function of its protein product. This principle can be extended to every gene in the genome.

The same is true for "regulatory" genes — genes that turn other genes or sets of genes on and off as needed. Typically, large regulatory networks exist. Variations within them do not change the organization of the networks, but rather, only speed and efficiency are affected. Permutations and combinations of such sequence variations in the constant set of genes can lead to a nearly unlimited number of variations in the individuals of a given organism's population. These variations are responsible for the slight differences in the size and shape of the body parts, and likewise the size and shape of the individuals of a species.

Genes that control development (homeotic genes containing a homeobox, see Genetics Primer) are also primarily responsible for the size and shape of organs and body parts. Normal sequence variations in the homeobox or the other regions of homeotic genes may also be responsible for the normal variations in the size and shape of a body structure in an organism, for example the nose of the human.

There is essentially no difference in the set of genes or the DG pathways among the different species of a distinct creature. The main difference would be in one or a few genes for trivial characteristics such as the size, height, coat color or thickness, which could be brought about by natural selection, geographic isolation, or genetic mutation. However, the difference between two distinct creatures is the presence of unique genes and/or unique body parts in one creature and not the other.

It is important to note that while the DG pathways among the individuals of a species (or among the various species of a distinct creature) is invariant, it is quite different in individuals from distinct organisms. The specificity of the DG pathway of an organism is extremely high, which is clearly distinguishable from those of all other organisms, each of which is rigid and unique; this is the true hallmark of an organism. Thus, each organism has its own characteristic DG pathway that contains a constant set of genes, while each of its genes can be present in any one of the myriad possible variant forms in the different individuals of the organism, or the different varieties and species of a distinct organism.

Source of individual variations

If the set of genes in a species never change functionally and its DG pathway is absolutely fixed, then what is the source of individual variations in a species? It is the mutations that lead to gene variants in the constant set of genes, and the recombination during sexual reproduction that mixes the different variants together that are the source of individual variations. Even if a population has all identical individuals to start with, mutational changes in the constant set of genes would lead to individual variations that would fall within a characteristically constant boundary. Consider a species that consists of only 10 identical individuals of each sex. As the population expands by reproduction, mutations will occur variably in a given gene in different individuals, producing variants of the same gene in the population. This will happen to all the genes in the constant set of genes in the genome. Furthermore, the different gene variants in the constant set of genes are mixed together during reproduction. Over geological time no two individuals (except identical twins) will have identical genomic sequence, although all the individuals will still have exactly the same set of genes, the same as that in the starting population. The DG pathway that was present in the starting population will also be absolutely unchanged even after geological time. This is the case in all species that exist.

Mutations can lead to defective genes resulting in developmental defects of existing body parts or cancer. They can produce many similar species of a distinct creature. Contrary to evolutionists' beliefs, such mutations cannot produce a unique organism from another unique organism.

If a creature is immutable, then what is the part played by natural selection in the scenario of organisms on earth?

Natural selection can only produce many similar species of a distinct creature (i.e., its natural varieties) within its closed framework, and never a new unique creature

We have now established that all artificial breeds of a given species fall within the confined boundary of individual variations of the species. Whatever breeds are produced by artificial selection, and however different "looking" they are from each other and from the parent stock, all of them still fall within this defined framework of the species. In a similar manner, we can see that natural selection can only produce natural varieties from the population of a given species, all of which have to fall within the closed limits of individual variations of the distinct creature. Natural selection has essentially no more powers than artificial selection, except that one operates under the direction of the human beings, and the other under the forces of nature. Artificial selection produces many artificial breeds and varieties of a species. Natural selection can produce many similar species of a distinct organism from one species of that organism.

The ability of natural selection to produce many similar species of a distinct creature has been the main phenomenon, the culprit, misleading Darwin and all the evolutionists into believing that it can lead to a new species, ultimately with new body parts. Although no natural variety has been known to contain a new body part, even a useless one with no selective advantage, the evolutionists believe, as did Darwin, that it is possible to arrive at new body parts in an organism by evolution, but only man did not and cannot witness that due to the long geological time it takes. It led them to believe that all organisms on earth, with multitudes of different organs and appendages, have evolved from one or a few original ancestral species such as some worms. But, natural selection cannot, even in a trillion trillion trillion years, produce even one new creature with even one new useful body part.

A phenomenon that collects the individual variations of a particular trait in one given direction, and magnifies certain existing characteristics in an organism, all within the closed defined framework of individual variations of the organism, had seemed to "change" or "plasticize" the organism. Natural selection has been only misunderstood and mystified to be capable of producing organisms with new body parts eventually from those that lack them through "lineages" of organisms over geological time. It is totally astonishing that such a mechanism of so little importance, confined only within every distinct creature on earth, has deceived the human intelligentsia for more than 13 decades. As we shall see in the next chapter, not only natural selection, but all other mechanisms that are said to be capable of evolving one distinct organism from another, are fundamentally incorrect.

If Darwin's theory of evolution is incorrect, then what is meant by the principle of adaptation that Darwin had proposed to be capable of aiding in the evolution of organisms?

If evolution can explain only the production of similar species within every distinct creature, then what happens to the idea of adaptation? It only means that adaptation is not a phenomenon that can contribute to the evolution of one distinct organism to another. Just as natural selection, adaptation is a phenomenon that occurs within the closed confines of every organism, aiding in the process of the formation of varieties of a species and the similar species of every unique creature. Subpopulations of a species do get adapted to environmental changes and do change in physical and physiological characteristics, just as Darwin had seen on the Galapagos Islands, but we must realize that all of these adaptations and changes occur only within the creature's closed framework — never outside of it. In the case of the giant turtles of the Galapagos, many similar species of the turtle organism have been produced. It would therefore make more sense if we call them varieties of the turtle organism.

True monstrosities required for evolutionary changes are nonexistent: Another new principle that shows organismal evolution never happened

In addition to individual variations which Darwin believed to be the basis of natural selection and evolution, he also believed that some rogues and "monstrous" variations occurred in the population of a species and served as the material basis of natural selection. But if we carefully analyze the monstrous variations he alluded to, we can see that true monstrosities potentially useful in evolutionary change are nonexistent in nature. We shall also demonstrate that all the genetic aberrations in the population of a species can only lead to defects in organs and appendages already existing in the species, and can never lead to true monstrosities with potential to serve as a material basis for evolution of unique organisms with new body structures. These are only developmental defects of existing body parts — useless for evolutionary change. In favor of this view, it can also be demonstrated that true monstrous outgrowths useful to evolution should be totally random — but the random DG pathway required to produce such new body parts is impossible to be generated or evolved within the genome of any organism.

In a bed of plants, such as rice or wheat, or in a population of an animal such as sheep, one that is conspicuously different from the rest is called a "sport" or a "rogue." It also refers to a mutant animal or plant, or part of a plant, that shows an unusual deviation from the normal or parent type; for instance, a red rose in a bed of white roses. On the other hand, the term monstrosity is used to mean an individual in whom a body part is either abnormally developed or located. The term is also used hypothetically to mean an individual in whose body a novel structure appears. Although Darwin said that sports and rogues contributed to evolutionary change, it is the last category that has been mistakenly assumed by even modern evolutionists to be the main cause for evolutionary change, aside from individual variations.

Our earlier discussions showed that the shape and size of organs of a breed can be varied only to the extent that the individual variations within the closed framework of the species allows. It cannot be endlessly extended in a given direction. Therefore, unless the monstrous variations that Darwin has alluded to occur among individuals, from which useful variations are selected leading to a new organ, evolution of one species into one or more new species with new body parts is an imaginary phenomenon. However, our discussion in this section will show that such monstrous variations do not occur in any organism. Even, for the sake of argument, if such meaningless monstrous variations did occur, we shall demonstrate that it is improbable for a useful new organ or appendage to be selected and evolved from such variations.

True monstrous outgrowths are imaginary. A new organ is built by a unique DG pathway. In the supposed evolution, first many random outgrowths must occur in the population of a species, so that a useful outgrowth, if any occurs, can be selected. Each random outgrowth must be produced by a specific DG pathway. One out of many such genetic pathways leading to mostly meaningless multicellular masses or outgrowths — which we can call random outgrowths — would be selected by the supposed natural selection.

The genetic network leading to a body part is extremely unique and is integrated with the developmental program of the whole animal. Errors occurring in the genes crucially involved in the development of a tissue or organ can cause an abolition, misplacement, duplication, or malformation of that organ.

In all these errors, there is no formation of a new DG pathway for any kind of multicellular mass either useful or useless. A set of genes within a DG pathway is either expressed or not as a whole, that is, the whole or substantial part of the DG pathway for an organ is either expressed or not. That is why when a developmental gene is mutated, a whole subordinate program is either expressed or not expressed, leading to the abolition or malformation of that part. Another mutation in a DG pathway gene that causes the expression of the whole subordinate program at an incorrect location of the embryo will thus grow that body part at that wrong location (see Figure 3.6).

Other than the defects in the existing tissues and organs — that is, abolition, duplication, misplacement or malformation — the errors or aberrations in the existing genetic programs cannot lead to new extraneous outgrowths. In addition to the developmental errors, when a mutation in one of the genes which are involved in the control of cell division occurs, it leads to uncontrolled cell division, resulting in cancer.

Mistaken monstrosities

What exist in nature that are mistaken to be monstrosities are only extreme normal variants

We can see that true monstrosities required for the evolution of a unique body structure should be new outgrowths either external or internal to the body — not the defective growths of already existing organs or appendages at normal or abnormal locations. But all the known aberrations in nature, for instance, the fruit fly *D. melanogaster*, only belong to this category of defective growths of already existing body parts. The only aberrational outgrowths occurring in all the living world are those produced by defects in DG pathways of body parts already existing in an organism. They may also include duplication, abolition, or abnormal development of an already existing part in the animal. These defects are the result of mutations, which are defects in single genes that are crucially involved in development, or defects in many genes that cumulatively affect the development of a particular body part.

Other kinds of mutations cause variations that have been mistaken to be some form of macromutations or monstrosities. Hugo De Vries[21] formed his conclusions of sudden or abrupt changes based on his work with a flower called evening primrose (*Oenothera lamarckiana*). He noted that in a field of white-flowered primroses, a red flower suddenly appeared. "Sports" similar to that have been noticed among animals. Through experimental breeding of white flowers he found that the red color was inherited. De Vries thought that such sudden or abrupt and often drastic changes might be capable of making new species. He was unaware of the unusual behavior of the genes in *Oenothera*.

Its genes tended to be shuffled through a configuration known as a Renner complex,[22] which leads to color variations. Neither a new gene nor a new DG pathway leading to a new outgrowth were generated, and the color changes had absolutely nothing to do with evolution.

What Darwin alluded to in his *Origin of Species* as sports and rogues are either extreme normal variants, or some trivial mutants such as that of the *Oenothera*. The aberrations in animals, wherein errors in development occurs with one body part growing at an abnormal location (homeotic mutations), are also misconstrued to be aiding in evolution. In the fruit fly *Drosophila* for example, legs grow in the place of antennae possibly due to a single gene mutation. All aberrations in all animals are abnormal growths of already existing body parts or tissues and not truly random outgrowths of parts that do not already exist in the animal; or they can lead to cancer. In my view, only completely new outgrowths occurring in an animal represent true monstrosities. Any kind of abnormal growth of an already existing body part — duplication, abolition, malformation, or abnormal location — does not represent a true monstrosity.

Medically reported monsters are only abnormalities of existing organs and other body parts

A monstrous characteristic should be an extraneous outgrowth that is anatomically different from any existing organ or appendage in the body. Only these can be the random characteristics that can be useful for Darwin's natural selection and evolution of uniquely new organs. True monstrosities have never been reported in the medical literature as some evolutionists claim.[23] What have been so far reported as "monsters" are only aberrations of existing organs.

The acardiac "monster,"[24] for example, is in fact due to reduced (or compressed) blood flow to the developing fetus from the placenta due to some obstruction. Thus, this is not even due to a genetic defect. Shunting between the blood vessels between twins can result in a situation where the arterial pressure of one twin overpowers that of the other early in fetal development. The defeated recipient then has a reverse flow from the cotwin. This results in a host of disruptions with deterioration of previously existing tissues as well as malformation of tissues that are in the process of formation. The extent of disruption may be broad enough leaving as the residuum an "amorphous" twin. There is every gradation, from amorphia to acardia to less severe degrees of disruption, with no one case being identical to another.

The two-headed monster is a form of conjoined twins. It is a developmental defect originating in the early embryo as it partially splits to form twins. Because the recurrence risk for the same parent is low, it may not be genetic.

These developmental defects are mistaken to typify the monstrosities that serve as the material basis for Darwin's natural selection or other mechanisms proposed as modifications to the theory.

TRUE MONSTERS ARE NONEXISTENT

In essence, monstrosities that the evolutionists since Darwin have believed in are either not at all instrumental in evolution or nonexistent in nature. Evolutionists believe that the so-called monsters must have appeared sometime in the past. They are hopeful of their appearance in the future. Their belief is based on an erroneous expectation that because life on earth has been there for a long geologic time, even sporadic occurrences of monstrosities that man has never seen might have occurred and could have been the cause of the evolution of new organisms with new body structures. But even if it is extremely rare, and man does not observe it in every species, it should be observable in at least one out of many species on earth. On the contrary, monstrosities have never been observed in any species.

Even if one organism may not have such extraneous outgrowths for a million years, how can it be that, in all the ~30 million different species that live on earth now, not even in one can we see such extraneous outgrowths? Statistically, if it does occur, we should not have to wait for millions of years to see such outgrowths in one organism. We can simply look at the millions of organisms that live on earth now to find if at least a few of them have such outgrowths. But none have any!

Evolutionists may say that only when an avalanche of physical mutagens such as X-rays, ultraviolet rays, and other cosmic rays should visit the earth, would such outgrowths occur. But, as we have seen in our probabilistic computations, no amount of radiation and mutation could bring about random DG pathways leading to extraneous outgrowths from which useful outgrowths can be selected and evolved. The odds against it are astronomical.

Monstrosities are purely imaginary and could not have been the cause of the multitudes of unique creatures with unique body parts that have come on earth. And the evidence that truly monstrous outgrowths do not occur in any species is consistent with my concept that the kind of aberrational DG pathways constructed by random genetic networks cannot and do not occur in any organism.

All our discussions have thrown light on the indisputable fact that the individual variations existing in nature are absolutely useless in evolution. Monstrous variations are totally nonexistent, but have been erroneously imagined by evolutionists to occur in nature. Therefore, the only two forms of variations in an organism that have ever been proposed to be capable of evolutionary change of one creature into another unique creature — individual variations and monstrosities — can never be the cause of such evolutionary change.

Where Darwin went wrong: Extrapolating the capability of artificial selection in producing breeds to natural selection in producing unique and distinct creatures

So far, we have established that the genome of a species has a tight boundary with a constant set of genes organized into a unique DG pathway, which is absolutely closed with respect to evolution of unique organisms. Sequence variations in the constant set of genes in the genome of a species are the root cause of individual variations. Although the number of such sequence variations are large, they can never produce a new gene or a new DG pathway for a new body part even over geological time. Natural selection and mutations in genes for trivial characteristics can only lead to similar species of a distinct organism, and never to a unique organism with new genes or body parts.

With this new background, one can see where Darwin went wrong. Darwin had extrapolated the power of artificial selection in producing breeds to the power of natural selection in an extended geological time producing unique creatures with new genes and body structures. But neither artificial selection nor natural selection can ever extend beyond the confines of the closed permanent boundary of a distinct creature. And this was Darwin's mistake.

Darwin's *Origin of Species* illustrates his misconception that individual variations are the basis of evolution and his error of extending artificial selection to natural selection — producing all the unique creatures on earth from one original ancestor

Darwin's *Origin of Species* shows his belief that individual variations are the basis for the evolution of one organism from another. In describing individual variations and the power of artificial selection,[25] Darwin says,

> The key is man's power of accumulative selection: nature gives successive variations; man adds them up in certain directions useful to him. ... The great power of this principle of selection is not hypothet-

ical. It is certain that several of our eminent breeders have, even within a single lifetime, modified to a large extent some breeds of cattle and sheep. ... Breeders habitually speak of an animal's organization as something quite plastic, which they can model almost as they please. If I had space I could quote numerous passages to this effect from highly competent authorities. Youatt ... speaks of the principle of selection as 'that which enables the agriculturist, not only to modify the character of his flock, but to change it altogether. It is the magician's wand, by means of which he may summon into life whatever form and mould he pleases.' Lord Somerville, speaking of what breeders have done for sheep, says: - 'It would seem as if they had chalked out upon a wall a form perfect in itself, and then had given it existence.' That most skillful breeder, Sir John Sebright, used to say, with respect to pigeons, that 'he would produce any given feather in three years, but it would take him six years to obtain head and beak.' In Saxony the importance of the principle of selection in regard to merino sheep is so fully recognized, that men follow it as a trade: the sheep are placed on a table and are studied, like a picture by a connoisseur; this is done three times at intervals of months, and the sheep are each time marked and classed, so that the very best may ultimately be selected for breeding.

Darwin was alluding to the normal variations within a population, and the ability to rapidly produce breeds. From his writings it is evident that he simply extended the mechanisms of artificial selection of breeds to the natural selection of species, except that natural selection takes more time:

Youatt gives an excellent illustration of the effects of a course of selection, which may be considered as unconsciously followed, in so far that the breeders could never have expected or even have wished to have produced the result which ensued — namely, the production of two distinct strains. The two flocks of Liecester sheep kept by Mr. Buckley and Mr. Burgess, as Mr. Youatt remarks, 'have been purely bred from the original stock of Mr. Bakewell for upwards of fifty years. There is not a suspicion existing in the mind of any one at all acquainted with the subject that the owner of either of them has deviated in any one instance from the pure blood of Mr. Bakewell's flock, and yet the difference between the sheep possessed by these two gentleman is so great that they have the appearance of being quite different varieties.' ... And in two countries very differently circumstanced, individuals of the same species, having slightly different constitutions or structure, would often succeed better in the one country than in the other, and thus by a process of 'natural selection', as will hereafter be more fully explained, two sub-breeds might be formed. ... On the view here given of the all-important part which selection by man has played, it becomes at once obvious, how it is that our domestic races show adaptation in their structure or in their habits to man's wants or fancies. ... Variability is governed by many unknown laws, more especially by that of correlation of growth. ... Over all these causes of Change I am convinced that the accumulative action of Selection, whether applied methodically and more quickly, or unconsciously and more slowly, but more efficiently, is by far the predominant Power.

Darwin could not have known what variations meant in molecular biological terms. Although he based his theory on extensive observations of individual variations and mutational aberrations, he was certainly not aware of the most fundamental aspects of genes and the genome, which are responsible for the variations. His assumption that selection of variations to produce a new breed should also bring about a new organism with new body characteristics or structures is baseless.

In his chapter, "Variations under nature," Darwin illustrated that natural varieties and species are continuous and difficult for even naturalists to distinguish. He said that many closely related species were mistaken by naturalists to be varieties of a single species. He argued that through "divergence of character" individual variations gave rise to varieties, varieties gave rise to different species, and each species gave rise to many other species — meaning that one original ancestor gave rise to all organisms on earth — all by natural selection. In other words, from one original species evolved all organisms that are classified today into higher taxonomic categories — genera, families, orders, classes and phyla.

> Certainly no line of demarcation has as yet been drawn between species and sub-species — that is, the forms which in the opinion of some naturalists come very near to, but do not quite arrive at the rank of species; or, again, between sub-species and well-marked varieties, or between lesser varieties and individual differences. These differences blend into each other in an insensible series; and a series impresses the mind with the idea of an actual passage. Hence I look at individual differences, though of small interest to the systematist, as of high importance for us, as being the first step towards such slight varieties as are barely thought worth recording in works on natural history. And I look at varieties which are in any degree more distinct and permanent, as steps leading to more strongly marked and more permanent varieties; and at these latter, as leading to sub-species, and to species. The passage from one stage of difference to another and higher stage may be, in some cases, due merely to the long-continued action of different physical conditions in two different regions; but I have not much faith in this view; and I attribute the passage of a variety, from a state in which it differs very slightly from its parent to one in which it differs more, to the action of natural selection in accumulating (as will hereafter be more fully explained) differences of structure in certain definite directions. Hence I believe a well-marked variety may be justly called an incipient species.

There is no doubt that varieties can be produced by natural selection starting from individual variations, just as breeds are produced by artificial selection. There is also no doubt that a species of a distinct organism can produce many similar species of that organism. But neither natural selection nor any other claimed evolutionary mechanism can produce a new creature with a new gene or body part. No amount of individual variations will produce a new body part. If we can prove that individual variations can never produce a new body part — such as a horn or a feather — then we can extend this and show that individual variations cannot also produce a distinct organism with a new body structure. Consequently, natural selection can never change a variety into a new species with a new body part.

This means that from a worm, even a centipede or a millipede could not have evolved, because the latter have legs; a winged insect could not have evolved from an invertebrate lacking wings; a fish could not have evolved from an invertebrate, because a fish has bones and vertebrate eyes; a bird could not have evolved from a reptile because it has wings; and so on. If a mechanism cannot produce an organism with a new body part from another lacking that body part, then it cannot explain the origin and diversity of the myriad organisms with multitudes of unique body parts.

In all these, we must clearly realize that the distinct organism that encompasses its own varieties that are now called species is the only real entity. The other "taxonomic categories" — the genus, family, class, order and phylum — are purely arbitrary distinctions, based on certain similarities among organisms, but not based on any real connections. On the whole, we may conclude that the evolutionary connections among them are based on beliefs induced and nurtured by Darwin's theory of evolution.

Darwin's misbeliefs that sports, rogues, and monstrosities also form the material basis of evolution are manifest in his *Origin of Species*

In the following we shall see several passages from the *Origin of Species* where Darwin speaks about sports, rogues, and monstrosities. These will demonstrate his belief that they, in addition to normal individual variations, could serve as the basis for evolution.

In his discussions on sports, Darwin wrote:[26]

> A long list could easily be given of 'sporting plants;' by this term gardeners mean a single bud or offset, which suddenly assumes a new and sometimes very different character from that of the rest of the plant. Such buds can be propagated by grafting, &c., and sometimes by seed. These 'sports' are extremely rare under nature, but far from rare under cultivation.

Darwin only pointed to the extreme normal variants in a population or some mutants as rogues.[27]

> When a race of plants is once pretty well established, the seed-raisers do not pick out the best plants, but merely go over their seed-beds, and pull up the 'rogues', as they call the plants that deviate from the proper standard. With animals this kind of selection is, in fact, also followed; for hardly any one is so careless as to allow his worst animals to breed. ... As a general rule, I cannot doubt that the continued selection of slight variations, either in the leaves, the flowers, or the fruit, will produce races differing from each other chiefly in these characters. ... In rude and barbarous periods of English history choice animals were often imported, and laws were passed to prevent their exportation: the destruction of horses under a certain size was ordered, and this may be compared to the 'rouging' of plants by nurserymen. ... At the present time, eminent breeders try by methodical selection, with a distinct object in view, to make a new strain or sub-breed, superior to anything existing in the country.

His thinking about monstrous characters are manifest in the following passage:[28]

> Domestic races of the same species, also, often have a somewhat monstrous character; by which I mean, that, although differing from each other, and from the other species of the same genus, in several trifling respects, they often differ in an extreme degree in some one part, both when compared one with another, and more especially when compared with all the species in nature to which they are nearest allied.

While discussing natural selection in the chapter "Variations Under Nature"[29] he spoke of the "monstrosity" which, as we discussed already, is clearly nonexistent in nature.

> We have also what are called monstrosities; but they graduate into varieties. By a monstrosity I presume is meant some considerable deviation of structure in one part, either injurious to or not useful to the species, and not generally propagated.

Although Darwin did not clearly distinguish the two types of variations — individual and monstrous — he has involved both in his natural selection mechanisms. On the one hand, what Darwin believed to represent monstrosities may in fact be developmental defects caused by some mutation resulting in a defect in an existing DG pathway. But these do not contribute to the evolution of a new body part. On the other hand, what Darwin imagined may represent a true monstrosity that evolution requires, but, as we already demonstrated, such monstrosities do not occur in nature.

If natural selection was capable of producing new species with new body parts, it needs true monstrosities to occur in the population of a species: not normal individual variations or developmental mutants of existing body parts, but real monstrosities with new and different extraneous anatomical structures. In the living world there are no examples of such monstrosities. In all his writings about artificial selection of breeds, no such monstrosity had been shown to play a role.

We have shown that mutations in the genes in a genome of an organism happen only on existing genes, either producing variants of these genes or making them defective, but cannot evolve them into new genes. Mutations are on existing characteristics producing a red rose in a bed of white roses, or the malformation or mislocation of an existing body part.[30] Further, normal variants of a species at the opposite extremes of the framework of individual variations might look quite different, appearing to be sports and rogues. But they are like any other individual of the species.

Rogues and sports cannot be causative of evolutionary change, and the developmental defects thought to be monstrous are also useless in evolution of new body parts. True monsters that are required for evolution are nonexistent. Thus, neither individual variations nor monstrosities can bring about or evolve an organism with a new body part from another that lacks it. Let us now remember that these are the only two sources that evolutionists have ever relied upon as the causes for evolutionary change.

NEO-DARWINISTS AND MODERN EVOLUTIONISTS STILL CONTINUE THE SAME MISTAKE OF DARWIN

As we saw in Chapter 2, the Modern Synthesis is supposed to have resurrected Darwin's theory from its eclipse. But did the Modern Synthesis really say anything that Darwin had not said before? No. It claims to have merely clarified some of the tenets of Darwin in more detail: that genetic variations arose by random (i.e., not adaptively directed) mutations and rearrangements; that populations evolve by random genetic changes, and especially natural selection; that most adaptive genetic variants have individually slight physical effects so that changes are gradual; and that diversification comes about by speciation. The main principles of Darwin that individual variations provide the material basis for natural selection, which leads to many new species starting from one species, was unchanged. The Synthesis merely "explained" it better.

Evolutionists since the Modern Synthesis also have not really made any viable change to the fundamental principles of Darwin's theory. Even after the elucidation of the structure of DNA and a good understanding of the structure of genes had been achieved, still the same arguments are continued. The same principles of genetic variations and selections have been applied at a deeper, molecular level. The neo-Darwinists and modern evolutionists argue that mutations cause genetic variations in a population evolving new genes, and evolving new organs and body parts by natural selection and random genetic change.[31] Even the most recent evolutionists working with DNA sequences argue the same.[32]

In essence, although there have been new explanations and modifications to Darwin's theory in the past century, they only discuss whether the changes in the population of a species are neutral or selective, and whether the changes are abrupt or gradual (see Chapter 4). Thus, the basic concept that changes occur in the individuals of a species, which are the basis of evolutionary change, still remains unaltered. It suffices to say that there is no change in the basic tenets of Darwin, that individual variations and monstrous variations form the basis of evolutionary theory.

We shall devote a whole chapter probing how the different kinds of mutations possible in a living organism cannot lead to a new gene or a new DG pathway for a new organ in any length of geological time (Chapter 4). There, we shall analyze even further the actual changes that take place in the genome, at the most fundamental level of genes and other sequences, and demonstrate why all evolutionary theories are incorrect.

EXAMPLE ANALYSIS OF RANDOM MUTATIONS IN THE GENOME, SHOWING THAT THE EVOLUTION OF A NEW ORGAN OR APPENDAGE IS IMPROBABLE

So far in the previous sections we have demonstrated two major principles. 1) The genome is closed for evolutionary change of one organism into a distinct organism. Individual variations and the production of varieties and similar species of a distinct creature — all the result of sequence mutations and rearrangements — occur within this closed genomic framework. Therefore individual variations cannot lead to evolutionary change. Remember that we do not consider the production of a similar species within the framework of a distinct organism as evolutionary change. 2) True monstrosities — thought to form a material basis for evolutionary change by Darwin and his followers — are nonexistent. Having unequivocally arrived at these principles, we can apply them to a few individual cases to clearly illustrate them. In the following we shall demonstrate, by systematic analysis of the supposed evolution of some individual organs, that neither individual variations nor random monstrous characteristics, even if they did occur, can lead to the evolution of a new organ, because of the required random processes.

A CLOSE SCRUTINY OF DARWIN'S MECHANISMS AT THE GENOME LEVEL DEMONSTRATES THAT THE ORIGIN OF DIVERSE CREATURES ON EARTH BY THESE MECHANISMS IS IMPROBABLE

A mechanism of evolution must be capable of evolving a new organ or appendage starting from organisms totally lacking them — such as bone, placenta or wing. Otherwise, it cannot be accepted as capable of explaining the origin of diverse creatures on earth.

When the whole scenario of life on earth at face value is considered, Darwin's theory that a species changes into another in response to a change in the physical and ecological environment by natural selection and adaptation, is very convincing. We have also seen that his theory is certainly correct in explaining the production of similar species of a distinct organism. But when we scrutinize the mechanism of change at the genome level, Darwin's theory of the mutability of one organism into a distinct organism breaks down. The sequence of events involved in the natural selection of a new body part in an organism raises several questions regarding the corresponding changes in the genome that cannot be answered. When the genomic events are carefully looked into, even a small new body part requires radical changes at the genome level. Such radical changes are improbable through random mutational and rearrangement processes within the genome of a species.

What is claimed in Darwin's theory concerning monstrous characters? The population of a species should include random outgrowths among which any outgrowth useful as an organ or appendage occurring at least in its rudimentary or primitive form will have a selective advantage. The individuals with such rudimentary organ or limb having selective advantage in a given environment will survive. In the first of several gradual steps through which an organ is supposed to evolve, at least some new genes and many new DG pathways should evolve in order to produce random anatomical outgrowths in different individuals of a species. From among these outgrowths, a primitive, rudimentary anatomical structure should have some selective advantage and should start the evolution of a more well-defined organ or body part. Our primary aim here is to show that since not even one new gene or new DG pathway can evolve within the genome of an organism by any known mechanisms, not even can the first rudimentary outgrowth evolve. We will demonstrate that the random genomic changes required for bringing about such random outgrowths never happen in any organism, and therefore no new organ could ever evolve by natural selection or any other evolutionary mechanism.

THE EVOLUTION OF A NEW ORGAN WOULD REQUIRE ENORMOUS CHANGES IN THE GENOME, INCLUDING THE GENERATION OF NEW GENES AND NEW DG PATHWAYS. THESE CHANGES ARE IMPROBABLE BY RANDOM MUTATIONS. THEREFORE, NEW ORGANS OR APPENDAGES CAN NEVER BE EVOLVED BY RANDOM GENOMIC PROCESSES.

In Darwin's theory, multicellular life started with one or possibly a few organisms. Because of the existence of heritable variations, natural selection and adaptation mechanisms enabled many organisms to "descend" from the first one by evolution. This process continued, expanding the variety of species and increasing their complexity over the course of geological time. In Darwin's mechanisms, the evolution of a new creature often implies the evolution of new organs and/or appendages. In the change from a reptile to a bird, a wing is supposed to have "evolved." In addition, an oil gland that secretes an oil which is spread on the wings must have come about. From the reptile to a mammal a breast must be evolved. More importantly, a new organ, the placenta, must be evolved. From an invertebrate to a vertebrate, the bones must be evolved, as well as the extremely complex vertebrate eye. Many of these new organs are developed as a consequence of having new genes that are not present in the supposed ancestral animal. In each of these supposed evolutionary changes, new genes and unique DG pathways must be brought about by evolution.

The evolution of a new creature possessing a new organ requires the availability of individuals with at least a primitive form of that organ in the population of the starting species. For these individuals with this rudimentary organ to have a better selective advantage, the rudimentary organ must have at least a rudimentary function.

Darwin's theory requires that random mutational processes lead to random physical changes, from which a useful organ or limb system is selected, fixing the corresponding genetic changes at the genome level. We already saw that it is improbable to arrive at a new gene or a DG pathway by random mutational processes within the genome of a species. We shall reinforce this by analyzing in the individual cases how it is improbable to bring about the new genes and the unique DG pathways required to construct each body part. First of all we have demonstrated that random outgrowths (an outgrowth whose development is dictated by a DG pathway not already present in the genome) do not occur in any organism on earth. If random outgrowths absolutely do not occur, where can the useful outgrowths come from that gave rise to all the organs and appendages that are found in the living world — starting from a worm-like creature believed to be the very first progenitor of all organisms? However, even assuming that random outgrowths can occur, we shall show that the probability for even one of them to be a useful organ is far too low to be meaningful.

What is a random physical change? It means random outgrowths, either external or internal to the body. For example, if a tail has to evolve in an animal that does not have a tail, then it has to occur purely as an outgrowth with at least some of its tissues, at least in a primitive form, among numerous totally unrelated random outgrowths in the body. All these need not necessarily occur in the same individual. But in different individuals in the population of a species, millions of diverse meaningless outgrowths should occur, before which an outgrowth useful as a tail can occur and be selected for some useful function. If we look at nature, there are at least several tens of thousands of different kinds of organs and appendages in the living world. And each one of them supposedly arose from a random outgrowth. Any useful part has to occur only by chance among myriad such meaningless random outgrowths. And that chance, even if outgrowths do occur, is vanishingly small.

Why the wing of the bird could not have evolved from the reptile

Evolutionists believe that birds evolved from reptiles by evolving wings. By evolutionary arguments, an individual reptile with no wing characteristics should have developed at least a small, most primitive wing among a great number of random physical variations in the population of the reptile. This reptile with a primitive wing, if capable of some flight, must have had a better selective advantage, and, because this characteristic is heritable, its offspring that express this characteristic would also have had a better selective advantage. A few individuals with this kind of a wing must have generated a line of descent to the first primitive winged reptile.

How do random physical changes appear? Any physical change in an organism can result only from random changes in the genes or the genome. While a random DG pathway must be able to develop a random body structure, an extremely large number of such DG pathways should be formed and be expressed for the correspondingly large number of novel structures to be developed. A wing with its feathers and the ability to fly should occur among such random variations. Unless the wing occurs in a form that is useful for flight, the animal will not be selected, and its genes and the wing-specific DG pathways will not therefore be preserved as a heritable change.

To evolve even a rudimentary wing in an organism, many new genes should evolve first. In addition, duplications and modifications of some existing genes should occur. Finally and more importantly, coordinate expression of all these genes, including both the preexisting and the newly generated ones, via the expression of new DG pathways developing the feathers and wings, must happen in order to build even the most primitive wing. This is because the wing is a unique structure, with many substructures to it, not present in any of the reptiles from which the wing is supposed to have evolved.

As we have discussed before, the probability of arriving at a given DG pathway consisting of only ten genes in a precise series or cycle, from an available one hundred genes, is 100^{-10} (10^{-20}). Thus, although superficially it appears that random genotypes would produce random variations from which a useful wing could be selected, such meaninglessly low probabilities indicate that it is simply not possible. The fact that we do not detect any such random outgrowths in any organism corroborates unequivocally that the said random variations are only imaginary. In short, the bird's wing could not evolve through descent with modification from an ancestral animal lacking it.

To the genome of an animal that lacks a wing, the new genes that code for the proteins of the feathers and the new DG pathway that develop the wings have no meaning. In other words, its genome is blind to the outside world and it has no way of recognizing that there is a medium called air and that if it developed wings, its host would be able to fly and be better able to live. This is what we should consider when we compute the probability for the genes and the DG pathway for a useful body structure to occur among the random mutations in the already existing genes of the genome.

In addition to feathers, birds have other unique aspects. For instance, the "uropygial gland" on its tail secretes an oil that lubricates and waterproofs the feathers. Birds rub the oil from this gland onto their beaks and use it to preen their feathers. This oil is indispensable to birds, without which the feathers would become brittle and useless in flight. Water birds also use the oil to keep their feathers water resistant, without which they would become nonbuoyant and sink.

Because evolutionists believe that the birds evolved from the reptiles, they erroneously say that feathers must have evolved from the reptilian scales. On the contrary, when we look at the highly specific structure and anatomic features of the feather, with many intricate substructures, it will become clear that it must be developed by an extremely specific DG pathway. A bird has thousands of feathers on its body, serving a variety of functions.[33] The structure of one, the contour feather, is shown in Figure 3.11. It has a central shaft consisting of the rachis, which supports the vane, and a naked base, the quill. The quill is anchored in the skin follicle. The vane consists of many parallel barbs, each with a row of minute projections called barbules. The vane is made rigid, because each barbule has tiny hooks that interlock with grooves

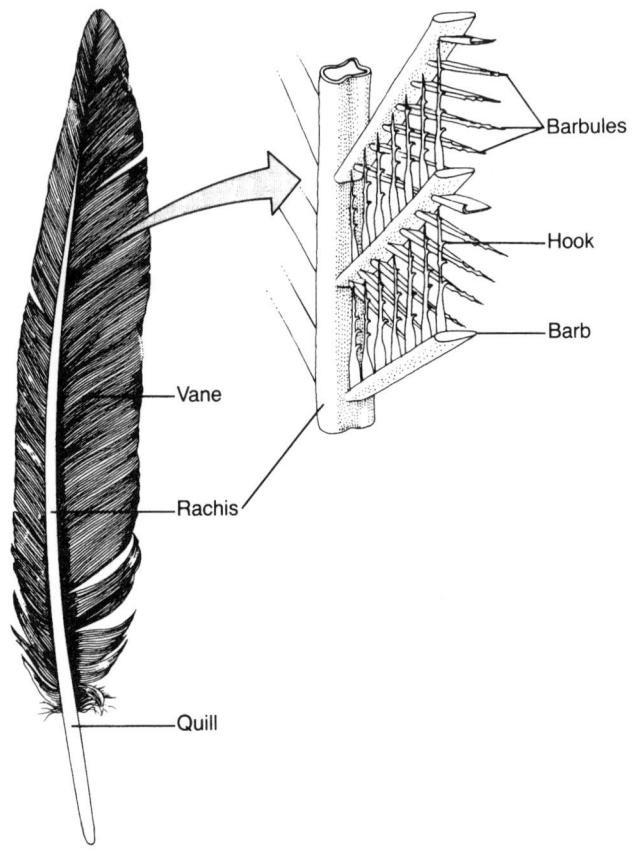

FIGURE 3.11. STRUCTURE OF A CONTOUR FEATHER. A flight feather from the wing, showing the microstructure of the vane. (From: Zoology by L. Mitchell, J. Mutchmor, and W. Dolphin. Copyright © 1988 by The Benjamin/Cummings Publishing Company. Adapted by permission.)

on adjacent barbules. If the barbules are detached and the vane is loose, a flight feather cannot function at all. Feathers develop as small folds (papillæ) in the skin. By elongating and sinking into the skin, a papilla forms a pit called the follicle. A feather develops at the base of the follicle from epidermal cells. The developing feather is nourished by blood vessels and nerves. A feather, with all its structures, develop in a protective coat. Color is added to the growing feathers by pigment cells in the epidermis. The feather takes shape as the epidermal cells die and become keratinized. Once it reaches full size, it unfolds and breaks out of the protective coat of dead epidermal cells, leaving a hollow feather shaft. The mature feather is a completely dead tissue. The follicle serves only to anchor it to the body.

Certainly the intricate developmental process of the complex structure of a feather requires a complex and specific genetic network involved in its development. Is it reasonable to say that such an extremely unique genetic network is produced by random mutations in the genome of a reptile? Does it not show that the probability for such a thing is practically nil even given a trillion trillion years? Whether or not these genes were present in the supposed reptilian ancestor, the unique DG pathway for feathers certainly was not present in its genome. It has got to be newly evolved. Only when a functional feather is evolved and the whole wing itself is formed, at least in a primitive but useful form for flight, will the bird be able to fly. That is, the selective advantage of flight would facilitate the first bird to be selected if and only when a functional wing is evolved. Until then, no part of the genetic network will be preserved in the genome of the supposed reptilian ancestor. It is an all-or-none law. How can such an intricately complex developmental genetic network be evolved all at once within the genome of the assumed reptilian ancestor, in which such a network is totally absent, by purely random genetic mutations? How can the genome recognize the evolutionary niche of flight, and produce such a DG pathway specifically? It cannot and it did not.

A fossil animal, *archeopteryx*, is cited by paleontologists and evolutionists to be an intermediate organism between the reptile and the first bird species. This animal, that had a fully developed pair of wings, also had a few characteristics of reptiles, such as claws at the end of the wings. Evolutionists argue that *archeopteryx* is one of the many successive intermediates with small improvements in its wings that any reptile must have gone through to become a bird, but the fossil record contains only this example.

Archeopteryx is indeed a fully developed bird like any other bird (see Chapter 10). Evolutionists' arguments that it is an intermediate between reptile and bird is seen to be incorrect when we consider our above arguments against the following fact. Out of the many supposed intermediate stages, no specimen of any other intermediate is found in the fossil record other than *archeopteryx*. Furthermore, the fact that 12 specimens of only *archeopteryx* have been found while none from any other supposed intermediates indicate that the specimens of others are not absent purely by chance. We can see that no other intermediates have ever existed, and the *archeopteryx* is not an intermediate between reptile and bird. It is a fully developed bird with some reptile-like characteristics. And this structure of the *archeopteryx* with "mixed" characteristics can be explained without involving evolutionary connections at all! (See Chapter 8.)

The evolution of placenta — an extremely complex and unique organ — is improbable through random mutational processes

When we analyze the supposed evolution of mammals from reptiles, a new mammalian organ, the placenta, must be evolved. Both anatomically and functionally speaking, it is a complex, highly specialized, and unique organ. The functional capacity of the placenta is far more complex than most organs in the body. It supplies beneficial nutrients to the embryo from maternal blood, prevents harmful material from entering the embryo, and excretes unwanted biochemicals from the embryo into the maternal bloodstream. A unique umbilical cord, with an artery and a vein, transfers blood between mother and fetus. In essence, the placenta is an organ with new tissues having highly specialized functions. It is derived from both maternal and embryonic tissues. Its development and function involves the expression of a large number of genes including a set of new genes, that are not expressed in other tissues and organs of the body, and by a precise genetic pathway unique to itself. One can imagine that it would take many structural and functional genes organized into a complex DG pathway to construct this complex organ.

It is possible that a number of these could be new genes, needed to bring about placental-specific functions, which are unique to mammals. There is a set of proteins called placental-specific plasma proteins (PSPP),[34] which appear to be mammal-specific, and absent in nonmammalian organisms such as reptiles. Furthermore, there are some proteins discovered in goats and sheep — produced by the free-floating embryo[35] — which act on the uterus inducing it to produce some of its own proteins. (Incidentally, this is the only way that the mother's system recognizes that there is a developing embryo in her uterus.) These proteins appear to prepare the uterus for the implantation of the embryo.[36] This is a mammal-specific process and therefore the proteins used in this process and thus the genes encoding these proteins may very well be mammal-specific. This sort of specific presence of some proteins is also found in primates. For example, chorionic gonadotropin is a placental protein hormone that exists only in primates,[37] not in reptiles or other animals (with the exception of the horse).

The assumed transition from an egg-laying animal to a placental animal is sudden without any intermediates in the fossil record, and without any transitional forms in the living world. How can random mutations in the genome come up with even one new protein that would be specifically useful in the construction of the placenta? Such a new protein is one out of a trillion trillion trillion and more possible protein sequences. Where do the new proteins constructing the placenta come from, if entirely new genes cannot be evolved from existing ones? How could these new genes and DG pathways have evolved within the genome of a reptile organism purely by random processes to specifically develop a functional placenta? As we discussed for the wing characteristics, random mutational processes within the genome should bring about a large number of random DG pathways leading to random outgrowths internally or externally to the body. One out of these numerous outgrowths, purely by chance, should be useful as the placenta — at least a very primitive one to start with. Even if, for the sake of argument, we suppose that random DG pathways are possible to occur in a genome, the probability that the unique DG pathway that could develop a placenta could occur among them is meaninglessly low. Therefore, the placenta could not have evolved from a reptile.

All mammals have mammary glands (breasts) which secrete milk for the nourishment for their offspring. Both the organ and the function are totally lacking in the reptiles, which lay eggs; the offspring is not nourished with any body fluid from the mother through any organs like breasts. When the evolution of even one organ is absolutely improbable from an organism lacking it, how can two organs — placenta and breasts — evolve in the same organism?

THERE ARE MANY DIFFERENT KINDS OF EYES IN THE LIVING WORLD, NONE OF WHICH CAN BE BROUGHT ABOUT THROUGH ORGANISMAL EVOLUTION

There are many different kinds of eyes in the animal world, and each has a specific structure and exhibits unique functional capabilities. Based on all the probabilistic and statistical methods we have so far used in order to understand if the specific genes or the genetic network for a new body part or structure can evolve within the genome, it is highly improbable to evolve any one of these eyes from organisms that lack any visual abilities. But yet, in the supposed evolutionary sequence, almost all the different kinds of eyes started with animals having no photoreceptors (the molecular or cellular unit that recognizes and responds to light). Seeing that there are so many varieties of eyes in the animal world which have no evolutionary connection, evolutionists Mayr and Salvini-Plawen interpret this to mean that they could have evolved many many times (at least 60 times, see below) in the animal world in independent evolutionary lines, and if so, they conclude that it must be very easy to evolve eyes. Their statements have truly no scientific basis and validity, other than their own beliefs based on the larger domain of evolutionary theory.[38]

Darwin was extremely confounded by the highest perfection of the eye. But he thought that if a gradation of eyes from simple to complex can be shown to exist in the animal world, that in itself could form a support to his theory. He wrote:[39]

> To suppose that the eye, with all its inimitable contrivances for adjusting the focus to different distances, for admitting different amounts of light, and for the correction of spherical and chromatic aberration, could have been formed by natural selection seems, I freely confess, absurd in the highest possible degree. Yet reason tells me, that if numerous gradations from a perfect and complex eye to one very imperfect and simple, each grade being useful to its possessor, can be shown to exist; if further, the eye does vary ever so slightly, and the variations be inherited, which is certainly the case; and if any variation or modification in the organ be ever useful to an animal under changing conditions of life, then the difficulty of believing that a perfect and complex eye could be formed by natural selection, though insuperable by our imagination, can hardly be considered real.

The presence of a gradation in eyes existing in all the animal world put together is true, but that gradation does not occur in a sequence correlatable with an evolutionary sequence of organisms. They occur in a random manner throughout the repertoire of organisms in the animal world. Even according to evolutionists' own account, it is very hard to correlate the gradations with the supposed ladder of animal evolution. Even after freely buttressing their theme, they could come up with sixty or more separate "independent" lines of eye evolution in the animal world. In fact, in my opinion, if all the buttressing is avoided, the number would go up to hundreds of different lines.

Complexity and uniqueness of different kinds of eyes

Just to indicate the complexity and uniqueness of the many kinds of eyes in the living world, let us examine three types of eyes — compound, pinhole, and vertebrate. Most insects and crustaceans have compound eyes.[40] A compound eye consists of an array of visual units called *ommatidia*.[41] Each of these gather information from a small portion of the eye's field of vision. An ommatidium is essentially a tube, with lenses at the outer end focusing light on photoreceptor cells located at the bottom of the tube. Light that enters neighboring ommatidia will differ in brightness, so that the field of vision is seen as a mosaic image of dots of light at different brightnesses.

Although the unrefined lenses of a compound eye produce coarse images, a compound eye has an excellent ability to detect motion. This feature is especially useful to fast-flying insects, such as a dragonfly watching for flying prey or a bumblebee scanning for landmarks near its nest. The eyes in such flying animals require that they are able to distinguish fast-moving images. This is made possible by the rapid "resetting" of the photoreceptor cells. Some insects can distinguish images (flashes from a light source) arriving at a rate of 330 per second.

Moreover, in some creatures the compound eye can make adjustments within the eye structures according to changing light intensities. In addition to color vision, the compound eye in some arthropods are sensitive to polarized light, thanks to a geometric arrangement of visual pigment molecules in the photoreceptor cells. With polarized light detectors, an animal can determine the position of the sun by the pattern of polarized light from only a small patch of blue sky. From the changing pattern of polarized light, the animal's orientation relative to the sun can be determined throughout the day. Honeybees and certain species of ants use this ability to point the way home when they are out foraging for food. Many other animals, including certain vertebrates, also have this capability and use it for orientation. Many birds and fishes and even certain mammals navigate long distances using patterns of polarized light.

Another kind of eye called pinhole eye forms images by admitting light through a small opening, or pinhole. The eyes of the chambered nautilus is an example. Light rays passing through the pinhole cast an inverted image on the retina at the back of the eye. The opening is small and therefore produces a sharply focused image; but since very little light is admitted, the image is dim.

Now, when we contemplate on the eyes of vertebrates, their single-lens eyes are quite different from either the compound eye or the pinhole eye. They are covered by a tough protective layer of fibrous connective tissue. The central transparent part of this forms the cornea, which admits light. The rest of this layer forms the "white" of the eye, or sclera. In many vertebrates, upper and lower eyelids protect the cornea and sclera. The eyes are closed by the contraction of a muscle in each eyelid, and a levator muscle in the upper eyelids opens them. Mucous membranes on the inner surface of the eyelids moisten and cleanse the eyes as they close and open.

The pupil is an opening in the center of a pigmented muscular diaphragm called the iris. Light first passes through the cornea, crosses a fluid-filled chamber, and then passes through the pupil. By automatically changing the diameter of the pupil as light conditions change, muscles in the iris control the amount of light that enters the eye. The autonomic nervous system controls the response of the iris muscles to changing light intensity.

The light, after traversing through the pupil, then passes through the lens. A lens consists of cells with large volumes of clear, fibrous fluid. The lens is suspended by suspensory ligaments extending from its margin to a surrounding muscle. The biconvex lens converges light rays as they cross its front and back surfaces. The chief function of lens is to project and focus images onto the retina at the back of the eye. The cornea is convex in shape and it also bends light rays at angles that aid focussing.

Here, the eye must adjust itself to focus on objects viewed at different distances. The eyes of many fishes achieve this by moving the lens forward and backward. On the other hand, the eyes of the human and that of many other vertebrates accommodate by changing the shape of the lens itself. To achieve this, many other structures and muscles are involved to reduce or increase the tension on the suspensory ligaments attached to the rim of the lens, allowing the lens to become more spherical and thicker or thinner and flatter.

Several kinds of nerve cells form three layers in the vertebrate retina. Before reaching photoreceptor cells in the third layer, light passes through two transparent layers. Photoreceptor cells are of two kinds called rods and cones. When light stimulates rods and cones, their membranes generate electrical potentials, which trigger release of neurotransmitter molecules at the first transparent cell layer, which in turn releases a neurotransmitter onto the second layer, which consists of ganglion cells. As a result, nerve impulses are generated in the ganglion cells, which then travel along the optic nerve to the brain, where most integration and interpretation occur in the visual areas of the brain.

Rods and cones have many functions making vision possible in bright, dim, and colored light. The different cones respond to different colors (red, green, and blue) because of differences in their protein components. Different combinations of these three populations of cones make it possible for many primates (including humans), birds, reptiles, amphibians, and fishes to see many colors. Colored oil droplets are present in the cones in birds and turtles, that filter certain components of light, allowing separation of colors before light strikes the retina. Most mammals, except for the primates, have few or no cones. Animals without cones are color-blind, while those with some cones can see subdued colors.

The improbability of evolving even one type of eye through descent with modification starting from an organism totally lacking even in photoreceptors

Starting from an animal lacking photoreceptors, how could even one type of eye evolve? Among the mutations that must occur randomly in the genome of such an organism, according to evolutionary theory, how do the genes which are specifically required for the biochemistry of the eye and for the construction of the structure of even a supposedly simple eye, evolve? It is clear that the genes that specifically construct an eye must be arranged in an extremely specific DG pathway to create all its substructures and the whole eye. One can see that the genome has no way of recognizing that there exists an entity called light surrounding the organism, and that if it brings about the necessary changes in it to evolve an eye, it would be useful for "seeing" things in its environment. Natural selection can operate only on blind, random changes. As we have seen in the supposed evolution of other organs above, the evolution of even one gene that is specifically useful in the construction of an eye within the genome of an organism is highly improbable. Then how can the large number of genes necessary in the structure and function of even one type of eye evolve? Even assuming they do, how could they be arranged within the genome in the right DG pathway to construct such complex structures and make them carry out such complex functions? How can all the different kinds of eyes, almost all of which are complex, evolve in independent lines of evolution, as claimed by the evolutionists, each starting from an organism which has absolutely nothing related to eyes in structure or function? The probabilities involved, even through gradual changes, are too low to be meaningful. Therefore, certainly, even one type of eye could not have evolved starting from animals lacking photoreceptors. In fact, as we shall see below, the scenario of the distribution of the different kinds of eyes in the animal world poses a great difficulty for evolutionists to evolutionarily relate the different kinds of eyes.

Single-lens eyes are also present in certain jellyfish, worms, spiders, mollusks, and vertebrates. Spiders usually have eight such eyes, arranged in two groups of four one above the other. Many insects also have single-lens eyes called ocelli. Squids and octopuses, which are invertebrates, have highly perfect eyes similar to, and even more advanced than, those of the vertebrates. However, according to evolutionary theory, vertebrates are not descended from these animals. Instead, vertebrates are supposed to have evolved from animals that had the compound, pinhole, or some other kind of eye of some other invertebrates. And this again is improbable.

There are hundreds of different kinds of eyes in the living world, evolutionarily uncorrelatable; staunch evolutionists themselves say that if one has to construct an evolutionary tree of different organisms in the living world based on their eyes, it would be a monumental task. This, we can see, is because these eyes were not evolved from earlier ancestors by descent with modification.

The large number of different eyes existing in the animal world are evolutionarily uncorrelatable, making evolutionists believe, albeit with much difficulty, that these evolved independently at least 40-65 times. However, an objective analysis shows that even a far larger number of unique eyes are randomly distributed in organisms and that the scenario is actually against evolutionary theory.[42]

Considering the genomic mechanisms, as we discussed, it is improbable to evolve a vertebrate eye from a compound or any other invertebrate eye, or even an invertebrate eye starting from an animal without any photoreceptor cells. However, because there are many different kinds of eyes in the animal world, which do not correlate with the supposed evolutionary tree, evolutionists believe that these different eyes have independently evolved — each one of them starting from a different organism that lacks even photoreceptors. We shall illustrate that there is a far greater number of such evolutionarily-uncorrelatable eyes than determined in these studies, and that these could not have originated by organismal evolution.

In 1977, evolutionists Plawen and Mayr published an article in which they reported an exhaustive and systematic study on the evolution of photoreceptors and eyes[43] — in order to verify Darwin's predictions that the various kinds of eyes are correlatable in one single evolutionary line, starting from a primitive ancestor. In fact, they found that there are at least 40-65 and perhaps more distinct types of eyes present in animals that cannot be connected by any meaningful phylogeny (a phylogeny is an evolutionary tree). Therefore, being staunch evolutionists, they concluded that each of these different eyes had independently evolved from animals possessing no eyes or photoreceptors. Only in some instances, they seemed to show possible sequences of eye evolution, but even in these cases, the authors believed that the evolution happens through a replacement of one type of cell or structure with another type, or acquiring new structures, quite easily and frequently — which based on our analyses is highly improbable. Some excerpts from this article would clearly illustrate the extreme difficulties in connecting the supposed evolution of eyes (italics mine):

> It requires little persuasion to become convinced that the lens eye of a vertebrate and the compound eye of an insect are independent evolutionary developments. But when we ask how many other times in the phylogeny of animals simple photoreceptors or more highly differentiated eyes developed independently, *we are up against methodological difficulties*. An organ that merely records light intensity can be quite simple, consisting of a slight specialization of a sensitive epidermal cell. *When such a photoreceptor is found in certain genera or families of a higher taxon (order, class, phylum) but not in others, it is sometimes difficult, and sometimes quite impossible, to decide* whether the multiple occurrence of the receptor is due to inheritance from a common ancestor or due to an independent response to similar selection pressure, because "there are so few ways of making an eye that the degree of convergence can be extraordinary." ... *The difficulty is hardly less great in many cases of highly complex eyes with a retina, a pigment layer, and various focusing devices (lens, iris etc.)*. The acquisition of these auxiliary structures, to enhance the effectiveness of the light sensitive cells, is apparently of such great selective advantage that it has occurred many times independently in the animal kingdom.
>
> ... *Although these various well-defined types of photoreceptors are sometimes transmitted from one group to a derived one, they have originated inde-*

> *pendently sufficiently often to keep them from supporting a far-reaching phylogenetic scheme. Their distribution in the various phyla of animals cannot safely be used as a basis for the construction of phylogenies.*
>
> *... Adopting the most rigorous criteria of homology, at least 40 different lines of photoreceptor differentiation must be postulated originating from unmodified cells with not-yet-determined special structure. These do not include photoreceptors, the homology of which is uncertain, or the ultrastructure of which needs to be investigated. Accordingly, and considering the ease with which eyes are apparently acquired during evolution, and considering furthermore that there is only a limited number of different eye types, it is possible, if not probable, that in at least some 20 additional cases similar eyes were acquired in certain phyletic lines independently by convergence. If true, this would mean that eyes evolved in evolution independently over 60 times. This number does not include those potential photoreceptors that have not yet been recognized owing to an absence of visible pigmentation.*
>
> *... The evolution of eyes raises also some rather general evolutionary questions. Natural selection is effective only when there is sufficient phenotypic variability on which it can operate. A selection for lenses, to offer one example, can be successful only when there is variation in the thickness of membranes, and a localization of such differences. The polyphyletic origin of lenses, and of other eye components, indicates that such variation must have always been abundantly available, as Darwin had postulated long ago. At the same time, so much variation in such a vital organ as the eye has always seemed somewhat unexpected.*
>
> *... A far greater puzzle is this:* Why should it have been "necessary," if we may use this expression, for eyes to evolve 40-65 times? One would think that the possession of eyes is of such great selective advantage that they would have evolved virtually simultaneously with the origin of Eucaryonta. In that case it would have been possible to trace back all eyes to the photoreceptive organ of that inventive ancestor. All the evidence, however, indicates that the earliest invertebrates, or at least those that gave rise to the more advanced phyletic lines, had no photoreceptors. This seems hard to understand because the possession of such organs should have been as advantageous in the pre-Cambrian as in the Ordovician or Devonian.

The authors also found that there are three structurally different types of photoreceptors, which have no correlation with major phylogenetic lines.

> *... there are at least three structurally different types (ciliary, rhabdomeric, and unpleated), and that there is an additional basic (diverticular) type of photoreceptor which originated from internal acilious parenchymous or ganglionic cells. None of these types in cellular differentiation can be strictly correlated with major phylogenetic lines.*

The distribution of various kinds of unique eyes — supposed to represent a gradation from optic nerve to vertebrate eyes — is not represented in the animal world as a gradation, but rather it is distributed in a nearly random manner. However, Plawen and Mayr tried to fit their observations with the evolutionary paradigm. The observations are fitted to their paradigm by "belief statements." For instance, just the existence of so many types of eyes is in itself a proof to the evolutionists that variations leading to lenses and eyes must have occurred, that eyes are easy to evolve, and that eyes did evolve in 45-60 different phylogenetic lines. With absolutely no other evidence, it is circular reasoning.

Even within an animal group, called a taxon, in which an evolutionary sequence of eyes is believed to be seen, the supposed evolutionary sequence is only superficial, and the authors unintentionally defend the evolutionary sequence and the evolutionary paradigm itself with their belief statements. It is also obvious that in most instances there is no such correlation:

> ... At least the aesthete eyes in chitons, the dorsal eyes in Onchidiacea, the distal retina in the eyes of Pecten, and the branchial eyes of Serpulimorpha are undoubtedly new acquisitions within their higher taxa.
> ... All these facts indicate that the differentiation of ciliated (epidermal) photosensitive cells has taken place several times resulting in different specializations (disks; lamellae; microvilli below or at the base of cilia; tubuli; numerous cilia as in Kamptozoa or Bryozoa larvae) and at different levels of organization according to functionally selected "need." There is no correlation whatsoever between type of ultrastructure and such phyletic groupings as Bilateria groups of Gastroneuralia (protosomes less lophophorates) or of Oligomera + Notoneuralia (deuterostomes plus lophophorates).

The authors' manner of stating observations mixed with assumed evolutionary connections illustrate how these statements are based on pure faith with absolutely no scientific validity, the only supporting entity being the evolutionary theory of Darwin.

> ... Photoreceptors are present in most animal groups, either through inheritance from an ancestral group *or by new acquisition*. ... When similar photoreceptors are found in the eyes of different higher taxa, no decisions can be made whether they are homologous or acquired by convergence, until the structure is traced back to the ancestral condition. *The eye may have been drastically modified during phylogeny. Eyes of similar structure may have evolved independently (e.g., mollusks and annelids), whereas rather different photoreceptors may have been derived from a common ancestral organ.*

Why are hundreds of different kinds of eyes evolutionarily uncorrelatable? Why even in the assumed sequences of eye evolution, there is really not a true sequence, that is, why are the different kinds of eyes randomly distributed in the animal world? Why are staunch evolutionists like Ernst Mayr and Salvini-Plawen puzzled, confused, and confounded to see this scenario? The answer to all these questions is that eyes have not evolved from animals not possessing eyes. The different eyes in the animal world are not the result of organismal evolution. An organism with a new body part cannot evolve from another organism lacking in that body part. This is why the authors found it extremely difficult to fit the scenario of eyes in the living world to the theory of evolution,[44] but they had to fit their data with the only theory for the origin of organisms that was available. The truth, however, is that the different eyes originated independently when the various organisms possessing them were born independently in the primordial pond (see Chapter 8).

IT IS FAR LESS PROBABLE TO SIMULTANEOUSLY EVOLVE MULTIPLE ORGANS IN THE SAME ANIMAL SPECIES THAN TO EVOLVE A SINGLE ORGAN

Under the evolutionary theory, many animals should have evolved more than one organ simultaneously. For instance, the evolution of the vertebrate eye and the bone are required in changing an invertebrate into a vertebrate.

We just saw that it is improbable to evolve an organ in an animal by random genotypic changes. The probability to evolve two organs in the same organism is the product of the probabilities for the evolution of each organ. When the probability for even one organ is too low to be meaningful, even less likely is the evolution of more than one organ in the same organism simultaneously. Thus, the probability of the evolution of the bone and the eye in the same organism is tremendously lower than when only the eye or the bone is considered. Similarly, consider the supposed evolution of the wing and an oil gland in the bird, and the placenta and the breast as well as hair in the mammal. The usual evolutionary argument is that natural selection selects a little here and a little there, and in geologic time its accumulative power brings forth order. But it is a totally misleading one. The requirement for the simultaneous occurrence of many subtissues of an organ in the same species for them to be coordinately selected, such as the many bones in a vertebrate or the many subtissues of a vertebrate eye, argues against this.

EVOLUTION OF NEW ORGANS BY MODIFICATION OF EXISTING ONES IS IMPROBABLE. EVEN IF IT IS TAKEN TO BE PROBABLE, IT WOULD ONLY EXPLAIN EXTREMELY RARE CASES IN LIVING ORGANISMS. WE ARE PRINCIPALLY INTERESTED IN UNDERSTANDING THE ORIGIN OF MYRIAD UNIQUE ORGANS AND APPENDAGES IN THE ANIMAL WORLD.

Evolutionists believe that existing limbs are modified to become new limbs, and that this provides a support for the evolutionary theory. In evolution, the fish's fins are supposed to be modified into the limbs of frogs. On the contrary, the structure of the fins of fish and the limbs of frogs are distinct, containing different numbers and arrangements of bones and tissues; the fins of fish and limbs of frogs are different and unique body parts. By our arguments of the uniqueness and rigidity of DG pathways it is clear that this is improbable. Similarly, the evolution of the front limbs of reptiles into the wings of birds is also improbable.

In other examples, evolutionists cite the resemblance of the mouth parts of some crustaceans to their leg parts and propose that the legs parts have become duplicated and modified to become the mouth parts. While I do not dispute the possibility of the duplication of a limb, the probability for the limb to be modified to be used for another purpose through descent with modification is extremely low. However, one may argue that the leg and mouth parts of crustaceans, such as the lobsters, are very similar; and that already there are three pairs of legs in a crustacean and if the front pair is duplicated, why can it not be used for collecting food? Even if it is granted that such a thing is possible, still it would explain only one in a thousand organs and limbs in the living world. But we are discussing the supposed diversification of all the living organisms on earth from an original species such as a marine worm which lacked any organ or appendage. For instance, we are discussing how the original limbs of the crustaceans came about and not how they have been modified. Or, how the wings of the fly or an insect originated in the first place and not how a pair of wings is converted into two pairs of wings. The main question is how the unique limbs and organs of the myriad organisms have originated. An understanding of this will lead to further understanding that the origin of all unique organs and appendages in the living world is not through an evolutionary process.

THE CONCLUSION: EXAMPLE ANALYSES ILLUSTRATE THAT THE THEORY OF EVOLUTION, IN ANY FORM, IS ERRONEOUS

In all the observations and conclusions we have derived from the examples we have discussed so far, we can see our principle that new organs cannot evolve from organisms lacking them is strongly corroborated. Along with our examples of wing, placenta, and eye, for the improbability of evolving new organs one could cite thousands more examples of organs and appendages throughout the living world and also show that they could not have evolved from their supposed ancestors. In conclusion, the concept that random genetic changes lead to random physical changes from which new useful structures are selected resulting in new organs and body parts is incorrect (see Figure 3.12). It is unfortunate that the theory of evolution has pulled and confined many a scientist into its fold, albeit they get confused and confounded in many instances such as in the above examples, only because there has been so far no theory that can explain these scenarios without evolutionary connection.

FURTHER OBSERVATIONS AND ARGUMENTS INDICATING THAT ORGANISMAL EVOLUTION NEVER CAN, AND NEVER DID, HAPPEN

THE GENOME IS BLIND AND CANNOT VISUALIZE THE EXISTING NICHES AND ENVIRONMENTS. THEREFORE, MILLIONS OF BIZARRE PHENOTYPES MUST BE PRODUCED IN A SPECIES FOR THE SELECTION OF ONE USEFUL STRUCTURE.

The niche (the physical and ecological environment occupied by an organism) for a winged animal did exist when there were no birds or winged insects. However, genetic changes do not "look" for niches — that is, the genome is "blind" to the environment. The genome of the reptile or the wingless invertebrate did not "know" that there was a medium called air in which the animal could fly if it developed a wing for its host. Therefore, in the supposed evolution, the genotypic changes should occur purely randomly, and the resulting physical changes, if advantageous, will be selected. The supposed occurrence of the very first primitive wing in a reptile (or in an invertebrate changing into an insect) should be one among millions of meaningless changes that should have occurred in different individuals of this species. This variation must first occur in at least one individual so that it can spread and get fixed in the population and continue to evolve. The very basis of evolutionary theory is that only out of random mutations those that produce useful characteristics are selected. Despite this, what the evolutionary biologists do not seem to realize is the fact that the genome is absolutely blind

> ### Why the Genome of an Organism Must be Evolutionarily Closed
>
> - The set of genes in a genome is constant.
>
> - The DG pathway of an organism is fixed.
>
> - Mutations cannot evolve new genes or new DG pathways.
>
> - Mutations can only lead to aberrations of existing body parts, genetic and congenital diseases, cancer, harmless alterations in characteristics such as skin color or pattern, body size, and other individual variations.
>
> - Mutations can change one species into a similar species but not into a distinct organism.
>
> - Individual variations are confined within the closed framework of every species of an organism.
>
> - True monstrosities required for evolutionary change can never be produced by any kind of mutations.
>
> - True monstrosities are nonexistent in living or extinct organisms.

FIGURE 3.12.

to the environment, and therefore an immense, almost infinite number of random mutations should occur in order to arrive at some useful structure. But our rich knowledge in the field of molecular biology and genetics indicates that this many mutations cannot occur in living organisms even in trillions and trillions of years.

For every new useful body part, one can imagine millions of bizarre structures that do not have any meaning. Questions arise concerning the "evolution" of genes leading to these variations. Again, the need for selection from random processes indicate that it is not possible to evolve even one such gene in the time that the earth has been in existence. If evolutionary mechanisms are correct, indeed a vast array of purposeless genes should be generated, among which a useful new gene should "occur" by chance and be selected. But there is absolutely no evidence that such a process is happening.

The genome size, although it consists of billions of nucleotides, is still too small for any such random mechanism to occur

Molecular biologists consider the genomes of multicellular organisms to be immense. The genome sizes run anywhere from 50 million nucleotide characters up to about 300 billion characters.[45] However, these numbers are still far too small for any of the random mechanisms leading to speciation to be occurring within the genomes. If the random duplications and tinkering with the genes occur within the genome by some genetic mechanisms leading to evolutionarily useful genes, then the genome sizes should be billions or even trillions of times larger than they are.

If Darwin's theory is correct, then evolution must be an ongoing process: Where are the incipient organs, appendages and species?

If evolution is an ongoing process, why don't we see reptiles in the process of evolving into birds currently? In responding to this, evolutionists might say that flying niches in the air currently are filled, and that is why no new birds are evolving from existing reptiles. This is an incorrect statement. There is nothing at present that can inhibit the evolution of new birds from reptiles. If organisms are even 1/10th as plastic as Darwin has discussed, the change of one creature into another with intermediate steps should be happening now among all the living organisms on earth. So reptiles should be evolving into incipient birds now, but the already evolved birds would compete more successfully with the currently evolving primitive birds.

Similarly, one can expect that invertebrates should be changing into winged insects and vertebrates, fish into amphibians, amphibians into reptiles, and reptiles into mammals. Why do evolutionists say that each of these transformations occurred at only one or a few times within very restricted geological times? Is it not reasonable to wonder that if they could change at one time into these different forms, then why couldn't at least one or a few species of each of them, if not a large number, currently be changing to organisms at higher steps in the evolutionary ladder? In fact, it should be a fairly continuous process. There is no genome mechanism which indicates that organisms evolve and then freeze. Similar arguments arise regarding incipient organs and appendages. Why do we not find any living organisms, or at least fossil organisms, with such incipient body parts?

If one reptilian species could grow a wing, then many reptilian organisms must be able to grow wings. Why should not mammals, such as rat or mouse, do the same? Why are fish not becoming amphibians now? If such evolutionary activities were possible then, some of it, at least rarely, must be seen now. On the contrary, we cannot demonstrably see any such evolutionary activity in any of the millions of creatures on earth today.

Darwin and other evolutionists have tried to fit the supposed mechanism of evolutionary change with the scenario of life on earth. There is no way the evolutionists can account for the lack of current evolution of fish into amphibians, amphibians into reptiles and reptiles into birds and mammals. Why should one think only about such a unidirectional pathway of evolution, where only the tip of the pathway is supposed to be active and not the starting points and middle points? This is because, in the first place, Darwin's theory is misguided and erroneous. The main reason for Darwin's theory to have been steadfastly followed is that the scenario of life on earth, containing certain deceptive similarities among organisms, misleads the human observer towards evolutionary change as the answer. Because the evolutionary theory is fundamentally incorrect, it can explain life on earth and the fossil record only superficially.

SPECIATION HAS NEVER BEEN DOCUMENTED

It has been the belief of evolutionary biologists that life on this planet is primarily the direct outcome of the process of speciation — the splitting of one species into two. In reality, true speciation has never been observed in any species on earth, either in experimental or in wild animals. Even among evolutionary biologists, it has been unanimously admitted that speciation has been seldom observed if ever and that we have little information about speciation. For instance, Guy Bush, who is a staunch evolutionary biologist, says the following in a recent article:[46]

> Although the importance of speciation is clear and convincing, the processes involved are, for the most part, unknown.
> ... Furthermore, speciation is usually a rare event, seldom if ever observed from start to finish. Our current concepts of speciation are therefore primarily based on *post hoc* reconstructions of past events, or derived from theoretical population genetic models usually based on classical Mendelian genetics, with all the inherent weaknesses and speculative nature of these approaches. The *post hoc* approach is, at best, subjective. ... One has only to peruse the literature to realize that although much has been written, little concrete information is actually available on the genetics of speciation.
> For instance, a genetic cornerstone of current speciation theory is "coadaptation." ... Unfortunately, the hard data on which the concept of coadaptation is based are not impressive. Specific cases

> ... are woefully lacking, and I am unaware of any unequivocal cases demonstrating that genetic regulations have been directly associated with speciation.
>
> ... In almost all cases, the observed facts are controversial and, upon close scrutiny, clearly open to conflicting interpretations.
>
> ... There is no question that speciation does occur as a result of geographical isolation. But how important has it really been? We only have score cards inferred from rather circumstantial evidence, which in turn has been derived from a few animals and plants that speciated sometime in the past. It is for this reason and others alluded to earlier that I remain skeptical of the universal application of allopatric speciation to all sexually reproducing eukaryotic organisms.
>
> ... It is clear that these nonallopatric models of speciation suffer from the same lack of hard data on the genetics of speciation that plagues the proponents of allopatric speciation. Whether any of these models reflect natural processes has not been conclusively established.
>
> ... Until we know more about the molecular machinery of adaptation — that is, what is and is not possible at the molecular level — our models of speciation must remain little more than speculation based on the subjective interpretation of equivocal data.

The highly speculative nature of the concept of speciation is clear. One cannot but wonder as to why one should still cling to the mechanism of speciation, even after so much skepticism of it. The reason must be that he first believes in Darwin's theory of evolution, and expects that the genetic mechanisms of speciation would be better explained someday in the future.

Grave inadequacy of evolutionary biologists in understanding the changes in DG pathways required concurrently in a supposed evolutionary change

Many evolutionary biologists themselves feel that genetic descriptions and mathematical models of changes in the shape of an organism during a supposed evolutionary change provide rather abstract pictures of the changes in developmental processes that might have transpired in evolution. These abstract descriptions do not tell us the mechanisms by which changes in DNA are translated into physical changes. For instance, Douglas J. Futuyma writes[47] (*D. melanogaster* and *D. simulans* referred to below are two different varieties of the fruit fly *Drosophila*),

> We have no idea of what we would have to do at a molecular or cellular level to transform *Drosophila melanogaster* into *D. simulans*, much less a fly into a flea. In all of biology, the mechanisms of development are the area of greatest ignorance, but they are central to major

> questions in evolution. ... Much of the progress in developmental biology bears only a tenuous, hypothetical relation to evolutionary studies. The mechanisms by which some genes exert their morphological effects are known, but chiefly through the study of rather drastic mutations; seldom does a geneticist determine the mechanism by which a gene difference between related species causes their difference in morphology. Similarly, few studies in experimental embryology describe the mechanisms that cause differences between related species. Geneticists and developmental biologists are fully occupied with the enormously difficult problems that are their proper province.

Futuyma's writing depicts that the understanding of the supposed evolutionary change at the level of DG pathways is superficial at best. Thus, we can see that what evolutionary biologists sincerely believe — that the developmental genetic program of one unique organism can be changed into that of another — is totally baseless; it is an erroneous expectation induced by Darwin's theory of evolution.

There is in fact a basic lack of understanding of the development of an organism in terms of its developmental genetic pathway. The genetic circuits which take the zygote through differentiation and development of specific organs and appendages via the precise genetic pathway that forms all the precise body parts in their respective positions at the appropriate times giving the specific shape and size of the developed individual, are not clearly comprehended. Still however, one can formulate certain basic principles based on the known facts in order to verify if evolution is occurring or not. Instead, unfortunately, most evolutionists tend to make vague statements that as evolution proceeds, the developmental program also should be changing correspondingly. The state of affairs is portrayed in the following passage, again from Douglas Futuyma's *Evolutionary Biology*:

> The development of an organism from a fertilized egg includes processes of CYTODIFFERENTIATION, whereby cells acquire different biochemical and structural features, and MORPHOGENESIS, the acquisition of the three-dimensional form of tissues, organs, and structures. For the most part, the mechanisms of cytodifferentiation and morphogenesis are not understood in detail. In this gap in our knowledge — between primary gene action and the development of complex phenotypes — lies much of what we do not yet understand about evolution.

Contrary to the beliefs of evolutionary biologists, we have now formulated a new principle that the developmental genetic pathway of an organism is fixed and have demonstrated that it is invariable into that of another by any mechanism within the genome of an organism even in very long geological time and even through a supposed lineage of organisms.

UNDERSTANDING HOW THE FIRST ORGANISM CAME INTO BEING IS OF UTMOST IMPORTANCE IN ORDER TO UNDERSTAND IF EVOLUTION OF ONE ORGANISM INTO ANOTHER CAN OCCUR AT ALL. BUT EVOLUTIONARY THEORY SIDESTEPS THIS ISSUE.

Darwin wrote at the end of *Origin of Species*,

> There is grandeur in this view of life, with its several powers, having been originally breathed into a few forms or into one; and that, whilst this planet has gone cycling on according to the fixed law of gravity, from so simple a beginning endless forms most beautiful and most wonderful have been, and are being, evolved.

However, he wrote little about how the one or a few original "ancestral" species were formed. Even under neo-Darwinism, little has been said concretely about how the genome of the original species came into being, especially considering that Darwin's mechanisms of natural selection start to act from the original species.

It is important to realize that the mechanism for the formation of even the simplest sexually reproducing multicellular animal species, which in itself is extremely complex, must be explained in order to understand how, or if at all, it could be changed into another species by Darwin's mechanisms. This is because, as we shall see later, the genome of even the simplest organism is no less complex than even the supposedly most highly evolved organism. But Darwin did not realize this, primarily because nothing about molecular biology was known at his time. In explaining how extremely complex organs, such as the eye, could have evolved he wrote,[48]

> How a nerve comes to be sensitive to light, hardly concerns us more than how life itself first originated.

Darwin was not concerned about the origin of the supposedly most primitive organisms or their organs, and was mainly concerned with explaining how the first organism, which he assumed came into being somehow, could possibly be changed into others. But understanding how the genome of even one supposedly primitive organism could have been put together, that is the origin of life itself, is of utmost importance to understand the origin of diverse creatures on earth. Because Darwin could not explain the mechanisms at his time, he assumed the "breathing" of life into the original organism. This was an innocent mistake that Darwin made, and it had the grave consequence of bringing forth a seemingly correct but erroneous theory. We shall offer a solution to the origin of life in Chapters 6, 7 and 8.

If evolution is an incorrect concept, then what are evolutionary biologists studying?

If it is evident from all our discussions that organismal evolution has not occurred, then what are the evolutionists doing? What are the mistakes that they make and why are they continuing to make them?

Evolutionists mistakenly believe that individual variations are the material basis for the evolutionary change of an original ancestor into all the numerous unique creatures on earth, and therefore are studying the various aspects of individual variations. They study the effects of geographical and environmental isolation on individual variations. They study individual differences at the level of the organism; and molecular sequence changes at the level of genes and proteins, called genetic polymorphisms. In our concept, these are nothing more than the many variants of the same gene. Furthermore, population geneticists study the various aspects of different populations of a species in different environments and the effects of various parameters in population change. But we can see that all of this research is carried out without realizing that it pertains to the changes of individuals and populations within the closed framework of every species, which can change only into the similar species of a distinct organism and never into a new organism.

In a similar manner, studies on mutations are also conducted with an erroneous belief that they evolve new genes. The presence of gene similarity among various organisms is shown as proof for evolution. Further, the presence of sets of similar genes even within a species is shown as another proof. First, these phenomena are clearly explainable without involving organismal evolution, which we shall deal with in Chapter 8. Second, such a "family of genes" is present not only in organisms supposed to be present at a "high" position in the evolutionary tree, but even in the lowest most primitive multicellular organisms such as worms, and even in single-celled organisms (see also Chapters 4 and 9). Does this not show that such a family of genes has not evolved through organismal evolution or descent with modification?

The main problem of evolutionists is that they think mutations are responsible for evolutionary change. But if we scrutinize what mutations are really capable of doing, as we will in the next chapter, one can see that they can never cause evolutionary change. All the mutations that do not cause a defect in a gene can only result in the normal variants of every gene in a genome — the fundamental cause of individual variations. Those that cause a defect in one or more genes can lead to one of the following conditions: 1) Mutation(s) can occur in a gene involved in a metabolic pathway, leading to congenital and genetic diseases such as phenylketonurea, galactosemia, albinism, and thalasemias. Sometimes a metabolic gene may be involved in the synthesis of a pigment for coloration of a body part such as the eye or the skin, and therefore when the gene is affected the normal pigmentation is affected leading to a color change. In fact sometimes a number of genes are involved in the pathway for the formation of a color pigment, and at each step in the pathway, the color of the pigment is different. Therefore a defect in any one of the genes in the pathway will stop color development at that step, producing a distinct color. Likewise, mutations in some genes such as that for the growth hormone can result in the alteration of the size of the animal, without altering its basic body structure — for instance, the various sizes of cats (e.g., domestic cat, cougar, tiger and ocelot). All the variations in such trivial organismal characteristics are produced by this category of metabolic mutations. Consequently, an absolutely immutable organism can occur in many different colors and patterns and in many different sizes. 2) Mutations can cause developmental defects such as those seen in the homeotic mutations, as we discussed earlier, e.g., antennepedia in the fruit fly. 3) Mutations can cause defects in one of the genes involved in the control of growth and division of the cell, leading to any one of many types of cancer. This is all the truth about mutations. Nothing more. They do not evolve new genes; they do not bring about any new body part; and they certainly do not cause evolutionary change.

But believing that they do, evolutionists are studying various kinds of mutations and mutational mechanisms. They study the variants of the same genes in different organisms by what is called "phylogenetic studies" because they think that these are the cause of evolutionary change.[49] Studies related to evolution have truly been based on faith in Darwin's theory and its seeming scientific appeal. The reason evolutionists pursue their misguided studies is that so far the truth about the incapability of mutations to bring about evolutionary change has not been revealed. Furthermore, because so far there has been no alternative theory for the evolutionary theory that can explain the origin and diversity of creatures without involving organismal evolution, scientists today have no choice but to work with evolution as the only reality, hoping that someday the theory will render itself free of all problems.

A phenomenon called "industrial melanism" is usually cited by the evolutionists as a proof of evolution, which in fact can be seen in every modern textbook of evolutionary biology. What is industrial melanism? Before the industrial revolution, the population of the peppered moth that lived in England consisted of mostly grey and a few black moths. The ratio of black to grey moths increased in soot-covered environments near major sources of industrial pollution. The moth population in such environments changed to mostly black (melanic) and were thus camouflaged among the soot-covered surfaces. This was because predators in polluted areas killed mostly the easily-spotted grey individuals. But in nonpolluted areas, the same predators killed mostly black moths, where they were easy to spot. When the industrial pollution is stopped and the soot-cover is reduced, one sees that the ratio of the black to grey moths reverses. Evolutionists offer this example for adaptation, and in general for evolution.[50] As we can see, the existence of the moths in two colors only means the existence of polymorphism with respect to color due to a mutation, as we discussed before. It should be noted that the mutation in the gene for forming color pigment is on an already existing gene in the genome. Where is the evolution of a new gene or a new DG pathway here? Is it not evident that during the industrial revolution, there was simply a change in the ratio of the population of the two already existing colored moths in the same species?[51, 52]

In general, evolutionists show examples of "evolutionary" changes apparently happening now as proof of evolution. When we probe these in the light of our new principles, we can very well see that these are really changes that occur within the closed framework of a species that they have mistaken to be evolutionary changes. We have seen that the population of a species can change within its characteristically defined closed framework through artificial or natural selection, such as in the peppered moth. Under both these circumstances, extremities of certain characteristics or traits that are not normally seen in a population under nature can be highlighted or brought about (Figure 3.9). To evolutionists, these changes within the closed framework appear to be evolutionary changes, *en route* to forming new species. Because of their mistaken belief that varieties lead to species, which lead to other very distinct forms, evolutionists always view any apparent change in a population as leading to a new organism. But this is an erroneous approach.

The mechanisms of natural selection and adaptation are now shrunk in their abilities to the production of varieties and similar species within the constant closed framework of a distinct creature. All the studies on evolution including those on molecular "evolutionary" processes, population genetics, and so on are therefore conducted unknowingly on the changes which occur within every immutable organism.

OUR ANALYSES ILLUSTRATE THAT NONE OF THE DISTINCT ORGANISMS ON EARTH ORIGINATED BY ORGANISMAL EVOLUTION

If not even a millipede or a centipede can be evolved from a worm that is lacking in legs, if a winged insect cannot be evolved from a wingless invertebrate, if no fish can be evolved from an assumed invertebrate ancestor lacking in bones, if a placental mammal can never be evolved from a reptile which totally lacks breasts and placenta, and if no bird can be evolved from a reptile without feathers or wings, how can any evolutionary mechanism explain the origin of these organisms? [53]

It is of extreme importance to note that the vertebrates are only a part of one phylum, out of the 35 phyla into which organisms are classified. It means that approximately 98% of all organisms are invertebrates. These are classified into 34 phyla, and thousands of subphyla, classes, orders, and families, the reason being they are extremely distinguished from one another — which really indicates that they are unique organisms.

The problem raised by the sheer uniqueness of creatures (known as the problem of the origin of higher taxa) — how the totally different organisms could have evolved from one or a few assumed original organisms — is a perennial unsolvable problem for evolutionary biologists by their own account.[54] Most organisms are totally unrelated and unique, whose bodily structures are either completely different or contain one or more unique body parts. All these organisms, except for the similar species within every distinct creature, could not have evolved by organismal evolution.

Is it not obvious that there is no organic relationship among most organisms, except for some which are currently defined as different species but in fact are only varieties of a single organism? It means that in the assumed evolutionary tree, we have severed the branches at various levels. The only remaining terminal branches, the twigs, represent the different species and genera within each distinct organism. Thus by our concrete and detailed analyses, almost all of the evolutionary tree is severed or broken down into pieces, showing that organisms have not come upon earth by descent with modification from one or a few original ancestors.

Conclusion

We have systematically disproved Darwin's theory in this chapter. We have shown that the major questions and difficulties faced by Darwin are still unanswered and unanswerable — even after 130 years of intense research in many different fields including paleontology and molecular biology — not because of smaller inconsistencies in the theory, but because the theory itself is fundamentally incorrect. Darwin felt that individual variations in the population of an organism, and monstrous variations that occurred among them, provided the material basis for natural selection and adaptation, producing many different creatures from one original ancestral organism. His followers, until today, have essentially believed in this theory, but the theory is wrong, because of the following principles we derived in this chapter.

1. No new genes can ever be formed in the genome of an organism even in a long geological time.
2. Likewise, no new DG pathway for a new body part can be evolved in the genome of a creature.
3. Based on 1 & 2, we can conclude that the genome of an organism is closed and locked with respect to evolution.
4. These principles taken together logically lead to the following conclusions:
    A) The set of genes in the genome of an organism is constant.
    B) The constant set of genes in the genome of a creature is organized into a unique and rigid DG pathway leading to an organism with a unique set of well-defined organs and appendages located in unique positions, and thus a uniquely shaped organism.
    C) Mutational changes of any kind can only produce defects in the existing set of genes in a genome or lead to normal variants of the same set of genes in the genome. For instance, mutations in developmental genes can only produce defects in the developmental genetic pathway, leading to misplacement, abolition, duplication, or malformation of an already existing body part in an organism. A second set of mutations lead to defects in genes participating in metabolic pathways, resulting in the absence of some normal metabolic products. A third set of mutations that causes a defect in one of the genes controlling cell division can result in uncontrolled cell division, leading to cancer. Thus all the defects in the genes of an organism can only lead to genetic and congenital diseases and cancer. None can lead to an extraneous outgrowth — a truly monstrous variation needed for evolutionary change; truly monstrous variations required for evolutionary change are nonexistent.
    D) Mutations that do not cause a defect in a gene can only produce variants of each gene in the constant set of genes in a genome. These variations in the constant set of genes are responsible for organismal individual variations. These four kinds of mutations are the only kinds possible in a genome, which put together cannot change the constancy of the set of genes or the rigidity of the DG pathway in a genome.
5. Consistent with the above conclusion, all the organismal individual variations in a species are confined within a closed constant framework, characteristic of that species. They can lead to the change of one species into many similar species within the confines of a distinct organism — without changing the constant set of genes and unique DG pathway of

the organism. However, they cannot change one organism into another distinct organism with new genes and/or unique body structures.

In summary, as we go up the assumed ladder of evolution, organisms at increasingly successive steps contain new or unique genes as well as unique body parts. The common belief among molecular evolutionists is that these new genes evolve by mutational mechanisms in the genomes of organisms from preexisting genes. To the contrary, we have demonstrated in this chapter that not even one new gene can evolve within the genomes of organisms by any mechanism even in long geological time. Furthermore, except for erroneous assumptions and inferences from the deceptive molecular scenario of life, the evolution of a new gene from a preexisting gene has never been demonstrated. In a similar manner, we have systematically proved that the belief of evolutionists that the genetic network of one organism can be changed into that of another by organismal evolution thereby evolving new body parts is erroneous.

The presence of the same and similar genes in entirely different organisms has been befogging our minds, and has been misleading us in discerning what is actually going on. It is no wonder, with this deceptive scenario, we have so far failed to see that there can be other explanations for this scenario without absolutely involving evolution. It is no wonder that we have so far failed to recognize that the gaps between organisms are true. There were never any transitional forms between organisms purported to be at successive steps on the evolutionary ladder because there is no evolutionary ladder. Because so far a coherent mechanism explaining all these scenarios without involving evolution is lacking, evolutionists are forced to find some mechanisms to connect organisms based on evolution. So they speak about many different mechanisms such as micromutations, macromutations, saltations, genetic drift, neutral mutations, punctuated equilibria, and such other mechanisms, none of which can explain the evolution of a new body part.

Evolutionists are bound to jump at me and show the production of one species of duck from another species of duck with a distinct variation in its color pattern or beak length, and say that this proves Darwin was right. The distinction is that Darwin was only right in an extremely limited sense. It is important to understand that all the changes in an organism cannot take it beyond a certain level. There can be a large number of artificial breeds and natural varieties (which are now termed species) within the confines of each organism. This scenario is sufficient to mislead and confuse one into thinking that because individual variations and natural selection lead into new varieties (i.e., similar species, which is true), varieties lead to completely new organisms (which is absolutely false). A frog might evolve into a toad, but it can never evolve into a rabbit, or anything that is not distinctly frog-like. The supposed ladder of evolution, from worm to millipedes, from invertebrates to vertebrates such as fish, fish to amphibians, amphibians to reptiles, reptiles to birds, and reptiles to mammals is imaginary because new body parts can never be produced by any form of evolutionary change.

Can anyone give a genetic mechanism for the change of one creature into another with a new body part? Can anyone give a mechanism by which a new DG pathway for a new organ be evolved even through a lineage of organisms? Other than showing the varieties of a species and simply stating that they are en route to evolving a new species, an evolutionist has no real way by which to demonstrate evolution!

The observed similarities among different species are not enough to prove that the mechanism of natural selection is responsible for earth's biodiversity. We can certainly understand how these observations might lead one to suspect such a mechanism at work, but suspicion is a far cry from proof. In any event, we have now seen that genetic changes actually can occur only within the sequences of an essentially fixed and closed genome, and can therefore produce only slight variations in the shapes and sizes of organs and appendages, and in their physical, physiological, and biochemical characteristics — leading only to varieties and similar species of every distinct organism. The characteristic gene set of an organism, however, is closed to any changes of a scale that might produce an entirely new organism.

Although we have examined these principles in this chapter, we must now analyze the individual mechanisms of mutations, and dissect and scrutinize them at the deepest possible level, to determine once and for all whether they are capable of evolutionary change from one distinct creature into another. These tasks await us in the next chapter.

4

Genetic Mutation and Rearrangement Cannot Cause Organismal Evolution

Darwin introduced his theory of natural selection and evolution in 1859, a full century before anyone had characterized the nature of hereditary material, and the mechanism by which heritable traits are transmitted. It was only a generation ago that researchers identified the most fundamental units of life in genes made of DNA, and discovered several mechanisms of genetic mutation and rearrangement in the genome. Indeed, the structure of genes in multicellular organisms was not unraveled until 1978, and only in the 1980s did we accumulate sufficient amounts of DNA sequence to begin a meaningful analysis of the origin of genes.

Evolutionary geneticists believe that the mechanisms of genetic mutation, collectively, enable the evolutionary change of one organism into entirely new organisms. Because similar genes and sequences are present in widely diverse organisms, and because genetic mutations and rearrangements superficially appear to be capable of evolving new genes from old genes, evolutionists assume that the collective sequence changes in a genome can lead to the change of one organism into another — even that new genes and unique body parts can evolve from old organisms that lacked them entirely. In this chapter we shall systematically analyze all the available scientific data for each of the mechanisms of genetic mutation, and we will find that none of these mechanisms — operating individually or collectively — can produce a new gene or a developmental genetic pathway. In short, these investigations will show us that the theory of evolution is fundamentally incorrect.

We saw in the previous chapter that the genome of a creature is a closed framework that cannot mutate into a new genome for another distinct creature. It follows logically that whatever mutations do occur in genes and sequences, they must occur within the confined framework of the genome of every distinct creature. The sequences may change and move around within the genome, but the constant set of genes in a genome and its fixed DG pathway cannot change. Mutations can produce individual variations — e.g., the size and shape of a human nose — and even many different varieties and similar species of an organism, but never a new organism. They can produce the normal variants of every gene, and defective genes, as well as defects in the DG pathway of an organ, leading to congenital and genetic diseases, aberrated growths of already existing body parts, and cancer. But none of these organismal changes has the potential to evolve one organism into a distinctly different organism with a new gene or a new body structure. The genome of every distinct creature has an innate flexibility, allowing all possible kinds of sequence mutation and rearrangement within its closed framework, but the genome of one distinct creature itself is absolutely immutable to that of another.

The whole theory of organismal evolution is founded on the assumed power of genetic mutations and rearrangements to evolve new genes and new genetic pathways of development. Note that except for the similar species of each distinct creature, multitudes of creatures on earth are distinct and unique and have new genes and/or unique body parts and structures. Therefore, if we can unequivocally prove that none of the mutational and rearrangement mechanisms are really capable of bringing about any new gene or new genetic pathway of development, we would essentially demonstrate that Darwin's theory — for that matter any theory of evolution — is incorrect. And that is what we shall do in this chapter. The final conclusion will be that evolution of one organism into another does not and never did occur.

In effect, any theory of evolution — purporting descent of organisms with modification producing unique creatures — is absolutely incorrect. We shall show in this chapter that all these mechanisms of mutation and rearrangement which occur in the cells of organisms are unavoidable biochemical properties of all living cells on earth, but are indeed totally incapable of organismal change beyond the production of similar species of a distinct organism, and as such have nothing to do with the assumed evolutionary change of one organism into another.

ANALYSIS OF THE KNOWN GENETIC MECHANISMS OF MUTATION AND REARRANGEMENT IN THE GENOME SHOWS THAT THEY ARE INCAPABLE OF BRINGING ABOUT THE EVOLUTION OF A NEW ORGANISM WITH A NEW GENE OR A NEW BODY PART

In Darwin's mechanisms, all the new genes and the developmental genetic pathways of all the future organisms with a variety of unique body structures and organs are assumed to be generated within the genomes of supposedly evolving organisms starting from the genome of an original primitive creature. This is supposed to happen by one or more methods of mutation. However, an analysis of the random processes involved will show that none of the known mechanisms can bring about even one new gene or developmental genetic pathway of one organism from that of another by evolutionary processes — even through an assumed long lineage of organisms. The critical point is that when such a process of building genes and developmental genetic pathways within a genome is improbable, then it confirms that Darwin's mechanisms are improbable.

What do we mean when we say that the genome of an organism is an evolutionarily closed framework? If, as we determined in the previous chapter, the genome of an organism cannot change into that of another organism, what do genetic mutations do to the genome and to the organism? The answer is: these mutations occur in the genome without changing the constant set of genes in the genome (i.e., without evolving even one new gene) and without changing the unique DG pathway of an organism (i.e., without evolving even one new DG pathway for a new body part or structure). Mutations only cause defects in existing genes and existing DG pathways — leading to congenital and genetic diseases and cancer — in addition to causing normal gene variants which are responsible for normal individual variations. Thus, the organism does not change — only its individuals change within a closed framework (sometimes producing similar species) and some of its individuals develop defectively — which is true even over any length of geological time. These mutations occur because of the basic biochemical propensity of DNA to undergo such changes, and therefore they occur unavoidably in every genome. In the following we shall fundamentally scrutinize every mechanism of genetic mutation and rearrangement and show that none of them has the capacity to evolve either a new gene or a new DG pathway for a new body part. Thereby we shall show, independent of the conclusions we derived in Chapter 3, that none of the mechanisms of genetic mutation can cause organismal evolution.

Several major categories of genetic mutation and rearrangement within the genome of an organism — sequence change or movement within the genome — have been discovered in this century, especially in recent years. In addition, some mechanisms have not been observed directly but have been proposed based on indirect genetic data. These classes of mutation are the following, and include all mutations occurring in the cells of all organisms living on earth.

1. Transposition
2. Gene duplication
3. Exon shuffling
4. Point mutation
5. Chromosomal rearrangements
6. Recombination
7. Crossing over
8. Pleiotropic mutation
9. Polyploidy

Even if new genetic processes are discovered in the future, we can still be absolutely certain, by our arguments on the random nature of mutation, that no possible hitherto unknown genetic mechanisms can be capable of evolving a new gene or a new body part.

As we discuss the various aspects of the assumed evolutionary mechanisms based on sequence mutations and rearrangements in this chapter we shall raise a lot of questions which cannot be explained by Darwin's theory of evolution (or any other theory of evolution). We shall discuss and answer all of them in my new theory (Chapter 8), which can account for almost all the questions we raise here.

Many mutational mechanisms are interrelated. For example, transposition, exon shuffling, and gene conversion can cause sequence duplication. Consequently, the discussion of one or more of these mechanisms may involve the others. We shall first discuss the basic details of every mechanism before we analyze their potential in organismal evolution. For some of the mechanisms, their inability to contribute to evolutionary change is explained separately under their description. For others, this is done collectively after describing all the mechanisms. In Chapters 7, 8 and 9, we shall discuss how these mechanisms are in fact expected and predicted in an immutable genome, that is, in organisms which never change beyond the level of species, based on my new theory.

The incapability of the transposon mechanism to evolve a new gene or a new DG pathway and to contribute to organismal evolution

Until two decades ago, the genome of an organism was thought to be quite stable. The genes were thought to be discrete elements with fixed positions along linear chromosomes.

It was found over 40 years ago that although most genes within a genome do stay put at their specific positions as stated above, a few of them could move from one place to another on the chromosomes.[1,2] (Use references 1 and 2 for further details below.) In 1947 Barbara McClintock discovered, through a series of genetic studies in the corn plant (maize), that some genes were jumping around the genome. The movement of genes from one location to another in the genome is due to what is called "transposable genetic elements." A transposable element can "pick up" a sequence on a chromosome and move it to another location on the same or different chromosome. Similarly functioning elements have been identified in a variety of organisms, ranging from bacteria to mammals, and the molecular details of some of these elements have been analyzed, but until the late 1970s the details of the maize elements themselves were not worked out. The widespread occurrence of movable elements was not accepted until long after McClintock's discovery.

What is a transposon?

A transposon is a DNA piece containing a set of genes for a few enzymes and a terminal repeat sequence at both of its ends, which has the ability to excise itself from one place in the genome and insert itself in another place. See Figure 4.1. The enzymes aid in cutting and re-inserting the transposon's DNA and the terminal repeats exist for the recognition of the transposon DNA by its enzyme system.

What are the physical effects of transposition during development of an organism?

It is well known that the Indian corn we find in grocery stores has a variegated appearance. The corn kernels have varying patches of purple color and colorless regions. The colorless patches are brought about by the action of a transposon in the cells during the development and maturation of the

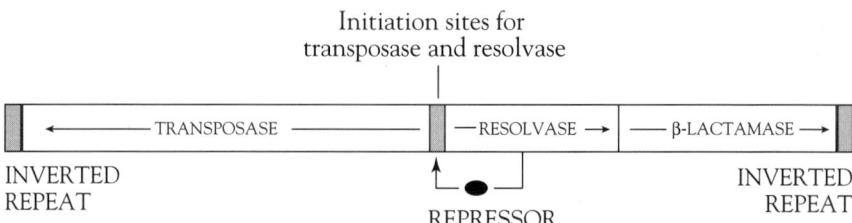

FIGURE 4.1. THE STRUCTURE OF A BACTERIAL TRANSPOSON. Bacterial transposon Tn3 is a DNA strand about 5,000 nucleotides long and carries three genes. Two of the genes encode enzymes, a "transposase" and a "resolvase," that catalyze transposition; the third encodes β-lactamase enzyme (a "passenger" sequence that has nothing to do with transposition). The product of the resolvase gene acts as a "repressor." It binds to a region between the transposase and the resolvase genes, keeping both genes turned off. When the repressor falls off the DNA, transposase and resolvase are produced, allowing transposition to occur. At the transposon ends there are 38-nucleotide "inverted repeat" DNA sequences which are symmetrical, reading the same in opposite directions on opposite strands. Such inverted repeats serve as recognition signals for transposition enzymes. [From "Transposable Genetic Elements in Maize," by Nina V. Fedoroff. Copyright © 1984 by Scientific American, Inc. All rights reserved.]

corn kernels. The purplish color in the corn is due to the synthesis of a purple pigment in the cells of the corn when they grow to form the kernel. A mutation in one of the genes leading to the purple pigment could affect the synthesis of the pigment in the pericarp, the layer of cells surrounding and protecting each individual kernel, rendering the kernel colorless. Such a mutation can be produced by the insertion of a transposon into one of these genes, which makes the gene inactive. However, because the transposon can also excise itself reverting the original gene intact, the synthesis of the pigment will now be resumed. Imagine what will happen if such a mutation and reversion of the mutation occurs in cells as they divide and grow — the mutation occurs in a cell, and, after several divisions of this cell, one of these cells reverts. A patch of colorless pericarp will be formed because of the cells which divided with the mutation. After reverting, the one cell with the normal gene divides into many cells, all of which will now synthesize the pigment, thereby forming a patch of purple pericarp.[3] See Figure 4.2.

GENETIC MUTATION AND REARRANGEMENT CANNOT CAUSE EVOLUTION   109

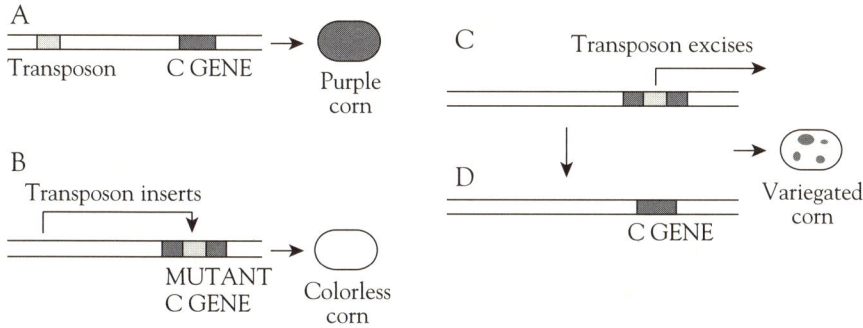

FIGURE 4.2. PRODUCTION OF VARIEGATION IN CORN KERNELS BY TRANSPOSONS. The C gene produces a pigment that gives Indian corn its purple color (A). A mutation in the C gene takes place when the transposon moves into the C gene (B). The mutation disables the gene, and the pigment is not made in cells with this mutation. Sometimes, the transposon excises away in some cells during kernel development (C). The mutation reverts when the transposon leaves, and the pigment returns (D). Frequent mutations and reversions in various cells within a developing corn kernel thus produces a variegated appearance. [From "Transposable Genetic Elements in Maize," by Nina V. Fedoroff. Copyright © 1984 by Scientific American, Inc. All rights reserved.]

WHAT ARE THE MOLECULAR EFFECTS OF TRANSPOSITION?

The molecular mechanism by which transposition is achieved

The typical mechanism of transposition, inferred from indirect evidence from studies in bacteria,[4] is shown schematically in Figure 4.3. The enzymes from the genes in the transposon cut the target site in the genome, and the transposon inserts itself. The ends adjacent to the inserted transposon are now directly repeated sequences. And when the inserted transposon is excised out of this site as a whole, it leaves the extra copy of the direct repeat sequences at the site of insertion. In general, when a transposon inserts at a host DNA site, it introduces a staggered cut, links the protruding ends to the transposon, and corrects these ends thereby generating direct repeats of the target DNA at the insertion site.

Transposition leading to duplication, deletion, and inversion of sequences

Transposons may be associated with various types of rearrangements.[5, 6] In *conservative transposition*, a transposon moves as a physical entity from a

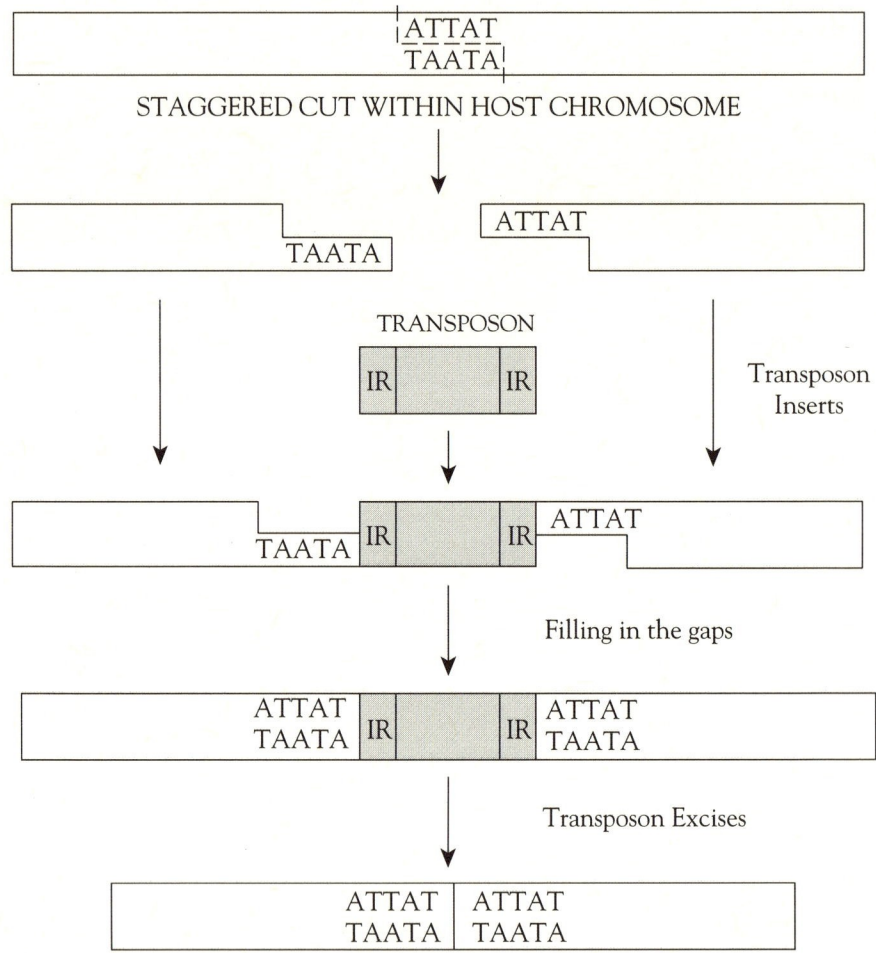

FIGURE 4.3. MECHANISM OF TRANSPOSON INSERTION AND EXCISION. A transposon makes a "staggered" cut at a site in the genomic DNA and inserts at that site. An enzyme fills in the single-stranded gaps, producing direct repeats flanking the transposon. When the transposon excises, a direct repeat remains as a "transposon footprint." This theory was developed to account for the observed flanking sequences. IR indicates the transposon's inverted repeat (see Figure 4.1). [From *RECOMBINANT DNA: A SHORT COURSE* by Watson, Tooze, and Kurtz. Copyright © 1983 by James D. Watson, John Tooze and David T. Kurtz. Adapted with permission of W. H. Freeman and Company.]

donor to a recipient site. What happens at the donor site after transposition is unclear. In *replicative transposition*, a copy of the transposon is created and inserted at a recipient site; the donor site remains unchanged.

When a transposon inserts a copy at a second site near its original location, it may result in rearrangements of the host DNA. *Deletion* of the genomic sequence between two direct repeats (between two closely placed transposons) occurs by reciprocal recombination between the direct repeats. Reciprocal recombination between inverted repeats can result in the *inversion* of the region between them. What we see in all these instances are breakage, discontinuity, deletion, or inversion of a sequence in the genomic DNA by the action of a transposon. This can cause a gene to become defective. As we shall soon see, the probability that any of these effects will lead to a new gene with a new function having a selective advantage for the cell or the organism is negligibly low.

Excision of transposons from a site can leave a "transposon footprint" at that site

When a transposon inserts itself at a location and then excises itself from there, it may leave a characteristic short sequence at that location, the "transposon footprint," which changes the original sequence within a gene. The locations of the insertions of many transposons in genes have been determined.[7] Insertions into any region of a gene have been described.

When transposable elements excise themselves from protein-coding regions, the number of nucleotides in a transposon footprint, within a given range, can be random. The excision of the transposons from a coding sequence of a gene can, however, lead to three possible results. The first is that the excision of a transposon leaves a transposon footprint of either three or six nucleotides (coding for one or two amino acids), thus restoring the correct "reading frame." That is, the protein-coding sequence around the excision will now be the same as the original coding sequence, except that there are one or two amino acid insertions at the site of the transposon footprint. In this case, these inserted amino acids may not affect the overall protein structure or its specific biochemical function, and therefore the changed gene is only a variant of the original gene. The second possibility is that the footprint does not restore the correct reading frame. In this case, the sequence of the protein from the point of the footprint will be completely changed which can lead to a defective protein. The third possibility is that the reading frame is restored but the protein is defective. In this instance, the sequence of the protein from the point of the footprint is restored to the original sequence, but the inserted one or two amino acids change the structure of the protein in such a way that it loses its biochemical activity entirely. In summary, the process leads to either normal variants of the same protein (i.e., the gene) or to totally defective, inactive proteins. However many times transposon footprints appear due to insertions and excisions, and in however many places, these are the only kinds of possible consequences.

Transposition can only activate or inactivate preexisting genes. They do not evolve new genes, even by their cumulative actions over geologic time.
The real question we have to examine in all these processes is: Where do we see the evidence for the evolution of a new gene? Did the transposition mechanism bring together short DNA sequences to make a coding sequence? Or, did it bring together parts of various genes and make or evolve a new coding sequence? Even if some short sequences are moved here and there, when we analyze this random process, it shows that sequences cannot be brought together to form a new, useful gene selectable by evolution. This is because even if a coding sequence for some protein is formed by such random combinations of sequences brought together by transposition, only one out of millions of such coding sequences may encode a protein with a biochemical function at all, and only one out of thousands or even millions of proteins with any function at all can have an evolutionary selection potential. When we do not have any evidence for sequences being brought together to form a coding sequence for even one gene by any transposon mechanism, how can there be the generation and expression of many such new genes so that the useful one can be selected?

It is of great importance to realize that the genes for the enzymes which produce the corn's color pigment already exist in the genome of the corn plant as a precise part of a DG pathway. Insertion or excision of transposons into and out of the sequences around the genes simply activate and inactivate them. There is no documentation that it can generate a new gene or new DG pathway for even a simple body structure. As has been discussed, the probability that a random function (i.e., a new biochemical function that does not necessarily have a useful purpose in the organism) can be useful in the evolution of a new organ is extremely low. While this is so, not even a gene expressing such a random function can be generated by the transposon mechanism by bringing together different sequences into a contiguous coding sequence. Similarly, as we shall see below, it is highly improbable that a new DG pathway for a new organ or limb would evolve by randomly moving around the sequences and genes preexisting in a genome. Thus, the superficial speculations of evolutionists, that transposons are responsible for the evolution of organisms from an original ancestor organism, are erroneous.

Excision of controlling elements can alter the quantitative properties of a given enzyme by leaving transposon footprints. But they cannot evolve a new gene for a new enzymatic function.

Molecular evolutionists such as Susan R. Wessler have speculated[8] that under life-threatening conditions the introduction of a few nucleotides in the regulatory or coding sequence of a gene may lead to genetic diversity (meaning that new genes and regulatory sequences may evolve through this process). But this has not been shown to be the case in any gene. Wessler herself comments in her article that although the properties of some transposons may suggest such a possibility, "these properties are also consistent with the alternative view that transposable elements are selfish DNA," that is, parasitic sequences that exist only to replicate themselves.

The introduction of amino acids into a protein by the introduction of one or two codons in the coding portion of genes does not alter the basic function of the protein, except where it results in a defective protein. For instance, transposon footprints are now well known to reduce the activity of an existing enzyme in the genome.[9,10] We can definitely see from all our discussions so far that, when an inactive or defective protein results, it may lead to 1) no effect with respect to evolution, such as in the case of the abolition of pigment synthesis in variegated corn, 2) congenital or genetic diseases (including the aberrated growths of already existing body parts, as we shall see below), or 3) the death of the cell or embryo. These observations are also consistent with my explanation in Chapter 3 that mutations can lead to variation in the level of a given enzyme's function (which is the basis of individual organismal variations) or to a defective gene unable to perform its biochemical function altogether, but not at all to a new function.

Random transposition of regulatory sequences cannot lead to new DG pathways

The introduction of a few nucleotides in a given regulatory sequence can lead to abolition (or misactivation) of the expression of a gene. When this happens, it may result in a defect in the development of a preexisting body part in an animal, but can never lead to a new DG pathway for a new body part. If complete abolition of gene expression does not happen, the mutation, however, cannot change the regulatory sequence in such a manner that it regulates another gene. It can only lead to the quantitative variation in the regulation of the same gene that it was regulating prior to the mutation, so that the production of the gene product may be increased or decreased. That is, the regulatory sequence is functionally still the same. The essential concept that these processes lead only to defects or variants of a given gene or regulatory sequence — and not to an entirely new protein function or a new regulatory sequence — indicates that transposons cannot lead to the evolution of a new gene or a new DG pathway.

What is fundamentally needed for the evolution of a new organ or body structure is the evolution of a new developmental genetic pathway that would dictate the development of that body part. But, from our analysis we can clearly see that transposons are incapable of evolving even the simplest DG pathway for even the simplest body structure in any length of geological time.

Where do we see the ability to construct a network of hundreds of genes in a DG pathway in all the actions of transposons? Nowhere indeed. A new DG pathway cannot be constructed within the genome of an organism through random processes; the transposon is not even capable of bringing together a small set of genes existing in the genome of an organism into a small, even meaningless network of genes by the processes of moving genes and other sequences.

The probability that even a few genes would be organized in a required order by this random process of shuffling the regulators and the genes preexisting in a genome is extremely small.[11, 12] One can then understand that the probability for the organization of a DG pathway for a new organ with a selective advantage for an organism by such random processes is virtually nonexistent.

It is also important for us to consider the fact that there is no documentation for the movement of a regulatory sequence of one gene as a new switch for another gene, or the movement of a coding sequence and putting it under a new promoter and regulatory sequence, which starts to function in a new situation. While this is the case, an extremely large number of random trials would be required to produce one combination of a specific regulatory sequence with an appropriate coding sequence that could be useful in a new DG pathway. Thus the transposition mechanisms do not have the potential to evolve a new DG pathway. They did not contribute to, and have absolutely nothing to do with, the supposed evolution of all organisms on earth from one or a few original ancestors.

Even if we assume that a gene's coding sequence can be precisely moved to be placed under the influence of another regulator by transposition, we can prove that such events can never lead to a new DG pathway. Let us take the hypothetical situation of having to arrange 10 genes, from a genome of 100 genes, in a given series that would construct a new organ. As we discussed in Chapter 3, the probability of doing this is 10^{-20}, which is meaninglessly low. Let us compare this with an actual genome of 10,000 genes and organizing the developmental genetic network of one new organ built by 100 genes.

Obviously, achieving this has a far lower probability (10^{-400}) than that for even the much smaller hypothetical DG pathway. It is also important to keep in mind that these mechanisms can act only on preexisting genes and sequences within the genome of the organism.[13] And among the mutations that occur in an individual, only those that occur in the germ cells (i.e., sperm and egg) of the individual can be effective in any possible evolutionary change in the genomic sequence.

The primary reason for the improbability of evolving a new DG pathway for a new body part is that this has to be achieved through random changes in the genomic sequences, expression of these changes into physical structures, and fixation through natural selection. Random generation of a large number of bizarre attributes by multitudes of new DG pathways should lead to the selection of the fitting ones. The complete developmental genetic network of a new organ, with suborgans and tissues (such as the bone, the placenta, the feather, or the eye), has to be achieved through such random processes. Because of the required randomness of the process, this could not be achieved even in trillions of years, which is far longer than the age of the universe itself. Furthermore, this requires that multitudes of bizarre phenotypes should exist in the populations of all living organisms. On the contrary, even given the tremendous diversity of life and a large population of each organism on earth, no such random phenotypes are seen in any living creature, nor do we have any evidence for these from extinct fossilized organisms.

ANALYSIS OF AN EXAMPLE ORGANISM: MUTATIONS OF THE FRUIT FLY INDICATE THAT TRANSPOSONS CAN HAVE NO EVOLUTIONARY CONTRIBUTION

In the case of the fruit fly *Drosophila melanogaster*, it is now estimated that perhaps as many as half of all spontaneous mutations that occur in them are not point mutations, but rather chromosomal rearrangements (probably causing gene inactivation or activation) brought about by the movement of transposons.[14]

One of the prominent visible traits studied by the first *Drosophila* geneticists was variation in eye color. The changes, originally thought to be point mutations, occur at a DNA site termed the *white* locus. The normal brick-colored eye is changed to various paler shades, including pure white, by a wide variety of changes at this locus (the chromosomal location of the gene or genes responsible for eye color[15]). Mutations in the proteins encoded at the white locus affect pigment production. The different intensities of eye color may correspond to the production of different amounts of a given pigment which result from various mutations.[16] Here we see only errors of the kind we discussed above in one or a few genes affecting the production of their products. It must be kept in mind that in all these mutational changes, there is nowhere seen anything about the evolution of either a new gene or DG pathway.

As it turns out, mutations in a locus called the *abl* can cause cancer in humans. In the fruit fly it is lethal, meaning that it kills the embryo during its development. There are many other lethal mutations that are possibly transposon-mediated in the fruit fly. As we saw in Chapter 3, some of the mutations caused by transposon movement in the genome of the fruit fly lead to defects in the development of an already existing body part — abolition,[17] duplication, misplacement, or malformation — such as its antennae. Thus, it is clear from our foregoing analysis that all the transposon-mediated mutations in the fruit fly fall under one of the categories of alterations or errors in existing gene functions: quantitative alteration of the synthesis of a protein or its activity, or abolition of the activity which leads to cancer, genetic or congenital aberrations, and killing the embryo *in utero*. They certainly do not contribute to the evolution of a new gene or new DG pathway.

Hybrid Dysgenesis and the improbability of evolution

In the fruit fly an interesting phenomenon known as *hybrid dysgenesis* is associated with transposable elements.[18]

Hybrid dysgenesis refers to a collection of related genetic abnormalities that arise spontaneously in offspring after certain varieties of fly are crossed. The defects occur in the germ cells, showing a large increase in the frequency of chromosome aberrations, gene mutations, and sterility. One of the factors responsible for hybrid dysgenesis appears to be a transposable element called P. It resembles the bacterial transposon in its basic structure. Within the P^+ fly, P does not appear to move and generate copies at new locations. When the P^+ fly is crossed with P^- females, the resulting hybrid progeny fail to produce many viable germ cells. The rare germ cells that do generate progeny often lead to sterile individuals in the second generation. Hybrid dysgenesis results from the mobilization of P elements after their insertion into P^- eggs. Probably due to the absence of a repressor protein for transposase within the recipient P^- eggs, the P elements begin to transpose, and many of the daughter P elements lethally insert themselves into vital genes. Such transpositions occur at noticeable rates only in germ cells, as opposed to somatic (body) cells. So the progeny of P^+/P^- crosses are normal except for their failure to produce viable sperm and eggs. Evolutionary geneticists expect that such activity can speed evolution. But this expectation is invalid in view of the following reasons.

The above mutations and transpositions seem to be random and large in number, disrupting many genes within the genome of the fruit fly. If these mutations disrupted the genetic networks, to the extent of generating some bizarre outgrowths, it would have supported the evolutionists' expectations. If the mutations had brought about some bizarre outgrowths, not existing in the normal fruit fly, then we could agree that this phenomenon contributes to evolution. But it does not happen. The disruption of the genome by random transposition and mutation of the vital genes resulted in the death of most of the progeny. Only in some cases did the cells survive, but for some reason the progeny became sterile. The latter event reflects the disruption of existing functions, namely fertility. Because only the genes responsible for fertility were affected in some individuals, and, as far as such sterile individuals are concerned, the mutations were not detrimental to life, they survived. In all other cases, mutations had hit one or more genes essential for life. The final conclusion is that no bizarre outgrowth has ever been seen by such transposition events. And without bizarre outgrowths, no evolutionary change can be expected.

Conclusion: Transposons cause genetic effects which are incapable of any evolutionary potential; these are passive parasitic processes occurring in immutable genomes

In conclusion, our analysis of all the activities of the transposons in the genomes of organisms clearly and decisively show that they can never evolve a new gene or a new DG pathway for a new body structure. The important point we need to understand for our discussion is that transposition either activates or inactivates the function of an already existing gene; it does not produce a new gene in any number of transposition events or in any length of geological time.

The properties of the transposons can be summed up as follows. A transposon can insert itself into, and excise itself from, any DNA sequence.

1. When the insertion or excision of a transposon is in a coding sequence of a gene, it can mutate the gene and produce an inactive protein or the same protein with exactly the same biochemical action but with reduced or increased activity.
2. When the transposon inserts itself within (or excises itself from) a gene's regulatory sequence, it can affect the expression of the gene either positively or negatively. If the gene was originally switched off, the mutation can lead to the switching on of the gene, and vice versa.

Consequently, none of the actions of a transposon can produce a new coding or regulatory sequence. Molecular biological research over the past decade has not shown any evidence that the coding or regulatory sequence of any of the thousands of genes whose details are known has resulted from the action of transposons. Transposon-mediated evolution is a conjecture without any evidence whatsoever, stemming from the superficial abilities of the transposons.

Even if the set of genes of an organism's genome is constant and the DG pathway of the organism is fixed, still all the described activities of transposons can take place in the genome. We can prove that even in a fixed and immutable genome we can have an entity like the transposon and all its activities without doing anything to the constancy of the set of genes in the genome or to its DG pathway, thereby doing nothing to the constancy or the fixity of the organism as long as it lives on earth. In fact, there is ample evidence which shows that this is indeed the case. Distinct transposons are present in different organisms, showing that this scenario could have occurred only if various organisms originated independently from a common pool of genes in the primordial pond (see Chapter 9).

THE BELIEF OF MOLECULAR EVOLUTIONISTS THAT TRANSPOSONS ARE CONTRIBUTORY TO ORGANISMAL EVOLUTION IS BECAUSE OF THE HIGHLY DECEPTIVE NATURE OF TRANSPOSONS

Although transposons are absolutely incapable of contributing to organismic evolution, they have been misunderstood to have had precisely such a function. The reason for this is that the activities of transposons are highly misleading and deceptive; molecular biologists are impressed by the superficial ability of transposons and believe that transposons have greatly contributed to the assumed evolution of diverse organisms from the original ancestor.[19] Even advanced molecular biology textbooks and research publications in reputable journals illustrate such beliefs. The following passage from the textbook *Molecular Cell Biology*[20] illustrates how deceptive are the activities of the transposons.

> In spite of the prevalence of mobile elements in the *Drosophila* chromosomes, it seems unlikely that mobile sequences play an essential role in either the early development of the fly or its later life cycle. For example, mobile elements are about 20 times more prevalent in *D. melanogaster* than in the almost identical *D. simulans*, a sibling species. ... Apparently the repetitive sequences are unnecessary to "build a fly." However, because of the dramatic effects the mobile elements can have on gene functions, they probably do participate in evolution.

Stanley Cohen and James Shapiro[21] state,

> ... transposable elements ... can bring together unrelated chromosomal-DNA segments to form a variety of structural rearrangements. Genetic rearrangements can have biological importance ... on an evolutionary scale.

Nina Fedoroff[22] speculates the following, with absolutely no evidence:

> Transposable elements may be of even greater significance in evolution. One can only speculate about such a role, but their properties appear to make them suitable agents for modifying not only the expression of genes but also the structure of genes and genomes. ... once the elements are activated they can promote many kinds of mutations and chromosomal rearrangements. It is as if transposable elements can amplify a small disturbance, turning it into a genetic earthquake. Perhaps such genetic turbulence is an important source of genetic variability, the raw material from which natural selection can sift what is useful for the species. Moreover, evidence is accumulating that in addition to turning genes off transposable elements can turn them on or amplify their expression. There is reason to suspect they can reprogram genes in more subtle ways as well, changing when and where in the organism a gene is active. This is indeed the stuff of remodeling and rebuilding, of organismic evolution.

In the textbook *Introduction to Genetic Analysis*, David Suzuki, Anthony Griffiths, Jeffrey Miller and Richard Lewontin[23] say the following, again with absolutely no evidence whatsoever.

> At present, it is not known whether transposons are elements that normally play a role in the day-to-day transactions of the genomes, as originally proposed by Barbara McClintock in the 1950s, or whether they are pieces of "selfish DNA" that exists for no purpose other than their own survival. ... At the evolutionary level, transposons may be important in the sudden leaps that characterize the fossil record.

The reason knowledgeable scientists speculate that transposons can contribute to evolution is clear when we consider that transposon activities are very deceptive. Since transposons can disrupt the coding sequence or switch off the expression of a gene and bring about many genetic aberrations leading to apparently big effects such as abolition or misdevelopment of an already existing body part, then it is tempting to assume that they are capable of far more effects over the course of geological time. But our foregoing systematic and thorough analyses demonstrate unequivocally that they can never contribute to the evolution of a new gene or a new DG pathway, and therefore cannot contribute to the evolution of an organism with a new body part.

Some molecular geneticists and evolutionists correctly believe that transposons are selfish genes, functioning only to propagate their own sequences within the genome

Some molecular evolutionists have proposed that transposons need not have originated for evolutionary functions. They feel that, although transposons can affect some physical attributes, or phenotypes, (as in the case of corn color), they need not have arisen for that purpose. For instance, Doolittle and Sapienza write from their molecular evolutionary standpoint that transposons have evolved within living cells with a single function: to maintain their own survival.[24]

> Although DNA sequences which contribute to organismal phenotypic fitness or evolutionary adaptability indirectly increase their own chances of preservation, and may be maintained by classical phenotypic selection, the only selection pressure which DNAs experience directly is the pressure to survive within cells. If there are ways in which mutation can increase the probability of survival within cells without effect on organismal phenotype, then sequences whose only 'function' is self-preservation will inevitably arise and be maintained by what we call 'non-phenotypic selection.' Furthermore, if it can be shown that a given gene (region of DNA) or class of genes (regions) has evolved a strategy which increases its probability of survival within cells, then no additional (phenotypic) explanation for its origin or continued existence is required.
>
> ... We do not deny that prokaryotic transposable elements or repetitive and unique-sequence DNAs not coding for protein in eukaryotes may have roles of immediate phenotypic benefit to the organism. Nor do we deny roles for these elements in the evolutionary process. We do question the almost automatic invocation of such roles for DNAs whose function is not obvious, when another and perhaps simpler explanation for their origin and maintenance is possible. It is inevitable that natural selection of the special sort we call non-phenotypic will favor the development within genomes of DNAs whose only 'function' is survival within genomes. When a given DNA, or class of DNAs, of unproven phenotypic function can be shown to have evolved a strategy (such as transposition) which ensures its genomic survival, then no other explanation for its existence is necessary. The search for other explanations may prove, if not intellectually sterile, ultimately futile.

Despite the beliefs of some geneticists and molecular evolutionists that transposons are possibly selfish elements, parasites that function to maintain their survival within the genome, still the fields of genetics and molecular evolution are filled with arguments that transposons could be great contributors to evolutionary change.

GENETIC MUTATION AND REARRANGEMENT CANNOT CAUSE EVOLUTION   121

In the final analysis, our discussions on transposons lead to the following conclusion. Molecular geneticists and evolutionists have believed that transposons aid in genetic rearrangements which contribute to evolutionary change. But they are incapable of contributing to the supposed evolutionary change because they are incapable of evolving new genes, regulatory sequences, or new DG pathways. They are parasites, and may result in some genetic waste in the genome. They can at the worst cause mutations leading to the defective development of already existing body parts in an animal. In addition to causing defects in existing genes, they can lead to sequence changes in a given gene without affecting its type of function, and thus can contribute to individual variations. But as we have shown earlier, individual variations of a species do not contribute to evolution. In summary, speculations that transposons can contribute to organismal evolution are incorrect. They cannot and do not.

PROGRAMMED GENOME REORGANIZATION IS USED FOR A FEW NORMAL CELLULAR AND ORGANISMIC FUNCTIONS

One kind of genome rearrangement occurs during the development of an animal, as part of the normal function of the genome, which can be called "genomically programmed" rearrangement. This does not happen in the germ cells, cells that produce the sperms and the eggs.[25] Since we are accounting for all possible genome rearrangements in this chapter, a discussion on this topic is relevant here.

Gene rearrangement in generating "antibody diversity" is a well-programmed genomic rearrangement, which is part of the normal function of the genome

Sequence rearrangement takes place in cells to produce antibodies tailored to defend against specific antigens.[26, 27] An antigen is an unwanted foreign molecule (such as a virus, bacteria, fungus, or toxin) in the blood stream. An antibody protein molecule recognizes an antigen, and deactivates it by binding to it. Because millions of different antigens exist in the environment, a similar diversity of antibody molecules need to be produced in the animal to counter their actions. Although the mechanisms of production of the immense number of different antibodies are still not completely understood, several cumulative mechanisms have been discovered. One such mechanism contributing to antibody diversity is the rearrangement of a few parts of the antibody genes, termed V, D, and J.[28] Various combinations of these parts, each of which exists in a few different forms, lead to enormous numbers of antibody molecules.

Each antibody molecule consists of two light (L) and two heavy (H) chains. Each chain consists of two regions: a variable (V) region and a constant (C) region. There are hundreds of variable-region genes for either light or heavy chains. There are only a few genes coding for C regions. In the context of the antibody molecule, a "gene" represents a DNA sequence coding for one of the parts that form the final antibody molecule. A light or heavy chain is constructed by physically joining a V gene to a C gene. Thus, any one of many V genes may be joined to any one of a few C genes. In essence, sequences are moved or recombined in order to generate the numerous antibody genes by the innumerable combinations possible from a few basic genes. This is a well-programmed, normal, genetic function of the genome of animals which is essential for their survival. This need not be related to any supposed evolutionary mechanism.

Programmed rearrangements occur even in single-celled organisms

Precise genomic rearrangements are used even in unicellular organisms to control gene expression.[29] The yeast S. *Cerevisia* can exhibit either of two sex "mating types," depending on the presence of either of the two genes, a or α at a specific site in its genome that specifies the mating type. It keeps the master copies of the two genes elsewhere in the genome, and puts a copy of one of the two genes at the mating-type site to switch from one mating type to another. See Figure 4.4.

Using a similar plan of DNA rearrangement, unicellular parasites, such as African trypanosomes, evade the host immune response by varying their surface features. The gene sequence present at the active locus determines the surface antigen. By substituting a sequence from any one of many silent loci, the sequence at the active locus can be changed.

The above phenomena are well-programmed cellular functions. They are precise rearrangements at one defined location in the genome. They are not random or indiscriminate. They are present in unicellular organisms. It is not at all logical to implicate such mechanisms in the supposed evolution of multicellular organisms through descent with modification. It is certainly reasonable to conclude that even if some hitherto unknown programmed rearrangements are found in the future in multicellular animals, still they cannot be implicated in evolutionary change, because they can never evolve a new gene or a new DG pathway.

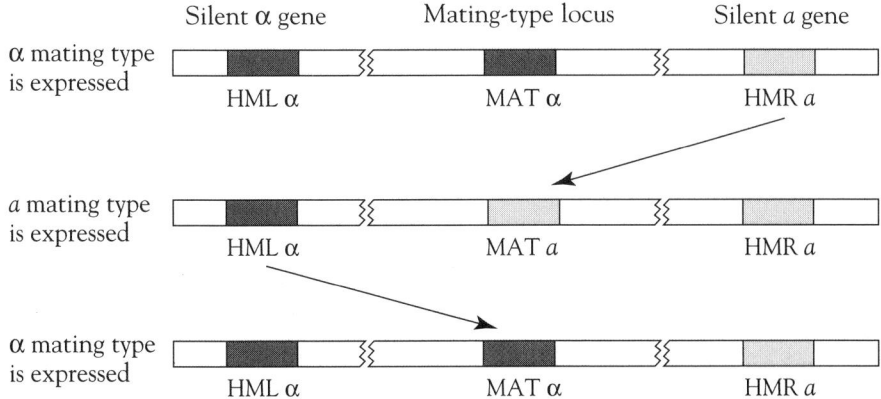

FIGURE 4.4. THE CASSETTE MODEL FOR MATING TYPE SEX INTERCONVERSION IN YEAST. There are two mating sex types of yeast, α or a, represented by two different genes called α and a. Mating type is determined by the α gene or the a gene being present at an active location called "mating type locus." The two genes are also present at two other locations (α at the HML site and the a at the HMR site), where they are always inactive. When the copy of one of these genes is inserted into the active mating type locus, it becomes active and the yeast expresses the corresponding mating type sex. The α or the a gene is inserted into the active site interchangeably, thus frequently changing the sex of the yeast cell. This mechanism is believed to be similar to transposon movement. [From RECOMBINANT DNA: A SHORT COURSE by Watson, Tooze, and Kurtz. Copyright © 1983 by James D. Watson, John Tooze and David T. Kurtz. Adapted with permission of W. H. Freeman and Company.]

GENE DUPLICATION CANNOT CONTRIBUTE TO EVOLUTION

One of the major mechanisms that evolutionists assume to contribute to the evolution of new genes is the duplication of an existing gene and its modification to produce a new gene. Just as any other mechanism of genetic change that evolutionists assume as capable of contributing to evolution, the gene duplication mechanism is also thought to be an ongoing evolutionary activity within the genomes of organisms. This is exemplified by the following quote from *Molecular Biology of the Cell*:[30]

> Evolution depends to a large extent on mutations that alter existing genes to create in their place new *alleles,* or variants, of these genes.
> ... The evolution of a complex organism, however, requires something more than the introduction of improved forms of existing genes. It requires the creation of new genes to serve new functions.
> ... Many of the proteins in a multicellular animal can be grouped into families: the collagens, the globins, the actins, the serine proteases, and so on. Proteins in the same family are related both in function and in amino acid sequence. There can be little doubt that each family has evolved from a single ancestral gene by a process of *duplication* and *divergence*. Different members of a protein family are often characteristic of different tissues of the body, where they perform analogous but distinctive tasks. The creation of new genes by diversification and specialization of existing genes has plainly been crucial for the evolution of complex multicellular organisms.
> ... Diploid species enjoy an important advantage: they have a spare copy of each gene, and this spare copy can mutate and serve as raw material for innovation. A haploid species does not have this easy means of taking the first step toward evolving a larger and more sophisticated genome.

We shall demonstrate below that this belief of the evolution of a new gene by the duplication of an existing gene within the genome is incorrect. Several examples of sequence and gene duplications have been discussed in the literature. When one analyzes these systematically, one can discern that, although the duplication of a sequence or a gene in a genome is certainly possible and demonstrable, the evolution of a new gene by gene duplication is highly improbable. Analysis of the scenario of the vertebrate plasma proteins — which are present with apparent duplications in all the vertebrates, but are completely absent, even in their unduplicated forms, in invertebrates (animals assumed to be the ancestors of vertebrates) — demonstrate that they did not evolve by gene duplication through organismic evolution. It can be shown in many cases that the presence of apparent duplications — purely inferred by the presence of similar sequences within the same protein or among different proteins with absolutely no other proof or evidence — could be in fact due to similar functional constraints in different independent proteins, and not due to real duplications. Lastly, the scenario of the globin genes, given as the best example of the evolution of a family of genes by gene duplication in organisms, exhibits discrepancies unexplainable by evolution. All the above scenarios are against evolutionary theory, and can be better explained by the new theory of the independent birth of organisms, described in Chapter 8. We shall analyze all the possible kinds of real and assumed gene duplication one by one and eliminate each of them as incapable of evolving a new gene. Furthermore, we shall illustrate by clear evidence in existing organisms that none of the genes which are claimed to have been evolved by the mechanism of gene duplication have indeed evolved by gene duplication from ancestral genes in ancestral organisms.

Genuine duplications due to the cellular need for large amounts of the gene product: Ribosomal RNAs and Histones

Some genes are duplicated many times and organized in large clusters within the genome, for example, the ribosomal RNA (rRNA) genes. Because ribosomes are the protein-making machinery, large amounts of ribosomes are needed in a cell. As a consequence, large quantities of rRNA are needed to construct the ribosomes. The multiple copies of the rRNA genes help in the synthesis of large amounts of rRNA. In a similar manner, large quantities of histones, used to package and protect the cell's DNA, are required in every cell, and as a consequence, multiple histone genes occur in clusters in every organism. These are exact copies of the same gene. Their presence is evidence that sequences can duplicate and be selected in organisms due to the basic functional demands of the cell.[31] Further, this kind of organization occurs even in unicellular organisms. Therefore, it is absolutely logical to conclude that this sort of gene duplication has nothing to do with the assumed evolution of multicellular organisms from unicellular organisms.

Sequence similarities in functionally-similar proteins, imposed by functional constraints, are mistaken to originate by gene duplications

Different proteins with similar functions sometimes have similar amino acid sequences. Such proteins are grouped under a "family" and are believed to have evolved from a single ancestral gene that in the course of evolution duplicated and gave rise to many copies.[32, 33] It is assumed that different mutations gradually accumulated in the various copies thus changing them into new genes, producing proteins with new functions (see Figure 4.5). An often cited example is the family of protein-cleaving enzymes called serine proteases. The digestive enzymes chymotrypsin, trypsin, elastase, and the blood-clotting enzyme thrombin are included in this family. About 40% of the positions in the amino acid sequences in any two of these proteins are occupied by the same amino acid.[34,35] Although these serine proteases cleave proteins, they have different target specificities and regulatory properties. Evolutionary geneticists believe that some of the amino acid changes that make these enzymes different were selected in the course of evolution, and resulted in the changed properties. A second set of functionally "neutral" amino acid changes survived because they did not affect the basic structure and function of the enzyme.

As we have seen in the previous chapter, it is improbable to evolve a new gene (even if it requires only a 10% specific change) by random mutations in the copy of an existing gene. There are other explanations for the presence of sequence similarity in different proteins and their genes. The different serine proteases can be totally independent proteins with no evolutionary connection. Because the basic nature of the biochemical reaction

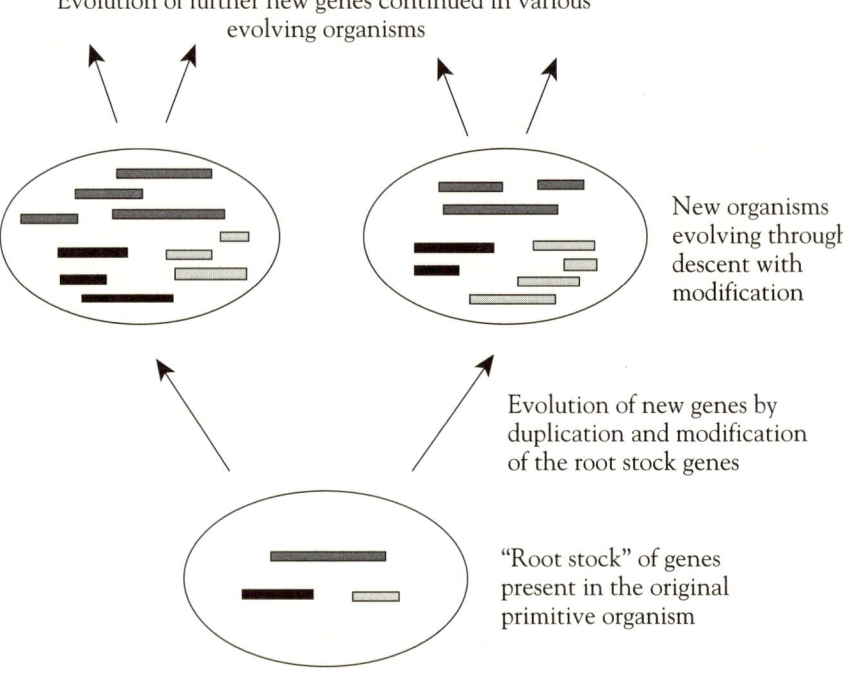

FIGURE 4.5. THE WRONG INFERENCE OF MOLECULAR EVOLUTIONISTS DERIVED FROM SEQUENCE SIMILARITY OF PROTEINS. Among today's living organisms, there exist many functionally similar genes with similar sequences. Wrongly connecting this scenario with the evolutionary theory of Darwin, molecular evolutionists simply believe that a "root stock" of genes existed in the original primitive organism, which evolved into many new genes as new organisms evolved from the ancestor. Every root-stock gene duplicated into many copies, and each copy gave rise to a new gene through accumulating mutations thereby evolving new families of genes with similar sequences, structures, and biochemical functions.

is very similar in these proteins, severe functional constraints can impose quite similar amino acid sequences in the proteins. Therefore, even if all these genes had originated independently, one can expect sequence similarity among them. In fact, evolutionary connection of these proteins is not at all needed to explain their similarity of sequence and function. In chapters 8 and 9, I provide an explanation.

In summary it is clear that although gene duplications can and do occur within the genome of an organism, the evolution of a new gene from the duplicated copy of a gene through organismal evolution is improbable. On the other hand, even if two similar genes can be shown to be the result of duplication of a gene and subsequent modification of a copy into a new gene, still we can show that such a process could not have happened through the closed genome of organisms while organisms were supposed to be evolving, because the probabilities involved do not permit this even in geological time. However, as we shall discuss in Chapters 7–9, such a process could have occurred in the fully open-ended gene pool of the primordial pond, allowing similar genes to have originated by gene duplication and modification.

Importance of distinguishing between the genuine and false phenomena of gene duplication

In dealing with gene duplications, it is important to distinguish between the real duplications and apparent, but false, duplications. Genuine gene duplications, for which the evidence is clear, include rRNA and histone genes. Apparent gene duplications are inferred from sequence similarity studies. Again, as we saw above, these need not be due to duplications within the genome. A third set of sequence similarities, reported by molecular evolutionists, appear to be due to considerable errors in judging similarities (and in the methods in identifying them). Sequence similarities occurring by chance may be misconstrued to be genuine similarities. We shall discuss later the reasons why the level at which chance similarity and genuine similarity can be distinguished is a fine line, which has to be carefully evaluated.

VERTEBRATE PLASMA PROTEINS: A STUDY DESIGNED BY MOLECULAR EVOLUTIONISTS TO SUPPORT THE CONCEPT OF EVOLUTIONARY GENE DUPLICATION ACTUALLY PROVIDES EVIDENCE TO THE CONTRARY

Vertebrate blood cells are suspended in a solution called plasma. It is a remarkable solution of substances which are mostly peculiar to vertebrates. It is astounding to see that many of the proteins in this blood plasma simply "appear" at the level of fish, with absolutely no evidence of these existing in protochordates (animals which are assumed to be the ancestors of fish). A typical mammalian blood plasma contains more than 600 proteins.[36] Furthermore, the blood coagulation system of proteins is specific to all the vertebrates and absent in all the invertebrates.[37]

Molecular evolutionists see vertebrate plasma proteins as excellent examples of evolution through gene duplication.[38] I shall demonstrate, by analyzing the available structure and sequence data on the genes encoding vertebrate plasma proteins, that these proteins provide evidence exactly to the opposite conclusion — that they could not have evolved by duplicating genes from an ancestral organism.

Two facts are clear from the studies of vertebrate plasma proteins carried out by modern molecular geneticists who happen to believe in Darwin's evolutionary theory. 1) Vertebrate plasma protein genes have absolutely no counterparts in the invertebrates. 2) Because several proteins in the vertebrate plasma contain multiple copies and varieties of a unit domain (a basic, shared amino acid sequence), evolutionists thought that these plasma proteins evolved from a single unit domain present in an ancestral animal. However, there exists no gene for the supposed "unit-domain" protein in any living organism for any of the plasma proteins so far tested, from which the "multidomain" vertebrate plasma proteins are believed to have evolved.

The protein albumin *contains three similar domains that are assumed to have originated by duplication of a single domain. But there is absolutely no evidence that such duplication occurred in evolution.*

In almost all the vertebrate plasma proteins, duplications of subsequences seem to be present. The details of most of these proteins were worked out first in the "higher" vertebrates, particularly the mammal.[39] *Albumin,* the most abundant protein in mammalian blood plasma, is primarily a transport protein. Brown determined the amino acid sequences of both bovine and human albumins (580 amino acids) and showed them to contain three similar "macrodomains" (Figure 4.6). On the basis of the degree of similarity between the three macrodomains, and from a consideration of the degree of difference of the human and bovine sequence, Brown concluded that there must have been a primitive gene in an ancestral organism coding for a protein about one-third the size of the mammalian type, and that a series of elongative duplications had led to the three similar macrodomains found in the mammalian albumin. The expectation was that this duplication should have occurred in some animals believed to be early ancestors of mammals, such as the fish.

Starting with this expectation and background, Russell Doolittle began a search for a "small" albumin representing the primitive macrodomain in the

FIGURE 4.6. THE DECEPTIVE APPEARANCE OF A VERTEBRATE PLASMA PROTEIN WITH THREE SIMILAR DOMAINS, MISLEADING ONE TO BELIEVE THAT IT HAS EVOLVED FROM A SINGLE DOMAIN FROM AN ANCESTOR. A typical protein (such as albumin) in the vertebrate plasma appears to have three domains with some sequence similarity (~40%). Purely based on this structure, molecular evolutionists assume that the tri-modular gene has evolved by the duplication of the single-domain gene which, they believe, could have been present in an ancestral organism. There is absolutely no evidence whatsoever, other than an inference from such an apparently repeated structure, that this has happened. In fact, such a tri-domain protein can occur with exactly the same probability as that for a protein without such repetitions (see Chapter 7 and 9) — showing that evolutionary connection is not required to explain the structure.

lamprey and other primitive fish.[40] Contrary to his expectation, the lamprey's major plasma protein was an oversized albumin composed of a single chain stretching about 1500 amino acids. The blood plasma albumins of kelp bass and guitar fish were also found to be far larger than the mammalian plasma albumin. The observation was puzzling to Doolittle, who then concluded that plasma albumin predated the evolution of vertebrates, and that the search for smaller versions must be pushed back to invertebrate creatures — from which vertebrates are assumed to have evolved.

The "Fibrinogen paradox": A case where molecular evolutionists expected to prove evolutionary gene duplication but where all the evidence goes against it

While molecular evolutionists were puzzled with the case of the plasma albumin, the study of many other plasma proteins resulted in more puzzles than answers for the evolutionists. The central protein responsible for the coagulation of vertebrate blood is *fibrinogen*. In mammals such as the human, the protein contains three pairs of nonidentical polypeptide chains. However, the three human fibrinogen chains — α, β, and γ — were found to have amino acid sequence similarities. By comparing amino acid sequences of fibrinogens from other species (rodents and bovine), Doolittle concluded that the β and γ chains themselves diverged about 600 million years ago by duplication. If one assumes a similar rate for the α chain, the duplication that gave rise to an α and non-α chain ought to have occurred about a billion years ago.

What is important to our discussion is that the structure of the fibrinogen of "lower" vertebrates, such as the lamprey, is similar to that of the mammals such as the human, and that each of the lamprey fibrinogen chains is similar to the corresponding mammalian chains. As a consequence, according to Darwin's theory of evolution, the gene duplications leading to separate chains predate the divergence of lampreys and other vertebrates, which is 450 million years ago. As said above, by extrapolating the rates of sequence changes, it was concluded that the β–γ duplication should have occurred about 600 million years ago, and the α–non α chains much before that. If this were true, then one ought to expect the presence of fibrinogen-related proteins among the protochordates and invertebrates. Most strikingly, so far no such proteins have been found in any of these organisms, even though a number of explorations have been undertaken. The absence of fibrinogen in the invertebrates is so puzzling and confounding to Doolittle that he calls it the "fibrinogen paradox."[41]

What we can infer from such a crystal clear scenario is that there has been no duplication of an ancestral gene to produce the α, β, and γ chains of fibrinogen; such an assumed gene duplication is unnecessary for the evolution of the lowest fish (lamprey) to the highest mammal; because essentially the same fibrinogen protein is present in all the vertebrates. And, it is totally absent in invertebrates, because they do not need these proteins for their blood coagulation. See Figure 4.7.

The complete absence of the thrombin-generation system *(blood coagulation system) as a whole in invertebrates and protochordates is clearly illustrative that these proteins did not evolve by gene duplication from an assumed ancestral protein through organismal evolution*

The thrombin-generation system is needed to convert fibrinogen to fibrin. The major serine proteases, used in blood clotting, are made as precursors: prothrombin, Factors VII, IX, and X. The amino acid sequences of all these proteins resemble each other. The inference of molecular evolutionists was that they have evolved as a result of gene duplications from an ancestral gene. All the four proteins have been found not only in "higher" mammals but also in nonmammalian vertebrates such as the lamprey, a creature supposedly at the bottom of the vertebrate evolutionary tree.

The puzzling thing about the vertebrate blood coagulation system, according to Doolittle, is that the whole system appears fully developed even at the level of the most primitive fish. Comparisons of the amino acid

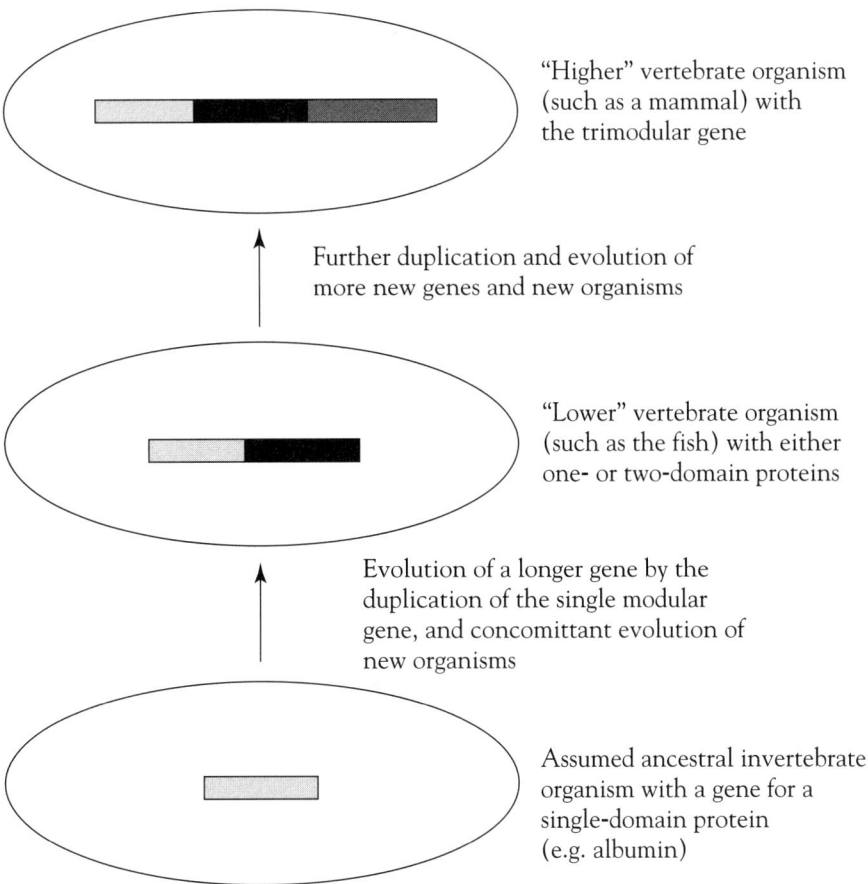

FIGURE 4.7. THE WRONG CONCLUSIONS DERIVED FROM THE STRUCTURE OF THE VERTEBRATE PLASMA PROTEINS. The presence of "trimodular" proteins (described in Figure 4.6) of the vertebrate plasma was first discovered in mammals. From this, molecular evolutionists simply assumed that these three modules evolved from a single module present in the genome of an ancestral vertebrate or invertebrate organism. But when molecular evolutionists searched for its presence in the "lower" vertebrates such as the fish, these proteins appeared in essentially the same full "trimodular" structure. Scientists began to think that the single-modular structure would be found in invertebrate organisms, from which they believed all the vertebrates had evolved. But to their disappointment, no such ancestral proteins could be found — either in the tri-, di- or single-modular form — showing that their original assumption is wrong.

sequences of these coagulation factors imply that their divergence occurred more than a billion years ago. Yet, there is no indication of a thrombin-generating system in protochordates and invertebrates. This is absolutely confounding molecular evolutionists, as illustrated in the following quote by Doolittle (italics mine):[42]

> The most astonishing thing revealed by our survey of vertebrate plasma proteins is that so many of them appear to be represented in even the lowest of fish. Not only that, but most of these proteins exist in fish in a general form not much different from that observed in mammals, in spite of *an obvious prehistory of elongation by contiguous(tandem) duplication.* The further paradox is that most of these same proteins have not been identified among the invertebrates or even the protochordates. When were they invented, and when did the elongations occur? Although we have some clues in the many cases of common ancestry, we have not been able to pin down the timing of these duplicating events with any precision. Moreover, the primal events whereby they were derived from other stock proteins remain mysterious in most of the cases.

We must note that the "obvious prehistory of elongation by contiguous (tandem) duplication" is only an assumption. It is neither obvious nor proved that these similar regions had originated by organismal evolution. On the contrary, as we shall see in Chapter 7, many similar sequence domains could occur in a random sequence purely by chance — thus showing that evolution is unnecessary to explain the presence of multiple similar domains in a gene.

The thrombin-generation system is involved in blood coagulation. It is interesting to note that in the coagulation system, the platelets are restricted to mammals. Inframammalian vertebrates (vertebrates considered to be lower than the mammals in the evolutionary ladder), including fish, have a nucleated white cell that plays a role equivalent to platelets.[43] This difference, taken together with the many other differences such as the unique presence of placentas and mammary glands in mammals and their absence in inframammalian vertebrates, indicates that mammals and inframammalian vertebrates have different or independent origins.

The cases of transferrin, fibronectin *and* ceruloplasmin *also disprove the assumed mechanism of the evolution of new genes by gene duplication through organismal descent with modification*

Transferrin is one of the more abundant plasma proteins in mammals which transports iron to the bone marrow and other parts of the body. All the vertebrate organisms examined have transferrins with molecular weights in the range of 70,000–90,000 daltons (a dalton, like the gram, is a unit of mass). In mammals and chickens, the protein seems to contain an internal duplication.[44] Evolutionary geneticists interpreted these very similar sequences to indicate that the original internal duplication ought to have occurred not too long before the divergence of birds and mammals, or about 250 million years ago. But transferrin occurs in the blood plasma of all vertebrates, including the "primitive" cyclostome fishes,[45] which implies a very early appearance of the internally-duplicated protein, and certainly an appearance long before the divergence of fish and amphibians (300 million years ago). Evolutionary geneticists suggested that a smaller prototype unit domain may be found in the invertebrates. Palmour and Sutton surveyed a large number of invertebrates to test this notion.[46] Of all the invertebrates they investigated only moths had a blood protein that even bound iron. The body fluids of amphioxus, sea cucumbers, horseshoe crabs, snails, clams, limpets, or assorted worms did not contain transferrin or anything similar to it.

Sometime later, iron-binding proteins were discovered in a spider and the crab *Cancer magister*.[47] In the spider, the iron was associated with two different blood proteins, of molecular weights in the range of 80,000-100,000 and 200,000–300,000 daltons. In the crab, only one iron-binding protein, of molecular weight 150,000, was found. This latter protein bound two atoms of iron per molecule. Surprising to evolutionists, the proteins that bind iron in invertebrates all had higher molecular weights than those of vertebrates. Furthermore, it is not clear that the vertebrate iron-binding proteins have any sequence similarities with those of the invertebrates. More importantly, there are differences in the presence and absence of transferrins in the different invertebrates. They are absent in almost all the invertebrates. From these findings we can determine that there is no evidence at all for the evolution of the multidomain protein transferrin from a unit domain protein assumed to have been present in the "lower" animals. It is clear that these findings clearly corroborate our conclusions that genes cannot evolve — and have not evolved — by duplication from an ancestral gene through organismal evolution in their genomes.[48]

The proteins such as *fibronectin* and *ceruloplasmin* also contain an apparent internal duplication (indeed we should call this internal subsimilarities, since we do not know if duplications actually have occurred). Like the other plasma proteins we have discussed, fibronectin and ceruloplasmin are present even in lamprey. Although it is not determined whether these two proteins are absent in invertebrates, it is quite possible, as in the case of other plasma proteins, that they are absent.

The immunoglobulins, the proteins that serve as antibodies, and the vertebrate immune system are totally absent in invertebrates

The immune response is a system in the blood of the vertebrate that is specific to vertebrates and absent in the invertebrates. The main feature of this system is the synthesis of antibodies capable of combining with foreign substances. The search for the origin of the system with a belief that its precursor should be present in the invertebrates has only ended up in disappointment. Those few substances which are capable of binding foreign materials in certain invertebrates have entirely different structures from the vertebrate counterparts.[49] On the other hand, proteins very similar to the mammalian immunoglobulins are present even in the lower vertebrates such as the fish and amphibians. The arguments that apply to the coagulation system and other proteins unique to vertebrates can be very well applied here to illustrate how it is impossible for the vertebrate immune system and its proteins to have originated from invertebrate creatures by organismal evolution.

The conclusion: The scenario of all the vertebrate plasma proteins illustrates that gene duplications never occurred in the context of organismal evolution

The most important aspect of these findings is that neither the different plasma proteins in the multidomain forms nor their supposedly unduplicated unit-domain forms have been identified in invertebrates or even the protochordates. Ultimately, of course, there is absolutely no evidence that such an ancestral domain exists in these "lower" animals, which has been puzzling to evolutionary geneticists (Figures 4.5–4.7). However, we can see that this situation is not puzzling at all based on the new theory of the independent birth of organisms. The viewpoint of molecular evolutionists that plasma protein genes have arisen by gene duplication has been developed purely by inference based on the presence of apparently duplicated regions in these proteins — in order to correlate it to Darwin's theory of evolution. It has to be noted that no duplication of a gene and the evolution of a new gene has been ever shown or proved in the genome of an organism. All these facts combined indicate that gene duplications have never contributed to organismal evolution.

Absence of vertebrate plasma proteins in invertebrates indicate that vertebrates could not have evolved from the invertebrates

There are several underlying truths revealed from the foregoing discussions showing directly that Darwin's theory is incorrect. The scenario of the vertebrate plasma proteins reveals that vertebrates could not have evolved from invertebrates, because the proteins of vertebrate plasma are absent in the invertebrates, even in a primitive form. It is improbable that these proteins were "invented" from out of thin air within the genomes of the vertebrates in a quantum leap when vertebrates were supposed to have evolved from invertebrates.

The existence of a much larger albumin protein in lampreys compared to that of the "higher" mammals may indicate that the evolution of lamprey to mammal is improbable.[50] Similarly the scenario of other blood coagulation proteins show unequivocally that vertebrates could not have evolved from invertebrates. The different sizes of some of the proteins in various vertebrates also indicate that the different vertebrates are not evolutionarily connected. How then did these proteins come about in vertebrates?

It is appropriate for us at this juncture to remind ourselves that the details about the vertebrate plasma proteins, illustrating that almost all of the 600 and more proteins present in the vertebrate blood plasma are absent in the invertebrates, serves as one example for the uniqueness of one type of organism compared to another. When we analyze the different invertebrate organisms, we can be very sure that they will also exhibit distinctness of proteins and genes among them. A number of examples are already known. However, what is known is only the tip of the iceberg. We can boldly predict that the kind of differences we have witnessed between the vertebrates and invertebrates will be shown among organisms that are believed to be even more closely-related through evolution.

Gene duplications could not have taken place in the genomes of organisms through organismic evolution: The globin gene example

It is the belief of molecular evolutionists that from one protein many different proteins can evolve by its duplication and modification. For instance, *Molecular Biology of the Cell* states,

> Once an amino acid sequence has evolved to form a useful and uniquely folded protein domain, the DNA sequence that codes for it can be duplicated and the duplicated copy modified to make additional, somewhat different proteins.

An example most often cited to be exemplifying the duplication of genes concomitant with the evolution of organisms is the blood cell protein hemoglobin. However, consideration of probabilities in duplicating and changing a given gene sequence specifically into another useful one, through descent with modification, shows that achieving this is improbable. In support of this conclusion, there are discrepancies in the presence and absence of the various forms of globin genes in the different organisms supposedly at different levels on the phylogenetic (evolutionary) scale.

In explaining the supposed evolution of the different globin genes through organismic evolution, the evolutionary argument goes as follows.[51]

> There exist different forms of the molecule hemoglobin in organisms at different levels on the phylogenetic scale in all vertebrates and in many invertebrates. By considering these different forms of hemoglobin one can reconstruct the events of evolutionary change of organisms at different levels on the phylogenetic scale. A molecule like hemoglobin became necessary to allow multicellular animals to grow to large sizes, where they could no longer derive their oxygen supply simply by diffusion through the body coverings; consequently, a similar molecule is found in all vertebrates and in many invertebrates.
>
> The most primitive oxygen-carrying molecule is α globin composed of a single chain of about 150 amino acids. In many marine worms, insects, and primitive fish, oxygen is carried by this kind of globin. In higher vertebrates, however, two kinds of globin chains make up the hemoglobin molecule. About 500 million years ago, during the evolution of higher fish, a series of gene mutations and duplications must have occurred. These events led to the establishment of two slightly different globin genes, coding for the α and β globin chains in the genome of each individual. In modern higher vertebrates, the hemoglobin molecules are composed of a complex of four of these chains: two α-chains and two β-chains. This structure is much more efficient than single-chain globins because the four oxygen binding sites in the $\alpha_2\beta_2$ molecule interact, causing a cooperative allosteric change in the conformation of the molecule as it binds and releases oxygen. This enables it to deliver a much larger fraction of its bound oxygen to the tissues.
>
> Still later, during the evolution of mammals, one of the two β-chain genes apparently underwent mutation and duplication once again giving rise to a γ chain that is synthesized specifically in the embryo (or fetus) to produce an $\alpha_2\gamma_2$ hemoglobin. This fetal hemoglobin has a higher affinity for oxygen than adult hemoglobin and is thus advantageous for the fetus. A further duplication occurred still later, during primate evolution, to give rise to a δ globin gene and thus to a second minor form of hemoglobin ($\alpha_2\delta_2$) found only in adult primates. Sometime during evolution an ϵ globin gene also appeared, resulting in an embryonic form of hemoglobin ($\alpha_2\epsilon_2$).
>
> The end result of the gene duplication processes that have given rise to the diversity of hemoglobin chains is seen in the arrangement of the genes coding for the different functional polypeptides chains that arose from the original β-chain. They are arranged as a series of homologous [similar] DNA sequences located within 50,000 base pairs

of each other on one human chromosome. Some duplicated globin DNA sequences in this region do not correspond to genes. These sequences, known as *pseudo-genes*, have homology to functional genes but have been disabled by mutations that prevent them from being expressed.

Evolutionarily expected scenario is different from that observed in the organization of globin gene family

What we have seen in the previous section concerning vertebrate plasma proteins clearly demonstrates that there is no evidence for the duplication of genes through organismic evolution and that invertebrates could not have evolved into the vertebrates. But the belief is contrary to the evidence.

If the evolution of various globin genes had happened starting from an ancestral gene in an ancestral invertebrate, the arrangement of the different forms of globins in the genomes of different animals also should conform to the supposed sequence of evolutionary events. But there are many discrepancies in these arrangements. The organization of clusters of genes in the globin "gene family" vary in different animals. Although the general organization of globin gene clusters seems similar in many organisms, the types, number, and order of genes vary randomly among them.[52] This discrepancy in the expected and observed genomic structures of genes contradict the evolutionists' concept that gene duplications occurred in the genomes of organisms and contributed to organismal evolution through descent with modification.

The red blood cell protein hemoglobin occurs in many forms in different organisms, e.g., α, β, γ, δ, ψ, and ε. The rabbit has four β-like genes: two embryonic, one pseudo (inactive), and one adult, lying in the order of their expression. In the mouse seven β-like genes have been found: two early embryonic, one late embryonic, two adult genes, and two pseudogenes. In the chicken, the cluster is smaller and does not seem to include pseudogenes, but rather only four functional β-like genes. The outside two are embryonic and the inside two are adult. There seems to exist only one adult gene in humans,[53] whereas there are two in mouse. Furthermore, the types of globin genes seem to vary in different organisms, for instance, we are not yet sure whether there are separate embryonic and fetal β-like globins in rabbit and mouse. A pseudogene in one species may be an active gene in another. For example, in goat, $\psi\beta_1$ of the higher primates is equivalent to an active embryonic gene.

If indeed the above different organisms evolved from a common ancestor, the organization of a given gene cluster in all these species should remain largely the same. Even if it did change, the changes should be evolutionarily correlatable, traceable, or justifiable. But there are discrepancies not explainable by evolutionary change.[54] A number of roundabout and unconvincing evolutionary hypotheses are usually proposed to account for such random variations. For example, the differences are suggested to be "recent" changes in the genomes of these organisms, without being able to trace them back to one possible ancestor.[55] It is not probable that the changes required to bring about the scenario of the distribution of the various globin genes in the different organisms can occur within the genome by random mutations. For example, to delete a gene precisely, and to duplicate a β gene, etc., let alone evolving a δ gene, is improbable. Then what is the answer to the scenario of this distribution? If it cannot be brought about by evolutionary changes, then by what mechanisms can it be brought about? This is answered by the new theory of the independent birth of organisms described in Chapter 8.

Duplications and modifications of globin genes could not have occurred in the closed genomes of organisms through organismic evolution

It is important to remember that the globin gene is only one of a large number of genes supposed to be included in the new DG pathway of an evolving organism. Different forms of globin gene have to be expressed specifically in the red blood cell at different times in the development of an organism. Furthermore, each form is supposed to have evolved through specific sequence changes, selected from random gene mutations. In addition, there are different sets of globin forms in different animals, such as the goat or the human.

Mammals have several unique mammal-specific hormones and proteins, as well as unique body parts and organs such as the mammary gland and placenta. All these are believed to have come about by the evolution of the mammal from the reptile. The mammal-specific forms of globin genes and their unique positions in the DG pathway of the mammal is only one of a large number of unique genes and their specific positions in the DG pathway of the mammal. When the DG pathway of even one organ cannot evolve through organismic evolution, how can all these come about when the reptile supposedly evolved into a mammal? Thus the scenario of the globin genes in the different organisms could not be derived by gene duplication while organisms were evolving.

As we shall see, the observed scenario in the case of the globin genes in the different organisms is most likely due to a duplication of genes and by other means in the primordial pond. In essence, gene duplication could have happened in the free genomes in the primordial pond, but not in the genomes of living organisms.

OTHER ARGUMENTS AGAINST THE EVOLUTION OF NEW GENES BY GENE DUPLICATION CONCOMITANT WITH ORGANISMIC EVOLUTION

The constancy of some organisms for a very long geological time while others are assumed to be rapidly evolving is contradictory to evolutionary theory. This also illustrates that genetic mechanisms such as gene duplications cannot contribute to evolution.

If Darwin's mechanisms are correct, the gene duplications and all other types of mutations should have happened in the genomes of all organisms equally over geological time, and should have led to their evolutionary effects equally. Therefore, organisms should be changing equally, in terms of their structure and function. However, the fact that fish remained fish for 300 million years (e.g., the crossopterygian fish), while all organisms "higher" than fish (the amphibians, reptiles, birds and mammals) supposedly evolved, indicate that this is not the case. What happened to the gene duplications and their supposed evolutionary effects in fishes and frogs that remained unchanged? According to the theory of evolution, they should have changed into creatures different from the fish and the frog they once were. But the fossil record shows them to be virtually unchanged since their appearance.

Evolutionists offer a speculative explanation called "balancing selection" for the observed constancy of organisms in the face of environmental change.[56] They say that an organism actively finds its niche and remains there, thereby balancing natural selection. This concept is shown to be incorrect when it is analyzed at the level of genomic changes. According to the evolutionary theory, environments are changing and new organisms are evolving. Then, how can an environmental niche be absolutely constant, so that an organism, despite all its random genomic changes, be virtually unchanged for hundreds of millions of years? Furthermore, the genomic changes must be random no matter what the environmental change may be. Under the evolutionary theory, there is no way a genome can remain constant for hundreds of millions of years on the face of random mutations that must constantly occur in the genome. When mutations are assumed to have changed the fish into rats, anteaters, and elephants, how can the very same kind of mutations not change many sharks, fishes, frogs, and scores of invertebrates into other organisms? That is, if organismal evolution by descent with modification is truly an ongoing process, organisms should not remain constant over geological time. How can any evolutionist explain this discrepancy?

These facts indicate that neither gene duplication nor any other mutational mechanisms are the cause for the presumed evolutionary change. It is appropriate to say that these changes occur passively in the genome of every organism, without ever changing the organism itself.

"The different genes of all organisms evolved from the 'root-stock' genes of the first original organism": A misconception

Molecular evolutionists assume that all the genes of all living organisms originated from a root stock of genes of the original organism. They believe that each gene in the original organism duplicated and the duplicated copy evolved by mutations into one or more new genes. For instance, Russell Doolittle states:[57]

> All living organisms must trace back to a common ancestor, and it is reasonable to think that some very early ancestor had a relatively small genome coding for a relatively small inventory of prototypic proteins. Most contemporary gene products are the result of past gene duplications and subsequent divergence resulting from gradual amino acid replacement. As a result, many proteins have already been grouped into families. There are sequences for scores of each of four major protease families and correspondingly large numbers of protease inhibitors. There are vast numbers of protein kinases, all apparently descended from a common ancestor, and we can anticipate a similar multitude of protein phosphatases.
>
> ... Common ancestry is the essence of evolution, and nowhere is Darwin's notion of 'descent with modification' more apparent than in the amino acid sequences unraveling before us.

If a large number of genes grouped into gene families "evolved" from a set of "root-stock" genes, which itself is large in number and no less primitive than the supposedly evolved proteins, then how were the root-stock proteins themselves evolved? It is imperative to analyze and understand how "a relatively small inventory of prototypic proteins" of "a relatively small genome" could have originated in "that very early ancestor." When we do such a scrutiny and analysis of this most crucial question, as we do in Chapters 7, 8, and 9, it becomes crystal clear that all the genes for all the proteins found in all living organisms must have originated directly in the primordial pond, and be selected from there into the genome of each organism directly. And, it thereby becomes clear that there is no need for a small inventory of proteins in a small genome of a common ancestor, and there is no need for their divergence into many further proteins while diversifying the common ancestor into multitudes of diverse organisms.[58]

Such families of proteins exist even in the unicellular bacteria, for instance, the family of bacterial activator proteins,[59] indicating the baselessness of the assumptions that such families were absent in the ancestral cell and then evolved through the evolution of multicellular organisms. In essence, the whole idea of the evolution of families of proteins, as multicellular organisms were evolving from the root-stock proteins present in a single, ultimate, common progenitor is absolutely incorrect.

It can be seen that the proteases, protease inhibitors, protein kinases and protein phosphatases, each of which Doolittle groups under a family of proteins, have common biochemical functions. In fact, most of the proteins, which are assumed to have evolved from an ancestral gene by gene duplication, and which have sequence similarities, exhibit similar biochemical functions — because of which they have been grouped under a family. However, even if these proteins had originated independently, they would have similar structural domains because they have similar functions. This would necessitate similar amino acid sequences, and similar corresponding DNA sequences within the genes. Thus, the inference, based on sequence similarities among proteins with similar functions, that their genes are derived from one another by evolutionary duplication and change is erroneous.

We know that various enzymes carry out their catalytic functions by binding to many common cofactors, metals, nucleotides, and other small molecules and receptors. As a result, many distinct proteins in nature have similar subfunctions. For instance, the nucleotide GTP or GDP is bound by a wide variety of different proteins. Calcium, magnesium, and iron is bound by widely different proteins. Each of several cofactors such as the NAD, NADH, FAD etc. (see Genetics Primer), is bound by enzymes that are totally unrelated. Numerous such examples can be given, wherein totally unrelated genes can have virtually the same functional domains. But why should one say that these enzymes (and their genes) are related by evolution, just because they have similar functional domains and therefore have sequence similarity? Is it not obvious that this is purely a conjecture, induced by Darwin's theory of evolution? Is it not clear that it is easy to fall into the evolutionary thinking, just by looking at the sequence similarity in different proteins, without looking for an explanation divorced from organismal evolution? Later, I will provide such an explanation.

If evolutionary mechanisms are correct, then duplication of sequences should be indiscriminate and defective genes should be far too frequent. But the contrary seems to be true.

If evolutionary gene duplications are occurring at all in the genomes of organisms, then for every gene that supposedly evolved by duplication, we must have thousands of random duplications of sequences (whether they contain a gene, part of a gene, or no gene) and their modifications, from among which the right one could be selected. One might say that all those other than the correct one would have been lost by natural selection. However, it is not possible to eliminate all of them by random deletion or by any other mechanism.[60] But when we see duplicated genes we see only complete and functional genes, except for one or two pseudogenes. If the process of gene duplication leading to the evolution of new genes does happen in the genomes of organisms and is contributing to organismal evolution as claimed by evolutionary geneticists, then the following should be true. 1) An extremely large number of duplicated but nonfunctional genes should be present in the genomes and 2) the duplications of a very large number of sequences in the genome in a totally random manner is certainly required for the duplication of every complete gene in the genome.[61, 62] Their absence in the genomes clearly illustrates that evolution by gene duplication did not occur and is not currently occurring.

Because there can be no directed evolution within the genome, there can be no directed duplication of only the coding sequences. If sequences duplicate within the genome, then it has to happen randomly. Only out of several such indiscriminate duplications can one, by chance, yield something useful. Any new gene must be expressed in order to be evolutionarily selected — even the genes that are not useful should be expressed and tested, so that the useful one can be selected. First of all, if this is the case, there should be a chaotic expression of a number of useless genes. Furthermore, such an indiscriminate expression of useless genes can be detrimental to the living system. Even if one could agree that the bad genes, or at least the expression of them, have been selected against, the process itself cannot be selected against. That is, the process must be ongoing in living organisms. In other words, if evolution did occur within the genome by gene duplications and other purported genetic mechanisms, then the genome must be sort of a mess. But instead, we see that the genome exhibits a highly programmed, streamlined processing of genetic information.

The probability that one good "selectable" gene evolves in an organism is extremely low. Therefore, the probability that several genes that would collectively express a body structure or a physiological function can evolve in a species is far too low. But one might say that each of these genes could evolve, one after another, and accumulate in a species. However, there arises a problem in the proposal that various genes belonging to the set of genes which can construct even a minimum unit (the smallest functional body structure in an organism), evolve sequentially, one after another, by gene duplication in the same line or lineage of organisms. We get into the problem of natural selection of incipient organs. Except for individual genes expressing independent units, in most cases multiple genes are required to express a unit. For example, unless all the genes that construct a minimally functional feather or a minimally functional eye have evolved or appeared in a genome, none of the individual genes for these structures will be selected, and, in fact, any such gene will be lost because it will be totally useless.

Therefore sequential evolution of genes is unacceptable, indicating that a new physical unit in an organism cannot originate by evolution.

A paradox unexplainable by evolution

A discrepancy, similar to that in the organization of the globin genes we discussed above, is found with the histone gene cluster.[63] Histones are the structural proteins that protect the chromosomes. Usually there are five types of histone protein: H1, H2A, H2B, H3 and H4. All eukaryotic chromosomes contain histones. Because the amount of DNA in the chromosomes is large, there is a need for large amounts of histone, and therefore the histone genes are repeated in the genome. Usually there is the same number of copies of each histone gene. But the repetition frequency of histones varies in different organisms. The pattern of differences does not correspond with the evolutionary hierarchy. For instance, the frequency of histone genes in the different organisms are: *D. melanogaster* (fruit fly) ~ 100, sea urchins ~300-600, *X-lavis* (frog) ~ 40, chicken ~ 10, mammals ~ 22. But more importantly the organization within the repeating unit poses a paradox to the geneticists.

The five histone genes (H1, H2A, H2B, H3 and H4) as one unit are repeated several times in a cluster in the genome. Each gene is separated from the next by a spacer sequence. The multiple copies of the five-gene unit within an organism are virtually identical. But between different organisms, the spacers differ in length and in sequence. Here is a discrepancy that geneticists call a "paradox."[64] According to them, the common organization of the repeating unit in the histone gene clusters among the different organisms suggests that it must have existed before speciation of the sea urchins. All of these clusters presumably evolved by duplicating the entire unit. Selective forces have acted to preserve the function of the genes. However, the geneticists ask, while allowing the intermingled spacers to diverge entirely between organisms, how did the spacers remain constant within each species? This is the paradox.[65] To circumvent this, geneticists believe superficially that some "corrective" mechanism must act within each species in order to maintain the spacers to have the same sequence.

Here we are discussing differences in genome organization among creatures with one or two examples. If we consider all the genes in the whole genome and all its other features, the total number of differences will be enormous. As we have seen, the rate of point mutation (10^{-9} to 10^{-6} nucleotides per generation, Chapter 3) is very small, and other kinds of mutations also do not occur rapidly. How then can such an immense number of differences come about in such a short geological time of a few million years through evolutionary change? The answer is that the wholesale differences are not the cause or the result of the evolutionary change of organisms, because, as we have discussed earlier, organismal evolution never happened. These differences could arise due to a fundamentally different phenomenon.

Deceptive scenario that misled molecular evolutionists to believe that gene duplication is a mechanism contributing to evolution

In concluding our analysis on gene duplication, one can see that the scenario is quite deceptive, because there do exist many seemingly duplicated genes in the genomes of different organisms. First, there are genuine duplications such as that of the rRNA genes; second, there are sequence similarities in different genes that appear to be the result of gene duplication within the genome. We can see that the presence of similar genes and repeated sequences led evolutionists to think that they arose by the duplication of sequences within the genomes of organisms while they were evolving from the original organism.[66] However, there is no evidence whatsoever that duplications have occurred through evolution of multicellular organisms, that is, through descent with modification. Actually the evidence is contrary to it.

It is clearly discernible from all our foregoing analyses that evolutionists have made the following false assumptions: 1) the presence of sets of similar genes and sequences is the result of duplication of an original root-stock set of genes and sequences from the original ancestral organism; 2) through such duplications, many new genes evolved; and 3) such duplications are to a great extent responsible for the evolution of the multitudes of creatures from an original ancestor. As unequivocal as we saw, gene duplication could neither be the cause nor the result of organismic evolution. The details of the scenario are clearly explained by the new theory of the independent birth of organisms.

EXON SHUFFLING

Introns and exons

Genes of all organisms except bacteria consist of short exons (coding regions) interrupted by long introns (intervening sequences). When a gene is expressed, its DNA sequence is copied into a "primary" RNA sequence. Then the "spliceosome" machinery physically removes the introns from the RNA copy of the gene, leaving only a contiguously connected series of exons, which becomes the "messenger" RNA (mRNA). This mRNA is now "read" by another cellular machinery called the ribosome, to produce the encoded protein. Thus, although introns are not physically removed from DNA, a gene's sequence is read as if the introns never existed.

What is exon shuffling?

The origin of introns in eukaryotic genes is one of the most important questions in molecular biology today. In one view, proposed first by Walter Gilbert[67] and extended by Colin Blake,[68] introns originated as a means of recombining and shuffling exons encoding distinct functional domains in order to evolve new genes. Thus, new genes are assembled from exon modules that code for functional domains, folding regions, or structural elements from preexisting genes in the genome of an ancestral organism, thereby evolving genes with new functions. Such "shuffling" of exons specifying discrete functions in genes would generate many new complex proteins with novel enzymatic functions, and the shuffling is mediated by the introns. Thus, in attempting to explain the origin of introns and the split architecture of eukaryotic genes, "exon shuffling" has been suggested as another mechanism for evolving new genes within the genomes of organisms.

Gilbert and Blake proposed exon shuffling when introns were first discovered in eukaryotic genes in 1978. Their aim was to explain the origin of introns in eukaryotic genes. However, except for an apparent initial support, subsequently there has been no real support for this hypothesis. At first it was discovered that a few proteins and their gene sequences seemed to follow this theme. However, even 12 years after it was proposed, there are only about a dozen modules and about a dozen proteins that have been shown to support this hypothesis. Extensive analysis of several thousands of proteins and genes have shown that only extremely rarely do genes exhibit the supposed exon shuffling phenomenon.[69]

Why the proposed exon shuffling mechanism cannot contribute to the evolution of new genes and new DG pathways, and therefore to organismic evolutionary change

As we discussed for the cases of transposition and gene duplication, if evolution has been ongoing, the proposed exon shuffling mechanism has to shuffle exons indiscriminately (that is, randomly) and, out of the random combinations, only those useful as genes should be selected. Several facts demonstrate that such a process is not happening.
   1. If this mechanism were happening, then the genome should be evident of this indiscriminate shuffling of exons, which is not at all the case. As

we saw, not even one in several hundred genes shows anything that even seems to be the result of exon shuffling. If genes had evolved (and are currently evolving) by shuffling a stock of exons, then the same exons should be repeated far more than observed in genomes of organisms. What we see in reality are extremely rare cases with superficial appearances of exon shuffling — in fact, compared against the whole repertoire of genes known so far, the number of genes even claimed to show exon shuffling is minuscule.

2. In fact, the simultaneous presence of similar DNA sequences encoding similar functional protein domains in some distinct genes is taken to reflect the prior shuffling of these part-DNA-sequence regions from a few preexisting genes encoding these domains.[70] As explained under our discussions on gene duplication, such a presence of a set of few functional domains in many different genes can be due to a constraint for the presence of similar biochemical subfunctions (such as binding a metal or a vitamin) in these various proteins. Also, the occurrence of discrete exons in different genes is seldom seen. Even exons slightly resembling other exons in other genes, without even a clear match at the boundaries of these exons, have been purported to be the result of exon shuffling.[71] All this shows that there is no need for an exon shuffling mechanism to explain the presence of some similar sequences in different genes.

3. The probabilities involved in achieving the right combination of exons forming a useful gene from the stock of exons present in the set of genes within a genome by random processes is exceedingly low. Consider that the recombination has to be indiscriminate, i.e., random in the genomic DNA. When we know that greater than 90% of the genome is junk (unused) DNA, and greater than 90% of a gene is introns, imagine the chance for shuffling of exons from various genes to bring about a useful new gene. Furthermore, let us not forget that an exon has three reading frames, and therefore it has to integrate in the right reading frame among the exons of a gene. In fact, if such random exon shuffling is occurring, the genome should be full of evidence of such a process — with many different exons of various genes being shuffled without achieving a new good gene for every good gene being evolved — which is simply not the case.

4. How the original stock of exons came into being in the original organism is not discussed under this exon shuffling proposal.

5. Even in such cases where there is a clear-cut repetition of exons in two different genes, exon duplication and usage in another gene could have very well happened in the genetic sequence pool of the primordial pond,

which will be explained in my new theory. This is not a widespread phenomenon at all, showing up in very few proteins (in about a dozen proteins out of the several thousands of proteins known to us so far).

All these reasons combined indicate that the proposal of the exon shuffling mechanism for the evolution of new genes is invalid. I would like to reiterate here that this is only a proposal for which there is no direct evidence. It is not at all an established mechanism, and it is in fact becoming more and more established that this is not occurring in the genomes of organisms.

Almost all the proteins that show any sign of exon shuffling are vertebrate plasma proteins and most of them are involved in blood coagulation. This phenomenon of coagulation is specific to all the vertebrates, and is not found in the invertebrates. Vertebrate blood plasma — the fluid in which blood cells are suspended — is a remarkable solution of substances mostly peculiar to vertebrates. Amazingly, many of the proteins in this blood plasma simply "appear" at the level of fish, with no evidence of these being in protochordates. A typical mammalian blood plasma contains more than 600 protein components.[72] Out of these almost none have counterparts among the invertebrates, and, as we saw above, albumin, fibrinogen, transferrin, thrombin, and factors VII, IX, X are totally absent in them. At the same time, if these proteins have evolved by either gene duplication or exon shuffling or both, they should have done so before the vertebrates appeared. In all the arguments of evolutionary biologists, there is a catch-22 situation: these proposed evolutionary activities of exon shuffling and gene duplication culminating in the set of proteins constituting the vertebrate blood plasma are a precondition to the evolution of the vertebrates themselves. That is, without these proteins there can be no blood plasma in the vertebrates. As a consequence, without the blood plasma, there can be no vertebrates. However, if all these were evolved in invertebrates, at least we should answer the following question that comes to mind: where are the invertebrate precursors of the vertebrate plasma proteins? If only a few of these proteins are missing and many are present in invertebrates, or, at least a few are present in invertebrates, then it could be a consolation. But almost none of them are present in invertebrates. All these facts clearly tell us that neither have these proteins evolved by mechanisms such as exon shuffling and gene duplication, nor have vertebrates evolved from invertebrates.

I have explained the reasons as to why and how introns originated at all in the genes of eukaryotic organisms in recent publications,[73] which are well accepted even by scientists such as Colin Blake, who is one of the proponents of the exon shuffling theory. My proposal is fully corroborated by all the molecular details from greater than 99% of genes we have known so far. Note that the exon shuffling theory is not supported by the details from even 1% of the genes.

Colin F. Blake writes in his recent article, "Proteins, exons, and molecular evolution," in the book *Intervening Sequences in Evolution and Development*, about my theory on why the genes are split into exons and introns.[74]

> Recent work by Senapathy, when applied to RNA, comprehensively explains the origin of the segregated form of RNA into coding and noncoding regions. It also suggests why a splicing mechanism was developed at the start of primordial evolution. He found that the distribution of reading frame lengths in a random nucleotide sequence corresponded exactly to that for the observed distribution of eukaryotic exon sizes. These were delimited by regions containing stop signals, the messages to terminate construction of the polypeptide chain, and were thus noncoding regions or introns. The presence of a random sequence was therefore sufficient to create in the primordial ancestor the segregated form of RNA observed in the eukaryotic gene structure.[75] Moreover, the random distribution also displays a cutoff at 600 nucleotides, which suggests that the maximum size for an early polypeptide was 200 residues, again as observed in the maximum size of the eukaryotic exon. Thus, in response to evolutionary pressures to create larger and more complex genes, the RNA fragments were joined together by a splicing mechanism that removed the introns. Hence, the early existence of both introns and RNA splicing in eukaryotes appears to be very likely from a simple statistical basis. These results also agree with the linear relationship found between the number of exons in the gene for a particular protein and the length of the polypeptide chain.

We shall see more about why genes are split by introns in Chapter 7. As we can see here, the proposal of exon shuffling is neither necessary to explain, nor can explain the origin of genes in the first place, let alone the origin of new genes within the genomes of evolving organisms.

Discredit to the proposed exon shuffling mechanism

Several molecular biologists have questioned the exon shuffling proposal, from a purely evolutionary view for both methodological and conceptual reasons. Scientists such as Russell Doolittle, who himself is a staunch evolutionary molecular biologist, have discredited the exon shuffling hypothesis.[76]

> Introns were discovered not very long after the notion was introduced that certain modular units in proteins — specifically nucleotide-binding domains — may have been shuffled about during the earliest stages of life on earth. .... Introns seemed a simple way of encouraging such combinations and recombinations. Still, the evidence for such a role was, for the most part, only circumstantial, and

a number of energetic if indirect efforts had to be made to buttress the hypothesis. Among these were many attempts to correlate the locations of introns with the boundaries of structurally independent segments, or domains. Unfortunately, it was mostly a subjective attack, with arrows boldly proclaiming the occasional success. In the cases of many proteins even the most ardent proclaimers conceded a lack of correlation.[77]

From all the foregoing, it is apparent that the proposal of the exon shuffling mechanism attempts to connect data that is very weak. This mechanism does not explain the evolution of the great majority of proteins. So, the proposal that all proteins with multiple exons in their genes had evolved by exon shuffling from a stock of exons is wrong. Obviously such a rare mechanism, which has not helped to bring about greater than 99% of all eukaryotic genes, cannot be responsible for the evolution of a multitude of new genes found in all living creatures.

In a recent paper,[78] Gilbert suggests that all the extant proteins have evolved from an original stock of 1000-7000 independent exons. If this is so, then where did the stock of independent exons come from? There is no answer for this question in Gilbert's proposal. Obviously such a stock could not come out of thin air; it had to have an origin. When we explain this origin, as we shall do in the new theory in Chapters 7 and 8, it becomes clear that there is no need for exon shuffling in order to bring about the multitudes of split genes in the living world. This is because when complete exons can occur in a long random sequence, complete genes with many exons and introns can also automatically occur. Indeed, in the background of all our discussions so far, it can be seen that except for the presence of some portions of similar coding sequences in a few different genes, there is nothing that would support Gilbert's proposal of exon shuffling.

A commentary made in *Science*[79] on Gilbert's recent work based on his assumed stock of exons,[80] adds credence to my arguments.

> Several specialists in protein evolution contacted by *Science* had serious reservations about Gilbert's paper, though only one wanted to express them on the record. The gist of their criticisms is that the way Gilbert and his colleagues attempt to detect common ancestry among exons and eliminate duplication is flawed. Russell Doolittle, a well-known protein chemist at the University of California at San Diego, was willing to comment for the record. Although Gilbert's team uses a standard mathematical method, it is "misapplied," he says, because it fails to identify the original exons correctly. It misses sequences that are known duplicates, and identifies repeats that are not related. Doolittle adds that he was disturbed to recognize several protein

sequences in the "distilled" set of ancient exons that were purported to be dissimilar but that today are known to be derived from a common ancestral molecule. That undermined the credibility of the Harvard group's model, says Doolittle, who adds that the value of the work has been "exaggerated."

A recent *Scientific American* article also quotes Doolittle on the work of Walter Gilbert.

> Even on its own terms, Doolittle thinks the Gilbert study is defective. He charges that the researchers did not adequately purge their 1,255-exon database of sequences that would skew the results. Several of the 14 exon matches found by the study involve related proteins that should have been excluded. In effect, those exons may match because they are the same exon, counted twice. Another matched pair — an exon from a keratin protein and one from an albumin — is misleading because the similarity is much more likely to derive from functional constraints on the proteins than from common exon ancestry. If fewer than half of the 14 matched pairs are truly significant, as Doolittle suspects, then according to the Gilbert's group's methods the upper boundary on the exon universe would rise to a half million or more. But Doolittle's point is even stronger. He contends that Gilbert's estimate is completely meaningless because unrecognized divergences or convergences could lower or raise the numbers unpredictably. ... By that time perhaps it will be clear whether they actually have identified the fundamental protein elements of the earliest organism or, as Doolittle suspects, a red herring.

In the new theory, I will show that myriad complete genes with the right exon-intron organization could exist in the primordial pond. In other words, there is absolutely no need for exon shuffling for the evolution of genes from an exon stock — if only such a stock could be shown to be possible to exist in the first place.

In conclusion, there is no doubt that there can be some sequences repetitively present in different genes. This can be entirely due to nonevolutionary reasons such as the constraint in functional similarity in totally independent proteins. Even if some of these arise due to recombination or shuffling of sequences, it could have originated in the primordial pond and not within the genomes of organisms. Thus, neither is there evidence that any exon shuffling has occurred nor is there any need for the exon shuffling mechanism to explain the genomic scenario of organisms on earth. This mechanism that is assumed to be happening within the genomes of organisms does not have any potential in contributing to evolutionary change. Figure 4.8 depicts this conclusion.

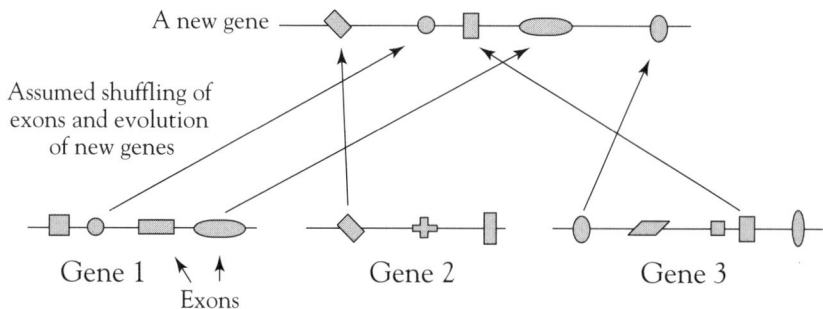

FIGURE 4.8. THE ASSUMPTION THAT SHUFFLING OF EXONS FROM PREEXISTING GENES IN AN ORGANISM COULD EVOLVE NEW GENES — "EXON SHUFFLING" — IS TOTALLY BASELESS. Exon shuffling leading to a new gene (as shown in the Figure) was suggested by Walter Gilbert and Colin Blake, when the split-gene architecture of eukaryotes was first discovered. The presence of the sequence encoding the same domain in a few different genes was originally thought to support this mechanism. But after 12 years of research with thousands of genes, it has been found that greater than 99% of the genes in the living world do not exhibit any pattern of shuffled exons. The observed patterns, in fact, could have originated independently (see Chapters 7 and 9). Thus, it is both unnecessary as well as incorrect to say that new genes evolve by exon shuffling.

Point mutations

A change of one nucleotide, either by deletion, addition, or substitution in the sequence of a gene is called a point mutation.

As shown in Figure 4.9, an insertion or deletion of a nucleotide in a coding sequence results in a "frame shift" mutation (see also Genetics Primer). The reading frame of the coding sequence is shifted by one nucleotide resulting in a completely changed amino acid sequence in the protein product. A mutation that results in the substitution of one amino acid for another is called a "missense" mutation. If a codon is replaced by another that codes for the same amino acid (e.g., a change from CUA to CUG, both of which code for leucine), it results in "samesense" mutations (also called "silent" mutations). If a codon that codes for an amino acid is replaced by one that codes for chain termination, resulting in the premature termination of the synthesis of the protein chain, it is called a "nonsense" mutation.

1. Deletion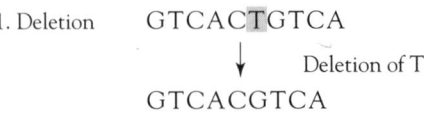

2. Insertion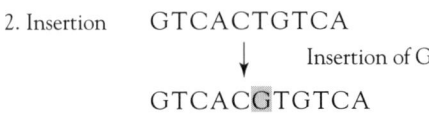

3. Substitution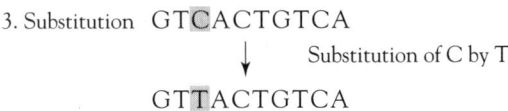

FIGURE 4.9. VARIOUS POSSIBLE POINT MUTATIONS. A point mutation can delete a nucleotide, insert an extra nucleotide, or substitute an existing nucleotide with a different nucleotide at any position. The change in the examples are highlighted.

Mutations can be brought about by chemical mutagens, physical agents such as ultraviolet radiation, and errors in the normal DNA replication process in the cells.

There is no question that point mutations can have powerful effects. But this power has been misunderstood to be capable of evolving new genes from old genes, even to the extent of evolving new genes with totally new functions. This power has been believed to be one of the major causes for the evolution of the multitudes of organisms from the original ancestor. For example, as we saw above under our discussions on gene duplication, after a gene duplicates into a spare copy of the gene, it is the point mutations which are supposed to modify the copy through many lineages of organisms into a new gene. When we carefully analyze what point mutations are capable of in living organisms, we shall see that their powers can never change a gene into a new gene with a new function by accumulating any number of mutations possible over any length of geological time.

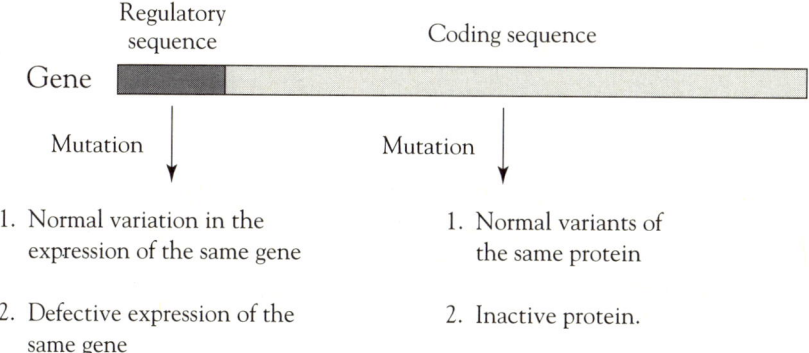

FIGURE 4.10. THE ONLY EFFECTS OF POINT MUTATIONS IN A GENE. A point mutation can occur in a regulatory or coding sequence. When it occurs in a regulatory sequence, it can produce either normal variations or make it defective. A normal variation in the regulatory sequence causes a quantitative variation in the synthesis of the same protein. When the regulatory sequence becomes defective, it will inappropriately switch the gene on and off. When the mutation occurs in the coding sequence of the gene, it can lead to either normal variants of exactly the same protein or to a defective, inactive protein. Because these are the only possibilities by any number of point mutations over any length of geological time, these mutations can never change a gene into a new gene coding for a new protein with a new function.

Figure 4.10 shows the three possible results of a point mutation. 1) It can lead to a normal variant of the same gene. 2) It can lead to a defective, i.e., inactive gene. 3) It can switch off the gene when the gene is normally switched on, and vice versa. This is all it can do, even if a creature lives for trillions of years. These mutations will only produce normal variants of the same gene leading to individual variations of the organism, and to some defective genes leading to diseases in some of its individuals. Even mutations in developmental genes (homeobox genes, see Genetics Primer) could only lead to the normal variations in the shape and size of an organ such as the human nose, or cause developmental errors. Thus, point mutations cannot contribute to the evolution of a new organism with a new gene or a new body part.

POINT MUTATIONS CAUSE CHANGES IN THE REGULATORY SEQUENCE
OR THE CODING SEQUENCE OF A GENE, ONLY LEADING TO EITHER
DEFECTIVE GENES (RESULTING IN A VARIETY OF DISEASES) OR TO
NORMAL VARIANTS OF THE SAME GENE (WHICH ARE THE CAUSE OF
ORGANISMAL INDIVIDUAL VARIATIONS)

Point mutations resulting in defective genes

It has often been demonstrated that point mutations can cause hereditary diseases, congenital abnormalities, and cancer. Many human genetic diseases are known in which an enzyme is either totally inactive or is defective in its catalytic or regulatory function.[81] The defective enzyme molecule may contain one or more "wrong" amino acids in its protein sequence due to a mutation in the DNA sequence coding for it. Replacement of a single amino acid at some critical position in the protein chain may destroy its catalytic activity. When the defective enzyme is a member of an enzyme system catalyzing an important metabolic pathway, the consequence may be a serious metabolic defect. If it is a member of the developmental pathway of an organism, it may lead to a serious developmental defect.

Point mutations can cause congenital diseases (*e.g.,.* phenylketonuria, albinism, galactosemia). In phenylketonurea for instance, replacement of a single amino acid in the enzyme *phenylalanine hydroxylase* destroys the enzyme's activity, and as a result the chemicals phenylalanine and phenyl pyruvate accumulate. An excess of phenyl pyruvate in the blood in early life impairs the normal development of the brain and causes severe mental retardation. Similar changes lead to other human genetic diseases such as galactosemia and albinism.

Globin, the protein part of hemoglobin, made up of two α and two β chains of amino acids, is involved in sickle cell anemia. In this disease, the β chains have a specific amino acid substitution, glutamic acid ⟶ valine, at the 6th position of the chain. Thus the change of a single amino acid could cause profound effects.[82] See Figure 4.11.[83]

Point mutations in genes can cause a variety of diseases which can be hereditary, such as thalasemias. Several classic cases are known. For example, a single point mutation in a splice junction of the β-globin gene is found to be the cause of one of the thalasemias.[84]

Point mutations can also result in many forms of cancers. For example, one form of human bladder carcinoma is found to be caused by a point mutation in a particular amino acid position in a normal protein. A point mutation affecting the 12th codon of the human *k-ras* gene changes the normal gene into a cancer-causing gene.[85] The sequence at the 12th codon is

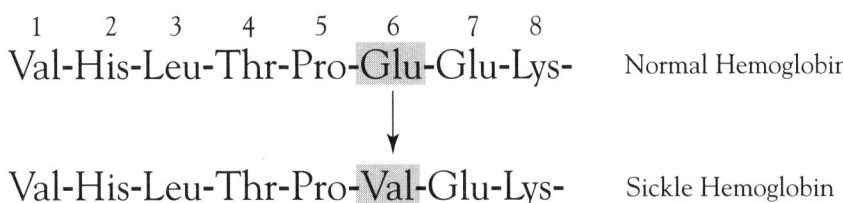

FIGURE 4.11. THE GENETIC MUTATION IN SICKLE CELL ANEMIA. Hemoglobin is made up of two protein chains called α and β. A mutation in the β-chain gene changes the glutamic acid in the 6th position to valine. This changed amino acid results in defective hemoglobin, which in turn results in defective red blood cells.

changed from GTC (coding for the amino acid valine) to GGC (which codes for glycine). This normal gene is one of the main switches for the cell division mechanism, whose action is needed to stop the cell from dividing; and when it is mutated and its function impaired, the "stop" or "do not divide" signal for the cell is lost, which triggers an endless cell division process leading to the cancer.

Point mutations can lead to normal variants of the same gene

Amino acid changes in a protein need not always lead to a defective protein. In fact most amino acid changes in an enzyme do not lead to a defective protein. We can call this a passive amino acid change, as shown in Figure 4.12. On the other hand, the change may increase or decrease the activity of the enzyme. But it must be noted that only the same enzymatic activity is quantitatively affected. The specific or basic biochemical function of the protein does not change. These changes therefore lead to normal variants of the same protein, that is, the same gene. In a similar manner, the changes in the regulatory sequence can lead to variants of the same regulatory sequence, regulating the same gene, but never change it to the extent that it can regulate a different gene. And any change in the junk DNA[86] between genes or within most regions of introns should, by nature, be necessarily neutral as far as the function of any gene or the whole genome is concerned — because junk DNA and introns do not have any function. As we shall see below, such changes can never evolve an entirely new function in the genomes of organisms.

1. Mutation which does not change the amino acid sequence.

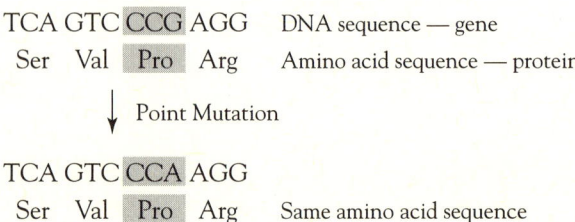

2. Mutation which changes the amino acid sequence but does not alter the function of the protein.

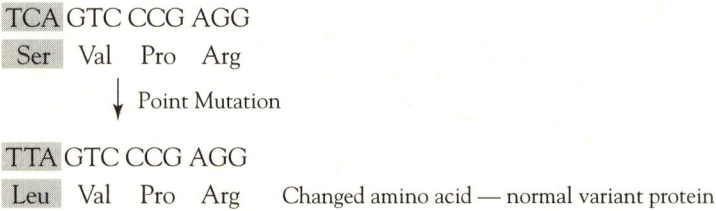

3. Mutation which changes the amino acid sequence but inactivates the function of the protein.

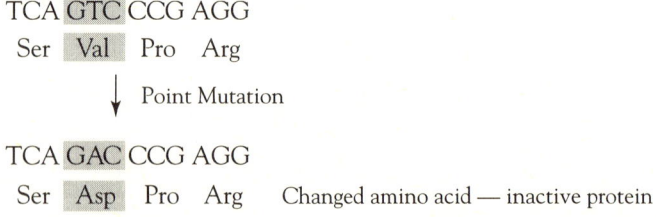

FIGURE 4.12. THE ONLY EFFECTS OF POINT MUTATIONS IN A PROTEIN. 1) A point mutation can change the codon into a "synonymous" codon which codes for the same amino acid, and thus does not change the protein sequence. 2) It can change the codon in such a manner to change the amino acid, although not changing the structure or function of the protein — and therefore resulting in a normal variant. This kind of mutation is most frequent. 3) A point mutation can also change a more crucial amino acid, which inactivates the protein.

Why point mutations cannot contribute to the supposed evolution of new genes

When we come to discuss the possibility of "evolving" a new gene by the cumulative effects of point mutations, we are left with few options. For the sake of argument, consider that a duplicate copy of a gene is changed keeping the function of the original gene. This has to be achieved through an extremely low mutation rate (10^{-9} nucleotides per cell generation[87]), and through purely random mutations. In the supposed evolution of an entirely new gene, a large number of possible sequences resulting from such random point mutations in the duplicate gene copy must be tried for its usefulness as a new gene. However, in all the positions in the gene sequence supposedly changing by point mutation, each one can be changed to any of the other three nucleotides by such mutation. The number of sequence permutations and combinations possible in a typical genome (~200 million nucleotides long) are far too large — $10^{100 \text{ million}}$. Because the mutations have to be random, the new genes have to occur among the permutations and combinations of these mutations only by chance. The probability that a gene is changed to any other new gene — leaving the original gene intact and mutating only the duplicated copy — even over a long geological time, is therefore practically zero.

We can demonstrate this by taking the frequency of mutations that can occur in one given gene as determined by molecular biologists.[88] Molecular biologists have estimated that one gene in a multicellular organism's germ cells accumulates one random mutation every 200,000 years. Let us assume that an average gene is only a mere 1000 nucleotides long. Let us also assume that it takes specific changes in only a given 10% of the sequence positions in order for it to change into a new gene. Furthermore, let us for argument sake assume that random mutations affect only the given 10% and not the rest of the gene. What this means is that, out of all the possible sequences resulting by such mutations in the given 10% locations, one must be a new gene. But how many sequences are possible by such random point mutations? The answer is 3^{100} (or $\sim 10^{50}$), because at the 100 locations, each nucleotide can change into any one of the other three nucleotides. If one nucleotide change takes 200,000 years as estimated by molecular biologists, how long would it take, on an average, for all these sequences to have occurred and tried in the cell for a new gene to have occurred among them? The answer is 200,000 x 10^{50} years. Ask if such a thing is possible on earth, when the age of the earth itself is under 5 billion (5 x 10^9) years. In this computation, we have assumed that the mutations are directed specifically to 10% of the gene — and not to all 100% of the nucleotides— in order to illustrate our point that, even when we do this, still it is improbable to evolve a new gene within geological time. In fact, computations show that when mutations occur randomly in the whole gene of 1000 characters, the time taken for the possible change of an old gene into a new gene with only a 10% specific change is $200{,}000 \times 10^{350}$ years.[89]

If the gene is split into exons and introns, it would take far longer to arrive at the desired gene by mutation. Let us assume that the coding sequence of 1000 nucleotides is split into many exons, separated by long introns, so that the complete gene length is 10,000 nucleotides. Again, mutations will occur all over the gene — in exons and introns — randomly, but the 10% expected change has to occur only within the exon sequences. For this to happen, ten times more mutations must occur all over the gene, so that the required mutations occur only in the exons (because mutations in introns are useless for evolution). Thus, it would take far longer for the new gene to evolve than that for the unsplit gene — that is $\sim 10^{450}$ years.[90]

Even if the particular changes required are only in 1% of the coding sequence for a given gene to evolve into an entirely new gene, i.e., 10 characters in the above example of a 1000 nucleotide coding sequence in a split gene of 10,000 characters, the time taken is estimated to be $200{,}000 \times 10^{45}$ years.

One might also argue that mutations in exons only should be useful for evolution. But we must remember that mutations will occur with equal probability in sequences of both exons and introns. Another usual argument is that most mutations in exon sequences are selected against, and only introns undergo random mutations. This argument would only mean that it is even more difficult to change the coding sequence of one gene into the coding sequence of a new gene.

The human mind is used to finding some purpose for the existence of everything. When we come to realize that the point mutations have nothing to do with evolution, then why are they there at all? We certainly can find an answer to this question. Even if a genome is immutable, as proposed in my new theory of the independent birth of organisms (Chapter 8), point mutation is an unavoidable property of the sequence, and is caused inevitably by the physicochemical forces within the cell (chemical mutagens and DNA copying errors) and outside the body (cosmic rays etc.). Because genetic selection permits the survival of only those individuals with functionally unaltered genes, the mutational mechanisms change the sequence only as long as the function of the gene is unaffected. When the point mutations change the sequence to the extent where the function of a gene is affected, then it becomes an error, leading to cancer and genetic disorders, but, never to a new gene with a different function. Thus, we can certainly be convinced that point mutations are only a secondary, inevitable result of the nature of DNA in the genomes of immutable organisms. They in no way contribute to any evolutionary change.

It is crystal clear that evolutionists have misconstrued the ability of point mutations in generating the variants of the same gene to be also its power to change one gene into an entirely new gene.

The generation of the variants of the same gene is very easy, but an entirely new gene is simply improbable, even over geological time

The generation or production of the variants of a given gene is a very easy phenomenon, which should not be confused with the ability to evolve new genes. When a given functional gene mutates, almost any mutation in greater than 90-95% of the nucleotide locations would accept changes without affecting the structure or the function of the protein, because with many codons for one amino acid (called degeneracy of the codon, see Primer), there is great margin for error as far as functionality is concerned.[91] Consequently, most nucleotide changes in a gene will lead only to its normal variants. See Figure 4.10. Only rarely will the nucleotide changes — in 5-10% of locations — affect the crucial amino acids in the protein such as those involved in the active site or binding site, or are essential for the structure of the protein.[92] Thus we can derive an important principle here: if the nucleotide changes do not lead to a defective protein which makes the protein nonfunctional, all other changes will lead to variants of the same gene. It is the ease with which the variants of the same gene are produced in organisms which has led the molecular evolutionists to believe that such changes can lead to entirely new genes. But, what is needed in bringing about a new protein (i.e., its gene) is a new protein structure, new active-site amino acids in precise locations, new binding sites, etc. When one carefully analyzes the probabilities for this, one can see that it is highly improbable.

Regulatory sequence mutations also can only bring about variants of the same gene or can only lead to defective (or incorrect) expression of a given gene — not to new genes or to a new DG pathway

If mutations in the coding sequence of a gene can never evolve a new protein, what can the mutations in its regulatory sequence achieve? Can they achieve a new regulatory sequence which will now regulate a different gene other than the one that it usually regulates? Again, the answer is, mutations in the regulatory sequence of a gene can only lead to the normal variants of such sequences (because even regulatory sequences exhibit considerable normal variations), or to defective sequences which can only lead to the destruction of its function. A gene can only be switched on or off, that is, in the normal case the switch is in either the 'off' position or in the 'on' position. Therefore, if the regulatory sequence becomes defective, the opposite of the normal effect will occur.[93] The normal variants of the regulatory sequence will still be switching on and off the same gene as before.[94] Thus, none of the mutations in the regulatory sequence can lead to the direct regulation of another gene which is not under its control.

What are the basic mistakes of evolutionary molecular biologists which lead them to misunderstand the abilities (or actually inabilities!) of mutations?

At this juncture, I would like to demonstrate some of the basic mistakes of modern molecular evolutionists. This mistake has indeed been made time and again by evolutionists since the advent of what is called the modern synthesis, even before the nature of the gene was known. But it is unfortunate that these mistakes are still being continued — even after the complete details of a gene have been understood. Surprisingly, molecular evolutionists simply believe that starting from a random sequence (with the length of a typical gene) within the genome, a new gene could be evolved by random mutations in the random sequence. Evolutionists such as Manfred Eigen have promulgated such an idea and others such as Richard Dawkins[95] and Bernd-Olaf Kuppers[96] have elaborated it. It can be quite easily demonstrated that their approach is fundamentally flawed.

They claim that they could easily demonstrate the evolution of a new gene by simulating their approach in the computer (using the sequence of letters in a sentence as analogy to the typical DNA sequence). First, they take a "target" sequence — the sequence they want to evolve at the end of their experiment. It is a fixed-length sequence. The target sentence that Dawkins chose is from Shakespeare's Hamlet, "METHINKS IT IS LIKE A WEASEL." Kuppers uses the phrase "EVOLUTIONARY THEORY." Then, they also choose a random sequence made from the 26-character English alphabet (plus the 'space' character) of exactly the same length as that of the target sentence. Mutations are allowed to take place in this random sequence, one at a time. After each mutation, the random sequence with the mutation is compared with the target sentence, letter for letter from the start to the end. If a mutation in the random sequence brings forth the same letter as at the corresponding location in the fixed target sequence, the computer is told to "select" that sequence, because, according to the evolutionist programmer, the random sequence is now that much closer to the sequence-to-be-evolved. No more mutation is allowed at that location. In the context of the evolution of a gene, according to these molecular evolutionists, this means that nature knows that that particular change is closer to the target sequence so that change should be preserved. Dawkins shows by this procedure that even with the slowest computer program he could achieve the evolution of the target sequence before he returns from his lunch; and that, with faster software, within a few seconds. He claims that such an evolution of a new gene by natural selection is as easy as and analogous to his computer simulation — which he achieved in about 40 mutations.

It is easy to show that the very basic idea of the evolution of a new gene by this process is flawed — because, as we have seen earlier, incipient genes, or small parts of genes do not have any selection potential and do not have any meaning. Therefore, in their experiments, first it is incorrect to specify a target sequence — which both Dawkins and Kuppers themselves very well agree (see below). Second, it is incorrect to select a single mutation that brings forth the same letter in the random sequence as that present in the target sequence. It is unreasonable, because, in nature, such a thing of selecting each of the supposedly good mutational changes can never happen. (The living cell or system — whether unicellular or multicellular cannot determine whether a mutation in a gene is good or bad unless and until a gene exists in a functional form, and therefore, unless and until the new gene is formed at least in its primitive functional form.) Each mutational step from random sequence to a functional gene — as purported in the evolution experiments — is unrecognizable by the living system, unless and until a functional gene is fully formed. This experiment puts the cart before the horse. Their experiments ought to be designed as follows: after every mutation, the mutated sequence as a whole should be compared with the target sequence. If and when the whole sequence matches with that of the target sequence, then the gene can be said to have evolved. We can even allow a considerable amount of variation to account for the codon and amino acid choice flexibility in the context of a gene. But this is not the approach they have taken. There is no doubt that their approach is incorrect.

If Dawkins had followed the protocol we have outlined here, which is actually the correct method, it would take him an average of 27^{28} (for his sentence length of 28 characters), i.e., $\sim 10^{40}$ mutations, to arrive at the target sequence of "METHINKS IT IS LIKE A WEASEL" from a random sequence. Needless to say that even for a computer which can carry out a trillion operations per second, it would take approximately one hundred million trillion ($\sim 10^{20}$) years to achieve this result. When we remember that the earth itself has been around only for the past 5 billion years, how could one say that even as small a gene as that of the above sentence can be evolved on earth by random mutations.

As I stated earlier, although the molecular evolutionists recognize their mistakes in some sense, they still do not and cannot do anything about it. They leave it at that, or they jump to another, again incorrect explanation. For instance, after describing in jubilant terms how a gene could be evolved by mutations, Dawkins writes about his computer simulation approach as follows:

> Although the monkey/Shakespeare model is useful for explaining the distinction between single-step selection and cumulative selection, it is misleading in important ways. One of these is that, in each generation of selective "breeding," the mutant "progeny" phrases were judged according to the criterion of resemblance to a distant ideal target, the phrase METHINKS IT IS LIKE A WEASEL. Life isn't like that. Evolution has no long-term goal. There is no long-distance target, no final perfection to serve as a criterion for selection, although human vanity cherishes the absurd notion that our species is the final goal of evolution. ... The "watchmaker" that is cumulative natural selection is blind to the future and has no long-term goal.

He then states:

> We can change our computer model to take account of this point. We can also make it more realistic in other respects. Letters and words are peculiarly human manifestations, so let's make the computer draw pictures instead.

One can see that Dawkins leaves the unsolved problem of approaching a "long distance" target as it is and then somehow proceeds with his discourse of a next argument. Although we are not going into the details, it is to be noted that his picture-producing approach is equally incorrect as that of his evolving the new gene using the Shakespearean example.

Similarly, Kuppers clearly realizes his problem while he tries to find support to the theory of evolution. He also describes a method as to how a new gene could be evolved starting by a random sequence by pretty much a similar approach as that of Dawkins — both of them seem to have followed the approach of another well-known molecular evolutionist, Manfred Eigen.[97] After describing the method and experiment, Kuppers writes the following:[98]

> The picture of sequence space, however, also uncovers a weakness in our simulation experiment. Unlike biological information, human language has no semantically hierarchical structure. There are, for example, no "half-meaningful" words. In this regard our experiment represents only a construction *a posteriori* of the evolutionary origin of information, that is, we started from a result that was already mean-

ingful (in this case, the target sequence) and show that with the help of a selection mechanism the statistical problem formulated in chapter 6 is soluble in principle. A construction *a priori*, however, seems impossible. If we could really simulate the process of the biological generation of information — for example, without giving in advance the target sequence and the evaluation scheme for the "letter mutants" — we would have solved a central problem in the field of artificial intelligence: the computer could generate information *de novo* in a self-organization process, simply consuming energy. Various attempts have been undertaken to make the simulation more realistic — for example, by omitting the definition of a target sequence and instead defining general criteria of fitness. The difficulties outlined above, however, cannot be put aside, as they are fundamental in nature.

Yes! They are certainly fundamental in nature. Even if molecular biologists such as Dawkins and Kuppers are well aware of the seriousness of the problems, they cannot but continue with such nebulous and self-contradicting arguments subscribing to the Darwinian world-view, because they have had no alternative so far that is totally outside of this world-view of descent with modification. In essence, whatever modifications have been made to Darwin's theory, they still purport the evolution of one organism into another, no matter what the mechanisms. That is where the fundamental problem lies! Such a philosophical constraint leads to the above-discussed erroneous and nebulous descriptions and self-contradicting arguments claiming that new genes could be evolved within the genomes of organisms. Other than such arguments, has anyone attempted to show even by genuine theoretical calculations that a new gene can be evolved? No! Almost always the molecular biology and evolutionary biology textbooks and treatises simply state the beliefs of evolutionists, something like: "Mutations of several kinds have the potential of evolving new genes, which are responsible for descent with modification." The reason is simple: although many evolutionists realize the basic mistakes in their arguments, they cannot do anything about it unless and until there is a theory that can explain the origin of the diverse creatures on earth absolutely without involving descent with modification. Such a theory has so far been lacking.

EXPERIMENTS DESIGNED TO ILLUSTRATE THE EVOLUTION OF NEW GENES FOR NEW FUNCTIONS BY MUTATIONS ACTUALLY DEMONSTRATE THE IMPROBABILITY OF EVOLVING NEW GENES

Some molecular evolutionists claim that they have directed the evolution of new functions in the laboratory.[99] But when we scrutinize their results, we can see that truly no new function is evolved. That new genes and new regulatory switches can never evolve in multicellular organisms even through lineages will be illustrated by a careful and objective analysis of these results.

Sometimes, an enzyme can have specificity for many closely related substrates, but with high specificity to one of them and poor activity towards others. In this case, the specificity of the enzyme can be affected by mutations in such a manner that the reaction with the related substrate is enhanced. For example, the bacterium Pseudomonas aeruginosa grows well on acetamide and propionamide, both of which are good substrates for the *amidase* enzyme, a product of the *amiE* gene. *Butyramide* is a poor substrate for this enzyme. By isolating mutants that grow on butyramide, Brown et al. could show that the microorganism synthesized an altered amidase with increased activity toward butyramide.[100] Similarly from this strain, another mutant could be isolated with increased activity toward *phenylacetamide*, a normally poor substrate for this enzyme. In the above studies, increased specificity of an enzyme towards a normally poor substrate is taken to be the evolution of a new function and a new enzyme. It is a mistake to take these observations as evidence for the evolution of new functions, in view of the following facts.

By objectively analyzing the results of experiments designed to demonstrate the evolution of new functions by mutations, we can show that although mutations can cause deleterious effects on proteins or change the specificity slightly, they cannot evolve proteins with entirely new functions. For example, extensive experiments have been carried out in bacteria in attempts to analyze the possibility of evolving new functions. Although scientists have reported that they could direct the evolution of new functions through selection of specific mutations, a close scrutiny of all the results shows that actually there has been no new function evolved in these studies. In the bacterium *E. coli*, the lactose operon (a well-studied system of genetic switches)[101] consists of three enzymes: β-*galactosidase* (*lacZ* gene), *permease* (*lacY* gene), and *transacetylase* (*lacA* gene). If there is a deletion in the *lacZ* gene, then *E. coli* cannot metabolize lactose. By genetic selection, a new β-galactosidase operon is claimed to have been evolved in a strain in which the *lacZ* gene has been deleted. The truth, as it was later found, is that there existed another operon which is closely related to the *lac* operon in the *E. coli* genome, whose function is still unknown. The authors suspect that this could be a duplicate of the *lac* operon that could have been silenced for a considerable time previously. A small number of point mutations in this related operon could allow synthesis of lactose-utilizing enzymes. The important fact is that if this related operon is also deleted from the genome, the bacterium cannot "evolve" another enzyme in spite of intense efforts by the scientists. Barry G. Hall, an evolutionist working on the problems of the supposed evolution of new functions, writes,

> The mutations that enable *lacZ* deletion strains of *E. coli* to utilize lactose are all in the genes of the EBG (evolved β-galactosidase) operon. All attempts to isolate lactose-utilizing mutants from a strain deleted for both *lacZ* and the EBG operon have failed.

It is therefore clear that what is claimed to have been evolved is not a new gene. It is a copy of the same gene, lying dormant in the genome of the *E. coli*, which through a few mutations became active again.

Even Hall, who proclaims that his experiments led to a new gene, cautions about its meaning in the context of the evolution of multicellular organisms:

> There are legitimate concerns about the degree to which we can extrapolate the results from laboratory experiments to natural populations. One special concern is the applicability of models based upon unicellular organisms to multicellular organisms, where specialized tissues and the isolation of the germ line from somatic cell lines may provide a strong buffer between the environment and selection acting on transmission of genetic information.

In summary, point mutations cannot aid in the evolution of a new organ in multicellular organisms for the following reasons. Point mutations do occur in multicellular organisms — in all the cells including the germ cells. However, an entirely new gene cannot evolve within the genome of an organism, nor can a new gene be integrated into preexisting genetic networks, in view of the exceedingly low probability in achieving this by random processes. All the mutations including point mutations in an organism occur within the closed framework of every distinct organism as we have predicted in Chapter 3. They cause diseases and defects or are detrimental to the system. All other mutations are functionally neutral or lead to variants of the same gene. These variants are indeed responsible for individual variations in the population of a given species. But it is improbable for any of these mechanisms to evolve a new organ. As we have ascertained, without the capability to evolve a new gene or a new organ, mutation as a general mechanism for evolution is meaningless.

Chromosomal rearrangements

We have discussed changes occurring at the level of the gene and protein. Now, we shall analyze what happens when part or whole chromosomes are involved in some rearrangements, and determine if there are any evolutionary consequences. Our analysis will show that there are in fact no evolutionary consequences at all due to any kind of chromosomal rearrangement.

Sometimes, a piece of a chromosome gets duplicated. The genes present in that duplicated portion of the chromosome now appear twice in the genome. During such duplication a gene may be broken at the middle. This can lead to a defective gene which can cause a disease in the organism. A repetition of a chromosome segment is known as *duplication*. If ABCDEFG represents an original chromosome sequence of genes, then a duplication of CD might be represented as ABCDCDEFG. The "bar eyes," an abnormality in *Drosophila*, results from a chromosomal duplication.

The transfer of a part of a chromosome to a different location is known as a *translocation*. A reciprocal translocation is one where there is an exchange of chromosomal material between two dissimilar chromosomes (for example between chromosome 3 and chromosome 17, instead of between the pairs of the same chromosome).[102] In general, translocations can result in gametes (sperm and eggs) with either duplication or deletion of chromosome material. When the gamete has a duplicated portion of the chromosome, it is called a Trisomy. The most common Trisomy in the human is due to translocations between the chromosomes 14 and 21 and is called Downs syndrome. The affected individual will have abnormal facial features, varying degrees of anomalies (cardiac, renal, etc.) and mental retardation. The female carriers of this type of translocation have a higher incidence of miscarriage and 1/3 of the viable offspring would have Downs syndrome.

Chromosomal rearrangements sometimes alter the number of chromosomes. *Fusion* of chromosomes occur when two chromosomes fuse together. When a chromosome splits into two pieces it is called a *fission*.

Several kinds of structural rearrangements of the chromosomes alter the positions of genes relative to one another without affecting the number and kind of genes.[103] Two breaks can occur in the same chromosome, and the segment between them can be rotated 180°. This results in the inversion of the region between the breaks. Inversions that overlap an already inverted region can also happen. Some inversions result in infertility.[104]

Sometimes, a portion of a chromosome can be deleted during the replication of the chromosome. If this happens in the germ cells (cells producing the sperm and egg), the offspring will be missing some genes. Many times, such a deletion can be fatal to the developing embryo, and therefore no individual will be born. But sometimes, the individual will survive with a diseased condition. Such a loss of chromosome material is termed deletion. A change from ABCDEFG to ABFG means a deletion of CDE. Deletions can be harmful in proportion to the size of the chromosomal portion deleted.

The effects of chromosomal rearrangements in humans are well-known. Even a small piece of extra chromosome or deletion of a portion or whole chromosome can cause serious congenital defects. An example for the deletion of a portion of chromosome is the disease called Cri-du-chat (or cat-cry) syndrome (also called LeJune's syndrome), in which the short arm of chromosome number 5 is deleted. The characteristics of this disease are: mental retardation, low-set ears, moon faces, abnormal larynx giving rise to high pitched cry similar to that of a cat.

An entire chromosome can also be deleted leading to a diseased condition, as in the case of Turner's syndrome, wherein a sex chromosome is deleted (either the second X in a female or the Y chromosome only can be deleted, because one X has to remain for viability; this condition is called 45XO). Characteristic features are sexual infantilism, and short stature. 96% of pregnancies involving this chromosomal defect end in miscarriage. An abnormality due to an additional sex chromosome leads to Kleinfelter syndrome (called 47XXY).

Some of the effects of chromosomal mutations have been misunderstood to contribute to sudden speciation. For instance, a case of translocation, known as the Renner complex,[105] is responsible for the mutation in the primrose *Oenothera lamarckiana* — which converts a white rose into a red rose.[106] As we discussed in Chapter 2, this was misunderstood by Hugo De Vries to be providing an instance of macromutation responsible for sudden speciation.

From our foregoing discussions, one can clearly discern that these chromosomal rearranging mechanisms do not contribute to new sequences and new genes. No new DG pathway can ever be generated by any of these chromosomal rearrangement mechanisms. Chromosomal aberrations can only lead to genetic diseases. In the extreme scenario these mechanisms produce reduced fertility or an organism with an incomplete genome. At the molecular level, these defects may be caused by defects in one or more proteins or in one or more regulatory pathways. None of the chromosomal mutations, even drastic ones, can lead to any drastic physical (phenotypic) effects that could be useful in evolutionary change. By what mechanism can chromosomal mutations lead to random outgrowths? None. Therefore, we can be sure, no mutations ranging from small single-gene effects to large chromosomal alterations could lead to new genes or new DG pathways for new body structures.

Recombination

Two DNA molecules can recombine between themselves producing two recombined molecules. Much of this process happens in the DNA of the germ cells, wherein the chromosomes from the two parents recombine to produce a set of recombined chromosomes carrying a mixture of characteristics from both the parents, which will be packaged into the sperm or the egg.[107] As we can see, the genome has a constant set of genes and the process of recombination only mixes the variants of the same set of genes that exist in the father's or the mother's chromosomes. This process definitely produces the individual variations in the population of an organism.[108] But the evolutionists misunderstand that this process is responsible for the production of new distinct creatures from an existing organism. They do not seem to realize that this process, being the cause of individual variations within an organism, can certainly lead to distinct varieties and similar species of the organism, but not to a new organism. This misunderstanding again stems from the concept that Darwin promulgated, which states that varieties of an organism are produced from its individual variations, which varieties in turn lead to new distinct species — which process can lead to entirely distinct new organisms. We shall see here from our analysis that the recombination process is undoubtedly the cause of individual variations, but certainly not capable of producing anything beyond the individual variations and similar species of an organism.

During meiosis in the germ cells (the process of producing the mixture of gene variants from the father and the mother chromosomes), parts of DNA molecules can recombine. This mechanism can bring together gene variants (alleles) present on homologous (paired) chromosomes.[109] For instance, if AB are two genes on one chromosome and *ab* are their alleles on the homologous chromosome, then recombination can result in chromosomes with A*b* and *a*B. Remember that A and *a* are the variants of the same gene A, and B and *b* are the variants of the same gene B. See Figure 4.13. The DNA molecules of homologous chromosomes recognize each other by some unknown mechanism during meiosis and become precisely aligned. Crossing over may then occur between the DNA molecules, producing the recombined molecules.[110]

The point of recombination between two homologous chromosomes need not always be precisely between two genes, that is between A and *b* in the above example. It can also happen within the sequence of A and *a* leading to a recombination of the variants of the same gene, producing a third variant. The latter is called *intragenic recombination*.[111] If two alleles for a gene sequence code for the amino acid sequences Asp-Lys-Arg-Leu and Pro-Lys-

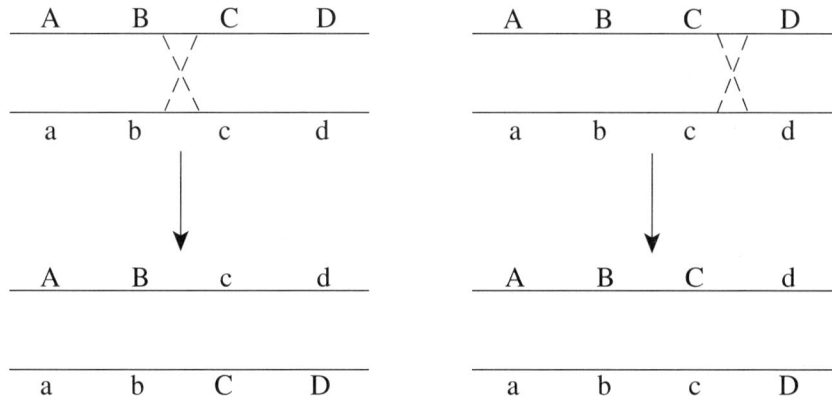

Figure 4.13. The mechanism of crossing over and recombination, mixing only variants of the same set of genes in the genome. When the sperm or egg is produced in the germ cells, different variants of genes are mixed by crossing over through DNA recombination. If A, B, C and D are different genes on one chromosome, and if a, b, c and d are their corresponding normal variants (alleles) on the "homologous" chromosome, then crossing over can produce ABcd on one chromosome and abCD on another, as in the first example in the figure. Note that only the normal variants of the same gene are recombined, and under no circumstances a new gene is produced.

Arg-Trp, recombination between them could give rise to the new amino acid sequence Asp-Lys-Arg-Trp. It has been suggested that a variation of the enzyme 6-phosphoglycerate dehydrogenase in Japanese quail (*Coturnix coturnix*) might have arisen in this manner. It is believed that intragenic recombination can result in different amino acid sequences compared to either parent sequence, and that this mechanism could generate new alleles at fairly high rates.

It should be noted that intragenic recombination can lead to only variants of the same gene, and never to the evolution of a new gene. The above recombined sequence is a new variant of the same protein. It is clear that the new allele, the recombinant gene, would still code for the same enzyme which would still specify the identical biochemical activity as the parent allele. It is improbable that it will specify an entirely new enzyme activity. Even if it does, the probability that it will be useful in a new DG pathway is exceedingly low.

The probability of several enzymes and structural proteins coming under a new DG pathway that would express even a bizarre phenotype is also too low to be meaningful. This is the central theme we have developed in all our arguments against the evolution of a new organism with a new body structure by any of these gene mutational and rearrangement mechanisms. Because this applies to any kind of recombinational event in the genome, we will not discuss the details of individual situations. Therefore, in essence, it is incorrect to say that a normal cellular mechanism, such as DNA recombination, can generate new genes or a new DG pathway developing a new body part over geological time either by natural selection or by any other mechanism that evolutionists propose. Because DNA recombination can lead to defective genes and normal gene variants, as well as bring together different alleles of the constant set of genes in a genome, molecular evolutionists have extended such abilities to the domain of belief that they can also create new genes and new body parts leading to new organisms.

According to the theory of evolution, a great deal of purely indiscriminate recombination should happen in the genome. If so, we should see evidence for it, and only one out of a very large number of indiscriminate recombinations can give rise to any new useful function. Such recombined sequences which are not useful simply cannot be eliminated from the genome, unless they are detrimental to the organism as a whole. Thus, we should see not only a few traces of recombinations but an extremely large population of sequences which are recombined sequences of existing genes, without losing the original genes (because individuals without essential functions do not survive). However, from the great molecular details we know of the genome, we certainly do not see any evidence for such a thing happening.

Recombination is a normal mechanism used in every cell division in meiosis. It is also used in the replication of viral genomes during their propagation.[112] There are many recombination mechanisms in the normal life of microorganisms such as bacteria and fungi. The normal recombination mechanisms, most fundamental in the life of organisms, can also cause mistakes resulting in mutations which can lead to some form of genetic abnormalities, including cancer in animals.

Thus the belief of evolutionists that these mutations are the material basis of evolution is incorrect. As we saw before, another belief of the evolutionists is that by bringing together and mixing alleles from various individuals of a species during meiosis, recombination leads to individual variations, supposed to be the material basis of evolution of new organisms. This is reflected clearly in the following writing of E. C. Minkoff[113] (italics mine):

Genetic recombination represents the proximate source of by far the greatest number of variations, even though mutations are the ultimate source of genetic variability. No two individuals are exactly alike (except perhaps some identical twins), and this uniqueness is usually the result of recombination rather than mutation. ... Because genes interact with one another, and because many characters are controlled by several genes, *genetic recombination may result in totally novel phenotypes*, as well as in new combinations of previously experienced character states. Look at the people around you and at their many distinguishing traits. Most of these traits have arisen from genetic recombination and from developmental (including environmental) controls, without benefit of new mutations. *The recombination of genetic material is thus an important aspect of the evolutionary process.*

There is no question that genetic recombination, in addition to mutations, is the basis of individual variations within each species of a distinct organism by its ability to mix the normal variants of the same set of genes. But individual variations do not contribute to the supposed evolution of new organisms. They are an innate property of every immutable organism.

Crossing Over

The phenomenon of crossing over between chromosomes cannot contribute to the evolution of a new gene or a new body part

The phenomenon of the crossing over of genes on a chromosome was discovered by Thomas H. Morgan and Alfred Sturtevant in the early twentieth century. They observed that during sexual recombination between the genetic material from the father and the mother, each chromosome from the father recombines with the same chromosome from the mother thereby mixing the variants of the genes from the father and the mother randomly. They observed that genes farther apart on a chromosome could cross over more frequently than the genes closer together. The germ cells contain two of each chromosome, one from the father and one from the mother. When the sperm or the ovum is made from a germ cell, only one complete set of chromosomes is put into it — representing a mixture of father and mother chromosomes.

We must note here that each set of chromosomes from the father and the mother, except the sex chromosomes (X and Y), contain the same set of genes, except that every gene is present in a different variant form on the two chromosomes of each pair. Rarely, one of the genes present on one chromosome may be defective (due to a mutation) while its counterpart on the other chromosome may be normal.[114]

The generation of the recombinant chromosome that goes into the sperm or egg occurs by "crossing over" of the chromosomal pairs. This introduces a great deal of variability in the genomes of the different individual sperm or eggs. Because the process of crossing over happens by recombination, we can see that the same principles which apply to the process of recombination as we discussed above applies also to the crossing over mechanism.

Evolutionists believe that the allelic variations (from the multitudes of combinations in which different alleles of the same gene may be paired in an organism) introduced by the crossing over phenomenon is a great contributor of the individual variations in a given species. We do not dispute it at all. But they believe, as Darwin did, that the individual variations are responsible for natural selection and the evolution of new organisms with new genes and new body structures — which is what we dispute. Figure 4.14 explains that the mechanism of crossing over can lead to only the variants of the same set of genes in the different individuals of an organism, which are indeed incapable of evolving a new organism with a new gene or body structure.

Only the variants of existing genes in the genomes of individuals in a species are mixed by means of crossing over. (And the many different similar species of a distinct organism all have the same set of genes.) It certainly explains the variations occurring among individuals in a species. Because there are almost unlimited ways in which the many variants of different genes from the population of an organism can be recombined or mixed in this manner, it can produce a nearly unlimited number of varieties and similar species, but never a new distinct organism with a new gene. Thus, we have to make a clear distinction between this phenomenon of the production of individual variations by crossing over and the phenomenon of the evolution of new genes, new DG pathways, and new organs. These are two distinct phenomena, which should not be confused with one another.

Unequal crossing over

Normally during meiosis genes on homologous chromosomes are aligned precisely. When crossing over occurs under such circumstances, it is an equal crossing over, meaning that different variants of the set of genes are swapped cleanly between the chromosomes. But the chromosomes are sometimes mis-

GENETIC MUTATION AND REARRANGEMENT CANNOT CAUSE EVOLUTION   173

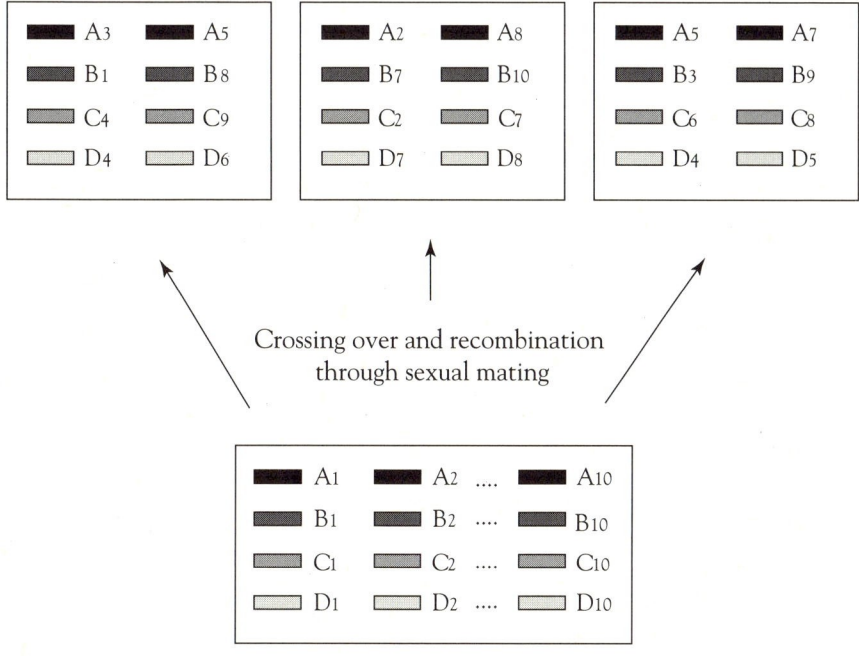

FIGURE 4.14. WHY THE MECHANISM OF CROSSING OVER CANNOT LEAD TO NEW GENES OR A NEW DG PATHWAY FOR A NEW BODY PART. Each gene in an organism's genome can exist in many different variant forms. Consider four genes, A, B, C and D. The variants of gene A are represented by A1, A2, ... A10; of gene B by B1, B2, ... B10; and so on. During crossing over in germ cells, any one of the variants from A, B, C, or D can be mixed to give a set of variants from all the four genes. Two sets of such variants are shown in the figure representing two chromosomes in an individual organism. Thus, the set of genes of each individual in the population of an organism is the same, although each individual may have different variants of the same gene. Note that crossing over simply assorts the variants of the different genes present in the population of an organism into various combinations of variants of the same set of genes. This process can lead only to individual variations, and never to a new gene or a new DG pathway for a new body part.

aligned. This results in unequal exchanges, giving rise to a tandemly duplicated region on one chromosome and a complementary deletion on the other. It has been suggested that a duplication may be advantageous, producing greater amounts of a gene product.[115] If deletions of coding regions or entire genes happen in this process, they are usually deleterious. Again, we cannot see any possibility for the evolution of a new gene or new DG pathway in any of these unequal crossing over mechanisms.

Pleiotropic Mutations

Pleiotropic mutations can cause errors in the existing developmental genetic pathways and produce aberrations of existing organs or limbs, but cannot produce new genes and new organs

Pleiotropy is defined as the multiple physical effects of a single gene. Because a single protein (i.e., its gene) may control an interconnected developmental genetic pathway leading to a body part, its inactivation may affect that whole pathway of development of that body part. That is, the activation or repression of a single gene may activate or repress the formation of the whole organ. This is true with almost any tissue, organ or body part. Consequently, a mutation in this particular gene will affect the function of many genes under its control. Such a gene could be called the master gene for the DG pathway of that body part. If this gene is activated due to a mutation in a cell in which it is normally switched off, it may trigger the development of that body part starting from that cell. In other instances, mutations in this master gene or in genes involved within the DG pathway of this body part may lead to either the abolition, duplication, or malformation of the body part.

Thus, we can see that pleiotropic mutations defined in the field of molecular genetics and molecular evolution is, under our discussions, nothing but mutations in the DG pathway of an organism. These mutations can be caused by almost any mechanism of genetic mutation, such as point mutation, transposition or chromosomal aberration — because a master gene is also a typical gene coding for a protein. It can affect a gene which is the master gene for cell division, in which case, the result would be uncontrolled cell division leading to cancer. As we have so far seen in this and the previous chapter, none of the pleiotropic mutations in the genome of any organism can lead to the evolution of a new body structure from an animal lacking that body part.

Many pleiotropic mutations which affect the development of the fruit fly D. *melanogaster* have been experimentally studied. Almost all the genetic aberrations of development in any animal organism must be due to such mutations. The observations that two pairs of wing can be produced in a fly normally having only one pair can be explained in a similar manner. However the important question is: Can we produce an aberrated appearance of something which is not, in some form, existing in the body of an animal? For example, can we produce a wing in an earthworm, even in the most rudimentary form, by pleiotropic mutations? It is clear that we can produce two pairs of wings in an animal which already has one pair of wings, but not even a single wing-like structure in an animal with no wings at all. We can even delete the wings from an insect and generate a wing-less animal through appropriate pleiotropic mutations, but can never generate wings in a wingless animal. We can boldly conclude that we can never produce any new organ or limb that needs new genes or new developmental genetic pathways in any creature by any pleiotropic mutations![116]

In essence, pleiotropic mutations can only cause errors in the developmental genetic pathways of existing organs, tissues and limbs, and construct some aberrated forms of these tissues and limbs in other parts of the body. Pleiotropic mutations do not produce the "monsters" thought to be useful in evolutionary change. They are simply straightforward errors caused in the normal expression of immutable genomes.[117]

THE NUMEROUS FRUIT FLY MUTATIONS LEADING TO A VARIETY OF MORPHOLOGICAL DEFECTS CANNOT LEAD TO NEW GENES OR BODY PARTS. THEY ARE ONLY ANALOGOUS TO GENETIC DISEASES IN HUMAN BEINGS, A CATALOGABLE AND DEFINABLE FINITE SET OF ALTERATIONS OF EXISTING GENES.

T. H. Morgan was the first to find mutations on the fruit fly *Drosophila melanogaster*'s chromosomes. He, and later Herman J. Muller, found a great number of variations in the fly — white eyes, vermilion eyes, ruffled hair, forked body, and flies with no wing, no antenna or no eyes, and so on, about one for every fifty thousand flies examined. However, when we analyze these mutations with our knowledge of how these can be produced, what do they show? They only show that these are due to errors caused by mutations of existing genes. Each mutation expresses a defined trait, a defect of each gene expressing itself as a defect in the organ the gene is used to build. Essentially, mutations in many genes in the genome of an organism can be expressed into such defects of an organ or a limb. But what do they mean? They mean only that they are similar to the genetic diseases in the human beings. There is a finite number of such mutations and defects that can be catalogued and counted in an organism.

What we are studying here is an entirely different phenomenon which can do nothing for the evolution of new genes, new organs and the evolution of new organisms: simply the effect of the various mutations of the genes that are specifically involved in the development of the particular organs and limbs, not the phenomenon of the evolution of new genes or new organs. Given all the facts about the properties of DNA, protein, genes and their potential to mutate, and so on, and given that distinct creatures are immutable, even then, there will exist such mutations leading to these morphological defects, because by definition, genes are responsible for constructing these organs, and mutations will necessarily cause these structural defects. Therefore, one cannot simply extend the phenomenon to eventually leading, by evolution, to new organs in the fruit fly. How will these defined mutations and morphological defects lead to new genes? How will these lead to a new organism with a new body structure starting from the fruit fly?

In treating this subject, it must be admitted that the evolutionists were looking for some mechanisms of changing one gene into another. And, when they found that mutations could cause such large morphological defects and aberrations, they were exuberant that this finding must support Darwin's theory of evolution based on individual variations.[118] What we need for natural selection to act and to produce new body structures are new kinds of aberrations (undefinable and uncatalogable and unlimited) caused by new developmental genetic pathways, and/or new genes. As we have reiterated several times, the fundamental requirement for the evolution of new organs are new developmental genetic pathways and/or new genes. When we consider this subtle but extremely important difference, it becomes apparent that the phenomenon of fruit fly mutations is one thing but the phenomenon of the evolution of new genes and new organs is a totally different thing.[119]

Polyploidy

Polyploidy, a mechanism which is not the cause for the origin of diverse organisms on earth

The cells of multicellular organisms termed "diploid" normally contain two sets of chromosomes (with two alleles of each gene). Rarely, and especially in plants, the chromosome complement doubles and becomes "tetraploid." This process is called polyploidy. In instances where the same chromosomes are doubled or tripled, it is called "autoploidy." In other, even rarer circumstances, and again mostly in plants, two different genomes from different species can combine and produce a new plant. This "hybridization" of two plant's chromosomes or genomes is called "allopolyploidy." When we analyze the situation carefully, we shall see that this is an extremely rare phenomenon which occurs almost exclusively in plants and which could not have been the reason for the origin of the myriad unique organisms on earth.

The structural and physiological effects of polyploidy within a given species (autopolyploids)

A mutation in a plant gene during the formation of the zygote (that is, during meiosis) may lead to the doubling of the chromosomes, and therefore, all the cells of the individual plant which develops from that zygote will contain a tetraploid set of chromosomes. For several reasons, these are usually sterile.[120] A few naturally occurring tetraploid plants have been described, such as certain fireweeds.

The doubling of the chromosome complement in a plant does not change the genes or their relative quantities. Therefore, polyploids are distinguishable from their diploid counterparts in only trivial characteristics. The tetraploid exhibits a thicker stem, a greater height, larger, thicker and relatively shorter and broader leaves, a darker green pigmentation, larger flowers and seeds. Tetraploids usually develop more slowly, probably due to a decreased rate of cell division. This is the most frequently noted physiological consequences of polyploidy. In animals, fish such as trout, salmon, and suckers are suggested to have been descended from ancestors that became tetraploid about 50 million years ago.[121] Polyploidy is sometimes found in parthenogenetic, that is, asexual forms of animals. The brine shrimp *Artemia salina*, for example, consists of asexual diploids, triploids, tetraploids, pentaploids, octaploids and decaploids, besides normal sexual diploids.[122]

"Hybridization" between two different species (allopolyploidy)

Hybrids between radish and cabbage have been produced artificially. The tetraploid hybrids represent a morphological type which is distinct both from radish and from cabbage. Some characteristics are intermediate, in others the influence of one of the parents predominates, and still others are peculiar to the hybrid.

In animals hybridization is extremely rare. It has been proposed to have occurred in some moth species. A well known case in animals is the hybridization between the horse and donkey producing the diploid sterile mule or jenny. The rarity of polyploidy in general can be seen from the following quote from *Evolutionary Biology* by E. C. Minkoff:

Grant has surveyed the frequency of occurrence of polyploidy among eucaryotes and reports that it is rare among animals, fungi, and most groups of gymnosperms, though it occurs regularly among angiosperms, pteridophytes (ferns and other vascular plants lacking true seeds), and one group of gymnosperms. The comparative rarity of polyploidy among animals has received various explanations that are not mutually exclusive but may reinforce one another. ... In those rare cases where polyploidy has been recorded in *Drosophila*, sterility has always resulted.

Why polyploidy has not been, and is not, the reason for the origin of multitudes of organisms on earth

The phenomenon of polyploidy occurs in plants and almost never in animals. We can clearly see that it only increases the frequency of existing genes from one organism or mixes the genes from two different organisms but never can evolve new genes. In my view, this phenomenon could have originated after the multitudes of organisms came on earth independently (by the mechanisms proposed in the new theory of the independent birth of organisms). Simply, the rarity of the polyploidy in animals will show that this phenomenon is not the cause of the origin of the diverse unique organisms on earth. Both its rarity and its incapability in evolving new organisms can be seen from the quotes of well-known evolutionists.

E. C. Minkoff writes in *Evolutionary Biology*:

> Speciation by polyploidy occurs at times among angiosperms and in certain other plant groups. Among animals and fungi, it is exceedingly rare and occurs only in some unusual circumstances, such as the parthenogenetic hybrid species of *Cnemidophorus* lizards.

Futuyma writes in his textbook *Evolutionary Biology*:[123]

> Allopolyploidy does not in itself create new morphological features of the kind that distinguish genera or other higher categories of classification, but allopolyploid stocks have often evolved into phyletic lines that achieve taxonomic distinction.

The noted evolutionist Dobzhansky writes that the gene mutation (as known in 1937) is the main source of evolutionary change although this process is extremely slow and takes geological time. Comparing the power of polyploidy to that of the gene mutation, he writes,

> It is highly remarkable, therefore, that alongside this slow method of species formation there should exist in nature a quite distinct mechanism causing a rapid, sudden, cataclysmic emergence of new species. It is remarkable, furthermore, that while the slow method seems to be encountered throughout the living world and may in this sense be called the general one, the cataclysmic origin of species is confined to some, though large, groups, mostly (so far as known) in the plant king-

dom. The latter method of species formation is connected with a multiplication of the chromosome complement, called polyploidy.

The foregoing passage indicates how even in the view of evolutionists, the mechanism of polyploidy is of little importance with respect to the mainline production of almost all animal organisms, whatever the mechanism may be. The evolutionists themselves agree that polyploidy cannot account for the production of even an extremely small fraction of the animal species.

In addition to the nine major mechanisms of gene mutation and rearrangement we have discussed so far, some minor mechanisms such as "gene conversion" have been predicted to evolve new genes. Again, by similar argument, we can show that no such mechanism can evolve distinct new genes within the genome.

Conclusion: All the known mechanisms of genetic mutation and rearrangement, even collectively, cannot contribute to evolution

Thus far we have accounted for all possible genomic rearrangements that can occur in a genome and have established that none of these can contribute to the evolution of a new gene or a new DG pathway for a new body part.

After an extensive series of analyses and observations, is it not obvious to us now that most of the mechanisms purported to be involved in evolutionary change ought to be operative even in an immutable genome for its normal functions of DNA replication (in cell division), repair of DNA, recombination, and crossing over (e.g., in meiosis)? Others, such as the point mutation, which do not participate in normal cellular mechanisms, can cause sequence changes in an immutable organism. Such sequence changes and errors caused by these mechanisms only lead to normal individual variations, diseases such as cancer, congenital and genetic disorders, and death. Some mechanisms, such as transposons, are parasitic, propagating themselves "selfishly" in the genome throughout the lifetime of an immutable genome without affecting the genome's constancy, except for causing disorders. Because each of these mechanisms is incapable of contributing to the evolutionary change of one organism into another with a new gene or a new body part, even collectively these mechanisms — all the known kinds of gene and sequence mutations and rearrangements — are no more potent than the individual mechanisms.

There is a certain amount of allowed flexibility for every genome within which the gene mutation and rearrangement mechanisms operate. This flexibility however, only works within the closed genome of the immutable organism.

The following principles apply equally to all the mechanisms of sequence mutations and rearrangements:
1. Mutations of any kind happen only on existing genes in a genome — without producing any new genes.
2. Movement of genes and other sequences affect only existing genes — either inactivating or activating the gene by integrating at a region within the gene's coding or regulatory sequence. They affect only the expression of genes, either positively or negatively, and play no role in evolution.
3. No new gene or regulatory sequence or DG pathway can be produced or evolved by any of these mechanisms of mutation and rearrangement within any earthly time.
4. By random processes it is improbable to generate a new metabolic or DG pathway within the genome of organisms.

Let us now ask a plain question that would clarify any of our remaining doubts why there are such mutations at all in living organisms, if they absolutely do not contribute to evolution. If, as we have determined in Chapter 3, the set of genes in the genome of an organism is constant and the DG pathway fixed, and, therefore, each organism is immutable or unchangeable into any other organism except its own varieties, what do all the mutations do and why do they exist at all? The only possible answer is that, given the biochemistry of DNA and proteins (the mechanisms of DNA replication and possible errors in this mechanism, etc.), point mutations are bound to occur even in a fixed genome. Likewise, the parasitic mechanisms of transposons, once introduced into the genome of an immutable organism (as described in Chapter 9), will simply persist, moving here and there within the genome, and, while doing so, inserting and deleting some sequences. What about DNA replication and recombination? Each is a normal, but error-prone mechanism absolutely required in all organisms, even when they are all totally fixed and unchangeable. The same arguments hold for gene conversion, gene duplication, pleiotropic mutations, and chromosomal rearrangements.

It can be said that there is flux in the genome in terms of its physical nature, but the functional nature of the genome is fixed, because no matter what happens in a genome due to mutations, the constancy of the genome in terms of its set of genes (except for some defective genes) or the fixity of its DG pathway (except for developmental defects) can never change. Whether individually or collectively, mutations have no power to cause evolutionary change. We should, however, not forget their power in producing varieties and similar species of an independent immutable organism. At the same time, it must be emphasized that they cannot produce anything beyond the varieties and similar species of a fixed organism.

Constancy of organisms over geologic time corroborates our prediction that changes in both "coding" and "regulatory" sequences of genes cannot contribute to evolution

The modern synthesis of Darwin's theory states that mutations in a genome are random — both in structural and regulatory sequences of genes — which are responsible for the change of one creature into one or more new distinct creatures. This means that mutations should be occurring in the genomes of all organisms equally. Therefore, all organisms should be changing into new organisms at fairly equal rates. That is, we should not see some organisms changing rapidly and some others not changing at all for long geological time. But it is well known that there are numerous organisms that are virtually unchanged for tens or hundreds of millions of years — called "living fossils," while others are believed to be changing continuously into other organisms.

This discrepancy tells us something: either 1) the said mutational changes are not occurring in the living fossils for some (unknown) reasons while they are occurring in the rapidly changing organisms; or 2) they are occurring equally in both cases and therefore the assumptions and conclusions of the modern synthesis of Darwin's theory are wrong.

The Port Jackson shark, a living fossil, has remained unchanged for 350 million years. During that period, the human species is supposed to have been evolving through a long lineage of several organisms from the fish — such as an amphibian frog, then a therapsid reptile, and then a mammal like a tree shrew and so on, and recently from a nonhuman primate. We can set up an experiment in order to analyze the questions we raised above. If genes of living fossils had undergone as many DNA base changes as the corresponding genes in the supposedly more rapidly evolving organism, then it would support point 2 above. If not it would support point 1. Molecular evolutionists claim that the α and β chains of hemoglobin have duplicated and evolved from a single hemoglobin chain in the common ancestor of the shark and the human, just before the two lineages had diverged from their common ancestor.[124] The sequences of the α chain and β chain of the principal hemoglobin of the Port Jackson shark and the human are known. Therefore, the number of sequence differences between the α chain and β chain in the Port Jackson shark can be determined (Table 4.1). A similar comparison of the α and β chains can be obtained from the human.

The two sets of comparisons show that the genes coding for the α and β chains of hemoglobin in the shark have the same (in fact a slightly higher) number of sequence differences as have the corresponding two

Type of change*	0	1	2	3	Gap	Total
Shark α vs. β	50	56	32	1	11	150
Human α vs. β	62	55	21	0	9	147

*The number of amino acid sites that can be interpreted from the codon table as due to a minimum of 0, 1, 2, and 3 nucleotide substitutions are given together with the number of gaps (expressed as equivalents of the number of amino acid sites).

TABLE 4.1. COMPARISON OF THE SEQUENCE DIFFERENCES BETWEEN THE GLOBIN α AND β CHAINS IN THE PORT JACKSON SHARK AND IN THE HUMAN. [From: Kimura, M., 1983, "The Neutral Theory of Molecular Evolution," in *Evolution of Genes and Proteins*, Nei, M. and Koehn, R. K., eds., page 208. Adapted with permission of Sinauer Associates.]

genes in humans.[125] It is clear that the mutations occur at the same rate, supporting point 2. Even if we accept that the human and the shark had evolved from a common ancestor, then the sequence variations are not responsible for the evolution of the human so far removed from the shark, because the same number of mutations has not at all affected the constancy of the shark for all these eons of time. This finding, generalized for all genes, demonstrates unequivocally that mutations in structural genes are not responsible for evolutionary change.

In more recent times, this realization seems to have changed the belief of some evolutionists that it is the changes in the regulatory sequences — not coding sequences — that are responsible for evolutionary change.[126] But, again they seem to miss the fact that similar changes presumed to have taken place in the regulatory sequences of the organisms that finally led to the human species must have taken place in the Port Jackson shark also. Nothing must discriminate the sequence changes, which are supposed to be random, between the Port Jackson shark and the ancestral animals that led to the human. Likewise, no kind of mutation can discriminate between the coding sequence and regulatory sequence of a gene. It is clear, then, that changes in regulatory sequences cannot be responsible for evolution of the human. If we can analyze the differences between two regulatory sequences supposedly evolved by duplication from one original sequence in the ancestral animal that led to the Port Jackson shark and the human (as in the case of the α and β globin genes), we can be certain, the difference between them in the shark compared to that in the human will be similar.

Consider the scenario, wherein the two animals originated independently with copies of the same α and β globin genes in them and wherein both the animals were immutable. Consider that the two genes (and any number of other genes) in both the animals underwent random mutations equally in them without functionally changing the set of genes, and therefore, without changing the organisms. This process would lead to the result that we just now saw — the number of sequence differences between the α and β globin genes being essentially the same in the shark and the human. Thus, this observation validates our concept that none of the mutational changes have any effect with respect to evolution.

Incapability of Other Seemingly Possible Mechanisms of Evolutionary Change

Addition of new genes into a genome by viral vectors: Putting the cart before the horse

In an attempt to find support for evolution, modern evolutionists put forth another mechanism that retroviruses can laterally transfer genes between organisms. Numerous distinctly unique creatures were found in the fossil record at the very start of the appearance of multicellular life (the Cambrian explosion, see Chapter 2). Evolutionists themselves agree that the rapid evolution of a multitude of unique organisms during the Cambrian explosion is highly improbable by any known genetic mechanisms. Because, even to evolutionists, evolutionary changes in genes are not rapid enough to bring about the multitudes of organisms within a short geologic span, some of the evolutionists have proposed "lateral gene transfer" by retroviruses to account for this phenomenon of rapid appearance of new creatures. The argument is that new genes can be transferred from one creature to another.

An important question arises here. The viruses only had the genes of the existing organisms on earth to work with at any given time. When there were only marine invertebrates, the genes for the vertebrate eye or the mammalian placenta did not exist anywhere. The action of viruses in laterally transferring genes are restricted to the pool of genes of all creatures existing at a given geologic time. Obviously, this kind of process, which has to work with previously existing genes, cannot transfer new genes that are nonexistent. Therefore, it cannot generate new organs either by generating new genes or by any other molecular mechanism. Only if an organism evolves a new gene, then viruses may transfer it into other organisms. But, as we now know, there is no way an organism can evolve a new gene.

Evolutionists such as Futuyma[127] believe that genes, occasionally transferred between unrelated organisms by viruses or other parasitic agents, occur rather frequently among different genera of bacteria,[128] and that there is evidence that transfers may occur between prokaryotes and eukaryotes, as well as among eukaryotes. However, Futuyma himself states that there is little evidence so far that transfers of this kind have occurred frequently in evolution.

GENERATION OF NEW SEQUENCES WITHIN A GENOME TO EVOLVE A NEW GENE IS HIGHLY IMPROBABLE

We have determined that all the known mutating mechanisms which appear to be capable of evolving genes within a genome actually cannot contribute to the evolution of new genes. The only mechanism that is left to us is the generation of new genes by the *de novo* generation of new random sequences *within a genome*. Statistically and biologically this is extremely improbable. We know of no mechanism that can generate new sequences in a genome. Even if we assume this is possible, to arrive at new genes responsible for developing a new organ such as the eye or the placenta is impossible from random sequences with lengths even as long as the genome itself. Indeed, if this were happening, it would constantly increase the size of the genome to thousands or even millions of times larger than those of the genomes of living organisms. On the contrary, the sizes of the genomes of living organisms, which fall between approximately a few billion up to a hundred billion nucleotides, are too small for such a mechanism to be occurring, and distinctively demonstrate that the probability for this mechanism is practically zero.

IMPROBABILITY OF THE VERY FIRST CREATURE HAVING CONTAINED ALL THE GENES AND DG PATHWAYS OF ALL THE FUTURE ORGANISMS IN A DORMANT FORM

There is another approach by which we can analyze this problem. Consider that, according to Darwin, the ~30 million species of organisms living on earth today and the approximately billion species which lived on earth and have become extinct so far, must have come from the first one original creature. Without a mechanism to generate new genes and DG pathways in the genome of an organism that will express a selectable attribute, we are forced to argue the following. The very first creature should have contained all the possible permutations and combinations of genes and their unique DG pathways, in a dormant form, to bring about all these millions and millions of organisms.

For example, in the genome of the very primitive marine invertebrate, there must have existed all the genes for the production of the numerous organs and body parts of multitudes of invertebrates, the vertebra of vertebrates, the fins of fishes, the limbs of land vertebrates, the wings of birds, insects, reptiles and bats, the sophisticated eyes, the brains of animals, and many other sophisticated organs. When we consider that in the biological world there must be several hundreds of thousands of genes, perhaps more than a million genes, the above argument, that the original creature contained the genes and the DG pathways for all the future organisms in a dormant form, is simply not valid.

ALL OF GEOLOGICAL TIME IS INSUFFICIENT FOR ALL THE GENETIC MECHANISMS OF MUTATION AND REARRANGEMENT TO EVOLVE EVEN ONE NEW GENE OR ONE NEW DG PATHWAY FOR A NEW BODY PART THROUGH ORGANISMAL EVOLUTION

Pointing to the slow rate of change in evolution, the noted evolutionist Dobzhansky wrote in 1937,[129]

> It has been pointed out that the sudden origin of a new species by gene mutation is an impossibility in practice. The argument employed to prove this thesis is simple enough. Races of a species, and to a still greater extent species of a genus, differ from each other in many genes, and usually also in the chromosome structure. A mutation that would catapult a new species into being must, therefore, involve simultaneous changes in many gene loci, and in addition some chromosomal reconstructions. With the known mutation rates the probability of such an event is negligible. The process of species formation is apparently a slow and gradual one, consuming time on at least quasi-geological scale.

Even in 1937 Dobzhansky had clearly comprehended the difficulty in changing one organism to another, but he believed that it happened in quasi-geological time. However, even geological time is not long enough. A new creature with a new organ cannot be evolved even in a trillion trillion years, a far greater period than the age of the earth and, in fact, the age of the universe itself.

Other Evolutionary Theories, Proposed as Modifications for Darwin's Theory, Are Also Incorrect

Neutral Theory deals with genetic observations occurring within the population of an immutable organism

According to neutral theory,[130] much of the variation observed at the molecular level (variation in DNA and protein sequences) is neutral; they are neither selected nor rejected. Much of the divergence among species at the molecular level has been caused by random fixation of selectively neutral mutants in the species (genetic drift). One prediction of the neutral model is that the rate of evolution should be higher in molecules (DNA and protein sequences) that are not subject to strong functional constraints than in those that are; in a weakly constrained molecule, fewer mutations will be selected against, and more mutations will have a neutral effect, because they are less likely to disrupt the molecule's function.

In neutral theory, the rate of change of an amino acid sequence by neutral mutations is denoted as the rate of evolution. However, we can ask: If amino acid changes do not change the protein's function to a distinctly new function, how can it be called an evolutionary change? It simply means a higher tolerance for amino acid variations in that protein.[131] Despite the sequence changes in the gene and the protein due to neutral mutations, the basic function of the protein is maintained to be absolutely the same.

We can show by Kimura's own account that neutral mutations can evolve neither new genes nor new DG pathways. Kimura writes,[132]

> I once stated that if the hemoglobins and other molecules of "living fossils" were shown to have undergone as many DNA base (and therefore amino acid) substitutions as the corresponding genes (protein) in more rapidly evolving species, this would support the theory. Since then, the amino acid sequences of the ß chain and the α chain of the principal hemoglobin of the Port Jackson shark have been determined. According to Romer, this shark is a relict survivor of a type of ancestral shark, which had numerous representatives in the late Paleozoic days, notably in the Carboniferous period (270–350 million years ago). So, this shark is well entitled to be called a living fossil. In Table 1, I present a result of comparison between the α and ß chains of the Port Jackson shark together with a similar comparison of α and ß chains of humans. In each comparison, the numbers of amino acid sites that can be interpreted from the code table as due to a minimum of 0, 1, 2 and 3 nucleotide substitutions are listed. From the two sets of comparisons, it is clear that genes coding for the α and ß chains of hemoglobin in the shark have diverged roughly to the same extent (or slightly more)

as have the corresponding two genes in humans by accumulating random mutations since the origin of the α and β globin by duplication.

One cannot say that the mutations found in the human are specifically responsible for the change of the common ancestor into the human, whereas those found in the shark did nothing with respect to evolution. It is unreasonable to advance such an argument. If Kimura is dealing with changes of gene and protein sequences which do not change the qualitative function of the gene, why then does he call his theory the "neutral theory of *evolution?*" It is clear that what Kimura deals with in his theory are the variations that occur to various extents in different proteins[133] due to the difference in their tolerance of amino acid sequence variations, all of which occur in genomes of immutable organisms. They certainly cannot evolve even one new gene or DG pathway useful for the evolution of a new organ.

Neutral mutations do not change the constancy of the set of genes in a genome. They have been totally misunderstood to have potential for evolutionary change. What they do is only to cause variants of the same gene and only to result in normal individual variants of the organism within the population of an immutable creature. In summary, this mechanism is incapable of evolving a new organism from another organism. This therefore is an incorrect theory, revolving around the neutral variations of fixed genes in the genomes of immutable creatures, and trying to connect them to evolutionary change.

"Punctuated equilibrium": An improbable mechanism

When a new species appears in the fossil record, it usually does so abruptly and then apparently remains stable for as long as the record of that species lasts. The fossils do not seem to exhibit the slow and gradual changes that are expected according to the modern synthesis. In order to account for this discrepancy in the fossil record, Niles Eldredge and Stephen J. Gould have proposed that a species stays stable for a long geological time and suddenly evolves rapidly into one or more new species.[134]

The theory of punctuated equilibrium states that evolution of new species takes place in isolated, small, peripheral populations rather than a large population living in a single large geographical area. Small populations may become geographically isolated from the main population. Such a small, isolated, peripheral population has the potential to change into a new species rather rapidly in geological time. The fossil record will then display a series of species suddenly replacing one another, instead of a gradual transition from each species to the next.

In the following, this theory is given in Gould's own words from his recent article in *Natural History* (italics mine):[135]

> The idea that we eventually called punctuated equilibrium had two sources. ... First, a statement about mode of change: Most new species do not arise by transformation of entire ancestral populations but by the splitting (branching) of a lineage into two populations. ... *Allopatric* means "in another place," and the theory argues that new species may arise when a small population becomes isolated at the periphery of the parental geographic range. Isolation can occur by a variety of geological and geographic contingencies — mountains rising, rivers changing course, islands forming. ... Most peripherally isolated populations never become new species; they die out or rejoin the larger parental mass. But as species may have no other common means of origin, even a tiny fraction of isolated populations provide more than enough "raw material" for the genesis of evolutionary novelty.
>
> Second, a statement about rate of change. ... Punctuated equilibrium gains its rationale from the idea, ... that most peripherally isolated populations are relatively small and undergo their characteristic changes at a rate that translates into geological time as an instant.
>
> For a variety of reasons, small isolated populations have unusual potential for effective change: for example, *favorable genes* can quickly spread throughout the population, while the interaction of random change (rarely important in large populations) with natural selection provides another effective pathway for substantial evolution. Even with these possibilities for accelerated change, the formation of a new species from a peripherally isolated population would be glacially slow by the usual standard of our lifetimes. Suppose the process took five to ten thousand years. We might stand in the midst of this peripheral isolate for all our earthly days and see nothing in the way of major change.
>
> But now we come to the nub of punctuated equilibrium. Five to ten thousand years may be an eternity in human time, but such an interval represents an earthly instant in almost any geological situation — a single bedding plane (not a gradual sequence through meters of strata). Moreover, peripheral isolates are small in geographic extent and not located in the larger area where parents are living, dying, and contributing their skeletons to the fossil record.
>
> What then is the expected geological expression of speciation in a peripherally isolated population? The answer is, and must be, punctuated equilibrium. The speciation event occurs in a geological instant and in a region of limited extent at some distance from the parental population. In other words, punctuated equilibrium — and not gradualism — is the expected geological translation for the standard account of speciation in evolutionary theory. Species arise in a geological moment — the punctuation (slow by our standards, abrupt by the planet's). They then persist as large and stable populations on substantial geological watches, usually changing little (if at all) and in an aimless fashion about an unaltered average — the equilibrium.

The theory can be valid only if the evolution of new organisms can happen in a peripheral population. (Note that Gould simply assumes that new genes, which he calls "favorable genes," can evolve quickly, in five to ten thousand years.[136]) But only varieties and similar species of an organism can be produced by such a mechanism. It can be certainly accepted that varieties can be produced faster in isolated peripheral populations than in the main population. Thus it can be seen that this process which produces varieties and similar species of a distinct creature has been misconstrued to be also capable of producing distinct new creatures — the very same fundamental mistake made by Darwin and his other followers.

Eldredge and Gould do not have any genetic mechanism for the production of a new gene or a new organ. If they say that their theory, as far as the genetic basis of evolutionary change or speciation is concerned, is the same as that of Darwin's theory, then we have already shown its improbability. Everything that we have considered and analyzed with respect to the improbability of evolving new genes, new DG pathways, and new organs and body parts in Darwin's theory can be directly applied to the model of punctuated equilibrium. Their fundamental expectation or assumption that "favorable" genes can evolve in peripheral, isolated populations leading to new organisms is itself incorrect, as is the rest of their theory.

Gould writes,

> Punctuated equilibrium is a specific claim about speciation and its deployment in geological time...[137]
>
> Punctuated equilibrium is not a theory of macromutation; it is not a theory of any genetic processes. It is a theory about larger-scale patterns — the geometry of speciation in geological time.[138]

Our unequivocal demonstration in this and in the previous chapter that genomes are a closed framework with respect to evolution — neither micromutations nor macromutations could effect speciation of one organism into another organism with a new gene or a new body part — shows that the punctuated equilibrium model is totally incorrect. Therefore, the problems of the fossil record are still fully unsolved as far as any theory of evolution is concerned.

Richard Goldschmidt's "hopeful monster" hypothesis: An incorrect concept

To explain the lack of transitional (intermediate) forms among distinct living organisms, a mechanism for sudden speciation was necessary for evolutionists. Authors such as Schindewolf and Goldschmidt proposed that the distinguishing features of major groups (higher taxa) evolved by single "saltational" mutations.[139] They said, for example, that the first bird hatched with essentially birdlike features from the egg of a typical reptile. Goldschmidt believed that a "systemic mutation" was responsible for the "hopeful monsters" which were responsible for the origin of new organisms. This mutation represented reorganization or repatterning of the chromosomal material in its entirety. This "macromutation" will produce, suddenly, a huge effect upon a series of developmental processes leading at once to a new and stable form, widely diverging from the former.

Based on the principles we have amply derived so far, we can clearly see that Goldschmidt's theory is also incorrect. No new attribute not already programmed in the genome can be produced in any organism by any type of genetic change. We have demonstrated that true monsters required for evolutionary change are nonexistent in reality.

As it turns out, Goldschmidt's theory had been rejected even by many reputed evolutionists such as Mayr, Charlesworth and Templeton.[140] According to them, it derives no support from modern genetics, because the evidence of morphology and genetics makes the hopeful monsters untenable. However, some recent evolutionists have started to give renewed credence to it. Because there is still no explanation for the sudden appearance of new organisms, or the origin of higher taxa, it is clear that these recent evolutionists are trying to seek assistance from Goldschmidt's theory — in a struggle to find a mechanism to fit Darwin's theory.

The renewed support of some current evolutionists for Goldschmidt's hopeful monster hypothesis is baseless

Although the Eldredge-Gould theory of punctuated equilibrium does not itself propose macromutations, Gould supports Goldschmidt's theory of the hopeful monster, as he says, at least in part. Clearly, this is an attempt to find support to his theory of punctuated equilibrium. But they do not have a genetic mechanism for this process and therefore, the problem of how these new species arise even in the peripheral population still remains unsolved. Gould writes,[141]

> As with ecologically rapid modes of speciation, punctuated equilibrium welcomes macromutation as a source for the initiation of species: the faster the better. But punctuated equilibrium clearly does not require or imply macromutation ... I do feel that certain forms of macromutational theory are legitimate, and I have supported them, though not in the context of punctuated equilibrium. I doubt that the initiation of species by macromutation has a high relative frequency, but

> even rare occurrences may produce important evolutionary results because major morphological shifts are themselves so uncommon.
> ... Few evolutionists recognize that Goldschmidt set his "hopeful monster" theory in this legitimate context. Goldschmidt was primarily interested in how development constrains and facilitates macroevolution. He defined hopeful monsters as phenotypic products of small genetic changes that impact early ontogeny. Cascading effects arise from potential alternative pathways of development already contained in inherited norms of reaction. Monsters may be hopeful because the regulative properties of development tend to channel perturbations along viable (if discontinuous) routes. ... Since genetic differences between hopeful monsters and normal forms are minor, breeding may not be impaired; under certain population structures, small populations of homozygous hopeful monsters may be established.

Unfortunately, the theory of punctuated equilibrium seems to be based on a lack of understanding of the genetic processes required for the evolution of new genes and new body parts. The authors do not seem to recognize what it would take to change the DG pathways to bring about an evolutionary change. As we have discussed, new organs cannot simply be developed due to some errors or perturbations in the existing developmental programs. The importance of the random processes and the improbability involved in the evolution of new genes and new organs cannot be ignored.

It is true that one single gene mutation can convert a pair of wings into two pairs of wings. It is a reality that with a mutation in one gene, legs grow in place of antennae in the fruit fly. There is no doubt that the development of a whole organ can be abolished by a mutation in one gene. We cannot deny that with a mutation in a single gene, or by reducing the activity of a single gene, the thyroxine gene, the whole metamorphosis of the salamander can be stopped. But, we must never forget, all this does not mean that these are monstrous variations. These are only aberrations of already existing genes. A mutation in any gene in the genetic program that develops the front leg of a reptile can never produce the wing of a bird.

Based on our discussions on the probability of the evolution of useful organs in random processes, it is improbable for even one useful organ to evolve by these processes even in trillions and trillions of years — let alone evolving them in five or ten thousand years as claimed by Gould.

Guy L. Bush[142] is another noted evolutionist who supports Goldschmidt's views. He writes,

> Goldschmidt's systemic mutations represented special alterations of the genome that changed the primary pattern of what he termed reaction systems controlling development. He was rather specific as to the

kinds of mutations that were likely candidates for macromutations. These involve chromosome rearrangements which result in chromosome repatterning of the genetic material. He also postulated that one or more macromutations could survive only under the right circumstances, such as the absence of strong selection pressure against the heterozygote and inbreeding. This view is consistent with the current theory on the fixation of chromosome rearrangements. Such mutations are what we would now recognize as regulatory mutations that alter the timing and pattern of expression of several genes simultaneously, producing a phenotypic change. ....

You might say that Goldschmidt, like contemporaries, was "half right." His view that chromosome rearrangements play a special role in evolution and speciation is now increasingly supported by evidence emerging from the discoveries on eukaryotic chromosomes and on the complexity of gene structure and organization by molecular biologists. Indeed, there is some indication that genes may even "talk" to each other by way of introns in the process of coordinating expression of multigene systems, and that the intervening sequence of one gene may serve as the coding region for another with which it interacts. Certain genes, such as those responsible for the immune system in vertebrates, even rearrange themselves in a specific manner within an organism in order to augment immune specificity. Thus, a single nucleotide sequence is used in two or more ways.

He further delineates, at length, the complexity in the molecular mechanisms of regulation. What he says in all this is that the molecular biology of the multicellular organism is so complex that a small perturbation or a mutation may affect many systems at once, which might help in evolution. On the contrary, any of the perturbations would only lead to either a genetic disease, congenital defect, or cancer, and not to the evolutionary production of an entirely new body structure. These mutations would not even lead to a meaningless outgrowth of something other than what already exists in the body. Even without such an aberrated outgrowth, where is the material basis for selection? If it is true that such outgrowths do exist in living organisms, can we show just one example in the myriad organisms living on earth today that has such an outgrowth?

While we clearly agree that the molecular control of the expression of genomes of multicellular organisms is highly complex, it does not mean that a perturbation would lead to something aberrated which would be useful in evolutionary change. It is clear that Bush does not seem to recognize what it takes genetically to produce something truly monstrous (as we discussed in Chapter 3). By our arguments, this is also an expectation based on the deceptive appearance of the molecular control mechanisms and without considering the probabilities involved in the random processes required in the supposed evolutionary mechanisms. It is wrong to expect that, because the genome organization and expression of the multicellular organism is complex, a mutation in it would somehow lead to monstrous outgrowths and therefore to evolution.

ALL THE MODIFICATIONS TO DARWIN'S THEORY ARE INCORRECT

In essence, all the three modifications to Darwin's theory of evolution that we discussed above are erroneous. Without the ability to bring about new organs or appendages, thousands of which exist in the animal world, no mechanism can be claimed to be an evolutionary mechanism. Imagine then, how can neutral mutations evolve even one new organ?[143] What mechanism(s) exist by which these mutations produce a new body part? In my opinion, these kinds of mutations can sway an immutable organism within the closed framework of individual variations depending upon the environment, and no more. Similarly, how can Gould and Eldredge's mechanism of punctuated equilibrium, or the hopeful monster mechanism, evolve a new organ or appendage? How can the perturbations of a preexisting developmental genetic program evolve an entirely new organ? How can they evolve the blood coagulation system with their many unique genes out of thin air from an invertebrate lacking these genes, or the placenta of a mammal with unique genes and biochemical functions? Similarly how can the systems like the sonar ears of bats, dolphins, and many insects be evolved by perturbations in preexisting developmental programs, which body parts are lacking in the supposed ancestral animals? Consequently, these three mechanisms have no meaning when we closely analyze the probabilities of evolving new organs which is the foundation for any evolutionary theory. In the final analyses, all our convincing explanations and evidence undoubtedly point out that all the modifications to Darwin's theory are as incorrect as the original theory itself.

CONCLUSION

In looking at nature, the scenario of life on earth deceptively points to evolutionary change because of 1) physical similarity among organisms, 2) gene similarity among their genomes, and 3) mutational and rearrangement mechanisms that superficially appears to change one genome to that of another. Furthermore, these mutational mechanisms lead to the individual variations within a population of a species, which, according to Darwin's theory, are the fundamental material basis for natural selection. But as we have demonstrated, the individual variations can never be the cause of evolutionary change of one organism into another distinct organism with a new gene or a new body part. Without this distinctive knowledge, one easily makes the mistake that individual variations would, over geological time, be the cause of the evolution of one creature into a new creature.

The reasons for Darwin's mistakes are clear. The proposal of Darwin's theory, which has strengthened the general belief in evolution for over a century, goes hand in hand with the misleading scenario to further its own deceptiveness. Thus, together, Darwin's theory as well as the scenario itself, make it difficult for anyone, including scientists, to come out of the domain of evolutionary thinking.

There has never been a proof for speciation or evolution. Only a great deal of speculation based on such a highly deceptive scenario exists. The whole literature on genetics and evolution is illustrative of this. Scientists tend to make speculative statements on the participation of these mutational mechanisms in evolution, without any hard evidence. Because the scenario is extremely deceptive, concealing the truth, the statements appear to be valid and to have some scientific foundation. What is given as evidence is always only circumstantial. Because of this, the statements depict faith rather than conclusive evidence.

For instance, Suzuki, *et al.*, say in their textbook *An Introduction to Genetic Analysis*,

> Nature has devised many ways of changing the genetic architecture of organisms. We are now beginning to understand the molecular processes behind some of these phenomena. Gene mutation, recombination between chromosomes, and transposition can all be reasonably explained at the DNA level. Far from merely producing genetic waste, these processes undoubtedly all have important roles in evolution.

Stanley Cohen and James A. Shapiro state in their *Scientific American* article "Transposable Genetic Elements,"[144]

> Natural selection, as Darwin recognized more than a century ago, favors individuals and populations that acquire traits conducive to survival and reproduction. The generation of biological variation, which gives rise to new and potentially advantageous combinations of genetic traits, is therefore a central requirement for the successful evolution of species in diverse and changing environments. ... The basic step in the creation of genetic variation is the mutation, or alteration, of the DNA within a gene of a single individual. Mutations involve changes in nucleotide sequence, usually the replacement of one nucleotide by another. This can lead to a change in the chain of amino acids constituting the protein encoded by the gene, and the resulting change in the properties of the protein can influence the organism's biological characteristics.

We have clearly demonstrated that individual variations can never contribute to the evolution of one creature into another with a new gene or a new DG pathway.[145] They can only lead to breeds of a species by artificial selection and varieties and similar species of an organism by natural selection. This is because the set of genes is functionally constant, and the DG pathway of an organism is rigid and can never be changed to that of another organism. Genes and regulatory sequences vary between the individuals of a species, but they are functionally constant, and specify the same set of biochemical functions and regulate the same set of genes.

Alberts, et al., write in their book Molecular Biology of the Cell,[146]

> As a genus distinct from the great apes, humans are only a few million years old. Each gene has therefore had the chance to accumulate relatively few nucleotide changes since our inception, and most of these will have been eliminated by natural selection. A great deal of our genetic heritage must have been formed long before Homo sapiens appeared, during the evolution of mammals (starting about 3×10^8 years ago) and even earlier. It is not very surprising, therefore, that the proteins of mammals as different as whales and humans are very similar. The evolutionary changes that have produced the striking morphological differences among mammals have had to do so with surprisingly few changes in the materials from which we are made.

How can such a conclusion stated in the last sentence above be derived?[147] Because molecular evolutionists take evolution to be a granted fact, and when they compare the observation that there are only a few differences in the genes of whale and human against this assumed fact of evolution, they infer that these few changes must be the cause of evolution of whale and human from the common ancestor. This argument is applied, of course, to the evolution of all the organisms from an original organism. As we have seen, none of these changes or variations can alter the fundamental function of the proteins and enzymes; they only represent variants of the same gene. They are the result of the tolerance of proteins for amino acid variations (see also Chapter 7). We must not forget that the regulatory sequence of a given gene also has tolerance for a great deal of sequence variation. These variations have been obviously mistaken to be the cause of evolution of organisms.

Arguments for evolution having occurred, based on the mutational mechanisms capable of changing gene sequences and rearranging genes and sequences within the genome, are seen to be the norm in biology and genetics textbooks. Another example is seen in Benjamin Lewin's writing in his book Genes IV:

> Genomes are usually regarded as somewhat static, changing only on the leisurely time scale of evolution. ... Genomes evolve both by acquiring new sequences and by rearranging existing sequences. New sequences may arise by mutation of existing sequences or may be introduced by vectors. ... Rearrangements may create new sequences and may change the functions of existing sequences by placing them in new regulatory situations. Rearrangements are sponsored by processes internal to the genome. ... Transposable elements can promote rearrangements of the genome, directly or indirectly.

Geneticists believe that the observed genetic change and evolutionary change are synonymous. They feel that because evolution is a process taking geological time, no one can witness it, and therefore, one can assume that the theory is valid, because the scenario of life on earth is apparently explained by the theory. But the truth is that the scenario is absolutely deceptive. All of these mechanisms are only illusory in "evolving" new functions within the genome of a given species and in contributing to the change of one creature into another.

Rearrangements are a deceptive phenomenon, seemingly capable of placing genes into new regulatory contexts that could contribute profoundly to the evolution of new phenotypes. But as we have seen unequivocally, no rearrangement even over geological time can lead to a new DG pathway for a new organ. The probability of random processes producing a new DG pathway for even the simplest new organ or body part is absurdly low to be evolutionarily meaningful. Proponents of evolution have simply misjudged — astronomically — the potency of rearrangement mechanisms as significant agents in evolution.

Mutational mechanisms are unavoidable in a genome because they are natural to the chemistry and biology of the DNA molecule and the gene. For instance, point mutations are due both to the error in DNA replication and to chemical and physical mutagenesis. Other mutations are side products of normal cellular functions such as DNA replication, recombination and crossing over. Yet other mechanisms of mutation are due to parasitic machineries such as transposons. Mutations by all these mechanisms introduce variations in genes without affecting the types of functions.

Every genome has the innate flexibility to allow all possible mutations and rearrangements within the constant confines of the genome, but none of these mutational mechanisms can mutate the genome of one organism into that of another, with a new gene or a new DG pathway for a new body structure. Mutations in a genome are unnecessary to the operation of the genome as a whole, but are inevitable because they are the natural and innate property of genes and sequences. In any event, however, it is beyond the power of mutations to produce evolutionary change.

A new theory of the independent birth of organisms, introduced in the next chapter and further developed thereafter, explains why such a scenario exists while organisms are not related by descent with modification.

5

A Prelude to the New Theory

While my research during the late 1970s and early 1980s involved primarily experimental molecular biology, and the biochemistry and immunology of nucleic acids, I was quite attracted to the elegance of Darwin's theory in explaining the diversity of life on earth. I marveled at the theory's compelling explanation of the role of natural selection and adaptation in the perfect fit of organisms to their living environments in every sense.

But Darwin's theory did not address the question of life's origin,[1] and that was what puzzled me most. Darwin simply declared that one or a few organisms were somehow formed on earth, and he then proceeded to explain how the multitudes of widely varying organisms could have evolved from them by descent with modification. I was also dissatisfied with other theories purporting to explain the origin of life. It must be admitted that at the molecular level, there is no real known mechanism that could explain the origin of life itself.

The theory of chemical evolution proposed by A. I. Oparin was very convincing, and the experimental research carried out by Stanley Miller and later by others, notably Sydney Fox and Cyril Ponnamperuma, provided an excellent insight into the general nature of the chemical and molecular processes by which the first life could have arisen from inanimate materials on earth. However, these experiments did not explain how the genomes of the first cells were formed, a critical omission considering that even the smallest bacterium contains a very large genome with several thousand genes. No one could explain how even a single gene could be formed from primordial genetic sequences. This notable gap in our knowledge prompted me to analyze genes and genomes specifically with respect to the possible methods by which enzymes and proteins could have arisen from the primordial genetic sequences.

From the published results of experiments simulating primordial earth conditions, I was convinced that the primordial genetic sequences had been random. If this were true, the primary question was: How could the very first genes have come about from these sequences, and code for proteins? Beginning in 1980, while at the National Institutes of Health in Bethesda, I spent a great deal of time studying genetic sequences trying to understand the problems of the origin of life. I became immersed in this problem and continued my extensive research in random sequences, comparing them with the sequences of living organisms. What I learned from these analyses convinced me that genetic material must have been abundant in the primordial pond, and that it must have consisted of random nucleotide sequences. Genes simply occurred by chance in these extremely long DNA sequences, from which they randomly recombined to produce the first cells.[2]

I then came to realize that, given a sufficiently large pool of genetic sequences in the primordial pond, almost any gene could have occurred in it. If this had in fact been the case, then complete genomes — for unicellular and multicellular organisms alike — could have formed by the random assembly of these genes. If multiple copies of the same gene or multiple genes for the same function existed in the primordial pond, then several genomes capable of forming various organisms could have been separately and simultaneously assembled in it. This hypothetical scenario, if proven to be true, would explain the absence of the so-called "missing links" between successive organisms in the assumed evolutionary pathway, which could not be explained by Darwin's theory.

Intrigued by the possibilities, and by their implications for evolution and modern biology as a whole, I began my investigations by devising methods to test the hypothesis using computer techniques. I analyzed the available genetic sequence information acquired from living organisms. A protein sequence database was available from the National Biomedical Research Foundation prior to 1980, and a DNA sequence database, which became a boon to my studies, was just being established in 1982. My proficiency in computer programming helped me to formulate the right questions to ask concerning genetic sequences, and the computer began to rapidly deliver answers. I thus left experimental molecular biology behind to investigate these problems, and soon found myself devoting all of my time to this work using computers at the Division of Computer Research and Technology within NIH.

From the outset I understood that, if my hypothesis were true, then the primordial pond must have contained the complete genes for any given animal or plant, so that its genome could be directly assembled from these genes in the primordial pond. I worked out several details about how these genes could have formed from the primordial genetic sequences, which I assumed to be random. Over the next several years of extensive analysis I verified that my original assumptions were correct — that genes could in fact occur in the large sequence pool of the primordial pond. Computer simulations helped me to show that eukaryotic genes — those of all animals and plants — could occur directly in the random genetic sequences in a primordial pond,[3] and that a vast number of genes could have assembled randomly into numerous genomes. I then determined the basic mechanisms by which different sets of genes could form distinct genomes from the gene pool of the primordial pond, leading to the independent birth of multitudes of organisms.

The extensive series of observations and research I carried out over the past twelve years form the basis for my theory of the independent birth of organisms. This can be briefly summed up in the following. It was discovered only in the late 1970s that eukaryotic genes were split into exons (coding sequences) and introns (intervening, unused sequences). This is one of the most important discoveries in genetics, and is crucial in understanding the origin of not only these genes but also of the genomes and indeed, the organisms. However, the reason for this split-gene architecture was not known for several years. In my work with random sequences, I found a reason why genes were split: If primordial DNA were random in sequence, and if genes simply occurred in the sequences, then the only way they could occur in the sequences was in a split form.[4] Then useful genes, complete with their split architecture and without any need to be evolved from shorter coding sequences, could simply be selected from among those available genes in the primordial sequences and assembled into genomes.[5] This indeed revealed several important facts to me: 1) the primordial sequences were random; 2) genes simply occurred in the primordial DNA; 3) the split nature of genes increased the probability of genes in the random sequences tremendously and made it possible for almost any gene to occur in the primordial pond's random sequences; 4) because the first genes occurred with a split structure, the first cells were eukaryotic, directly assembling their genomes from the primordial pond; and 5) if the full complement of genes to make a genome of a multicellular organism were available in the primordial pond, then these principles enabled the direct assembly of these genes into not only a eukaryotic single-celled organism, but also into a seed cell, an egg like a zygote, which could give rise to a multicellular organism.

All these reasons enabled complex cells to appear, for instance a eukaryotic cell with a nucleus and other organelles, selecting and assembling all the required genes directly from the primordial pond. I could in fact show that the probability of forming the genomes of multicellular organisms is not too different compared to that for a unicellular organism. Likewise, the probability of independently assembling different genomes for many multicellular organisms is not very different from the probability of forming a genome for one multicellular organism, however anatomically complex an organism was, from worm to human. This is because there is not much difference in the complexities of the genomes of organisms at extreme ends of anatomical complexity. Therefore, if sufficient numbers of genes were available to make one viable genome in the primordial pond, it would inevitably enable the assembly of numerous genomes simultaneously, and consequently multitudes of diverse organisms. Since I made the original assumptions in formulating the new theory of the independent birth of organisms, the results and principles derived from my extensive computer studies involving simulations and sequence analyses over the next several years, along with a few already known principles, made it all too clear that the multitudes of diverse organisms on earth must have originated separately in the primordial pond and that the logic of my theory must be correct.

The theory I finally formulated is summarized briefly in the following.

1. Primordial chemical reactions on earth, approximately several hundred million years ago, produced a primordial pond with enormously large amounts of long DNA sequences, comprising the universal sequence pool (USP). Genes coding for different functional proteins were available in this pool in abundance and variety. Multiple copies of the same gene were also available. This pool of genes is called the universal gene pool (UGP).
2. Various organisms arose by independent births in the primordial pond by the independent assembly of genes from the USP into various genomes. The quantity of unique genes available in the primordial pond was more than sufficient to form a vast number of genomes capable of developing into multitudes of complex organisms.

3. The complexity of the genomes is not too different among various multicellular organisms, from worm to human, all of which in turn are not far removed from those of unicellular eukaryotes. Therefore the probabilities for the assembly of the genomes for these organisms are not widely varied.
4. At the time of the birth of organisms, an individual of one organism could be developed from a "seed cell," which contained the complete genome, analogous to the zygote of today's animal or plant. This seed cell, through a maturing process, had the potential to grow fully into an offspring and then into an adult. This process is analogous to the fertilized egg of a frog or a bird developing into an adult.
5. The immense number of genes in the UGP could be independently assembled to form genomes comprising common genes as well as unique genes — because all the genomes were derived from the same common pool of genes. Therefore, although different organisms differing in unique characteristics were born independently, they could have the same and similar genes, as well as unique genes, with basic similarities in their cellular morphology, biochemistry, and physiology.
6. An organism is developed from a single cell, the zygote, through a complex network of on-off switches in all the genes of its genome, which we call the developmental genetic pathway (DG pathway). The DG pathway for the construction of each organism was assembled independently of those of others in the primordial pond by random processes during the independent assembly of genomes.
7. The DG pathway, once formed in an independently-born organism, is unchangeable. That is, it is simply not possible to change the DG pathway of one organism into that of another distinct organism.
8. The gaps in the fossil record among distinct creatures are real, and are caused by the inherent discontinuities among the randomly assembled genomes. Each of the genomes is unique due to its unique DG pathway, and sometimes due to its specific set of genes not present in other organisms. This leads to the uniqueness of the different independently-born organisms. The "missing links," predicted in Darwin's theory, are therefore imaginary.
9. Once an organism was born it was, like its DG pathway, immutable and fixed forever. Every immutable creature is endowed with an innate ability to change within a closed and constant framework of individual variations, beyond which it cannot sway. Gene mutations can enrich the individual variations by producing normal gene variants. Refinements are possible from these individual variations in an organism, and there-

fore varieties and similar species can be produced within the closed framework of a distinct creature (by natural selection and mutations) and artificial breeds (by artificial selection). But they can never produce a distinctly new organism. In addition to producing normal individual variations, gene mutations can also lead to defined genetic diseases.

10. The constant framework of the individual variations of a species of a distinct organism can tolerate only a fixed range of physical conditions. If conditions change beyond this fixed range, the species will become extinct.
11. At the time of the birth of organisms, "random perfection" of organisms filtered the meaningful organisms from among the myriad mostly meaningless independently-born organisms. Those creatures that fit well with the physical environment survived while others perished. Among the physically fit immutable organisms, ecological fitness occurred by chance. This process resulted in creatures that were perfectly fit in both the physical and the ecological environments. In this connection, the concept of adaptation and speciation put forth by the evolutionary theory of Darwin can only explain the production of natural varieties and similar species of every distinct organism.
12. The chemistry of the primordial pond developed over a long geological period, and enabled the primordial pond to become fertile and conducive for the simultaneous birth of different organisms. This led to the sudden outburst of multitudes of organisms at the beginning of the appearance of multicellular life on earth. However, the fertility of the primordial pond continued for a long period afterwards, during which time a large number of additional new organisms were born independently by the same principles, leading to the sudden appearance of new organisms in later geological periods.
13. The births of organisms from the primordial ponds ceased when the ponds became barren millions of years ago. Extinctions can occur, but no new organisms will ever be formed (although numerous varieties and similar species can be formed from each independent organism). The number of organisms on earth can only decrease by extinctions and can never increase by any natural mechanism.[6]

The next several chapters will explain my theory in detail, with abundant supporting evidence and corroborations for its principles and conclusions.

6

The Primordial Pond: Universal Sequence Pool and Universal Gene Pool

Because the new theory of the independent birth of organisms depends on the synthesis of very long DNA molecules in the primordial pond, our first task shall be to determine whether conditions in the primordial pond were ripe for that process.

Over the past several decades, scientists studying "chemical evolution" have advanced a great many experimental and conceptual details regarding the primordial pond. The aim of this research was to demonstrate that living cells could have been assembled from nonliving matter in that environment. Molecular evolutionists believe that chemical evolution first led to the evolution of one or a few microorganisms from inanimate matter, and that these in turn gave rise to a primitive multicellular creature that, as Darwin proposed,[1] then evolved and diversified into all the creatures found on earth. The basis for the field of chemical evolution, and the details uncovered therein, are quite valid despite our earlier demonstration that the theory of organismal evolution is fundamentally flawed. I shall therefore retain the term "chemical evolution" in our discussions here.

First, we shall demonstrate the high reactivity and chemical complexity of the primordial soup. Next, we shall illustrate that prebiotic chemical processes could have synthesized vast amounts of long DNA molecules with random sequences. These prebiotic chemical processes[2] could also have led to the primitive but complex molecular machineries made up of prebiotically synthesized RNAs and proteins, not coded by DNA, capable of reading the messages contained in DNA sequences. These machineries would have decoded and expressed the vast number of genes existing in the long DNA molecules, leading to the repeated systhesis of authentic, genetic machineries coded by DNA.[3] These more efficient machineries would have enabled DNA duplication in the primordial pond. Together, these prebiotic chemical activities and DNA-coded machineries would have synthesized vast amounts of unique DNA molecules, and would have multiplied them over geological time. This creates a "universal sequence pool" (USP) of DNA molecules. We shall also discuss how these activities would have led to complex living cells.

The primordial pond and the primordial soup

The high reactivity of elements on primitive earth led to an abundance of extremely complex macromolecules in the primordial soup

The earth and living beings are one and the same in terms of their basic elemental composition

We are quite sure of one thing. By whatever mechanism life came about on earth, living beings are built of the same elements as those of the earth. This is true whether one accepts that animals and plants originated on earth by evolutionary mechanisms, or by some other entirely different mechanisms. Organisms take up chemicals from the earth on a day-to-day basis and build the complex macromolecules that they need for their everyday growth and activity. Plants do this directly from the earth, and animals do this indirectly by eating the plants; all the biochemicals that any living being consumes comes from the earth. The only apparent demarcation between the earth and the living beings is that the latter have elements organized into complex structures with different shapes, sizes, and functions, whereas the former has elements which are randomly distributed or organized in simpler complexes like sands, stones, and crystals. We shall see in the following how the elements that constitute living beings came from those of nonliving matter and organized themselves into life.

What is the primordial pond?

In the 1920s, the Russian biochemist A.I. Oparin and the British biologist J.B.S. Haldane theorized independently that a "chemical evolution" must have taken place on the primitive earth before the formation of the very first living cell. Haldane speculated, "When ultraviolet light acts on a mixture of water, carbon dioxide, and ammonia, a variety of organic substances are made, including sugars, and apparently some of the materials from which proteins are built up. Before the origin of life they must have accumulated until the primitive oceans reached the consistency of a 'hot dilute soup.' "[4] Amazingly, these predictions were later proved to be correct.

Earth, along with other planets of the solar system, is supposed to have condensed from primordial dust and gas some 4.6 billion years ago. The early atmosphere on the earth was very different from that of today. The core of the proto-earth melted under gravitational force, releasing enormous energy. Lighter rocks rose to the surface and cooled gradually. These eventually became the continents. Several gases such as oxygen escaped out of the molten core. Gradually, an envelope of water vapor, nitrogen, methane, ammonia, carbon dioxide, hydrogen, and other minor components surrounded the earth. In time, the water vapor cooled, rained, and gradually collected as ponds and oceans on the earth, dissolving these gases in the process. Under the fiery and fierce conditions of the hot molten earth, the small molecules in these ponds started to react with each other. All these molecules, dissolved in the water, reacted with one another to form more complex molecules using the energy of heat from volcanic lava, ultraviolet radiation from the sun, and electrical energy from lightning. Among the myriad other chemicals, amino acids, sugars, fatty acids, and nucleotides were formed. These reacted among themselves and formed macromolecules — among others, polypeptides (proteins), nucleic acids (DNA and RNA), fats, and sugars. Over a long period of geologic time, these complex macromolecules formed in extreme abundance.[5] This rich soup of brewing organic molecules is called the primordial soup, and the ponds and oceans containing them are termed the primordial ponds.

We shall see in the following sections how, over geologic time, the small molecules and large macromolecules interreacted among themselves leading to precellular machineries, proceeding inevitably towards the formation of living cells in the primordial pond.

The formation of complex molecules in the primordial pond by random chemical reactions

There are about 100 elements on the earth. Most elements react with a large number of other elements. Furthermore, each of the reactions may not occur at the same rate. The number of chemical reactions and the number of molecules formed from all the possible reactions of these elements are enormous. The set of all possible reactions among the chemicals must occur randomly, the only limitation being the availability of the chemicals and the inherent reaction rate of each reaction. Because the primordial pond must have been abundant in these elements, there was no dearth of reactants. Elements gave rise to molecules, which reacted with other elements and molecules giving rise to more complex molecules. It is logical to expect that this process increased the complexity, variety, and the numbers of the molecules in the primordial pond. A great many kinds of molecules, small and large, must have been stable in the primordial pond.

Prebiotic chemical experiments and evidence for complexities

Almost thirty years after the predictions of chemical evolution by Oparin and Haldane, Stanley Miller and Harold Urey of the University of Chicago simulated the primordial earth conditions[6] by subjecting gases (hydrogen, methane, ammonia, and water vapor) to the energy of electrical sparks in a closed laboratory flask. They succeeded in producing several organic compounds including amino acids. Purines and pyrimidines (precursors of nucleic acids) were not produced in their initial experiments, partly because the gas mixture was too rich in hydrogen. Subsequent experiments by Miller, Cyril Ponnamperuma, Sidney Fox, and others simulated conditions closer to the primordial earth by removing much of the hydrogen. This succeeded in producing a host of biological compounds including nucleotides (adenine, uracil, porphyrins, ATP, etc.) and numerous straight-chain carbon compounds (formaldehyde, urea, deoxyriobse, ribose, etc.). In fact recent experiments have proved that a wide variety of biologically important molecules could be produced under these conditions.

In an attempt to understand the process by which the compounds found in living cells could have been formed, Sidney Fox's group carried out a variety of experiments in which they tried to generate polymerized compounds of small molecules.[15] When amino acids were heated in the absence of water, they formed "thermal proteinoids," macromolecules similar to proteins. Very interestingly, they contained a variety of catalytic activities: e.g., hydrolysis, decarboxylation, amination, deamination, and oxidoreduction. Even a thermal proteinoid with a common hormone activity was produced. Several other properties of proteins found in nature are shared by proteinoids.

Such primitive proteins in the primordial ponds might have catalyzed a great number of reactions, albeit with poor efficiency.

In recent chemical evolution simulation experiments, starting from gases similar to Miller's experiments, Cyril Ponnamperuma has obtained a mixture of short random protein chains, 10-12 amino acids long. This mixture also was found to have a number of catalytic activities. A graduate student working on cytochrome C protein in a neighboring laboratory took the random mixture from Ponnamperuma and found, to his surprise, that it had cytochrome C activity. This incidence indicates that if one tries to test randomly for any known enzymatic function, the mixture might contain it. I asked Ponnamperuma if this meant that one can demonstrate any catalytic activity that we know in nature from this mixture, and he replied, "Almost."[7] This strongly supports the conclusion that the primordial pond must have had a great deal of catalytic activities.[8]

The production of long oligonucleotides in the primordial pond

It is possible to visualize the widespread occurrence of protein-like material in the primordial soup, some of which might have catalyzed the formation of *oligonucleotides*, or DNA chains. In fact, oligonucleotides of varying lengths have been synthesized under variously presumed prebiotic conditions in the laboratory by many scientists. Scientists have synthesized up to 20-nucleotide-long oligonucleotides under primordial earth conditions starting from gaseous mixtures. Cyril Ponnamperuma says that obtaining longer chains is not at all a problem — the short chains can be linked together by the action of catalytic activities found in the simulations.[9] There can be no doubt that proteinoid-like catalytic activities could aid in the process of building long DNA molecules in the primordial pond.

Keeping in view that a variety of enzymatic activities existed in the primordial pond, it is highly probable that the four nucleotides of DNA (A, C, T, and G) could have linked randomly. If sufficient quantities of the nucleotides were available, then random linking among them must be logically expected because all four nucleotides are similar in structure. It is also reasonable to expect that, over geological time, very long DNA molecules could have formed. The stability and the length of DNA molecules in today's living beings indicate that the DNA formed in the primordial pond must likewise have been stable. Furthermore, as stable DNA molecules accumulated, they could be joined end-to-end by appropriate catalytic activities giving rise to longer and longer DNA molecules. There are a variety of extremely efficient DNA-linking enzymes in today's living organisms. It is quite conceivable that catalytic activities akin to these were present in the primordial soup, and aided in the formation of long DNA molecules.

The formation of prebiotic, primitive genetic machineries

Because of their highly charged nature, the prebiotically-synthesized proteins and nucleic acids (DNA and RNA) could have become associated with one another. Ponnamperuma has demonstrated that protein-nucleic acid interactions are quite possible under primordial soup conditions. Over geological time, the primitive ribosomes and spliceosomes (the machinery which edits out the intervening sequences of the genes in the RNA molecule) might have formed from the protein-nucleic acid associations. It should be remembered that these proteins and RNAs of the primitive ribosomes (and other complex machineries) were not DNA coded. They were the random polymers of amino acids and nucleotides, chemically synthesized in the primordial pond. Rare associations of proteins and nucleic acids into "nucleoproteins" could carry out flashes of activities such as duplicating DNA (DNA replication), copying DNA into RNA (transcription), editing and splicing RNA (RNA splicing) and decoding RNA into protein (translation) in the primordial soup.

From current knowledge, it is reasonable to expect that such molecular associations leading to functions were possible in primordial ponds. Recent experiments have shown that under appropriate conditions, viral protein and genetic components will assemble themselves spontaneously into fully-infective virus particles. In fact, such a "packaging" system is routinely used to package foreign DNA into virus particles in recombinant DNA experiments. The concept of self-organizing systems is scientifically well founded (see below), and can be directly applied to the formation of nucleoprotein machineries, which in turn could have paved the way for the DNA-coded cellular genetic machineries.

In my opinion, complex cellular machineries must have been established in the primordial pond before the very first cells were formed. There is no need to assume that the first simple cells evolved with simple machineries which subsequently evolved complex machineries inside the cells. On the contrary, complex machineries must have developed first in the open pond, which then must have aided in the formation of the first cells.

Extremely large amounts of DNA with random sequences were possible in the primordial pond: the universal sequence pool

The primordial replication of DNA and the switch-over to DNA-coded genetic machineries

The four nucleotides (bases) on a strand of a double stranded DNA pair with those in the other strand following a specific rule (see Genetics Primer). The adenine (A) in one strand pairs with thymidine (T) on the opposite strand, and likewise guanine (G) pairs only with cytidine (C). Following this basic rule, DNA replication is carried out in living cells by a specific enzyme called DNA polymerase, producing two identical copies of DNA. It is very probable that primitive but similar catalytic activities could have occurred in the primordial pond. As we discussed before, there is evidence that almost all kinds of catalytic activities could have been present in the primordial pond. Complementary strands for single DNA strands may have been assembled by such DNA-polymerizing activities, using short random "primers" of about 10 nucleotides which could have been available abundantly in the primordial soup. Similar activities could have copied the DNA strands into RNA strands (transcription). In time, complex physico-chemical interactions between RNAs and proteins could have resulted in primitive protein-synthesizing systems (translation). Thus, the primitive transcription and translation mechanisms must have come into existence before such systems were coded from DNA sequences. We have to bear in mind that such systems which arose in the prebiotic primordial soup must have been feeble in their activity, before the consistent and reproducible DNA-coded machineries came into being. But even a very low level of such primitive activity could have been sufficient to bring about the switch to the highly reproducible, authentic DNA-coded machineries. In the following we shall see how this could have happened (see Figure 6.1).

It is plausible that one or a few of the myriad DNA sequences in the primordial pond, by chance, might have had the message for coding a protein with DNA polymerase activity. (One of our main principles is that genes for almost all possible enzyme activities could have occurred within the vast primordial DNA sequences.) Similarly, the messages for the transcription enzymes and the more complex translation machinery could have been present in random DNA. All these might have been expressed from DNA sequences by the primitive non-DNA-coded enzymes and nucleoprotein complexes of the primordial pond. Once the DNA-coded machineries were synthesized and assembled, they were capable of carrying out these activities in a far more streamlined and efficient manner. In fact, such a switch from non-DNA-coded machineries to DNA-coded machineries must have been inevitable, over a period of time, even though the non-DNA-coded machineries were primitive and slow.

Enormous quantities of unique DNA sequences as well as duplicated copies of the same DNA sequence would have been produced in the primordial pond

As we have discussed earlier, even before DNA-coded machineries arose, immense amounts of DNA molecules could have been synthesized in the pri-

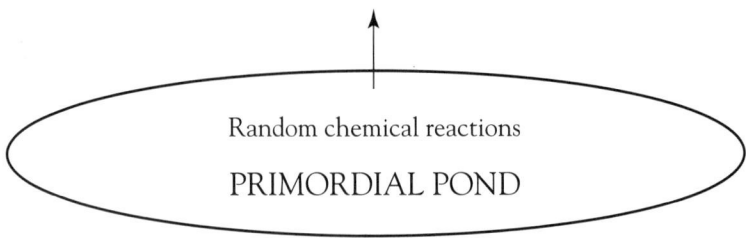

FIGURE 6.1. FROM PRIMORDIAL CHEMICAL REACTIONS TO LIVING CELLS — FROM INANIMATE MATTER TO LIFE. The history of the formation of living cells from inanimate matter in the primordial pond.

mordial ponds. After the DNA-coded enzymes had evolved, the replication of the DNA must have become more efficient and produced an abundance of the authentic DNA-polymerizing enzymes. This would have enabled the duplication of unique DNA sequences over and over again. In time the primordial soup would have attained high concentrations of DNA molecules, containing both unique and duplicated sequences.

A point I want to emphasize is the length of DNA sequences. It is very possible that small DNA molecules with random nucleotide sequences, existing in the primordial ponds, were linked to form very long DNA molecules, even to the extent of millions of nucleotides. It may be difficult for us to envision such long DNA molecules floating in the primordial ponds. However, imagine the length of the DNA in chromosomes of living cells. The smaller chromosomes contain about 50 million nucleotides, and the larger ones contain about 400 million nucleotides, in one single DNA molecule.[10] In the chromosomes of cells these sizable DNA molecules are stabilized by the binding of basic proteins and other small molecules. Similar processes happening in the primordial ponds, with many kinds of molecules available in it, may have stabilized the primordial chromosomes.

One might doubt whether all these enzymatic activities, discussed above, could have occurred in the primordial pond before the DNA-coded proteins were synthesized. However, we must remember that when the primordial soup attained highly complex chemical activities, it must have contained almost all the catalytic activities that we can think of. These would not have existed in the form of well-defined enzymes, but must have been the cumulative result of random protein molecules exhibiting similar catalytic activities over a period of time. Such a cumulative activity can be as effective in building complex molecules over geological time, as the well-defined enzymatic activities in today's living cells do in minutes or days.

In essence, the possibility for enormous amounts of DNA material and sequences to be produced in the primordial pond seems quite conceivable, and, in fact, absolutely reasonable. Their quantities would have been sufficient for the coding sequences (in the form of split genes) for almost all the enzymes that we can think of to occur in them by chance (as demonstrated in Chapter 7). Once the DNA-coded genetic machineries took over the expression of these "genes," then all these authentic and reproducible enzymes could be produced in abundance.

The huge amount of DNA present in just one individual animal is an indication of the amount of DNA in a primordial pond

The amount of DNA in one human cell (the genome size) is three billion (3×10^9) nucleotides.[11] Since there are approximately ten trillion (10^{13}) cells in a human body, the amount of DNA in one individual is 3×10^{22} nucleotides ($3 \times 10^9 \times 10^{13}$). This is a tremendous amount of DNA sequence. Because the same sequence present in each cell is repeated in all ten trillion cells, they could, hypothetically, be scrambled to produce a DNA, 10^{22} nucleotides long, with unique sequences.

Currently, there are 5 billion people (5×10^9) in the world. Thus the amount of DNA in the human species alone is 1.5×10^{32} nucleotides ($5 \times 10^9 \times 3 \times 10^{22}$). Now consider the tens of millions of species[12] that live on earth and in the seas, and the flora of all the forests. Consider that the amount of DNA in each cell of many amphibians and plants is 50 to 100 times larger than that of human beings. Further, the mass of all the plants on earth is about 10-100 times larger than that of all the animals on earth. Keeping in mind that approximately 30 million species are living on earth today, the total amount of DNA on earth today is at least a billion times more than that in human beings alone, i.e., approximately 10^{41} nucleotides ($1.5 \times 10^{32} \times 10^9 = 1.5 \times 10^{41}$ nucleotides). We must keep in mind that we have not taken into account the fossils, whose amount is enormous. If we consider the fact that only a small fraction of all the carbon, hydrogen, nitrogen, and oxygen on earth is present in the form of living things, we can understand that a lot more nucleic acids than we estimated were possible in the primordial pond. There must be at least a million times more organic material on the surface of the earth than in living beings. And this, in the primordial soup, would have given rise to DNA. Although millions of small and large ponds must have existed on the primitive earth, many of these ponds would have had approximately 10^{30}–10^{35} nucleotides of DNA material and a great deal of biochemical complexity at one time or another.[13] For our discussions hereafter, we shall assume this total amount of DNA to represent the size of the universal DNA sequence pool in a typical, biochemically-rich primordial pond. Some variations in this quantity should not really affect our concepts.

The stability and multiplication of DNA in the primordial pond

Today we can multiply DNA extremely efficiently in the test tube using DNA polymerases isolated from living cells. Further, the DNA molecule is highly stable even by itself in the test tube, indicating that such a stability is the inherent nature of the DNA molecule. Also, protective mechanisms similar to those existing in the chromosomes of living cells could have existed in the primordial pond. One would logically expect that DNA must have been abundant and very stable before complex cells were developed.

In this connection, there is still more support for the stability of DNA molecules in the primordial pond. It is significant that the DNA from mummified bodies has been stable for thousands of years and could even be cloned using recombinant DNA techniques. In fact, DNA is routinely boiled in the laboratory. The DNA is quite stable under these conditions. This shows that DNA could have been stable even at the boiling temperatures of the primordial pond. The best evidence in favor of this fact is that there exist many microorganisms that live at high temperatures. For instance, a fungus,

Thermophilus aquaticus, lives at 90° C. Its DNA is polymerized by a DNA polymerase (called Taq polymerase) that functions at this high temperature within the cell and in the test tube.

Humans have only discovered a few of the methods for DNA synthesis and replication in the laboratory that Nature can carry out.[14] Many more chemicals and conditions than can be created in the test tube could have existed in the primordial broth. It is extremely important that we should not underestimate the potential of the primordial broth, which must have contained millions of highly reactive organic chemicals of different types, sizes, and structures, brewing with all kinds of molecular catalysts.

Future primordial soup experiments could demonstrate the synthesis of even longer oligonucleotides

At first glance, the molecular complexities of the primordial pond may seem farfetched. However, from the time Oparin proposed the chemical evolution theory, it has been consistently shown that the primordial broth was very rich in its variety and complexity of chemicals. Even the solutions resulting from the simulated primordial-soup experiments contained biological molecules.[15] One may ask why extremely long DNA molecules, on the order of millions of nucleotides, have not been demonstrated in experiments simulating primordial ponds. My feeling is that Miller's experiments have been extended only to a limited extent. We have not carried out the further necessary experiments to demonstrate all the aspects of the primordial pond, especially that one can make very long DNA molecules. What has been achieved by the investigations so far is the basis of our conclusions, but there is much more to be learned.

I am confident that if we incubate large amounts of nucleotides along with proteinoids, given sufficient reaction time, oligonucleotides and long DNA duplexes may be formed. Similarly, the random mixtures of peptides, obtained in Ponnamperuma's experiments, may also contain such polymerizing activities. The underlying reasons for not attempting to synthesize long DNA molecules in chemical evolution simulation experiments may be that no theoretical framework existed that guided or forced one to look for long oligonucleotides. Therefore, the fact that long oligonucleotides have not been so far demonstrated in chemical evolution experiments does not mean that they are not possible.

DNA or RNA?: The question of which is the first genetic molecule is immaterial for the new theory of the independent birth of organisms.

Because RNA has "self-splicing" catalytic activity, and has a central role in translation, it has been speculated that it preceded DNA in precellular chemical evolution and that there was an "RNA world" before a "DNA world."[16] In my view, RNA need not be expected to have preceded DNA in life's prebiotic history. It is quite reasonable that the precellular non-DNA-coded translation machineries could have been formed from the RNA and protein molecules available in the primordial pond. Their messages could not have come from the DNA. The fact that RNA had some catalytic activities does not mean that RNA originated first in the primordial pond, after which DNA came into picture. As a matter of fact, several scientists have already argued against the RNA world theory.[17]

DNA is an extremely stable molecule that can be immensely long. Neither of these characteristics is shared by RNA. Both RNA and DNA could have occurred simultaneously and exhibited their functions — RNA by itself or in conjunction with protein had several catalytic activities serving to build the prebiotic genetic machineries; DNA had its own function of passively carrying the message for the RNAs and proteins that formed the DNA-coded genetic machineries, among the vast number of other genes.

Some form of translational machinery, far more complex than a single enzymatic function, must have come into existence in the primordial soup to switch from non-DNA-coded to DNA-coded genetic machineries. Consequently, it is quite possible that transcriptional activity, a far simpler function (of RNA polymerase) could have been present in the primordial protein mixture. Sequences for a vast number of unique proteins and RNAs existed in the universal sequence pool. Transcription of DNA sequences by prebiotically synthesized primitive RNA polymerase must have resulted in a large number of RNA molecules which included ribosomal RNAs and messenger RNAs (for proteins). The protein molecules translated from some of these mRNAs could have become associated with some DNA-coded ribosomal RNA molecules.

Therefore, although RNA had some catalytic activities, DNA must have been the genetic material that contained the messages for all the proteins and other cellular machineries for any living cell. These must have been initially decoded by the catalytic activities of the random polypeptides and the nucleoproteins in the primordial soup. The USP contained all the genes for the proteins and RNAs required for the construction of the genetic machineries, among the vast number of other genes, thus enabling the switch to authentic and reproducible genetic machineries.

Some have speculated that only the RNA molecules had messages for proteins, and that these RNA molecules were later reverse-transcribed into DNA molecules.[18] But it was the long random DNA sequences which contained all the genes necessary for living cells, as we shall establish in the next chapter, only because of which life was possible.

Self-organization of membranes in the primordial pond

Present-day living cells are surrounded by a membrane composed of fats and proteins.[19] One can make artificial membranes in the laboratory with properties similar to those of the natural membranes by simply mixing fats and water together and agitating. Such artificial membranes form closed vesicles under suitable conditions. These vesicles can enclose a distinct population of macromolecules and could form a spatially isolated functional unit. It has been suggested that the first cell was formed when phospholipid molecules in the prebiotic soup spontaneously assembled into such membranous structures, enclosing a self-replicating genetic system that could regenerate itself as well as the system of enclosing membranes.[19] In essence, the properties of artificial membranes that can be easily formed in the laboratory are quite similar to those of the membranes of living cells, and therefore the concept that, in the primordial soup, membranes can self-assemble and can enclose a self-replicating genetic system leading to living cells is quite valid.

Organization of genes from the USP to form a self-perpetuating system — the origin of the first cells

The discussion so far concerning primordial simulation experiments indicate that the universal DNA sequence pool could have been vast and could have had a random sequence. Genes could occur in it sporadically. If we consider the universal sequence pool as a single long sequence and walk on it from one end to the other, we will find genes only rarely and will walk most of the time on "junk" DNA. In the vast, total universal sequence pool, however, an extremely large number of genes and regulatory sequences would be present. We shall call this the universal gene pool (UGP).

A living cell is composed of a certain minimum number of genetic machineries, cellular structures, and a cell membrane surrounding everything. In a crude sense, if genes for all the proteins that would produce all these machineries and cellular structures were organized into well-defined genetic pathways, and were enclosed in a membrane-bound system, then it would be a fully self-perpetuating system — the living cell. As we shall see in the next chapter, vast numbers of genes, including not only those for all these machineries but also a great number of other biochemical functions, must have occurred in the UGP, enabling the assembly of these genes into a genome for complex cells.

The paramount significance of all these considerations is that all the genes for the genetic machineries and cellular structures could have existed in the universal sequence pool. What is required is their assembly in the right combination and in the right metabolic pathways. In my view this certainly could be expected. Random recombinations among genes and regulatory sequences would have happened in the primordial pond for a very long geological time. Several reasons ensure that the recombinations would inevitably lead to living cells. For example, 1) more than one sequence in the USP could specify a given enzymatic or structural function, and 2) multiple copies of the same gene could occur in the UGP, because multiple copies of long DNA sequences were produced in the primordial pond. Many such reasons lead to the inevitability of living cells. The same arguments can be applied to the processes leading to seed cells, i.e., cells capable of developing into multicellular organisms (see Chapters 7 and 8).

Sydney Fox has stated that chemical evolution must have culminated in the evolution of the first cells, otherwise it would have reached a dead end.[20] It is clear that the processes leading from inanimate matter to the living cell is a necessary prerequisite for the appearance of life on earth. In my opinion, random chemical processes must have led to vastly long random DNA sequences. Given that genes for all the cellular machineries and structures occurred in these sequences, it is reasonable to expect that random recombinations among DNA sequences led to genomes of viable cells. The next chapter will demonstrate, by probabilistic analysis and computer simulations, that an immense number of unique genes must have occurred in the universal sequence pool.

Conclusion

Eons ago, in turbulent ponds scattered across a still hot and partially molten earth, organic chemicals were being synthesized far more fiercely and ferociously than in today's laboratory flask. All of these organic chemicals were boiling, broiling, hydrating, dehydrating, complexing, condensing, precipitating, breaking, and recombining in a dizzyingly random dance of chemistry gone mad! We can only begin to imagine the great richness and fertility of the organic reactions. Use all the words for the processes that chemists today use to synthesize organic chemicals in the laboratory — condensation, diazotization, amination, hydration, halogenation, pyrrolysis, and so on. All these and more occurred spontaneously, and prolifically, in the millions of ponds that were sometimes mixing and sometimes isolated. Reactions ensued not just for a few hours or days, or even years, but for several millions of years. Consider that Miller synthesized myriad small organic chemicals, including small DNA and protein materials, by sustaining likely primordial conditions in a laboratory flask for just a few days. Then imagine the probable outcome of nature's own experiments, conducted in all of earth's many ponds under an incomprehensibly broader range of conditions, and sustained for several millions of years. And what a wonderful outcome it was: small molecules and large macromolecules that could construct those beautiful self-replicating things we call living cells.

The biochemical complexity of the primordial pond increased tremendously over geological time. The DNA, RNA, and protein molecules, among others, were prebiotically synthesized by random physicochemical reactions among elements and molecules of the primordial soup. They would have led to primitive machineries that carried out transcription, splicing, and translation. These non-DNA-coded machineries would have "expressed" the "genes" present in the vast universal sequence pool for the first time. The expression of these aided in the switch to, and the takeover by, the DNA-coded genetic machineries of such functions as DNA replication and the transcription and translation of genes. These randomly combined machineries, with the right conglomeration of genes from the USP, would have produced a self-replicating system leading to the first living cell.

The importance of all these considerations is not only that they could form complex single-celled organisms, but that if appropriate genomes were organized in such cells, these cells could directly give rise to multicellular organisms. Such single cells with the potential to yield individual organisms — analogous to the single-celled fertilized eggs of today's multicellular organisms — we call "seed cells." In the coming pages, we will see that the first genes and the first cells must have been highly complex — not simple as traditionally believed. This concept is important in developing the new theory that myriad seed cells could have been assembled, and multitudes of complex organisms could have been born directly from the primordial pond, deriving their basic genetic codes, genes, and cellular machineries from a common universal gene pool.

7

THE ABUNDANT OCCURRENCE OF GENES IS INEVITABLE IN THE PRIMORDIAL POND

Genes are fundamental to all living things — carrying all the instructions for the embryonic development, growth and functioning of organisms. Genes also are responsible for the perpetual self-replication of all organisms. The development of every organism is governed by genetic instructions, and this is as true today as it was eons ago, when life first appeared on primordial earth. Where there is life as we know it, there must also be genes. We therefore feel safe in concluding that, wherever life came from, its first manifestation must have bloomed from genes.

The existence of even the simplest living cell requires a certain minimum number of biochemical functions.[1] We know that these biochemical functions are facilitated by proteins, and that proteins' "marching orders" are stored, disseminated, and replicated only as genes. The primordial pond therefore must have contained an abundance of genes for even the simplest life form to have originated in it, since no life form could come into being without a certain critical number of complete, unique genes.[2]

Enormous quantities of DNA material must therefore have come into existence in the primordial ponds, although the total amount possible in a typical pond is within a finite limit (approximately 10^{35} nucleotides). How likely was it that the minimum number of genes required for a living entity actually existed in this amount of DNA material? A superficial probabilistic approach to this question, based on the lengths of typical protein molecules, suggests that not even one gene could have occurred by chance. But a more thorough analysis that accommodates fresh insights into gene characteristics shows that an abundant number of genes must have in fact existed within the primordial pond's genetic sequences.

Certainly it is notable when the seemingly preposterous and impossible are revealed to be inevitable. The notion that genes could have occurred abundantly in the Universal Sequence Pool (USP — the primordial genetic sequences) has seemed absurd since the first characterization of genes as the agents of organismal reproduction and inheritance. But a new and closer look at the nature of genes will show that, in fact, a great many genes occurred within the random DNA material in the primordial pond. This new inquiry is further enriched by asking, at every step: What really constitutes a gene within the context of life?

The probability of finding a genome book in a random sequence: An incorrect analogy

Finding the complete works of Shakespeare in a random stream of English letters: an incorrect analogy to finding a genome in a random stream of DNA nucleotides

The occurrence of a Shakespearean work in a single stretch in a random stream of English letters is improbable

Molecular geneticists and evolutionists attempt to show that it is improbable to find the sequence of the genome, of even the simplest living entity such as a bacterium, in a random stream of DNA characters. The genomic DNA of a bacterium such as *E.Coli* has approximately 5 million characters. The probability for this genome to occur in a random DNA sequence is $4^{-5\,\text{million}}$. This is a meaninglessly low probability. Therefore, it goes without saying that the genome sequence of even the simplest form of life such as a bacterium cannot occur as a complete entity in the primordial pond based on probability.

Evolutionists often use this analogy to ascertain their view that life could not have originated as a probabilistic result.[3] We shall see below that this kind of approach of looking for the "genome book" as a whole in a random sequence is truly incorrect. But more importantly, molecular geneticists and evolutionists are convinced that even a gene, let alone a genome, could not occur in a random sequence. Their reasoning runs as follows. The probability for a specific 200-nucleotide coding sequence, as we discussed before, is 10^{-120}, and the *expected mean length* of the random sequence (in which this long coding sequence should occur by chance) is 10^{120} nucleotides. Compare this with the total length of the DNA sequence, if the mass of the whole universe (which is 10^{80} H atoms) is converted into a single DNA molecule — leading to a DNA of $\sim 10^{78}$ nucleotides in length (taking approximately 100 H atoms to constitute one nucleotide). So while the expected mean length of a random sequence in which the specific 200-nucleotide sequence can occur probabilistically is 10^{120} nucleotides, no DNA longer than 10^{78} nucleotides could ever be formed even if the mass of the whole universe were converted into a single long DNA molecule. Evolutionists take this as a proof that a coding sequence of even 200-nucleotide length cannot occur in a random sequence, let alone a complete gene, which is far longer. Based on such arguments, they believe that somehow genes had to evolve from shorter coding sequences, which, as we have determined in Chapters 3 and 4, is improbable.[4]

Evolutionists such as Jacques Monod have considered that life was improbable, but that since life is a reality on earth now, first some very simple organism must have originated as a freak accident — and that the genes and sequences required for the formation of even the simplest living cell originated not as a definite probabilistic outcome but as an unlikely accident.[5] And that one accident gave rise to all other life on earth by means of evolution.

To support their view that genes could not have occurred by chance in the primordial genetic sequences, it is customary for molecular evolutionists to cite an analogy.[6] It is clear to anyone that it is extraordinarily unlikely for a book such as Shakespeare's Hamlet to appear perfectly, in one piece, within any reasonable length of random English letters. And the genome of even a bacterium contains far more "letters" than Hamlet, so molecular evolutionists believe that even a simple genome could not have occurred in any random DNA sequence. Typically, a molecular evolutionist first shows that the probability for finding even one word in a random stream of English letters is very low — therefore, it is improbable to find a sentence. Without even attempting to probe any other details of the gene, evolutionists would then jump to conclude that it is improbable for a given gene to occur in a random sequence. Because genes and genomes are a reality, they simply assume that one original genome must have somehow evolved in the past. Period! Having stated that — without even attempting to find out how such a thing could have happened — they are overtly concerned about a truly nonexistent problem: How, given one organism, all other organisms evolved by descent with modification.

A "genome book" is not analogous to a "Shakespearean book." Even a gene should not be directly equated to a sentence.

We can show that it is an incorrect analogy to equate a "genome book" to an "English book" in discussing the origin of the genome. Although a genome sequence can be generally considered a book, the mechanism of its origin is absolutely different from that of a real book. I take a totally different approach to show that genes could occur inevitably in random DNA sequences, and you don't need a DNA molecule as massive as the universe. In fact, DNA molecules possible in a primordial pond are quite sufficient for this. We must analyze several distinct aspects of a gene and the structure and function of its protein product in order to answer this question. However, before getting to these details of genes, let us discuss briefly the correct analogy in the context of genes.

First, it is not necessary that a genome sequence should occur in a single stretch in a random sequence. It is sufficient if individual genes are scattered through the random sequences, from which a genome can be assembled by biochemical processes in the primordial pond.[7] Here we can equate the gene to a sentence in Shakespeare's book. But even the sentence cannot be equated to the gene directly. The gene contains exons which have the functional portions of a gene, and introns (intervening sequences) which are meaningless. We can thus equate the words in a sentence to the exons, and the spaces between the words to the introns. Then we can show that sentences can certainly occur in a random stream of English letters with lengths meaningful to our discussions. This analogy, as crude as it is, will help us to understand the observed structure of genes in living organisms. However, even to show that a gene's word, an exon, can occur in a random DNA sequence, it involves several principles concerning the fundamental nature of genes, proteins, and their functions, which we shall discuss later.

Hamlet may consist of approximately one million characters with 10,000 sentences. If we can show that a sentence in Hamlet can occur probabilistically in a stream of random letters, it would apply to any other sentence in Hamlet. Within a random character stream, fairly longer (say ten times) than that required probabilistically for one typical sentence to occur, almost any sentence must be possible to occur. If we want the 10,000 consecutive sentences in Hamlet to occur in the same order as found in the book, then we would need 10,000 times the length of the random sequence in which we expect one sentence to occur.

When we consider words as exons and spaces as meaningless introns, complete sentences are highly probable in a random character stream.

The probability for a 20-character English sentence is one in 27 (for each letter in the alphabet plus a space) raised to the power of 20 (number of characters in the sentence), or 27^{-20}. This is roughly equivalent to 10^{-29}. Therefore, the length of a random sequence in which this sentence can be expected to occur in a single stretch at least once is about 10^{29} characters (the expected mean length). If we use a computer to type a random sequence of letters, 10^{30} long, the sentence will likely be found somewhere in it.[8] However, if we employ a new method of looking for the sentence — that is, look for only words of the sentence in the same order as they occur in the sentence and ignore the nonsense streams of characters between words — then the probability of the sentence increases tremendously. Here the spaces between words are equivalent to the nonsense strings of letters.

On this view, it is certainly true that the gene is very similar to such an English sentence. It has exons which are analogous to the words, and introns which are analogous to the spaces. As with the sentence, the probability of a gene occurring is far greater when its complete coding sequence is split into shorter coding pieces (exons) by meaningless introns than if the gene's complete coding sequence had to occur in a single stretch.

Let us take the first word of the sentence, "To be, or not to be." Start with the first character in the random stream of letters and look for the occurrence of the first word, "To." Once we succeed, take the next word "be," and keep looking for its occurrence starting from the next character in the random sequence, ignoring the nonsense passing by between the consecutive words. By this process, one would see, it is inevitable that the sentence would be found fairly soon in a random character stream. We can see in Figure 7.1 that the sentence "To be or not to be" is found in about 3500 characters (i.e., within one page[9]), whereas in a single stretch it can be found in about 27^{18} (6×10^{25}) characters (i.e., about 10^{22} pages).

The probability that a four-letter English word occurs in a random sequence is 26^{-4}. The expected mean length of a random stream in which this word can be expected to occur is 26^4 (456,000) characters long. Therefore, if we have a 3,000,000-character random sequence of letters, far more than the expected mean length, we can expect to find almost any four-letter word in that random sequence. In a random sequence of

AVTQMOIBIYUTTYRXBVGHSFRETYPNMKJBZXCVBFGTWRWEDDFALH
OILPMNKJUVBGHYFQSZVDFTRYOPMMJLAJSHJGFRTYQREFFGFBNBMI
ALKEIUQJURYTWSDHTRHFMNZBXVCHQYTNVHSKFYWURIOPMCVHY
HDFQIOREUYSKJGHADGLZXMNRBCNVYQNEUCBNRTYVBNFYURHJY
NBBNZCXJKWOPIKIUQWYR**TO**HVBCNMZJSGHFGTWRERUUIOPPMKJH
HGAWRYBCGDFHNXCYRQZCVNBIOVYZNSGHENMBKHJJYQXHFJGHII
YOPPZNCVJHFGJJMMBJHQOVNMZBXVTRRQEWFHKPLOIQAZSGJHUIO
THKLPMLOK**BE**UBNJCGTQRWETRYIIPOCNMKSALIDERQWTYHCBZVX
ASPMQIZUXEMCCUVIEASRTTYPOIVLASFGUEYRTHNBCVXVAQWHGK
JFGURYTOPZDFGKHLUITYWRERYVNGFASDFLJIWKERHVNXJWIWZAQ
WSDDSXCMKOPLJHEDRFTGYHUBVFEWSXMUBTCWQA**OR**WAGJLNVX
ZWSAQEDCBHYNMKOLPIUWQERVFCXNBMJHGFTIOLEWQAJUIMJTED
WSAQZXCVBNMKIJHYTEWQSDFHLOPKYQAMOECIMTBUNHFSKOPMQ
PMZALYBECMIQMXNCVHVKITUYQASLKJGNHVMCHFDZCERWTQYTU
IOMBJGPOQWUNXGDFFHSYRTYAZNGHKITYTQRHSDNJHKLJPGKJFJHS
LQEIWUZV**NOT**IHLGKJAFDJGDFKIUUTWTYERUJGHDNMHKJHKJKJKQ
ZATMLPOKJIMNBVZXUYTRWDKJHGFAASOJHSAWQHJKLYUIOPQWED
FGHBVCDEWSQAZXMJIKOLPMYHTGRFEDSWQAZXCFRDLOPNJHGGVJ
JFHJJYGHRTEDWQEXSSDXGFVOPLKMJHJJBHNGHQWSRFGGYUHIOKP
OLMKJNBHVCZSSZWAWDFGOPNGYHY**TO**RDFFGVGBHYHGUHJQWSC
FVBMNIYUTRERJDFHOIURTYIERUQOBMNZBXCBNCGHFTRYGJHJJGH
GFGSGFSKSKHJJKTKMQIOYUTCBTREOPZWSETFBGJGNUJBDFEDLOPT
EGJALJQEPBMNZWAX**BE**CVBIYOYQWTREHJLMBNQACVIMJXKFGODR

FIGURE 7.1 FINDING AN ENGLISH SENTENCE IN A RANDOM SEQUENCE OF ENGLISH CHARACTERS BY A NEW METHOD. A random stream of English characters was simulated in the computer. The sentence "To be or not to be" was searched by looking for each word separately. First the starting word "To" was located, and then the nonsense characters in the random sequence were ignored until the second word "be" was found, and so on.

letters ten times the expected mean length for an eight letter word (approx. 10^{12} characters), almost any word of eight characters would be found. This is because my research has shown that a random sequence with six times the expected mean length for a given word of specified length will contain more than 99.9% of all possible words of that specified length and all the words under that length.[10] This applies to a complete sentence too. But there is one important difference. Because we search word-pieces of a sentence in a stretch, the probability of a sentence does not depend on the length of the sentence. It depends only on the probability of the longest word!

For a sentence made up of 10 words, with the longest word being eight letters long, the probability of finding the complete sentence is nearly the same as that of finding this word. The longest word will almost certainly occur in a random sequence of length ten times the expected mean length for that word. For instance the expected mean length of a random sequence of letters for an eight-letter word such as "question" to occur is about 10^{11} characters, assuring that a random sequence of 10^{12} characters (ten times the expected mean length) will contain this word. However, all the other words in the quote, "To be or not to be. That is the question," being shorter, will also occur in this same random sequence. We can also expect them to occur in the same order they occur in the sentence. So to find any complete sentence, one only needs a random sequence long enough to ensure that the longest word is found. The difference between taking the sentence as a whole and splitting it into its word-pieces in computing its expected mean length is truly immense: if a 10-word sentence is 50 characters long with its longest word of eight characters, the expected mean length for the sentence as a whole is about 10^{71} characters (27^{50}); but to find the sentence in word-pieces by our new approach, the expected mean length is only 10^{11} characters (26^{8}). What a stupendous difference between the two! Even if there are ten words each eight characters long in the sentence, still the expected mean length for this sentence by our new approach is only ten times the expected mean length for one eight-letter word (10^{11}), which is 10^{12} characters.

As a result, an important principle we can derive from these analyses is that if the expected mean length of a random sequence of English letters for a sentence (with the longest word of eight characters) to occur is 10^{11} characters, then within a sequence 100 times longer (10^{13} characters), we can find almost any sentence with words eight characters or fewer. For instance, in the same random sequence of letters in which we found the above sentence "To be, or not to be. That is the question," we can find another sentence, "Love sought is good, but given unsought is better." The number of words does not matter and thus the total length of the sentence does not matter. Only the length of the longest word matters!

Another interesting thing is that in the same random sequence in which we find all the sentences of Shakespeare's Hamlet, we can find any sentence from any other work — for instance, any of the novels of Charles Dickens or Ernest Hemingway. In fact, for all the ~ 10,000 sentences of

Hamlet to occur in the same sequential order in the random sequence — assuming the longest word is 10 characters — it only requires a random sequence of ~10^{18} characters (10,000 times the length of the random sequence for one sentence to occur). In the very same random sequence, any book of the size of Hamlet would be found (with the longest word of 10 characters). With the longest word of 14 characters and with a maximum of 100,000 sentences, it takes a random sequence of length ~10^{25} to contain all the works ever written and will ever be produced — with the sentences of each book in the same order as they were written.

Incidentally, the expected mean length for a eight-character DNA "word" (4^8, or about 10^4 DNA characters) is far shorter than that for an English word of the same length (10^{11} English characters). This is because the alphabet has 26 letters to choose from, while DNA has only four kinds of nucleotides. To obtain a three-letter English word, one has to walk 20,000 random characters, whereas to obtain a specific three-letter DNA word, one has to walk only about 200 DNA characters.

We can conduct a computer experiment to verify our predictions. Let us simulate a random stream of 26 letters and the space, 3 billion (10^9) characters long (this length is 10 times longer than the expected mean length for a six-character word, and therefore, probabilistically, in this random sequence, any word of length six characters will occur). Let us look for the occurrence of some sentences from different English works (or some quotable quotes) in this simulated random sequence, with the only rule that the longest word in all these sentences should not consist of more than six characters. We shall also note the location where each sentence occurs for the first time in the whole random sequence. [11]

Figure 7.2 shows the location of the occurrence of these sentences. As we predicted, all the sentences we searched for occur within the random stream of English characters. The number of words and the length of the sentences are immaterial. The sentence G, containing 24 words, is three times longer than the sentence B, which is also found in the random sequence. As long as the longest word is under six characters, the upper limit we have set for this search, sentences with almost any length are found somewhere in the random sequence. Indeed, any sentence from any English work, and in fact any sentences that we can construct now or any of our descendent generations will ever construct in the future, will occur in this very same random sequence, as long as its longest word is six characters or fewer in length. This analogy powerfully illustrates that if we look for any sentence in word-pieces, it will occur in the random sequence — no matter how many words the sentence contains, and no matter how different are the sentences.[12]

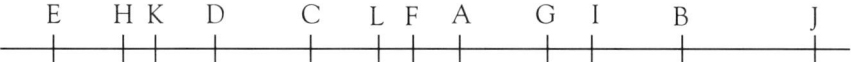

A — If music be the food of love, play on, give me excess of it.
— *William Shakespeare, Twelefth Night*

B — Chaos often breeds life, when order breeds habit. — Henry B. Adams

C — There are people who laugh to show their fine teeth;
and there are those who cry to show their good hearts. — Joseph Roux

D — It takes a clever man to turn cynic,
and a wise man to be clever enough not so. — Fannie Hurst

E — Die, my dear doctor! That's the last thing I shall do! — Lord Palmerston

F — God heals, and the doctor takes the fee. — Benjamin Franklin

G — There is no slave out of heaven like a loving woman;
and of all loving women,
there is no such slave as a mother. — Henry Ward Beecher

H — The mind is its own place, and in itself
Can make a Heaven of Hell, a Hell of Heaven. — John Milton

I — Love is the wisdom of the fool and folly of the wise. — Dr. Samuel Johnson

J — Home is the place where, when you have to go there,
They have to take you in. — Robert Frost

K — Do not do unto others as you would that they should do unto you.
Their tastes may not be the same. — George Bernard Shaw

L — Man is the head of the family,
woman is the neck that turns the head. — Chinese aphorism

FIGURE 7.2. FINDING MANY DISTINCT ENGLISH SENTENCES IN THE SAME RANDOM SEQUENCE OF ENGLISH CHARACTERS BY THE NEW METHOD. A random stream of 3 billion English characters was simulated in the computer. Many quotes, each of whose longest word contained six characters, were chosen. Each sentence shown in the figure was searched following the method described under Figure 7.1. The location of each sentence is marked on a line depicting the three billion character random stream of letters.

This is the same with genes. The probability of a gene increases far greatly when it is split into exons among which meaningless strings of introns can occur. No matter what information a gene may contain, and no matter what sequences exons may contain, all of a gene's exons will certainly exist in the long random DNA sequences of the primordial pond. Furthermore, if we can find one gene in a given random DNA sequence, then we can find any given gene in the same random sequence. There are indeed many more reasons that increases this probability to an even greater extent in the case of genes. We shall go into these particular details in the following sections.

It is evident from these extensive considerations that the traditional Shakespeare analogy is thus incorrect in the context of the genome. As we shall see later, the genes of the genome are inevitable in the vast USP, and the genome is certainly probable to be assembled from these genes randomly in the primordial pond.

The first genes were split genes and the first cells were eukaryotic cells

Computer analysis of DNA sequences reveals that the very first genes in the primordial pond were split into coding (exon) and intervening (intron) sequences

Traditional theory says that bacteria are more ancient than the eukaryotic single-celled organisms

Scientists have traditionally regarded bacteria (prokaryotes) as more primitive and more ancient than the single-celled eukaryotes because they are smaller and less complex. Eukaryotic cells are larger and contain a nucleus (the membranous sac within which all the cell's chromosomes are housed) and organelles that compartmentalize machineries for a number of specialized functions (see Genetics Primer). Examples of organelles are mitochondria (powerhouses that unleash the energy contained in food), and chloroplasts (which enclose a plant cell's photosynthesis machinery). The nucleus is the hallmark of the eukaryotic cell. In comparison, prokaryotes neither have a nucleus nor any organelles. They are also thousands of times smaller than typical eukaryotic cells. They almost always contain their genome (all their DNA material) in a single chromosome, while the eukaryotes typically have many chromosomes. Correspondingly, the size of the genome in a prokaryote is much smaller (~1–5 million nucleotides) than those of eukaryotes (several millions to billions of nucleotides). Prokaryotic genomes do not have "junk" DNA between genes; eukaryotic DNA has a significant amount of it. Finally, prokaryotes do not have introns within their genes while eukaryotic genes almost always contain them.

Scientists have traditionally theorized an "evolutionary tree" in which bacteria appeared first, followed by the more complex eukaryotic cells. Some scientists have suggested that eukaryotes were formed by "endosymbiosis," the unions of many different primitive bacteria.[13] The result of these unions, they believe, are eukaryotic organelles, such as mitochondria and chloroplasts. To them the nucleus of the eukaryotic cell was also formed as a result of engulfing one bacterium by another and subsequent modifications.

One ought to remember that, beyond speculation, there is absolutely no evidence whatsoever that the nucleus of the eukaryotic cell originated as a bacterium. Although there exists some resemblance between the mitochondrion and bacterial cells, the origin of the nucleus in the eukaryotic cell is still considered to be a total mystery. It is often said that the single most striking hallmark of a eukaryotic cell is its nucleus, whose origin is an absolute enigma. Thus, we must remember that it is purely an assumption in the evolutionary literature that the prokaryotes came first and that they combined to evolve single-celled eukaryotes. Likewise, it is purely an assumption that from the unicellular eukaryotes evolved one or a few multicellular organisms, and that from the first multicellular creature evolved all organisms on earth.

A New Hypothesis: The single celled eukaryotes are the first to have come on earth. The prokaryotic genes could only have originated from the genes of the eukaryotes by losing introns.

It is amazing to see the clear demarcation between the prokaryote and the eukaryote in almost all respects. When one views the vast difference in the structure of the genes in the prokaryote and the eukaryote, the question that would certainly come to mind is: why should there be introns in the eukaryotic genes and not in the prokaryotic genes? It is even more astounding especially when one thinks that, although the introns have no function at all, the proportion of introns in a gene is greater than 90%. Why should the genes of eukaryotes mostly consist of useless DNA material and how could the prokaryotic genes avoid them?

I addressed this question of why and when the introns originated in living organisms by asking the most fundamental question: How could genes have come into existence at all in the primordial pond? It was through computer simulation and analysis of DNA sequences that I recently demonstrated that genes must have randomly existed in primordial DNA which was present in large quantities in the primordial pond.[14] I have shown that only split genes (with exons separated by introns) could occur in a random sequence, and that a contiguous coding sequence of a gene is highly improbable in a random sequence — that is, introns are inevitable if genes occurred purely by chance in long random sequences in the primordial pond. Eukaryotic genes today resemble those primordial sequences: mostly junk, punctuated by short exons. Logically, therefore, all the eukaryotic genes must have come directly from the primordial pond, because almost all the eukaryotic genes occur with the typical split structure, whereas almost no prokaryotic gene occurs with such a structure. Since contiguously long prokaryotic genes were absolutely improbable to occur in the primordial sequences, the prokaryotic genes could not have directly come from the primordial genetic sequences; the contiguous genes of prokaryotes could only be derived from eukaryotic split genes by losing introns.[15] Let us now look into the computer analysis I carried out to demonstrate that eukaryotic cells were the first to appear on earth.

Probabilistic analysis of coding sequences using the computer shows that split genes, typical of eukaryotes, must have been the first genes in the primordial DNA sequences, and not the contiguous genes of prokaryotes

I began by asking why the unused stretches of DNA in genes, the introns, exist in eukaryotes and where they came from. The eukaryotic gene and the prokaryotic gene are clearly distinguishable based on their structure: The eukaryotic gene is split into many exons while prokaryotic gene is contiguous in the coding sequence. However, the length of almost all the exons (the split pieces of the gene's coding sequence) of eukaryotic genes are almost always shorter than 600 nucleotides. In prokaryotes, the length of the coding sequence of a typical gene does not have such an upper limit. It can be as high as nearly 10,000 contiguous nucleotides. However, the overall lengths of the complete coding sequences are roughly the same for both prokaryotes and eukaryotes. This information can be derived from the fact that the lengths of proteins are similarly distributed in both of them. The question I asked was: Why should the coding sequence of the gene be split into short pieces which are limited to 600 nucleotides in the eukaryotes whereas the prokaryotic gene is not at all split?

In order to find why eukaryotic genes are split into exons and introns, first of all we must try to find why and wherefrom the meaningless introns originated. Logically, the origin of introns should be very closely related to that of exons. Since the characteristics of all coding sequences (i.e., exons) are common to all organisms, the origin of exons could be explained by analyzing the earliest evolution of the structure of coding sequences in general.

It is reasonable to assume that the coding sequences for proteins were derived from the preexisting DNA sequences in the primordial soup, and not by construction from shorter coding sequences. If primordial DNA contained random nucleotide sequences, the next question is: Was there an upper limit in the coding sequence lengths, and if so, did this limit play a crucial role in the formation of the structural features of genes?

Consider that we are given a long random sequence and certain parameters concerning the sequence. The first is that it contains a random distribution of all four nucleotides (A, T, G, and C), which can be read in triplets such as ATG, CTT, GCA and so on. Out of the 64 possible triplets, or *codons*, three specify the stop signal for protein synthesis because they do not code for any amino acid — and these are called the *stop codons*. When codons are randomly distributed, what we get in a DNA sequence are very short *reading frames* (RFs) — linear sequences in which reading sequentially one codon at a time will lead to a contiguous chain of amino acids before being stopped by a stop codon.[16] The RFs range from zero nucleotides long to approximately 600 nucleotides (200 codons). Also, the shorter the reading frame the more frequently it appears. This kind of distribution is called a "negative exponential distribution" (see Figure 7.3).

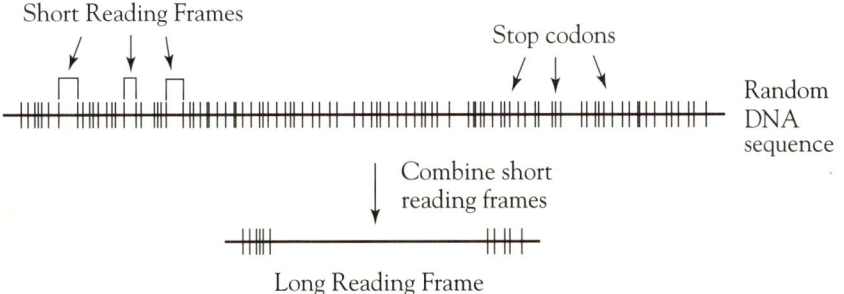

FIGURE 7.3. ONLY VERY SHORT READING FRAMES CAN EXIST IN A RANDOM DNA SEQUENCE. In a random DNA sequence, the stop codons are also randomly distributed. The property of this distribution is such that only short reading frames — the DNA sequence between consecutive stop codons occurring in the same phase when reading triplets — with a maximum of 600 nucleotides occur in a random sequence. Even if we walk for tremendous lengths in a random DNA sequence (e.g. thousands of trillions of nucleotides), still we do not find reading frames longer than about 600 nucleotides. The only way in which we can obtain a long reading frame is to combine some of the available reading frames in the random sequence. The top line depicts a random DNA sequence, and the tick marks denote the stop codons.

The zero RF means that two stop codons occur next to each other, and in a random sequence, stop codons appear frequently and close to each other. Greater than 95% of all random RFs are shorter than 100 nucleotides. Only one out of 1,000 RFs reach 500 nucleotides long. The important thing about such a distribution is that, whether we walk on a random sequence 1000 nucleotides long or one million nucleotides long, the reading frame length characteristics do not change much. This is true even if we extend our walk to 100 million nucleotides or 100 billion nucleotides.

With an understanding of these characteristics, let us ask the important question that pertains to the origin of life on earth. Life depends fundamentally upon the function of many thousands of proteins. A majority of these proteins are much longer than 200 amino acids, many reaching lengths over 3000. And to make a 3000-amino-acid protein from a DNA triplet code, you need 3000 codons, or 9000 nucleotides. Suppose we are given a long random DNA sequence of a billion characters with the characteristics we discussed above: the DNA has an inherent feature of restricting or limiting the length of the RF under an approximate maximum of 200 codons. Even if we increase the length of the random sequence to 1000 trillion characters (10^{15}), the upper limit increases very slightly, to only 220 codons. So, we know that because of the stringent length limit in the RFs, no protein can be synthesized that is longer than about 200 amino acids, no matter how long we walk in the many trillions of characters. Under such circumstances, if we are asked to come up with a RF of 500 codons or 1000 codons, how shall we arrive at it? There are only a few possible ways that one can think of under the given conditions.

1. We can take a fairly long reading frame, say the longest in all the thousand billion nucleotide sequence we are given, and eliminate the first stop codon. It obviously lengthens the RF only very slightly, say from 200 to 210 codons, because only too soon we arrive at another stop codon. Let us try to eliminate that also, but again we face the same problem. So, even to arrive at 400 codons from a 200 codon RF by this method, we have to specifically eliminate approximately 50 consecutive stop codons.
2. Another possible method is to take a fairly long RF, and recombine it with those which are as long as possible out of the available nearby RFs. This is done in a contiguous manner so that the recombined sequence forms a much longer contiguous RF. By this process, clusters of many stop codons are automatically eliminated at a stretch.

It seems that in order to arrive at a 1000-codon-long RF, instead of eliminating 200 consecutive stop codons individually, it may be easier to link about 10 or 15 different RFs together. The more important thing is that if such a thing had to happen in a real life situation, what was more likely? Perhaps the more reasonable approach would be to analyze what had actually happened and then to infer from it that that method was the more efficient or probable one under the given circumstances.

When we look at the real situation in split genes existing in eukaryotes, it becomes clear that the second method we discussed above is precisely what had been used in order to lengthen the RFs. Now, given the properties of the random sequence and the requirement for synthesizing long proteins from such a random sequence, we can reasonably conclude that we have solved the theoretical question as to how to arrive at them. In living cells, this is the linking of exons into genes. Instead of linking the DNA sequence itself, the DNA message is read into an RNA copy and the introns, containing clusters of stop codons, are spliced out of the RNA. The RNA molecule prior to splicing, which is a full copy of the gene with its complement of exons and introns, is called the "primary RNA."

Figure 7.4 describes how a gene coding for a typical protein could have simply occurred in the long primordial random DNA sequence, with no evolution from shorter coding sequences. We can confidently conclude that in the primordial pond, long random genetic sequences existed, and long proteins were synthesized from them by first linking the short RFs (exons) and making them into a long contiguous coding sequence for a protein.

What we have so far done is to simply look at the architecture of the typical eukaryotic gene and correlate it with what is possible from a random DNA sequence. By doing this, we found that the structure of the eukaryotic gene is precisely the same as that which can be obtained from a purely random DNA sequence in the primordial pond. However, we can analyze the eukaryotic DNA sequences compiled in the DNA sequence databanks, and verify if indeed it contains a random nucleotide sequence. Further, we can correlate the differences between the gene architectures of eukaryotes and prokaryotes by comparing them with purely random DNA sequences. We can begin to do this by directing the computer to generate a random DNA sequence of one million nucleotides in which all the four nucleotides A, T, C, and G have equal probabilities at any given position. We can then find out the lengths of all the possible reading frames in this sequence in all the three reading phases (note that there are three possible "phases" of reading a DNA sequence in steps of three nucleotides). The statistics of these random RF lengths can then be compared with those of the actual eukaryotic DNA sequences in the databanks.

When we plot the lengths of the reading frames against their frequencies in the million-nucleotide random sequence, the shortest RF, zero length, happens to be the most frequent. As the length of the RF increases,

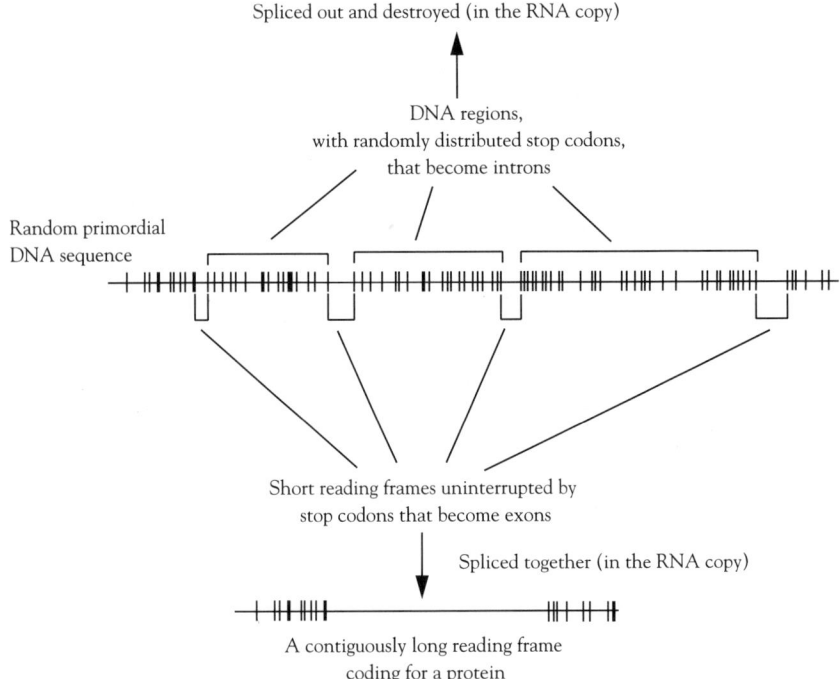

FIGURE 7.4. HOW WAS A GENE SELECTED FROM THE RANDOM PRIMORDIAL DNA SEQUENCE? In a random primordial DNA sequence, a random distribution of stop codons led to very short reading frames (with an upper length limit of 600 nucleotides) and long DNA regions with clusters of stop codons. The only way a gene longer than 600 nucleotides could originate was to select some short reading frames and splice them together consecutively (in the primary RNA copy, not shown in the figure), by editing out the intervening regions containing many stop codons. Such a splicing resulted in a long reading frame which could then code for a long protein. In today's biology, the short coding pieces which are spliced together are called the exons, and the intervening pieces, the introns.

the frequency decreases and the curve tails off to near zero frequency at a length of about 600 nucleotides (see Figure 7.5 C, left). As one can see, the reading frames are almost all smaller, usually much fewer than 600 nucleotides long. The curve is typical of a negative exponential distribution.

I then isolated the eukaryotic DNA sequences in GenBank,[17] and plotted a similar curve for reading frame lengths against their frequencies. The curve resulting from the eukaryotic DNA looks remarkably similar to the one obtained from the random DNA sequence, that is, a negative exponential

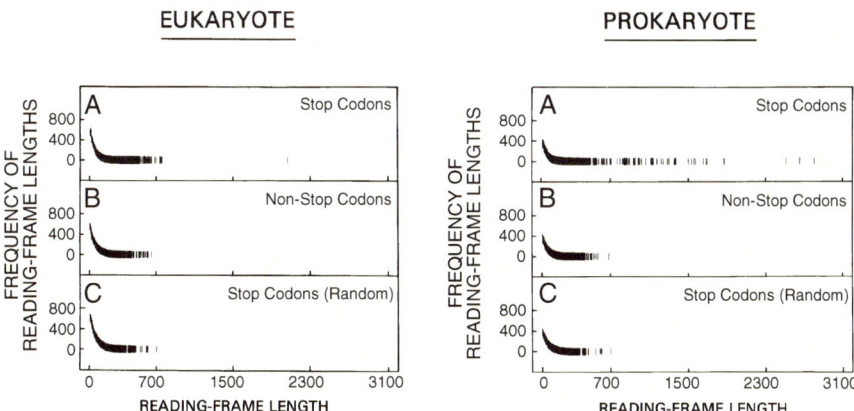

FIGURE 7.5. DIFFERENCES IN THE FREQUENCY DISTRIBUTIONS OF READING FRAME LENGTHS IN EUKARYOTIC AND PROKARYOTIC DNA SEQUENCES. The lengths of reading frames in the eukaryotic and prokaryotic DNA sequences (from a total of about one million sequence characters) in the DNA sequence databank (GenBank) were computed using the computer. The frequencies of these lengths were plotted against the reading frame lengths. (A) Distribution of reading-frame lengths between stop codons found in actual DNA sequences. (B) Distribution of RF lengths between non-stop codons in actual sequences. This is a control study that shows how the non-stop codons are distributed. (C) Distribution of RF lengths between stop codons in a computer-generated random sequence (about one million nucleotides long).

distribution. Again, the frequencies of reading frame lengths tails off at a maximum of about 600 nucleotides (Figure 7.5 A, left). Both these characteristics indicate that the eukaryotic DNA is indeed random in its sequence. This agreement between the lengths of protein-coding sequences generated randomly and those that actually exist in eukaryotic organisms tells us clearly that genes must have occurred in random sequences in the primordial pond in short pieces interrupted by noncoding sequences — and that eukaryotic genes were directly selected from such genes. As we predicted before, the coding sequences became longer by linking or splicing the best of the consecutively available short coding segments (reading frames) in the primordial DNA into longer split genes. Note that all the codons other than the stop codons also are randomly distributed in the eukaryotic DNA (Figure 7.5 B, left).

In essence, according to my interpretation, long genes were generated from random sequences of the primordial DNA by turning the long noncoding sections between the short original coding sections into introns, which could be "spliced out" from a primary RNA during the formation of messenger RNA. The genes that occurred in the USP, therefore, were split into coding (exon) and noncoding (intron) sequences, just like in today's eukaryotic DNA sequences. The successive splicing of the consecutive exons would produce a long contiguous coding sequence, theoretically of any length. Therefore, a gene could occur at a stretch in a random sequence, but in short successive coding pieces interrupted by meaningless intron sequences. For some reason, this architecture was maintained in the DNA, and the editing occurred only in the RNA copies of the genes.

The prokaryotic gene must have been derived from the typical eukaryotic split gene by losing introns

The story of the prokaryotic DNA is different. The curve of its reading frame lengths plotted against their frequencies shows that the frequencies only tail off to zero after a few thousand nucleotides (Figure 7.5A, right), indicating a nonrandom distribution of stop codons. It indicates that the prokaryotic DNA sequence is unlike the random sequence. Probabilistic calculation shows that it is highly improbable for the long reading frames typical of the prokaryotic genes to occur even in an unreasonably long random sequence.

How then could the long prokaryotic genes arise? Once the eukaryotic cells were formed, some of these cells containing split genes could have begun to lose introns.[18] This would have resulted in the protein-coding sections of DNA being joined up without introns, forming the genes of prokaryotes. So rather than being primitive forms of life, bacteria may be indeed more advanced than eukaryotes as far as the structure of their genes is concerned.

On the contrary, it is also possible in the primordial pond that the split genes from the UGP might have lost their introns before the first cells came into being, from which the prokaryotic genomes could have been directly assembled, without a need for a nucleus (see below). Thus it is possible for the prokaryotic genome to have been derived directly from contiguous genes in the open primordial pond.

A small digression is appropriate here. Everything so far has been reasonable, and clearly correlatable with what must have actually gone on in the primordial pond. But while all the processes in the primordial pond are supposed to be random, the process such as that for linking the RFs, as that in the RNA splicing process, appears to be quite directed. How is it possible? In analyzing this, we are actually coming to the fundamental question of how any molecular or genetic process or mechanism can come into existence at all in the primordial pond. As we saw in Chapter 6, random chemical processes link small molecules into macromolecules. Among many different kinds of macromolecules, there are random interactions. Some of these macromolecular interactions form more complex structures with many macromolecules in them. Now, there can be numerous such distinct kinds of structures possible (such as the combinations of many RNA and protein molecules, called ribonucleoproteins) when we consider the vast quantities of molecules even in a small portion of the primordial pond, and a very long geological time.[19] Such large structures will randomly interact with one another, but some will have stronger and more stable interactions than others. This sort of argument can be extended to show that while most interactions will lead to nothing interesting or biologically meaningful, some would lead to something biologically useful — such as the enzyme DNA polymerase interacting with the DNA nucleotides. Such are the processes that lead to genetic machineries and ultimately a living cell. In essence, it is possible, by purely random processes, to build complex machineries and living cells, which only on the surface appears to be a directed process. The RNA-splicing process that we are discussing now is just one such machinery that was the outcome of the random process. It has thus aided in bringing together purely random sequence pieces into coding sequences encoding biologically meaningful proteins, not only to build single celled organisms, but also to construct the more complex multicellular animals on earth with their repertoire of long proteins.

The very first cells were highly complex eukaryotic cells with a nucleus

By asking these fundamental questions, it is possible to find out not only why and wherefrom the meaningless and useless intron sequences originated in the genes of eukaryotic cells, but also why the nucleus of the eukaryotic cells originated. Moreover, we can extend this question and ask why the prokaryotic cells do not have a nucleus. If we can find clear reasons for this demarcation, it will uncover the most fundamental aspect of the history and biology of living cells.

Probabilistically the intron sequences are expected to be very long. This is because long reading frames (even within the upper limit of 600 nucleotides) do not occur often in a random sequence. Furthermore, the right combination of splicing signals also do not occur frequently in a random sequence.[20] This in fact is what we find in eukaryotic genes. So, the primary RNA, from which the splicing machinery must edit out the introns, is very long — sometimes as long as one million nucleotides. When the introns are removed, the resulting mRNA is far shorter — a maximum of approximately

10,000 nucleotides is so far known in the living world. By correlating these distinctions with the biology of transcription, splicing, and translation, we can formulate some reasons why these processes are compartmentalized in the nucleus and cytoplasm of eukaryotes. This analysis reveals the most fundamental reasons for the origin of the nucleus itself in the eukaryotic cell.

The nuclear boundary originated in the first cells in the primordial pond in order to segregate the extremely long primary RNA, with very long useless introns, from being unnecessarily translated by the ribosomes

The chromosomes reside in the nucleus. The gene transcription (of DNA into RNA) occurs in the nucleus. The primary RNA is spliced within the nucleus before entering the cytoplasm, thereby "editing out" the very long useless introns within the nucleus itself. Therefore, the ribosomes outside the nucleus never see the primary RNA; they only find the mRNA containing only the protein-coding message. Why is there this kind of compartmentalized operation? Why is the primary RNA not transported first out into the cytoplasm and then spliced? Why can't the ribosome be present in the nucleus, translating mRNAs there?

When we consider what will happen if the protein-synthesizing machinery is not separated from the primary RNA in the eukaryote, we can see that the ribosomes present in the same environment would start to translate them. Because the primary RNA molecules contain all the translation initiation signals, the ribosomes would not be able to distinguish between the primary unspliced RNA and the spliced mRNA. At any given time within a cell, thousands of genes are transcribed and translated into proteins. Imagine, then, what would happen if there is no nucleus separating the ribosomes and the primary RNA. A sort of a chaotic mess would certainly result. The ribosomes will attempt to unnecessarily and wastefully translate almost all the primary RNAs, each of which is ten times longer than its mRNA counterpart. This will tremendously reduce the probability of the genuine, spliced mRNAs being translated. Such an enormous and unnecessary waste is not a simple burden on a cell.

It would clearly be advantageous to separate the unspliced RNA from the protein-synthesizing machinery, allowing only the spliced RNA to be translated. It is most probably for this principal reason that the nucleus originated in the first cells in the primordial pond. It is the best way to specifically present the clean, edited mRNA copy of the gene to the ribosomes. Therefore it can be seen logically that the first cells were typically eukaryotes (see Figure 7.6). The fact that splicing always occurs in today's eukaryotic cell within the nucleus before the RNA is transported to the cytoplasm corroborates this concept. Moreover, nuclear compartmentalization does not

Separation of Unspliced RNA and Ribosomes in the Very First Cell by a Nuclear Boundary

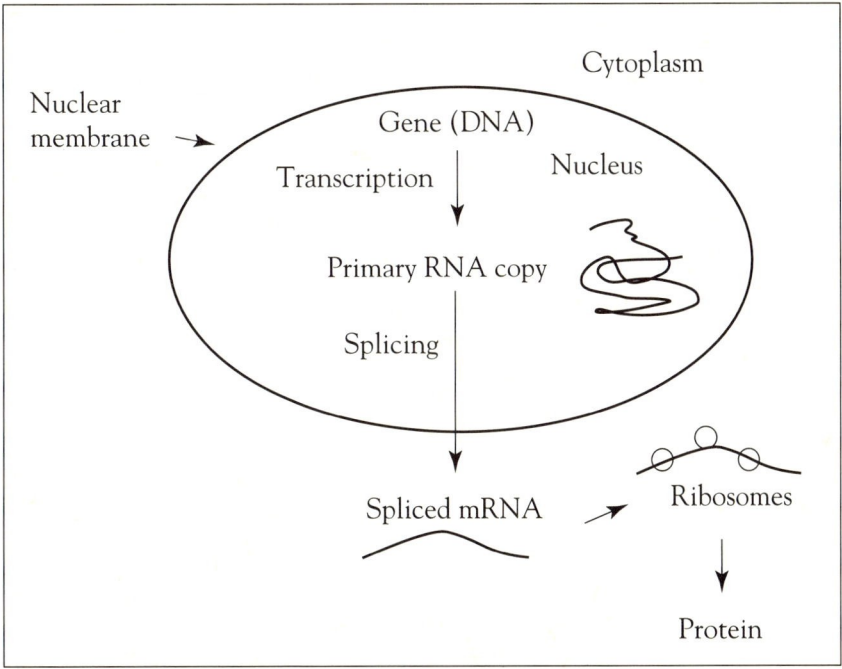

FIGURE 7.6. WHY DID A NUCLEUS ORIGINATE IN THE VERY FIRST CELLS? According to my new theory on the origin of introns, the key reason for the origin of the nucleus in the cell is to keep the ribosomes from translating the unspliced primary RNA. Without the nuclear boundary, there is nothing that would stop the ribosomes from acting on the primary RNAs of all the genes that are expressed in a cell, which would create a colossal waste. There is no other known function for the presence of the nucleus. Therefore, the nucleus must have originated in the first cells when they were formed from the genomes assembled directly from the split genes in the primordial pond. Thus, the very first cells were typical eukaryotic cells.

seem to have any major function other than separating the ribosomes from acting on the long primary RNAs. The fact that the RNA synthesis temporarily stops when the nuclear membrane itself dissolves during the DNA replication phase during cell division indicates that my concept must be correct.[21]

The nucleus that came with the unicellular eukaryotes was lost by the prokaryotes

When we look at the structure of the typical gene in prokaryotic cells, it is found already in the spliced form. That means, logically, there is no need for the separation of DNA from protein-synthesizing machinery (ribosomes) by a nuclear membrane, as in a eukaryotic cell. Therefore, it seems that when introns were lost from the genes of the primitive single-celled eukaryotes, the cells also lost their nuclear boundary and became nonnucleated cells — the prokaryotes (see Figure 7.7). The presence and absence of the introns in the eukaryotic and prokaryotic genes, and the corresponding presence and absence of the nucleus in their cells are logically correlated — the introns are the primary reason for the existence of a nucleus. If no introns are present in the RNA directly transcribed from DNA, then there is no need for segregating the ribosomes. In fact, it is amazing to see in the prokaryotes that, while the mRNA is still being synthesized from the DNA, the ribosomes start to bind and translate the still-growing mRNA.

Another explanation for the absence of a nucleus in the prokaryote is that the intron loss from the split genes could have happened in the primordial pond, instead of inside the eukaryotic cells. From a large population of intron-less genes in the primordial pond, an assembly of a complete set of genes with the ability to form a living cell could have formed the prokaryotic genomes, and thus the prokaryotic cells, without need for the nucleus (Figure 7.7).

Interestingly, *prokaryote* means "before nucleus" and *eukaryote* means "with nucleus." These terms were derived from the traditional belief that prokaryotes were the first cells, and that from the prokaryotes the eukaryotes evolved by gaining a nucleus. These terms can be changed to reflect the reality based on our new concepts — *eukaryote* can be retained to represent cells with nucleus and *postkaryotes* or *akaryotes* can be used to represent prokaryotes which came later.

THE PRESENCE OF STOP CODONS IN SPLICE JUNCTIONS: A STRONG SECOND LINE OF SUPPORT TO THE EUKARYOTE-FIRST THEORY

The intron is the entity that is "edited out" from a gene during the splicing process. The junctions between the intron and the exons on either side are

THE ABUNDANT OCCURRENCE OF GENES IN THE PRIMORDIAL POND 243

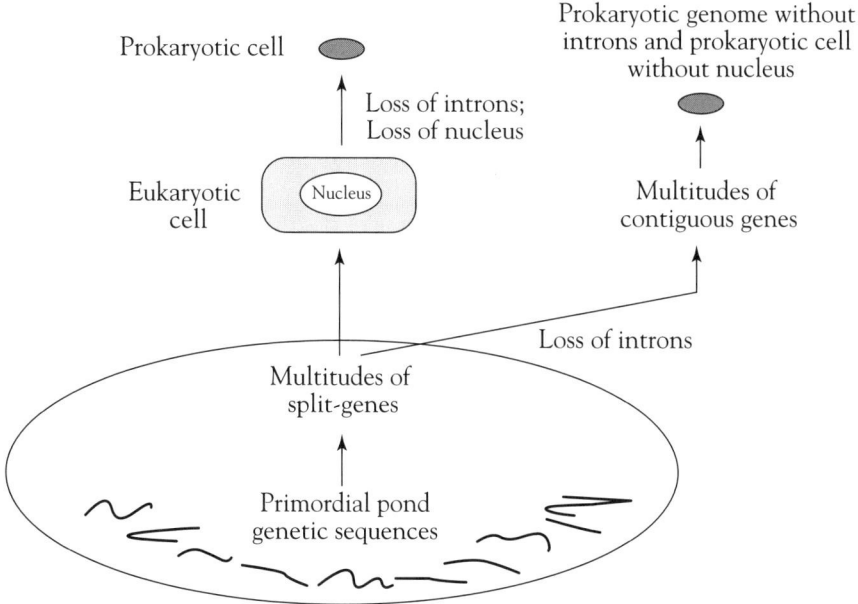

FIGURE 7.7. THE ORIGIN OF THE EUKARYOTIC CELL AND THE PROKARYOTIC CELL IN THE PRIMORDIAL POND. Genes could occur in the primordial pond only with a split structure of exons and introns. The conglomeration of genes gave rise to a genome that could develop a viable cell. The nucleus within the cell had originated when the first cells came into being from the primordial pond due to reasons described under Figure 7.6. From the split genes, the introns could be lost by copying back the genes' spliced RNAs (mRNAs) into DNAs by an enzyme called reverse transcriptase — thereby creating intron-less genes. This coupled with the loss of a nucleus in a eukaryotic cell would lead to a prokaryotic cell. Alternatively, the intron loss could happen directly from the split genes in the primordial pond, where a cell without a nucleus could be constructed.

called the splice junctions — the one on the "left" side of the intron is called the 5' splice junction and the one on the "right" side is called the 3' splice junction (this terminology stems from the convention that a printed DNA sequence is read from left to right, from what is termed the 5' end to the 3' end). These splice junctions between exons and introns are short but highly conserved, meaning that nearly the same sequence occurs at every junction in all the genes in almost all the organisms (see Genetics Primer for details). A sequence of nine nucleotides is highly conserved at the 5' splice sites. The

3' splice sites also exhibit a highly conserved sequence of four nucleotides, preceded by a region rich in C or T. These short conserved sequences are an essential part of the process of exon splicing and provide a specific molecular signal to the RNA splicing machinery in order to identify the precise splice points. Why are they present at all in genes? How did they originate? Understanding the mechanism by which these signals originated may reveal their biological meaning and throw light on the mechanisms of origin of the gene itself.

If my theory on the origin of introns in eukaryotic genes is correct — that the split genes originally occurred in the random primordial DNA sequences and were randomly selected in the assembly of the genomes of eukaryotic cells — then splice junction sequences must also have been selected at that time. If, as we have so far discussed, splicing mechanisms came into being primarily for removing stop codons thereby lengthening the coding sequences, it is possible that the splice junction signals were closely associated with stop codons.

Figures 7.3 and 7.4 show the distribution of stop codons in a random DNA or RNA sequence. We can see that the stop codons are frequently clustered together, and only rarely they are somewhat farther apart. The reading frames, the sequences between two successively occurring stop codons, are thus most often short. The longer a reading frame, the rarer it becomes, the longest being approximately 200 codons. If a gene encoding a fairly long protein of a 1000 amino acids or more has to exist in a random sequence, then, under these circumstances, it can occur only in short pieces which could then be spliced together into a single, long contiguous message. This is what we discussed before to understand why genes are split. Let us look into this process a little bit deeper and ask whether there is any connection between this and the presence of the splice junctions in present-day genes.

If exons represented reading frames and introns represented sequences with clusters of stop codons in the random sequences of the primordial pond, then logically there must have been a system which distinguished between the exons and the introns. This system must have been primarily able to distinguish between what is a reading frame and what is a stop codon. Suppose we are asked to read a random sequence, starting from a given point, and to identify the reading frames and to connect a few relatively long reading frames occurring consecutively. How can we do it? The most obvious way is to read from the starting point, looking for a stop codon. Once it occurs we know that that is the end of the reading frame. More likely than not, there will be a cluster of stop codons soon after the first occurrence; we can skip this cluster, and when we find another fairly long reading frame, we can continue to read again. Remember that in a random distribution of stop codons, even a reading frame of 50 nucleotides is relatively long.

When we encounter a stop codon, we mark it as the end of the reading frame (the exon) and the beginning of the intron. Therefore, every intron begins with a stop codon (see Figure 7.8). If genes were indeed selected by this process from random sequences in the primordial pond, then such a phenomenon of having a stop codon at the intron beginning must exist in the genes of all organisms. If the structure of the genes of organisms did not change much over the eons through which each organism has lived, then, today in living organisms the same phenomenon should exist. We can test if our predictions are correct by simply analyzing the gene sequences of today's living organisms available in the DNA sequence databanks (e.g., GenBank).

I examined the codon frequencies around the 5' splice junctions, which showed amazingly that all three stop codons occur with very high frequency on the intron side, just one nucleotide from the splice point.[22] Out of a total of 1030 introns I examined, 726 contain stop codons at the second nucleotide position in the intron.[23] The expected frequency of a random occurrence of stop codons is 3/64, which is only 55 out of 1030 sites. My theory of the stop codon scanning mechanism would predict stop codons immediately after the 5' splice points, and, in fact, they do appear there. These facts strongly corroborate my hypothesis that introns in eukaryotic genes originated directly from the random primordial DNA sequences — and that the mechanism that identified genes consecutively selected its successive exons by looking for stop codons while reading a random sequence from 5' to 3'. Furthermore, it is also clear that the splice junction sequences which contain these stop codons must have originated due to these reasons, and serve as molecular signals for the exon-splicing process. Thus, by fundamentally analyzing the process by which a gene could have originated in random primordial DNA sequences eons ago, I could find a reason why there are "conserved" splice junction sequences in the genes of today's living organisms, as well as the meaning of these sequences.

Our predictions on the origin of introns and the split structure of genes are further strengthened by the fact that the stop codons exist in the splice junction signals of only the protein-coding genes which are copied into the messenger RNAs (mRNAs) for proteins. Genes coding for the other kinds of RNAs which are not messengers for proteins (such as the ribosomal RNA used to build the ribosome and the transfer RNAs which aid in translation of the mRNA) also contain introns and use the splicing process. But these do not have stop codons in their splice junctions. The consistent presence of stop codons in the splice junctions of only the protein-coding genes illustrates that stop codons have played a role in the origin of splice junction signal sequences only in protein-coding genes, the only places where stop codons are meaningful.

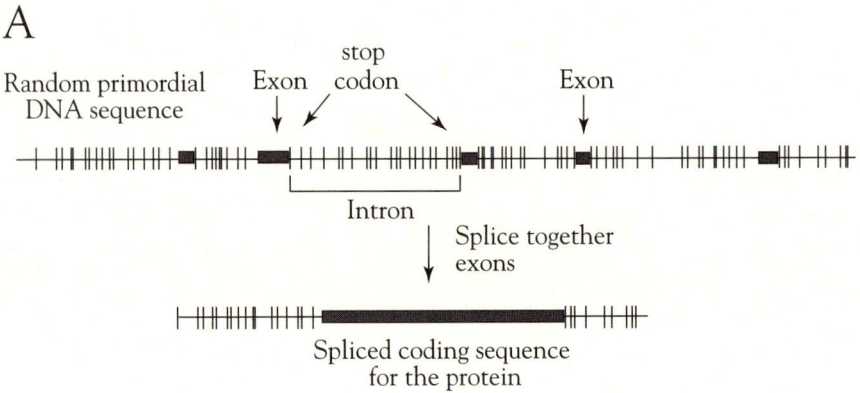

FIGURE 7.8. WHY STOP CODONS SHOULD BE PRESENT AT THE BEGINNING OF ALMOST ALL INTRONS. (A) In selecting the exons of a split gene from a random primordial DNA sequence, whatever machinery that did this should be capable of searching for stop codons (tick marks) to identify regions without stop codons (in the primary RNA copy, not shown), which are reading frames. In doing so, the first encountered stop codon in such a search will be marked as the beginning of the intron. This process will lead to the presence of a stop codon at the beginning of almost all introns. Sometimes, all of a reading frame is chosen to be an exon, because of which the end of the previous intron will have a stop codon. (B) The beginning and the end of the intron are part of what are called the "splice junction sequences." The stop codons are shown with a grey background.

In addition to the splice junction sequences which aid in splicing the exons of the primary RNA of the protein-coding genes, there are a few other sequences which are involved. For instance, a short stretch of conserved nucleotides present in introns near the 3' splice junction has been shown to aid the splicing process (see Genetics Primer). This is called the "lariat signal." Nomi Harris[24] and I carried out a systematic study of this signal sequence in numerous genes and found a consistent presence of stop codons in it. The polyA signal, a sequence which aids in the translation of the mRNA, also contains stop codons in it. Thus, the evolution of the whole RNA processing mechanism seems to make these stop codons the focal points for RNA processing.

It is truly amazing that all our predictions about the structure of genes present in the primordial sequences eons ago can be verified from the gene sequences of today's living organisms. It indeed means that our overall predictions about the manner in which genes could originate on earth — the presence of random genetic sequences in the primordial pond, origin of split genes directly in these random sequences, and so on — are all correct. There exists unmistakable and undeniable evidence for our concept that introns and split genes originated directly in random sequences in the primordial pond. There were absolutely no contiguous genes (as found today in prokaryotes) in the primordial pond. Each gene in the primordial pond occurred with a split structure of exons and introns, just as they are found in the genes of animals and plants living today. In essence, our predictions regarding the origin of the genes of complex organisms, the animals and the plants, directly in the primordial pond, and their structure, are clearly and unequivocally provable from genetic analyses of today's living organisms.

My theory as to how the split genes originated in the primordial pond is well accepted by scientists. For instance, Colin F. Blake from the University of Oxford in England, who is a well-reputed scientist in protein structure and function and who is a proponent of the exon shuffling theory, states that the problem of the origin of the split gene is comprehensively explained by my theory in his article "Proteins, exons, and molecular evolution," in the recent book, *Intervening Sequences in Evolution and Development*.[25, 26] In fact he states that this theory not only explains the origin of the introns and the split architecture of the eukaryotic genes, but also gives a reason for the origin of the splicing machinery (see his quote in Chapter 4, page 147).[27]

WHY THE FIRST GENES IN THE PRIMORDIAL POND MUST HAVE BEEN LONG, NOT SHORT, AND WHY THE FIRST CELLS MUST HAVE BEEN COMPLEX, NOT SIMPLE

Our foregoing discussions indicate that the very first genes must have been quite long, just as they occur in today's eukaryotic cells. They need not have been short coding sequences which later duplicated or recombined to form longer genes. It is clear that these genes must have occurred in the vast USP and were selected for the assembly of the genomes. Only if long DNA molecules, large numbers of genes, and all the cellular machineries for transcribing, splicing, and translating genes were available in the primordial soup (much before the first cells came about), then, the formation of complex cells from them was possible. Logic not only predicts but demands this. My studies indicate that as a rule, only if splicing had come into being first, then long genes capable of coding for long proteins could have been possible. Because there was a stringent upper limit of about 200 codons in the length of all the exons available in primordial sequences, without the splicing of these exons, no protein longer than 200 amino acids could be coded. Looking at the fact that all living entities — even the simplest bacteria or the bacterial viruses — have a majority of proteins longer than 200 amino acids, in fact as long as 3000 amino acids,[28] it is obvious that splicing must have originated in the primordial soup, before the very first cell could ever start to live. (In other words, no living cell exists with all its proteins under a length of 200 amino acids. Therefore, it can be said that no living cell could come into being only with proteins shorter than 200 amino acids.)

By simple logic, when a complex system such as that of splicing could come into being, then it is at least equally probable for the origin of the simpler systems of transcription and translation in the primordial soup. All these basic systems must have come into being before the organization of the first living cells. Without these machineries a living cell cannot be assembled. Thus the very first cells could have been very highly complex. They were comprised of a full complement of split genes, a fully complex splicing machinery, and a nucleus to house all the chromosomes. It is possible that they also contained other organelles such as mitochondria and chloroplasts, in which they compartmentalized some special functions. When a very highly complex cell with a nucleus could be formed by a large genome containing an extremely large number of genes from the USP, it is quite probable to form an organelle with a far smaller set of genes directly from the primordial pond's genetic sequences. I am convinced that this is what had happened.

It is truly not necessary to be bogged down by the traditional feeling that simple things should appear first on earth and only then complex things could be derived from them. In fact in our new view, once long DNAs existed in the primordial pond to the extent that complete genes were available, complexity will inevitably ensue. It is the inherent nature of this concept that, once such a thing is possible in the primordial pond, there is absolutely no need for simple molecules and cells to come about and then evolve into complex molecules and cells. It is an all-or-none-law that if complete genes existed in the random primordial DNA sequences, then complex machineries and complex cells should be formed; if complete genes did not exist, then no living cells will ever be formed.

Let us shed our traditional beliefs of always going from simplicity to complexity. Let us open our minds to the rationale of complexity first and simplicity next.

THE EVIDENCE THAT THE VERY FIRST CELLS MUST HAVE BEEN TYPICALLY EUKARYOTIC SUPPORTS GREATLY THE THEORY OF THE INDEPENDENT BIRTH OF ORGANISMS

The possibility that eukaryotic genomes and eukaryotic cells could arise directly in the primordial pond strongly corroborates the theory of the independent birth of organisms from the primordial pond. With regard to the nature and organization of split genes and intergenic sequences, the genome of the multicellular eukaryotes is quite similar to that of a unicellular eukaryote. As we shall discuss in Chapter 8, the structures of the genes and genomes of all the animals, simple or complex, are quite similar. The number of genes and the complexity of genetic networks even in the unicellular eukaryote are not far different from those in multicellular animals. Consequently, assembling a genome of a multicellular eukaryote is not far more difficult than assembling the genome of a unicellular eukaryote. Multitudes of eukaryotic genomes and cells can therefore arise directly in the primordial pond independent of each other.

What are the implications if we follow the traditional belief that prokaryotes somehow originated on earth, from which single-celled eukaryotes evolved, and in turn from which one or a few multicellular organisms came about — which then were the basis for Darwin's theory of the diversification of organisms? We would have absolutely no way of showing that the genes and genomes of various animals and plants can directly arise from the primordial pond's genetic sequences through a prokaryotic origin. Thus, only because we have destroyed this belief and have shown that the genes of eukaryotes are the ones that directly originated in the primordial pond first, from which the prokaryotic genes must have been derived, we are able to show that the genomes of various animals and plants could have directly originated in the primordial pond. This is a critical finding for the theory of the independent birth of organisms.

Note that the distinction between the gene structures of the eukaryotes and prokaryotes was discovered only in 1978, and sufficient amounts of gene sequence data became available only in the 1980s. Thus, it was perhaps impossible until the 1980s to determine the ultimate structures of genes, correlate them with the primordial genes, and to solve these fundamental problems. Only because of the availability of genetic sequence information, I was able to show that the eukaryotes originated first, and that "simpler," intron-free prokaryotic genomes are in fact more advanced and more efficient than those of eukaryotes. In my opinion, without this fundamental finding about the genes which are central to life, it would be impossible to show that multicellular animals and plants could have directly originated in the primordial pond.

All our foregoing analyses also show that an assumption that we made earlier in our discussions is absolutely correct. The analyses of the distribution of all the codons including stop codons prove that the DNA sequences in the genomes of eukaryotes are random. Furthermore, my study regarding the splice junction signals in eukaryotic sequences also indicated their random distribution.[29] The results of all these analyses culminate in the conclusion that the eukaryotic genome is truly a random sequence in which genes are embedded — showing that our original assumption, that the genes and the genomes of organisms directly originated in the primordial pond's random DNA sequences, must be correct. It also shows that the random character of these sequences has changed little over the hundreds of millions of years of the life of these organisms.

The inevitability of the occurrence of multitudes of split genes in the universal sequence pool

As we saw above, the foremost principle that makes possible the independent birth of myriads of organisms is that the very first cells were typical eukaryotic cells whose genomes were directly assembled from split genes in the primordial pond. This makes it possible for "seed cells" with different genomes to arise directly from the primordial pond which could independently give rise to different organisms. But this is possible only if complete split genes, with all their characteristics found in the living animals (we shall hereafter call such genes "real" or "actual" genes), were available in the primordial soup. We shall show here that genes are not only probable but they are inevitable in the universal sequence pool (USP) of the primordial pond. So far we have shown that eukaryotic genes arose directly from the primordial DNA sequences. We have not yet shown that real genes, which are assumed to be highly evolved, could have indeed existed in the vast primordial DNA sequences with all their sequence and structural characteristics — and coded for the proteins of the living organisms with all their complex biochemical functions. In fact I shall illustrate in the following that the split genes present in animals and plants living today could have actually occurred randomly in the universal sequence pool in abundance.

Can a real gene with all its functional characteristics simply exist in the universal sequence pool?

We have been discussing that the universal sequence pool of random DNA sequences totalling approximately to 10^{30} nucleotides could have existed in the primordial ponds. Let us now ask: Can a real gene (either contiguous like in the prokaryotic genome or split as in eukaryotic genome) occur in a random sequence 10^{30} nucleotides long? If we superficially analyze the probabilities based on the expected mean length of a random sequence in which a gene coding 300 amino acids (900 nucleotides) can occur, without taking into account the various features of a gene, then this notion would seem preposterous. By such an approach one is led to see that the probability of a gene for a contiguous 900-nucleotide coding sequence to appear in a random sequence is one in 4^{900}, and that the expected mean length for such a random sequence (approximately 10^{540} nucleotides) is far too long to have any meaning. This is because, as we saw before, nucleotide sequences longer than 10^{78} cannot exist even if all the matter in the universe is converted into a DNA molecule. However, if we analyze this problem deeply, by taking into account the many structural and functional aspects of genes, several reasons are unearthed which explain why it is highly probable for a typical eukaryotic gene to have occurred even in a far smaller USP of the primordial pond than that we have estimated to be approximately 10^{30} nucleotides in length. These principles offer tremendous support for the theory of the independent birth of organisms. These analyses also indicate that many structural and functional features of the typical eukaryotic genome in today's living organisms originated directly from the universal sequence pool.

We shall uncover and demonstrate several important principles that cumulatively make it a realistic, in fact, an inevitable probability for eukaryotic genes to occur abundantly in the universal sequence pool, permitting the assembly of genomes which give rise to the birth of myriads of creatures. These principles are: 1) split genes occur in a random sequence with far more probability than contiguous genes; 2) the degeneracy of codons (many different codons are functionally interchangeable) greatly increases the probability for genes; 3) a codon with higher degeneracy will occur more frequently in a random DNA sequence, and this also increases the probability of a given amino acid sequence to be encoded in a random DNA sequence;[30] 4) a very high degeneracy of amino acids (many amino acids in proteins are functionally interchangeable) tremendously increases the probability of gene occurrence in the USP; and 5) the phenomenon of "negative exponential distribution" of sequence waiting intervals makes the occurrence of eukaryotic split genes in the USP, with realistic gene lengths found in today's living animals, extremely probable.

TREMENDOUS INCREASE IN THE PROBABILITY OF A GENE BY THE SPLITTING OF ITS CODING SEQUENCE INTO SHORTER PIECES

THE PROBABILITY FOR THE OCCURRENCE OF A SPLIT GENE IS FAR HIGHER THAN THAT OF A GENE WITH AN UNSPLIT STRUCTURE

As we saw before, the length of a random sequence in which a specific gene of 900 nucleotides can be expected to occur by chance is 10^{540} nucleotides, a size which cannot exist. Therefore it is even more improbable for the occurrence of contiguous genes with coding sequence lengths of 3000-6000 nucleotides, required to code for proteins 1000 or 2000 amino acids long. (Prokaryotic genes contain contiguous coding sequences of up to about 10,000 nucleotides. For this reason, it is absolutely out of question that the prokaryotic genes could have occurred in the primordial DNA.)

We must remember that although eukaryotes code for proteins as long as those of the prokaryotes, the eukaryotic genes are split into small pieces (exons). The pieces have an upper length limit of approximately 600 nucleotides (200 codons). Thus, we are dealing with far shorter coding sequences than that is required for the actual lengths of protein sequences in eukaryotes. In fact a majority of exons are much shorter than this 600-nucleotide limit, on the order of 100–200 nucleotides. The mean length of the random sequence in which the coding lengths of exons, in contrast to the complete coding sequence, would occur would then be reduced tremendously.

Let us take a short hypothetical eukaryotic gene (70 nucleotides in length), containing five exons with their lengths 17, 10, 9, 20 and 15 nucleotides in the order of their occurrence (see Figure 7.9). As we discussed before for the English sentence, the probability of finding this gene is essentially the same as the probability of finding the longest exon (20 nt), that is, the least probable one of all these exons. Thus, the expected mean length of the random sequence for finding this gene would be approximately 4^{20} (10^{12}) nucleotides. Note that it is not 4^{70} (10^{42}) nucleotides, the expected mean length for the contiguous coding sequence of all 70 nucleotides. This calculation shows that there exists a tremendous difference in the probability

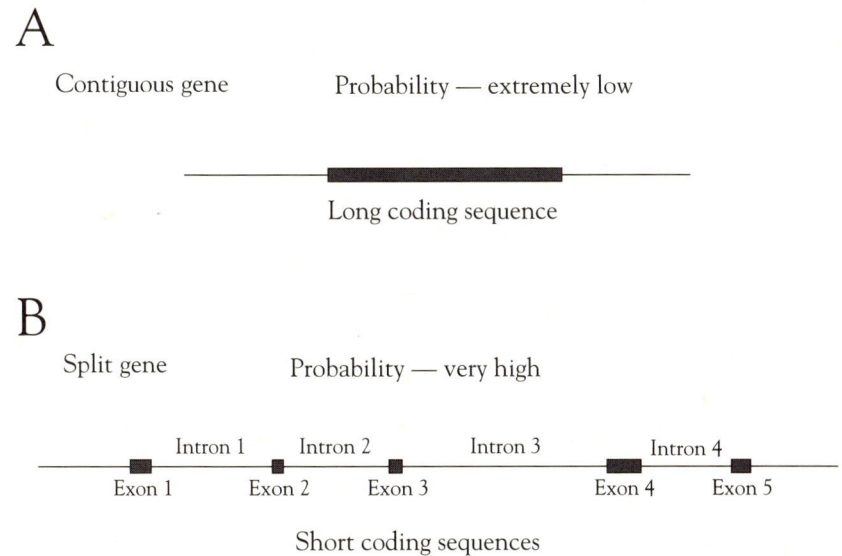

FIGURE 7.9. THE PROBABILITY FOR A GENE IS EXTREMELY HIGH WHEN IT IS SPLIT INTO MANY EXONS. (A) The probability for a contiguous gene, say 70 nucleotides long, is 4^{-70}. The length of the random sequence in which it is expected to occur once is 10^{42} nucleotides. (B) In contrast, when this gene is split into many exons, say 5 exons with 17, 10, 9, 20 and 15 nucleotides consecutively, the probability for the gene is essentially that of the longest exon (20 nucleotides). The probability for this split gene is thus 4^{-20} and its expected mean length is reduced tremendously to 10^{12} nucleotides.

of a split gene (1 in 10^{12}) compared to that of a contiguous gene (1 in 10^{42}). The difference is magnified greatly when we consider long, real genes.

In almost all the eukaryotic genes, the longest exon is approximately 600 nucleotides. The expected mean length for this is 10^{360} nucleotides. Look at the expected mean length for a 5000-nucleotide contiguous coding sequence, which is 10^{3000}. What a stupendous increase in the probability for the gene, that is associated with the decrease in the expected mean length, when it occurs in pieces. However, a 10^{360}-nucleotide sequence is still far longer than any of the realistic limits of the USP in the primordial ponds. But, as will be discussed in the coming sections, there are four additional phenomena that tremendously reduce this expected mean length and thus greatly increase the probability of finding a real gene in the primordial pond's universal sequence pool with an approximate total length of only 10^{30} nucleotides!

THE DEGENERACY OF CODONS IN GENES, AND THE DEGENERACY OF AMINO ACIDS IN PROTEINS: THE TWO CRUCIAL PHENOMENA THAT IMMENSELY INCREASE THE PROBABILITY OF EXONS

THE CODING SEQUENCE OF A GENE FOR A GIVEN PROTEIN IS NOT AT ALL RIGID. IT IS HIGHLY VARIABLE FOR MANY DIFFERENT REASONS

The genome is expressed largely through the set of proteins it encodes

How does a genome express itself into an organism that carries out a variety of physical, physiological, and mental functions? At the most fundamental level, an organism is manifested mainly by the characteristics of its proteins. The genomic DNA contains the coded messages for proteins for the most part as a "read-only memory," and expresses different subsets of proteins at appropriate times and locations in the body during the development of the animal. It is the proteins that have unique three-dimensional structures and specific functions which enable the building and functioning of organisms.

The important characteristics of a protein, with regard to its role in a living cell or animal, are only its structure and function — not its precise sequence.

Living things need only a set of biochemical functions mostly carried out by proteins. To achieve this, life does not need proteins with rigid sequences. It is crucial to remember that it is the biochemical *functions* that are fundamental to life and not the exact protein sequences.

What, therefore, are we looking for in a protein when we are trying to understand the life of an organism? Not its size, shape, amino acid sequence, or length, but its basic enzymatic or structural function. That is the bottom line. We do not care what the different parameters of a protein are, as long as it specifies its given function. The structural and functional parameters of a protein required in an organism may be rigid, but a protein does not require a unique amino acid sequence or size to exhibit a given structure or function. As it turns out, trillions and trillions of varying sequences of a protein can specify exactly the same biochemical function. We shall return to this point in more detail later.

This background information lets us take a look at the primordial pond's universal sequence pool from a high vantage point and ask: Are there nucleotide sequences in the USP that code for a protein that would specify a given enzymatic or structural function? Remember that we are not asking if the USP contains an exact, invariant, DNA sequence coding for an exact amino acid sequence of a protein. First, if many entirely different genes can code for the same biochemical function, then any one of these is acceptable as an authentic gene for this structure or function. Second, the sequence of a particular protein that specifies a particular function is not that rigid, that is, if the amino acids at several positions of the protein are highly degenerate, then the nucleotide sequence at these corresponding coding positions can be almost random. Keep in mind that in this case we are alluding to a specific protein with many sequence variations. These two principles make it possible for a vast number of gene sequences to specify exactly the same biochemical function. This radically increases the probability of finding a gene in the USP for a given structure or function. Furthermore, we need not find a gene as a contiguous sequence; it is enough if we find it in very small exon pieces. This again catapults the probability of a gene's occurrence in the USP to the level of inevitability.

THE DEGENERACY OF THE CODON, AND THE HIGH FREQUENCY OF DEGENERATE CODONS IN A GENE: A PHENOMENON WHICH TREMENDOUSLY CONTRIBUTES TO THE PROBABILITY FOR THE OCCURRENCE OF A PROTEIN-CODING GENE IN A RANDOM DNA SEQUENCE

The degeneracy of the codons contributes to a high probability of a given gene in a random sequence

There are 64 codons, of which 61 code for amino acids (see Genetics Primer). This means that most of the 20 different amino acids (18 in fact) are coded for by more than one codon. For example, the amino acid *serine* is coded by six codons, TCT, TCC, TCA, TCG, AGT and AGC, and the amino acid *proline* is coded by four codons, CCT, CCC, CCA, CCG. Only *tryptophane* and *methionine* are coded by single codons, TGG and ATG, respectively. Because the probability of any codon occurring in a random sequence is 1/64 (0.015), the probability of *serine* (which has six codons) at a given position in a protein is six out of 64 (0.09). This is an increase by six fold in just one codon position in a random sequence. Figure 7.10 shows

Amino acid Sequence:	Arg–	Pro–	Tyr–	Ser–	Gly–	Met
Degenerate Codons:	CGT CGC CGA CGG AGA AGG	CCT CCC CCA CCG	TAT TAC	TCT TCC TCA TCG AGT AGC	GGT GGC GGA GGG	ATG
# degenerate codons	6	4	2	6	4	1
Probability:	.09	.06	.03	.09	.06	.015

Examples of possible DNA sequences:

    CGTCCTTATTCTGGTATG
    CGCCCCTACTCCGGCATG
    CGACCATATTCAGGAATG
    CGGCCGTACTCGGGGATG
    AGGCCCTATTCTGGTATG
    AGACCTTATAGTGGAATG
    CGTCCCTATAGTGGCATG
    CGACCGTACTCCGGTATG

FIGURE 7.10. THE DEGENERACY OF CODONS INCREASES THE PROBABILITY FOR A GENE IN THE PRIMORDIAL POND. When more than one codon codes for an amino acid, they are called degenerate codons for that amino acid. The degenerate codons for every amino acid in the example polypeptide are shown. Any codon for the first amino acid can be combined with any codon for the next amino acid, and so on, and any DNA sequence formed by this combination can code for the same polypeptide chain. Out of the numerous coding sequences possible from such combinations, a few examples are given at the bottom. The number of such possible sequences can be calculated using the formula $CD_1 \times CD_2 \times ... CD_n$, where $CD_1 ... CD_n$ specifies the codon degeneracy at each position. In this case there are 6 x 4 x 2 x 6 x 4 x 1 = 1152 possible DNA sequences that can code for exactly the same amino acid sequence.

how an amino acid sequence *arg-pro-tyr-ser-gly-met* can be coded by an extremely large number of different DNA sequences.

The importance of this degeneracy is realized when we consider that the probability of an amino acid sequence is the product of the probabilities of the individual amino acids in a random sequence. That is, the probability for the sequence *met-met* is 1/64 x 1/64 (2.25×10^{-4}), and the probability for the sequence *ser-ser* is 6/64 x 6/64 (8×10^{-3}). The sequence *ser-ser* is 36 times more likely to appear than *met-met*. The difference between the probability of *met-trp-met-trp* (both *met* and *trp* are each coded by one codon) and *arg-ser-leu-arg* (all coded by six codons) is 1296. This clearly illustrates how greatly the probability of a given gene is increased by the occurrence of degenerate codons in a gene which code for proteins hundreds of amino acids long. The difference would be truly stupendous. Because 61 codons in DNA code for all the 20 amino acids in proteins, there is an average of three possible codons at every position coding for the same amino acid. Thus, if we compute based on an average of three degenerate codons at every position, the increase in probability for a 200 amino acid protein — due to the degeneracy of the codons alone compared to invariant codons — is well over 10^{100}.[31]

The frequency of occurrence of a degenerate codon in a random DNA sequence is proportional to the extent of the degeneracy

If we analyze a random sequence, all the 64 codons are randomly distributed, and *Ser* codons will occur six times more frequently than *Met* codons. Thus, in a long protein coded by a random DNA sequence, for every 61 amino acids, on average, six *Ser*, six *Arg,* and one *Met* will occur. Therefore, there will be an unequal proportion of the 20 amino acids in that protein. This means that the degenerate codon positions are the most frequent and the nondegenerate codon positions are the least frequent in a random DNA sequence. These frequencies are predictable.

If genes were available in the random sequences in the primordial pond and were selected in the first living cells, then their coding sequences would also have been random. This would be reflected in the frequency of the 20 amino acids in the population of proteins coded from the random sequence. Therefore proteins coded from the random DNA sequence would contain the predicted unequal proportion of the 20 amino acids. If it is true that the genes of the living cells were directly selected from the primordial pond's random sequences as I have proposed, then this can be tested by checking if the predicted proportion of the 20 amino acids exists in the natural proteins of today's living organisms. In fact, astonishingly, the frequency of amino acids in natural proteins is nearly the same as expected in a population of proteins coded from a random DNA sequence (Table 7.1). This is determined

	Predicted frequency of amino acids in proteins	Actual Frequency of amino acids in proteins
Met	1.6	2.1
Trp	1.6	1.4
Asn	3.3	4.9
Asp	3.3	5.2
Cys	3.3	2.3
Glu	3.3	6.7
Gln	3.3	4.6
His	3.3	2.3
Lys	3.3	5.7
Phe	3.3	3.7
Tyr	3.3	3.1
Ile	4.9	4.5
Ala	6.6	6.9
Gly	6.6	7.4
Pro	6.6	6.1
Thr	6.6	5.8
Val	6.6	6.2
Arg	9.8	5.4
Leu	9.8	9.2
Ser	9.8	7.9

TABLE 7.1. THE PREDICTED FREQUENCY OF THE 20 DIFFERENT AMINO ACIDS CODED FROM A RANDOM DNA SEQUENCE IS ESSENTIALLY FOUND IN PROTEINS OF TODAY'S LIVING ORGANISMS. Because different numbers of codons code for each of the 20 amino acids, the frequency of the amino acids coded by a random DNA sequence is predictable, as shown in the table. Interestingly, essentially the same frequencies of amino acids are found in the proteins of all animals and plants living today, except for small variations and exceptions. The frequencies are computed from the National Biomedical Research Foundation's protein sequence database. (Courtesy of Dr. Daniel Haft, Protein Information Resource, National Biomedical Research Foundation, Washington, D.C.)

by counting the occurrence of each amino acid in a large number of protein sequences from the National Biomedical Research Foundation's Protein Sequence database. In fact, this has been shown previously.[32,33] Moreover, when all the 64 codons are counted in the eukaryotic DNA sequences contained in the GenBank database, I found that all the codons, except for a reduced frequency of CG-containing codons, are generally distributed as expected in a random sequence. This is further evidence that the DNA sequences of genes in today's living eukaryotic organisms are random.

It is clear that the probability for the occurrence of a given exon in the random sequences in the primordial pond is enormously increased thanks to codon degeneracy and the frequency of degenerate codons in random sequences (see Figure 7.11).

A VERY HIGH DEGENERACY OF AMINO ACIDS IN PROTEINS INCREASES GREATLY THE PROBABILITY OF THE OCCURRENCE OF GENES IN THE UNIVERSAL SEQUENCE POOL

Another extremely important phenomenon that contributes to the probability of finding a gene sequence for a protein in the USP is the degeneracy of amino acids in proteins. In fact, the contribution made by this phenomenon is comparatively far greater than that made by the degeneracy of codons.

Many amino acid sequences can specify exactly the same function, so long as those particular amino acid positions specifying and controlling the protein function remain constant or change only to certain other specific amino acids — such positions are very small in number in the whole protein.[34] In most positions, there can be a variety of amino acid changes tolerated by the protein without affecting the protein's function. A typical protein functions using its active site, the small portion of the protein that actually carries out the biochemical function. It usually consists of only a few amino acids, but they are positioned in a precise three-dimensional configuration — usually at different places in the linear protein chain (see Genetics Primer). All the rest of the amino acids in the protein chain only aid to bring about the overall three-dimensional structure to the protein so that the active-site amino acids are projected in the required structure. As we shall see below, most amino acids in the protein chain can vary a great deal without affecting the protein's overall three-dimensional structure, or the active site.

If the change of an amino acid at a given position in a protein to another amino acid will not affect the function of the protein, then that amino acid position is said to be degenerate (as opposed to the degeneracy of codons). Sometimes an amino acid can be changed to one of several amino

				AGG
				AGA
			GCG	CGG
			GCA	CGA
		TAT	GCC	CGC
Codons:	ATG	TAC	GCT	CGT
Amino acid:	Met	Tyr	Ala	Arg

FIGURE 7.11. THE AMINO ACIDS HAVING MORE DEGENERATE CODONS OCCUR MORE FREQUENTLY IN PROTEINS. The frequency of each of the 20 amino acids in a protein in living organisms is proportional to its number of codons in the codon table. For example, the amino acid Met occurs only once for every 61 amino acids in a protein, whereas, Tyr occurs twice, Ala occurs four times and Arg occurs six times. At each of these positions, a corresponding number of degenerate codons can code for that amino acid. The total number of variable codons for every 61 amino acid positions is 235. This increases the codon degeneracy from 3.05 (61/20) to 3.85 (235/61). This phenomenon considerably increases the probability for a protein in a random DNA sequence in the primordial pond.

acids without affecting the protein function. If an amino acid can be changed to only one other amino acid without affecting the protein function, that is, even if the degeneracy at a given amino acid position in a protein sequence is low, it still makes the nucleotide sequence of the gene, coding this amino acid position, more nearly random, especially when combined with codon degeneracy. This seems to be the norm in most regions of most of the natural proteins.

In order to understand amino acid degeneracy, let us look at the sequence of -Met-Glu-Pro-Arg-Ala- as an example (Figure 7.12). Consider that at the Glu position another amino acid Trp or His can occur; at Arg, Leu can occur; and at Ala, Gly, Ser or Gln can occur. Now the sequence becomes -Met-{Glu/Trp/His}-Pro-{Arg/Leu}-{Ala/Gly/Ser/Gln}. The notation used here means that any of the amino acids occurring within the { } is acceptable at that position. In reality, more amino acids than shown in this example can occur at a given degenerate amino acid position, making the corresponding nucleotide sequence at that position almost random.

Protein sequence: Met–Glu–Pro–Arg–Ala
                           | | |
Allowed amino acid   Trp   Leu Gly
variations: His   Ser
                                          Gln

Examples of possible amino acid sequences:
Met–His–Pro–Arg–Ala
Met–Trp–Pro–Leu–Gly
Met–Glu–Pro–Arg–Ser
Met–His–Pro–Leu–Gln

FIGURE 7.12. THE AMINO ACID DEGENERACY OF A PROTEIN. Only at very few positions (~10%) in the overall three-dimensional structure of a protein are particular amino acids required for function. In all the other positions, each amino acid can be varied to a number of other amino acids without changing the function. The ability of the protein to change in its amino acid sequence without changing its structure or function is called the amino acid degeneracy of the protein. In the hypothetical example shown, the second, fourth and fifth positions are degenerate positions, and the first and the third are invariant positions. At the second position, for example, any of the amino acids Glu, Trp or His can occur. Met at the first position and Pro at the third cannot be changed to any other amino acid without disrupting the protein's function. The number of various possible amino acid sequences that can specify the same protein function can be calculated by using the formula $AD_1 \times AD_2 \times ... AD_n$, where AD_1, AD_2, etc. denotes the amino acid degeneracy at each position. In this case, it is $1 \times 3 \times 1 \times 2 \times 4 = 24$ possible amino acid sequences.

The high degeneracy of amino acids in proteins: The structure and function of a protein are tolerant to a great deal of sequence variation

There are two main approaches to studying the degeneracy of amino acids in a protein. The first method is to study the same protein in different organisms traditionally considered to be evolutionarily related. The traditional idea is that through evolution, a protein changes in amino acid sequence due to mutations while retaining its biological activity. In the second method, genetic approaches are used to introduce amino acid changes at specific positions in a protein sequence, by changing the nucleotide sequence at corresponding positions in a cloned gene. From these variations, functional sequences are identified based on biological tests. Several proteins which can be expressed in bacteria or yeast by using appropriate genetic manipulations have been studied by this approach.

Both methods reveal a list of different amino acid sequences possible in a given protein which have the same function. These can now be compared and analyzed to identify sequence features that are essential for biological function. If a particular property of an amino acid is important at a given position of the sequence, then it will be revealed by discovering what kinds of amino acids are permissible at that position.

These studies have revealed that proteins are surprisingly tolerant of amino acid variations. For example, Miller and associates, while studying the effects of approximately 1500 amino acid changes at 142 positions in *lac repressor*,[35, 36] found that about half of all variations do not affect the biochemical function of the protein. Widely differing substitutions were allowed at some positions, while no substitutions or only conservative substitutions were allowed at other positions.

The invariant or conserved amino acids play important roles in maintaining both structure and function in proteins. Usually, amino acids that are directly involved in protein functions, such as binding or catalysis, will be highly conserved. For example, when the amino acid Asn, which binds DNA, is changed to Asp in the λ repressor protein,[34] a substantial loss of its activity occurs. Also, in addition to amino acids that are directly involved in function, those that are required for structure stability can also be equally important.

Examples of amino acid degeneracy in known proteins that do not alter structure and function

An example of the allowed sequence variations for a short region in λ repressor is given in Figure 7.13.[34, 37] Out of 17 positions only three are completely invariant. Three others are relatively invariant. The rest tolerate a wide range of amino acid variations. It is to be noted that the invariant or highly conserved amino acid positions are buried inside the protein in its three-dimensional structure, while most of the highly varying positions are situated on the outer surfaces of the protein. This may indicate that most of the structural information in this region of the protein is carried by amino acids that are buried.

FIGURE 7.13. AMINO ACID DEGENERACY IN A SHORT REGION OF λ REPRESSOR PROTEIN.[34] The sequence of λ repressor in a short region of 17 amino acids is shown. The amino acids that can be varied at the different positions are shown above each position. Only positions 2, 4 and 10 are invariant. At all other positions, each amino acid can be changed to any one of the amino acids shown above, without altering the structure and activity of the λ repressor. [Adapted from Bowie, J.U., et al, *Science*, 247:1306, with permission. Copyright 1990 by the AAAS.]

An extremely large number of amino acid sequences will specify the same structure and function of a protein

The essence of our discussion in this section is that the amino acid sequences in proteins can be highly degenerate, and only at certain key positions specifying structural or functional information, such as binding or catalysis, amino acids are invariant. This would tell us that a stupendously immense number of different protein sequences can specify the same structure and function. The number of possible different amino acid sequences that can code for the 17-amino-acid portion of the λ repressor protein shown in Figure 7.13 can be computed by simply multiplying the number of variable amino acids at each position, which is 5.6×10^{11}. Imagine this when proteins in living systems which are far longer are considered; for instance, the number of possible sequences in a 500-amino-acid protein, with a similar degree of amino acid variation, is about 10^{400}. Remember that any one of these amino acid sequences of a protein can specify exactly the same biochemical function.

This phenomenon can be coupled with codon degeneracy and discontinuity of genes made from short exons. All these principles cumulatively have a tremendous impact on the probability for the occurrence of genes in the USP. The expected mean length for the occurrence of a gene correspondingly reduces tremendously from the 10^{360} nucleotides (that we estimated for the longest exon of about 600 nucleotides) to well under the length of the USP available in the primordial pond, that is, 10^{30} nucleotides. Therefore, the USP is vast enough for the occurrence of a gene in it. It is clear that what originally appeared to be highly preposterous — the occurrence of complete exons in the primordial genetic sequences — must have been indeed inevitable.

The tremendous cumulative effects of codon degeneracy and amino acid degeneracy on the probability of the occurrence of genes in a random sequence: An example analysis

We can compare the probability of an exact DNA sequence coding for a specific amino acid sequence in a given protein to the probability of a DNA sequence when all the possible codon and amino acid variations are allowed in it in such a way that it would still code for a protein with the same biochemical function.

Let us take a DNA sequence, 24 nucleotides long, that can code for a stretch of eight amino acids in a protein (shown in Figure 7.14A). What is the probability for the occurrence of this sequence in a purely random DNA sequence? It is one in 4^{24} or about one in 10^{15}. The expected mean length for finding this DNA sequence is about 10^{15} nucleotides. Because of the degeneracy of the codons (see Figure 7.14B), the number of possible 24-nt-long DNA sequences that can code for the given eight-amino-acid sequence is very large (10,368 possible sequences). Therefore, the probability of finding any one of them increases heavily, to about one in 10^{11}. This is a 10,000-fold difference in the probabilities due to the allowance of codon degeneracy.

Now, considering the additional variability allowed by amino acid degeneracy (Figure 7.14C), the number of possible amino acid sequences that will specify the same protein function is increased by about three million.[38] And this, combined with codon degeneracy, results in the probability of finding this 24-nt "gene" in a random DNA sequence becoming only one in 5840. And correspondingly, the expected mean length in which any one of these 24-nt DNA sequences can be found is also 5840 nucleotides. What does it mean? It means that if we walk only about 5840 nucleotides in a random DNA sequence, we can find a 24-nt DNA

A. The probability of an invariant DNA sequence of a gene

Protein: Ser-Ile-Ala-Arg-Glu-Ile-Tyr-Glu $P = 3.5 \times 10^{-15}$
DNA: TCC ATA GCT CGA GAA ATC TAT GAG $EML = 2.8 \times 10^{14}$ nts

B. Effect of only Codon Degeneracy on the probability of the gene

Protein: Ser-Ile-Ala-Arg-Glu-Ile-Tyr-Glu
DNA: TCC ATT GCT CGA GAA ATC TAT GAG
Degenerate TCT ATC GCC CGT GAG ATT TAC GAA
Codons: TCA ATA GCA CGC   ATA
         TCG   GCG CGG
         AGT   AGA
         AGC   AGG

# Variable codons:	6	3	4	6	2	3	2	2	$P = 3.6 \times 10^{-11}$
Probability:	.09	.04	.06	.09	.03	.04	.03	.03	$EML = 2.7 \times 10^{10}$ nts

C. Effect of Codon Degeneracy and Amino acid Degeneracy on the probability of the gene

```
              Arg              Arg  Asp
              Lys        Arg   Gln  Gln
              Asp   Lys  Lys   Glu  Ser
              Gln   Gln  Gln   Ser  Thr
              Asn   Asn  Asn   Thr  Tyr
              Glu   His  His   Cys  Gly
              His   Ser  Ser   Gly  Ala
              Tyr   Thr  Thr   Ala  Met
              Thr Lys  Gly  Gly    Trp  Trp
              Cys Cys  Met  Met    Leu  Leu
Amino Acid    Gly Met  Leu  Leu    Val  Phe
Degeneracy    Ala Leu  Ser  Val Val  Ile  Ile
```

Protein: Ser-Ile-Ala-Arg-Glu-Ile-Tyr-Glu
DNA: TCC ATT GCT CGA GAA ATC TAT GAG

# Variable Codons:	40	14	10	39	41	3	46	39	$P = 1.7 \times 10^{-4}$
Probability:	.62	.21	.15	.61	.64	.05	.72	.61	$EML = 5840$ nts

FIGURE 7.14. THE TREMENDOUS INCREASE IN THE PROBABILITY OF A GENE IN THE PRIMORDIAL RANDOM SEQUENCE BY THE CUMULATIVE EFFECTS OF CODON DEGENERACY AND AMINO ACID DEGENERACY. The probability for a fixed DNA sequence coding for an eight-amino-acid portion of the protein λ repressor is compared with that when codon degeneracy as well as amino acid degeneracy are allowed. (A) The probability and expected mean length for the fixed 24-nucleotide sequence; (B) when codon degeneracy alone is permitted; and (C) when both codon degeneracy and amino acid degeneracy are allowed. The variable codons are not shown in C. The expected mean length of the fixed 24-nucleotide DNA sequence is 2.8×10^{14} nts; with codon degeneracy alone it is 2.7×10^{10} nts; and when amino acid degeneracy is also taken into account, it is a mere 5840 nucleotides. (P = Probability; EML = Expected Mean Length.)

sequence that will code for an eight-amino-acid sequence which will still specify the λ-repressor function.[39,40]

So if the DNA sequence of this eight amino acid protein were rigid, the length (expected mean length) of DNA needed to find this particular sequence would be 2.8×10^{14} nucleotides. But when we consider the degeneracy of codons and the degeneracy of amino acids, the length needed to find this protein will be only 5840 nucleotides!

In the above example, the effect due to the split structure of gene in the increase of the probability of a gene has still not been taken into account. As we shall see later, when the coding sequence is split into exons and introns, the probability of the complete gene is the same as the probability of the longest exon. Since the longest exon in all the genes in most living organisms is 600 nucleotides (with some exceptions), the probability of genes of any length is the same as the probability for a 600-nucleotide exon. Taking into account of the overall amino acid degeneracy of proteins, we can compute the expected mean length for a 600 nucleotide exon coding for a 200-amino-acid portion of a protein to be approximately 10^{20} nucleotides. It is interesting to note that this amount of DNA is less than that found in the body of just one human individual. From these extensive series of observations, is it not amazing that what appeared to be highly preposterous at the beginning of our discussion, that complete exons can occur in the USP, must in fact have been an inevitable outcome in the primordial pond?

INTRON LENGTHS AS FOUND IN THE GENES OF TODAY'S ANIMALS AND PLANTS ARE PROBABLE IN THE GENES OF PRIMORDIAL POND ONLY DUE TO THE NEGATIVE EXPONENTIAL DISTRIBUTION OF SEQUENCE WAITING INTERVALS

"NEGATIVE EXPONENTIAL DISTRIBUTION" OF SEQUENCE WAITING INTERVALS: A POWERFUL PHENOMENON THAT ENABLES THE OCCURRENCE OF SPLIT GENES WITH REALISTIC LENGTHS IN THE UNIVERSAL SEQUENCE POOL

We have unequivocally demonstrated that exons are possible, and in fact inevitable, within the total length of the USP. But a gene contains many exons, the splicing of which results in a long contiguous message coding for a specific function. Therefore, our next question is: Can we find multiple exons of a long gene closely spaced in a random sequence, as we find in the genes of today's living animals? We shall see that the answer is a definite yes — the phenomenon of negative exponential distribution is responsible for this, which enables the occurrence of any two given sequences very closely to each other in a random sequence. Because of this phenomenon, the exons of a given gene will be found closely spaced in the primordial pond's long DNA sequences, and the length of the complete gene including introns will be similar to those of eukaryotic genes in today's organisms.

Although we saw that a coding sequence split into smaller sections is far more probable in the USP than if it is not, we have not asked how far away the consecutive pieces of the coding sequence (exons) can be separated, and whether these distances are biologically meaningful. In other words, are the lengths of introns in the primordial genes comparable to those in the present-day living organisms? Consider, for instance, a gene with five exons. We can search for the sequence of these five exons consecutively as they occur in the gene in a long simulated random sequence, without paying attention to the nature or the length of the sequences that intervene them. Suppose we find that the complete gene with all its five exons occurs in a random sequence of 50 million characters. The intron lengths average 10 million nucleotides. Is this meaningful? Certainly not — because real genes are usually under 200,000 nts long, the longest known so far being a little over 1,000,000 nts. Therefore, only if a gene conforms to these lengths, would it be biologically meaningful.

Here is where the phenomenon of negative exponential distribution of *sequence waiting intervals* comes into the picture. Although we have discussed this kind of distribution before in our analysis of the exon-intron structure of eukaryotic genes, it would be advantageous at this juncture to describe some of the principles involved in a more basic manner.

The sequence waiting interval

What is a sequence waiting interval? In the sequence ATGTACGTAC TCTAGTGCTAGTA, the first occurrence of G is at the 3rd position (see Figure 7.15). The second occurrence is at 7th position, and the third at the 15th position. The interval between the first occurrence and the second occurrence (four nucleotides) is a waiting interval of G for its successive repetition. Similarly, the second waiting interval is thus eight nucleotides. The sequence TG occurs at the 2nd and 16th positions, and the waiting interval is 13 nucleotides; this means we have to wait 13 nucleotides for the second occurrence of TG from its first occurrence. This can be applied to any specified sequence, and also for the waiting intervals between different sequences. For instance, the waiting interval between ATG and TAG is 10 nucleotides.

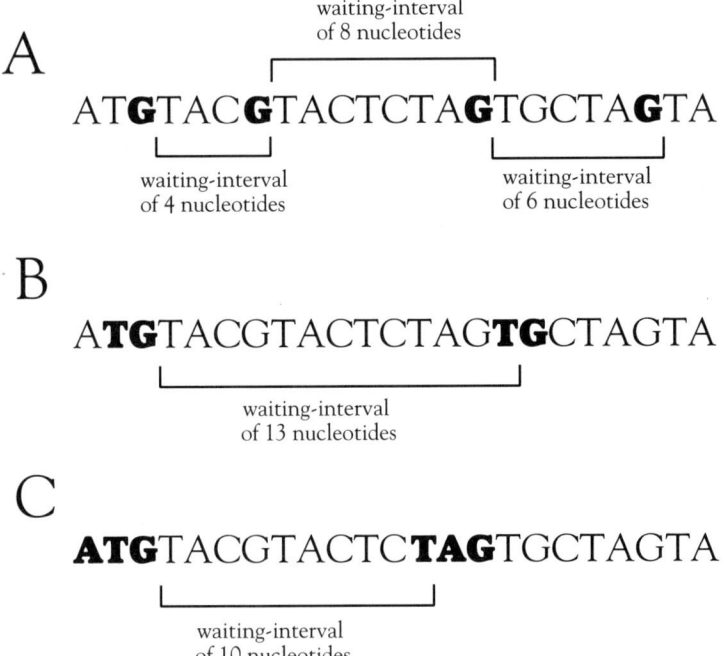

FIGURE 7.15. WHAT IS A SEQUENCE WAITING INTERVAL? The distance in number of nucleotides from one occurrence of a given subsequence to the next consecutive occurrence is called the sequence waiting interval for that subsequence. The subsequence can be either a nucleotide or a sequence of nucleotides. (A) The waiting intervals between the successive occurrences of the nucleotide G. (B) The waiting interval between the successive occurrences of the dinucleotide TG. (C) The waiting interval between the successive occurrences of ATG and TAG.

The properties of sequence waiting intervals for a given individual nucleotide: A simple example

Based on probability theory, we can derive several principles concerning random sequences. When one carefully considers the successive repetitions of a particular sequence element, one finds that the shortest waiting interval is the *most frequent* of all possible intervals. At first glance, this seems counterintuitive. However, when we look at it closely, we shall see that this is to be expected simply based on the property of distribution of waiting intervals.

In order to easily understand the properties of sequence waiting intervals, let us consider the distribution of one of the four nucleotides in a random sequence. If, for convenience, we assume that each of the four nucleotides occurs with equal probability, then the probability of occurrence of a given nucleotide, say A, is 1/4. The probability of a nucleotide not being A (i.e., being C, G or T) is 3/4. Based on this, the probability of two successive 'A's being separated by any number of nucleotides that are not A can be calculated as shown in Figure 7.16.

By looking at this figure, we can see that if we consider only the series of subsequences bounded by two successive 'A's in a random sequence, the most probable in the series is AA, the next most probable is AXA (where X is any nucleotide other than A), the next most probable is AXXA, and so on. The greater the waiting interval separating the two successive repetitions of A, the less probable it is that this waiting interval will occur in a random sequence. If we plot the probability that we thus calculate in this table against the corresponding waiting interval, the plot (Figure 7.17) decreases in an "exponential" manner. This distribution is what we call the "negative exponential" distribution. Therefore, AA is the most frequent when compared individually with any other element in the series AX_nA, where n can be any number from one to infinity and X is any nucleotide other than A.

Sequence series	Waiting intervals (t)	Probability of t
AA	0	1/4
AXA	1	$(3/4)^1 \times 1/4$ = 3/16
AXXA	2	$(3/4)^2 \times 1/4$ = 9/64
AXXXA	3	$(3/4)^3 \times 1/4$ = 27/256

FIGURE 7.16. PROBABILITIES OF INCREASING WAITING INTERVALS BETWEEN TWO SUCCESSIVE REPETITIONS OF 'A'. The sequence length between two successive 'A's separated by non-A nucleotides (G, C, or T - denoted by X in column 1) is called the waiting interval between the two successive 'A's. As this increases, the probability decreases.

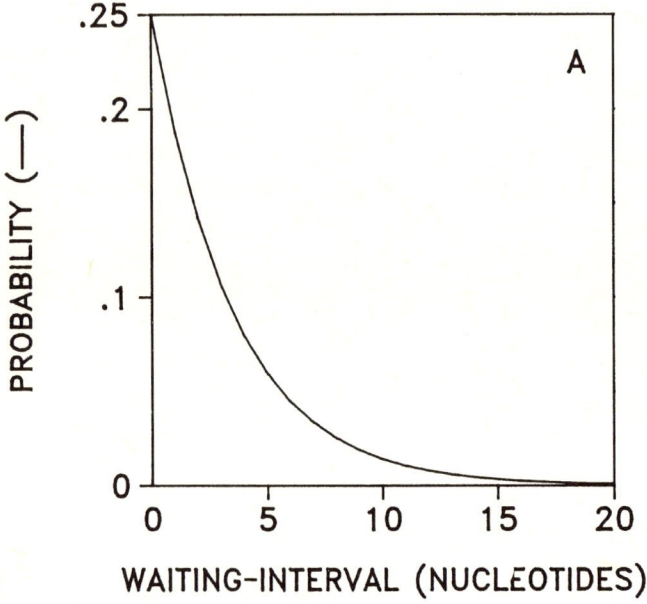

FIGURE 7.17. THE NEGATIVE EXPONENTIAL DISTRIBUTION OF THE PROBABILITIES OF WAITING INTERVALS. The probabilities of the waiting intervals between the successive repetitions of a given nucleotide was computed as shown in Figure 7.16. These probabilities were plotted against the waiting intervals, which follow a negative exponential distribution curve.

This phenomenon is also true for the successive repetitions of a sequence of nucleotides, and can be worked out by simply replacing A with the particular sequence. For example, the waiting intervals between the successive repetitions of AGTA will also be negative exponentially distributed. Furthermore, the successive waiting intervals between two different sequences, for example AGTA and CTGC, also will be distributed in a similar manner. We can in fact replace these sequences with any sequences we want and with much longer sequences, and still the property of distribution will be the same. Now consider that the two sequences represent two different exons of a gene. In a very long random sequence, such as the universal sequence pool in the primordial pond, we can expect numerous occurrences of these exons. Short distances between the consecutive occurrences of these two exons is more probable than longer distances. This favors our concept that genes found in primordial soup will have exons separated at fairly close distances as much as found in real genes.

Because a random sequence exhibits this sort of a distribution between any two given subsequences in it, we can use this distribution characteristic to test if a given DNA sequence is random. We can therefore test whether our following assumption is true: that the primordial sequences were random and that in these DNA sequences the genes of the living cells simply occurred and were directly selected into genomes. If this is true, then the DNA sequences of today's living organisms should be random. In fact when I did this negative exponential distribution analysis on the DNA sequences of the eukaryotes available in the GenBank database, it indicated that today's eukaryotic genes are almost random in sequence.[41] This is one of the best results in favor of my theory that the eukaryotic genes were derived directly from the random DNA sequences of the USP in the primordial pond.

The meaning of negative exponential distribution of sequence waiting intervals

Based on our above discussion, if we collect the waiting intervals between the successive occurrences of a given subsequence in a random sequence and plot these lengths against their frequency, we will obtain a negative exponential distribution curve. Figure 7.18B shows a distribution curve for the waiting intervals between the successive repetitions of the nucleotide A in a random sequence simulated by the computer. It can be seen that the shortest waiting interval is the most frequent as predicted; and the frequency decreases exponentially with the increase of the waiting interval.[42] When we construct a curve for an actual gene, the human globin gene, we see that it is quite similar to that of the random sequence (Figure 7.18A).

The next question is: What does this negative exponential distribution mean when we look at a linear sequence? Because shorter waiting intervals are more frequent, there is significant clustering of a given subsequence in some regions of the random sequence while it is rare in some other regions. An example is given for the tetranucleotide ATGC in Figure 7.19. In this example, I took a purely random DNA sequence simulated by the computer, and then marked wherever a ATGC occurred. Quite clearly, there are regions where many ATGCs are clustered while there are other regions where there are no ATGCs at all for a long distance. In fact, if we look at all the various possible subsequences of a given length, such as each of the 256 possible tetramers, we shall find that there are many subsequences that would be very heavily clustered in some regions. When scientists look at such clustered regions in actual genes of living organisms without realizing these sta-

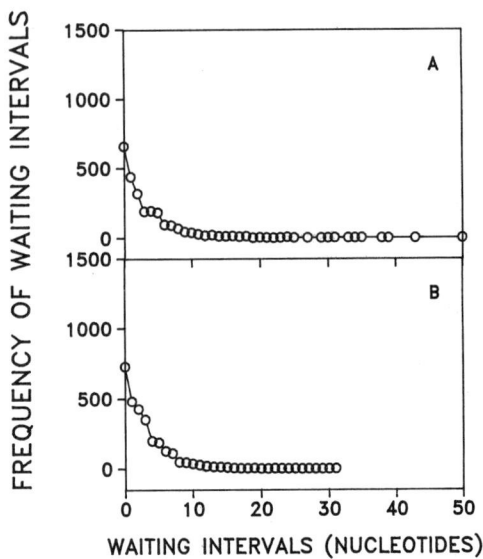

FIGURE 7.18. THE DISTRIBUTION OF WAITING INTERVALS BETWEEN SUCCESSIVE REPETITIONS OF 'A' IN A NATURAL DNA SEQUENCE. Starting from the first nucleotide in a natural DNA sequence (the hemoglobin gamma gene sequence), the location of the first and the subsequent occurrences of the nucleotide 'A' were noted. The waiting intervals were computed, and these distances were organized as frequency distributions, i.e. how many occurred with a waiting interval of 1, how many 2's and so on. The frequencies of the waiting intervals were then plotted, giving a negative exponential distribution. Shown in the figure are the distribution from (A) the hemoglobin gene sequence, and (B) a random sequence simulated by the computer.

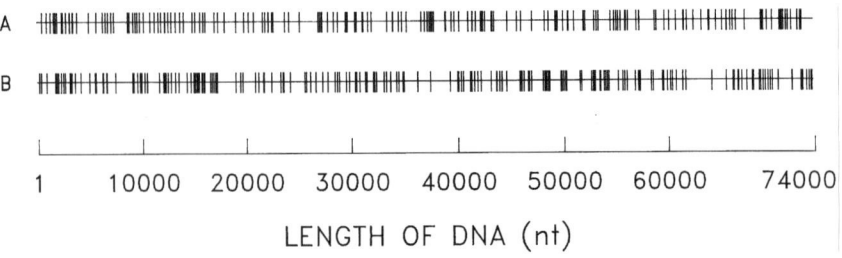

FIGURE 7.19. LINEAR DISTRIBUTION OF ATGC IN THE HUMAN GLOBIN GENE SEQUENCE AND IN A RANDOM SEQUENCE. The successive occurrences of ATGC in the human globin gene sequence of 73,360 nt (A) and a random sequence of the same length (B) were plotted serially. The tick marks represent the occurrence locations of ATGC in the sequences.

tistical realities, it makes them believe that the sequence is nonrandom with respect to the distribution of the given subsequence — and that it must have evolved by molecular evolutionary processes. But as we just saw, it is simply the property of the negative exponential distribution of sequence waiting intervals in a long random sequence.

Searching for the occurrence of the genes of today's animals and plants in the universal sequence pool: Computer simulation experiments

Let us now discuss how it is possible that long split genes with many exons and introns found in the genomes of animals and plants living today could have randomly occurred in the universal sequence pool. We have estimated the length of the universal sequence pool to be approximately 10^{30} nucleotides. Assume that a particular gene Z occurs somewhere in this vast random sequence. We can take a long random computer-generated sequence as the hypothetical USP that occurred eons ago in a typical primordial pond and search for the gene in this sequence. In fact we shall obtain many occurrences of the gene Z, each split into many exons — the length of the complete gene Z in each occurrence varying considerably due to the introns. Our aim will be to find the shortest possible ones and show that these lengths are comparable to the genes of today's living animals and plants. If we achieve this, then we essentially show that such genes were available in the USP and could have been selected in building the genomes of various organisms directly in the primordial pond.

The principle of negative exponential distribution of sequence waiting intervals is true with any number of consecutively occurring subsequences, for instance exons P, Q, and R. The distances between P and Q, and Q and R will keep decreasing as we search for numerous consecutive occurrences of P, Q, and R with gaps of random sequences (introns) between. Remember that we are searching for many occurrences of the same P-Q-R gene — the same coding sequence split into the three exons but the lengths of introns (as well as their sequences) may vary widely in the different occurrences of the P-Q-R gene — in an extremely long random primordial sequence of 10^{30} nucleotides. Statistically, the longer the random sequence we search, the more often a given gene will occur, and the shorter the introns, leading to a shorter complete gene.

We now know from our previous analyses that actual exons of eukaryotic genes, almost all of which are under a length of 200 codons, can exist in the primordial USP (taking into consideration the degenerate amino acid positions and the degeneracy of the codons). What we want to know next is: Can the consecutive exons of a gene occur at close distances in the universal sequence pool, with introns of a few thousand to about a few hundred thousand nucleotides, as we see in the genes of today's living animals and plants? The computer simulation studies that Ganesan Ramalingam[43] and I have conducted, which are described in the following, demonstrate that such a thing is clearly possible.[44]

At this moment, it is worthwhile to recall the manner in which we searched for an English sentence in a random stream of letters that we discussed earlier in this chapter. We can follow essentially the same approach here to search for the occurrence of the gene (with all its given exon sequences interrupted by random intron sequences) starting from the first nucleotide in the long random sequence. In our search for an English sentence, we searched until we found only one occurrence of the sentence. However, here in our search for genes, we shall find several occurrences of the same gene (split into many exons) consecutively occurring in a long random DNA sequence. Once the first complete gene is found, we shall repeat the search again for the same gene starting from the next nucleotide onward in the simulated random sequence. In these searches, we have fixed the protein sequence (with its degeneracies) and the order of exons to remain constant. The length and the sequence of the introns will vary, thereby the total length of the gene will also vary. We shall repeat this process several times, record the length of the gene each time, and compute the frequency of occurrence of the gene versus the gene lengths.[45]

The first occurrence of the gene found in the random sequence may be meaninglessly too long. But, the idea is that as more occurrences of the gene are found, the average length of the gene would statistically decrease. If we keep track of the shortest gene found so far in our continuing search in this manner, the more occurrences of the gene we find, the shorter the gene length will become. And the shortest gene found will be far shorter than the expected mean length.

A realistic probability for finding the genes of today's animals and plants in the primordial pond's USP

For practical reasons, it is not possible to simulate the vast lengths of the universal sequence pool (10^{30} nucleotides), and carry out computer simulation experiments to verify if we can find a given actual gene in it, because the USP is far too long for the present day computers to deal with. However, we can do experiments in which we search for short hypothetical genes in a small hypothetical USP to the extent that our computers will allow. Our conclusions can then be extended to real gene situations — without loosing any conceptual meaning and validity whatsoever.

Let us search for a short hypothetical split gene sequence, coding for the protein sequence shown in Figure 7.13, in a random DNA sequence simulated in the computer. Although this is part of the bacterial λ-repressor protein, we shall use this as our hypothetical protein because this is one of the most well-studied proteins with respect to the variability of amino acids, and because the fundamental characteristics of proteins are the same whether it is a bacterial protein or an animal protein. We shall split this portion of the protein into two, three, or four exons at various positions for our studies, and treat each one as a separate gene in order to analyze the increase in probability by splitting a gene. In searching for the gene, its consecutive exons must be found sequentially in the hypothetical USP. However, as discussed before, the introns can be of any length, and we are not interested in their sequence. For the moment, let us not worry about finding the splice junctions either.

At this juncture, it is important that we discuss in greater detail the process of how we find the gene of interest in a random sequence. As we saw earlier, the length of the random sequence in which a given gene is found depends upon the length and the probability of the longest exon. However, when consecutive exons are searched for in a long random sequence, by the time we find the first occurrence of the longest exon, there would be many occurrences of all the shorter exons. As it turns out, although all the other exons may occur at the very start of the random sequence and several times before the longest exon occurs, almost all of them will also occur very close to the longest exon in the same order as found in the gene. For instance, let A, B, C, D, and E be the exons of the gene, and let D be the longest exon. A, B, and C can occur several times early on in the search before D occurs, but they can occur close to where D occurs in the order ABC. E being another short exon, it may also occur several times before D, but it will also occur soon after D occurs. Therefore, what we do first is to search for the exons A, B, C, D and E in order in the random sequence. Then, starting from where the last exon E occurs, we search backwards for the occurrence of D, C, B and A. Figure 7.20 illustrates this process. The pattern of the exons ABCDE thus finally obtained will be the shortest possible gene in the sequence. These considerations show clearly that the gene itself with all its exons will occur in one piece within a fairly short span of the random sequence compared to that probabilistically expected for the longest exon in a random sequence.

In fact, due to this new approach in searching for a given gene, the probability and expected mean length for the gene now entirely depends not on the length of the longest exon, but upon the length of the second longest exon in the gene — because, from the longest exon that occurs only once per search, all other exons proximal to it are then located. It is of great

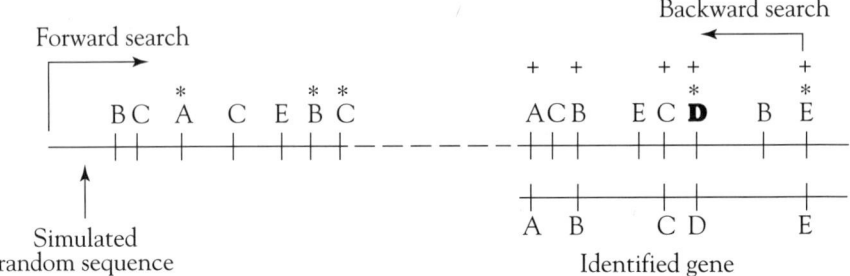

FIGURE 7.20. THE METHOD OF IDENTIFYING THE GENE OF OUR INTEREST IN A RANDOM SEQUENCE. Let A, B, C, D, and E be the exons of the gene and let D be the longest of them. In a long random sequence, all the shorter exons can occur multiple times before the longest exon occurs. In order to find the shortest possible and the first pattern of ABCDE, we search for the first occurrence of A, then the next occurrence of B, and so on until the last exon E is found. Then, we search backwards from E for the first occurrence of D, then C, B and lastly A. This approach ensures that the first shortest occurring gene in the random sequence is identified. The exons with * above them are the ones that we find in the forward search. The exons with + are found by backward search. In reality, many more As, Bs, and Cs will be found than shown in the figure before we reach the longest exon D; and the length of the random sequence before D may also be considerably longer than that shown in the figure. The gap - - - - indicates the long sequence in which many As, Bs, Cs, and Es can occur.

importance to understand and remember this, for this feature heavily reduces the length of the gene. Now, the even more important and extremely crucial consideration is: when we carry out a large number of searches for several occurrences of the gene of our interest, using the above approach of finding a gene — first locating the longest exon and then locating the surrounding exons — the length of the gene will certainly reduce tremendously even further because of the negative exponential phenomenon.

Let us now return to our experiment. The first occurrence of the hypothetical gene in the random sequence may be quite long. Once it is found, continue from the next nucleotide in the random sequence and search for the gene again. Repeat this process 1000 or 10,000 times, each time recording the length of the gene — from the start of the first exon up to the end of the last exon (and the point in the simulated random sequence at which the gene occurs). We can now plot the lengths of the genes obtained against their frequency. For clarity of viewing the results, we can group the occurrences of genes in terms of their lengths in increasing steps, so that we can easily see if the short ones are more frequent and the longer ones are less and less frequent as we predicted.

Figure 7.21 shows the results of our experiment wherein we split the 51-nucleotide gene (coding for the 17-amino-acid portion of the λ repressor) into two exons. The search for 1000 occurrences of this gene, which, due to degeneracy, can be one of many possible sequences, was carried out in a contiguously long random sequence (the hypothetical USP). The X-axis shows the gene lengths, and the Y-axis depicts the frequency of the occurrences in each length group. The shortest length of the gene among all the 1000 occurrences was 54 nucleotides, and the longest was 67,493 nucleotides. It is amazing to note that while the expected mean length of the gene when it is split into two exons is approximately 3,814,500 nucleotides,[46] the shortest gene was only 54 nucleotides. It is also worthwhile to note that the total length of the exons itself is 51 nucleotides, which means that the length of the intron in this case was just three nucleotides; in the case of the longest gene that occurred, the intron length was 67,493-51 = 67,442 nucleotides. The figure illustrates that the shortest occurrences of the gene are the most frequent. The longer genes are progressively less frequent, as we predicted. We can also note that as much as 70% of the occurrences were shorter than the expected mean length of the gene as computed by the combined probabilities of all the exons.[47]

It is crucial to note that even the longest gene is far shorter than the expected mean length of the gene computed using the probabilities of the exons. It is only because of our backward search method that the length of the gene itself is tremendously short. If we did not use the backward search method, then the gene length would be nearly the same as that of the expected mean length of the longest exon. Because of this, it takes a considerable length of the random sequence, from the start of each search to the location where the gene is found. This length of the random sequence from the start to the location of the gene is another parameter we can compute and make a plot. Such a plot is depicted in Figure 7.22 for the gene split into two exons. Here too, the same phenomenon of the negative exponential distribution is seen. The shortest search length of the random sequence in which the gene was found is 5712 nucleotides and the longest was 8,974,631. The expected mean length is 3,814,500 nts.

Figure 7.23 illustrates the length distribution of the gene split into three exons, when 10,000 of its consecutive occurrences were searched in a random sequence. The shortest gene found in this case is 53 nucleotides (indicating that the length of the two introns together in this case was two nucleotides), and the longest gene is 3442 nts. As can be seen from Figure 7.24, the shortest search length of the random sequence in which the gene

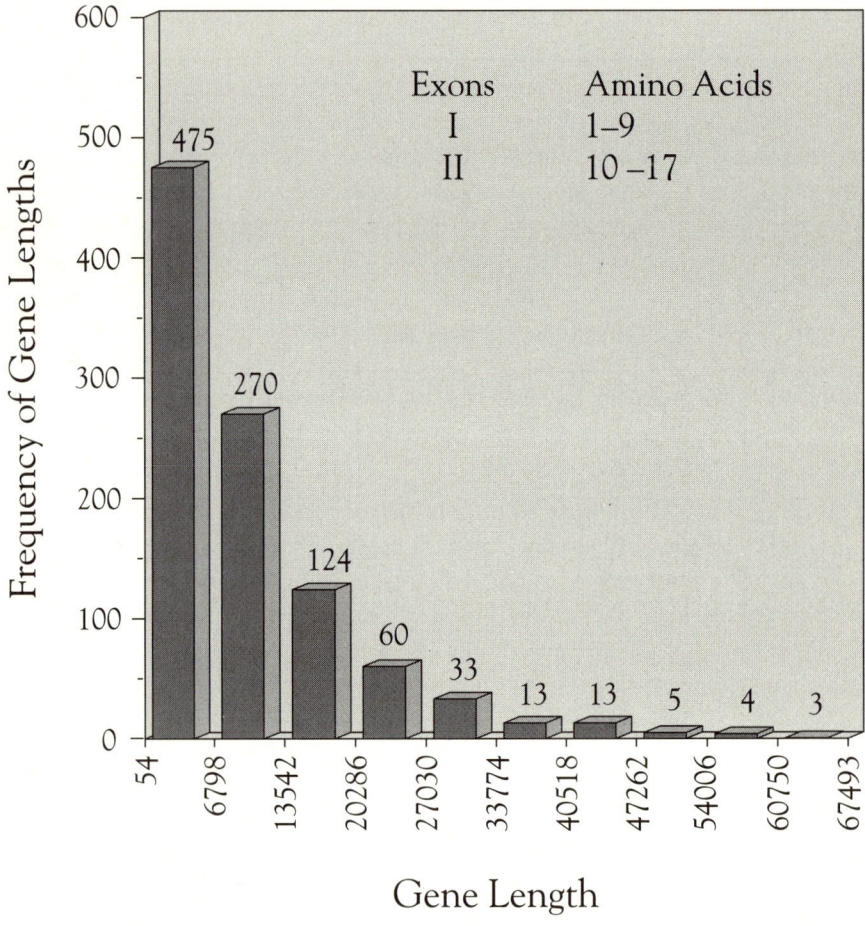

FIGURE 7.21. SEARCH FOR THE SHORTEST HYPOTHETICAL GENE SPLIT INTO 2 EXONS. The 17-amino-acid portion of the lambda repressor protein shown in Figure 7.13 was split into 2 exons (exon 1: amino acids 1-9 and exon 2: 10-17). The possible DNA sequences coding for this split gene were searched in a simulated random sequence taking into account all the codon degeneracies and amino acid degeneracies and using the backward search process described in Figure 7.20. One thousand consecutive occurrences of the gene were located in a long random sequence. The length of the gene (from the start of the first exon to the end of the last exon) in each occurrence was computed. They were grouped into ten equal gene-length ranges from shortest to longest. The frequency of the gene's occurrence in each range is plotted against the length of the gene. For example, 475 occurrences of the gene were shorter than 6798 nucleotides.

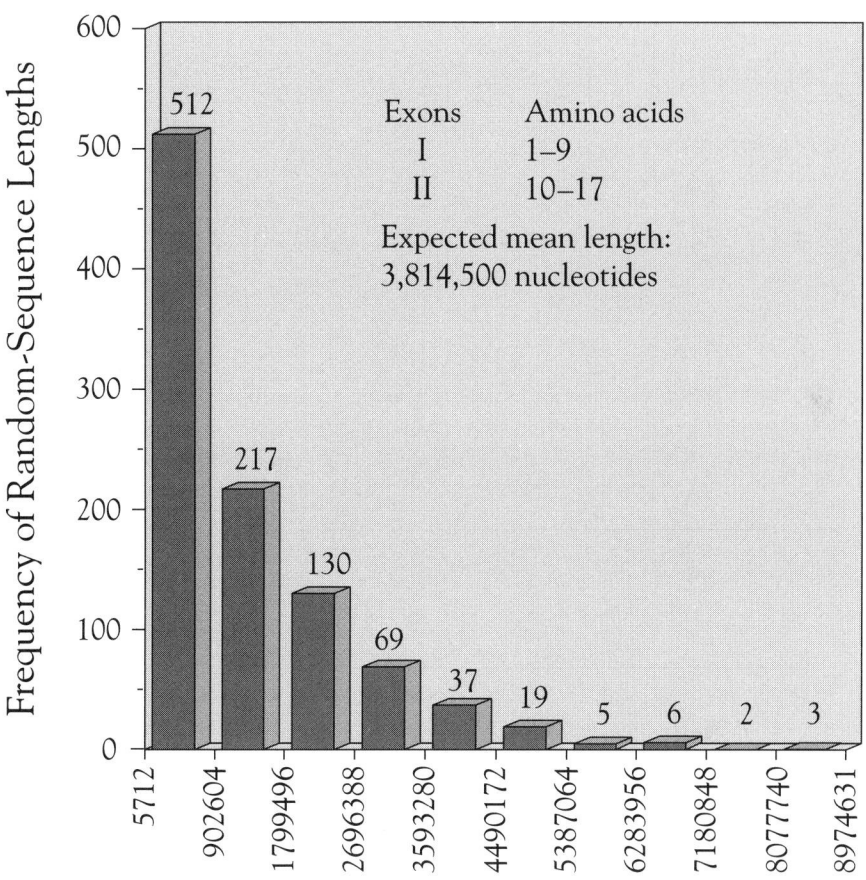

FIGURE 7.22. THE LENGTH OF THE RANDOM SEQUENCE IN WHICH THE HYPOTHETICAL GENE SPLIT INTO TWO EXONS OCCURS. In the search for many occurrences of the gene described in Figure 7.21, the length of the random sequence at which the gene occurred was computed for each search (this length is measured from the start of the search to the start of the gene). The lengths were grouped in increasing order and were divided into ten equal ranges from the shortest to the longest. The frequency in each range was plotted against the lengths. The origin of the X-axis shows the shortest length of random sequence in which the gene occurred. The expected mean length of the random sequence was computed as described in the text.

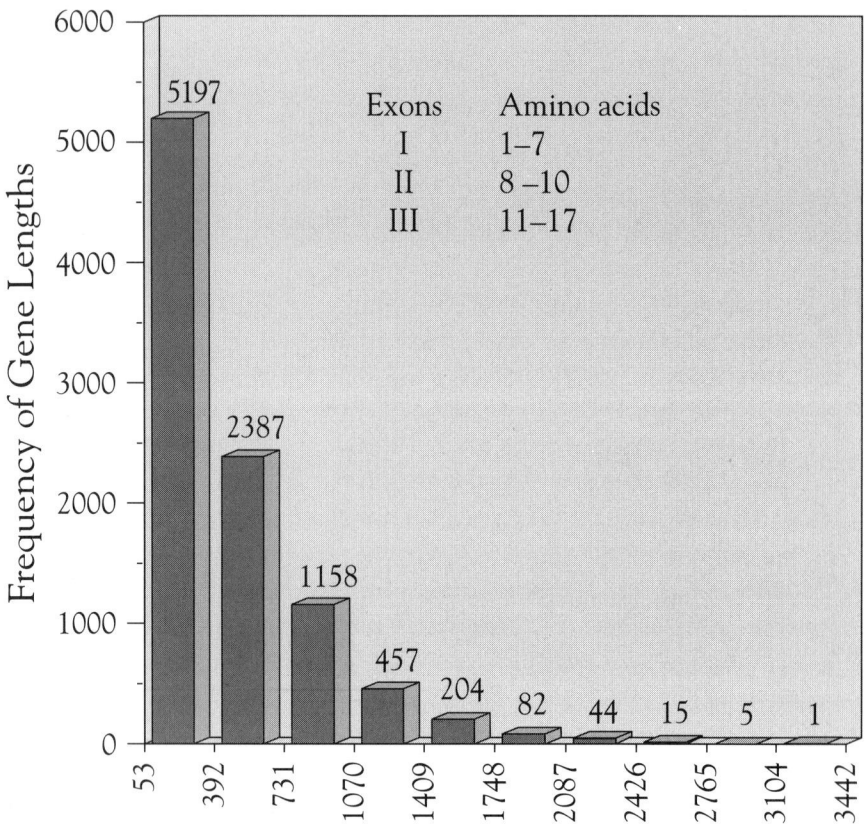

FIGURE 7.23. SEARCH FOR THE SHORTEST HYPOTHETICAL GENE SPLIT INTO 3 EXONS. The results of an experiment similar to that described in Figure 7.21, except that the gene was split into three exons, are plotted.

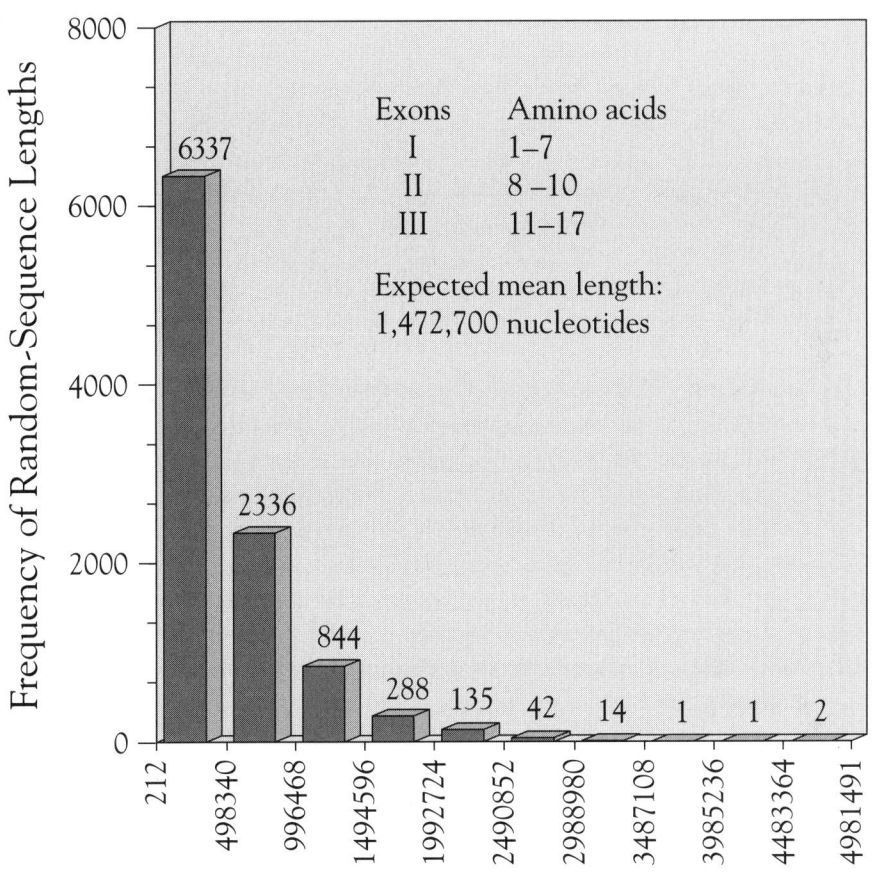

FIGURE 7.24. THE LENGTH OF THE RANDOM SEQUENCE IN WHICH THE HYPOTHETICAL GENE SPLIT INTO 3 EXONS OCCURS. The results of an experiment similar to that described in Figure 7.22, except that the gene was split into three exons, are plotted.

was found is 212 nts, the longest being 4,981,491 nts. Again, it is interesting to note that the expected mean length of a random sequence in which this gene can occur is 1,472,700, as computed from the probabilities of the three exons. One can see that when the gene is split into more exons, it is easier to find the gene in a random sequence.[48]

A very important observation that we should make here is that a particular gene occurs 1000 times in the random sequence that is 1000 times longer than its expected mean length. Most of these genes do occur within very short lengths (much shorter than the expected mean length) because they are compensated by the fact that some genes are far longer than the expected mean length of the gene. The implications of these observations is that in the universal gene pool of the primordial pond, a given gene can occur many times giving rise to essentially the same protein with respect to its structure and function. Indeed, we see this phenomenon at least in some cases of animals from the details we know today: the collagen gene is known to occur with distinct exon-intron structures in various organisms. Their exon-intron structures are so different that they cannot be said to have evolved one from another through organismal evolution. However, this scenario fits very well with our predictions that different collagen genes should have occurred independently in the universal sequence pool and could code for essentially the same protein, "collagen." All these results abundantly show that our predictions concerning the occurrence of genes in random sequences are absolutely correct.

In our experiments in which we illustrated that we can find the genes of today's living animals and plants in random sequences, we have not included the splice junction sequences at the junctions of the exons and introns. We can certainly consider an experiment that includes the splice junction sequences. Although we have not done such an experiment, it is certain to yield similar results.[49] In this case, only more iterations have to be done in order to obtain the shortest gene length, but this would not alter the results because splice junction sequences are very short and occur with a great deal of variation, and they also occur quite randomly in eukaryotic DNA sequences as expected in my theory, and shown by my published experiments.[50]

The essential theme we must derive from all these experiments is simple. When one contemplates the length of the random sequence in which a real gene can randomly occur probabilistically, it seems to be meaninglessly long. However, when we take into account the several underlying features of proteins and genes, in actuality we find the gene in a tremendously short random sequence. This is what we illustrated when we showed that the 51-nucleotide hypothetical gene's expected mean length was 5×10^{30}

nucleotides, whereas we found it in a mere 55 nucleotides. We can scale this up to show (see below) that within the finite sequence of the primordial pond, genes of any actual living animal or plant can indeed occur.

THE PROBABILITY OF ANY SEQUENCE IS THE SAME — WHETHER IT HAS MANY REPETITIONS IN IT APPEARING TO BE HIGHLY NONRANDOM OR IT IS A PURELY RANDOM SEQUENCE. THEREFORE, PROTEINS WHICH APPEAR TO BE HIGHLY EVOLVED CAN OCCUR IN THE PRIMORDIAL DNA SEQUENCE WITH THE SAME PROBABILITY AS A RANDOM PROTEIN SEQUENCE.

Molecular geneticists and evolutionists always speak about the nonrandomness of a protein sequence or a gene sequence when they see a sequence having many repeated amino acids or nucleotides. Their idea is that such a repetition is not probable just by chance, and therefore, should have come about by evolution through duplication of the repeated sequences and so on. But interestingly, we can prove that as far as probability is concerned, it really does not matter whether a sequence is purely random or it has many internal repeats in it.

There are many instances where there are such internal repeats in the sequence of a protein (and therefore in the sequence of the gene). A classical example is the sequence of *collagen*. It has a nearly perfect repeat of Gly-X-Pro-Gly-X-Pro, where X is any amino acid. Looking at this, molecular biologists and evolutionists have speculated that a primordial exon coding for a primordial collagen peptide should have duplicated and evolved the long collagen protein. But we can prove that such a gene can exist in a random sequence, with absolutely the same probability as that for a protein which has no sequence repetition at all.

Another example of a repeated sequence is the mammalian protein *albumin*. Also, many of the vertebrate coagulation proteins such as *transferrin*, *fibrinogen*, and *fibrinectin* have internal sequence repetitions. All of these could occur in the vast sequence of the USP with the same probability as that for proteins which have no sequence repetitions.

From our analysis we can derive the following important principle. As far as the probability of a protein (or a gene) in the primordial pond is concerned, what matters is only the length of the longest exon. The sequence of the exon does not matter. The repetition of a sequence within the exon, or the repetition of the exons within the gene does not matter. The length of the protein does not matter. The sequence of the whole protein does not matter. The protein domain for binding of cofactors or metals, or for catalytic activity does not matter. What does this mean in our context? Any protein with apparently high nonrandomness could have occurred in the primordial

DNA sequences and could have been selected in the assembly of the various genomes — directly performing a specific biochemical reaction in the cells and the bodies of organisms. Remember it did not have to evolve, because it was available as such in the random DNA sequences of the primordial pond. And thus, for the millions of years an organism lives (or lived) on earth, the protein performed the very same biochemical function, with only modifications of its sequence changing into the normal variants of the protein.

We can conduct some simulation experiments to prove our concepts. Consider a short hypothetical gene in which a sequence is repeated several times (as shown in Figure 7.25). We conducted the same kind of experiments with this sequence as we did for the unique sequence of the lambda repressor protein gene. We then compared the frequency of occurrence of the hypothetical gene with that for the repressor gene. The results showed that the frequencies were not at all different. In fact, we can simply compute the probability for a purely unique sequence which is indeed a piece taken from a random sequence generated by the computer and compare this probability with that for a sequence which has repetitions. For instance, the probability for the protein sequence His-Asp-Gln-Val is the same as that for His-His-His-His. The effect of repetition is the same when we consider the DNA sequence or the protein sequence. Let us take the example of ATGCTA or TTTTTT. The probability for A, T, C or G is the same as that of consecutively occurring Ts, when they are taken from a large pool of all the four nucleotides. Only when the pool is very small does it affect the probability for the next occurring nucleotide in a sequence built from such a pool. When the pool is large, taking one nucleotide or amino acid does not deplete the pool to any significant extent.

We then conducted an experiment where the same exon is repeated several times (Figure 7.25). Again the probability does not change whether the exons of a gene are different or the same. We also conducted an experiment with an actual protein sequence, a portion of the highly repeated sequence of the collagen protein. The results again showed that the collagen gene can occur in a random sequence with the same probability as that for a gene which codes for a protein with no repetitions at all. All these experimental results show that repetitions can occur within a gene or in different genes simply as a probabilistic outcome without any evolutionary implication.

The influence of evolutionary theory over biologists is quite interesting when we consider their belief that proteins in living systems are highly evolved through hundreds of millions of years of organismal evolution. We see the protein domains which carry out biochemical reactions, called the active sites. We see other domains, which, in order to aid the particular biochemical reaction, binds with some other chemicals called cofactors (vita-

A

    Sequence without repetition: His-Asp-Glu-Tyr   ⎫ have the same
    Sequence with repetition: His-His-His-His   ⎬ probability
                                                                                                                 ⎭

B

    Many exons repeated in a gene

    Exon 1:   Cys-Asp-Glu-His

    Exon 2:   Cys-Asp-Glu-His

    Exon 3:   Cys-Asp-Glu-His

C

    Collagen gene

    Exon 1:   Gly-X-Pro-Gly-X-Pro-Gly-X-Pro
    Exon 2:   Gly-X-Pro-Gly-X-Pro-Gly-X-Pro-Gly-X-Pro

FIGURE 7.25. THE PROBABILITY OF A GENE IS THE SAME WHETHER IT IS A UNIQUE SEQUENCE OR HAS REPETITIONS IN IT. (A) A protein sequence with or without repetitions has the same probability in a random DNA sequence. (B) The probability of a gene in a random DNA sequence is the same regardless of whether many exons of the gene have repeated sequences. (C) Similarly, the collagen gene in animals, which has the repetition of Gly-X-Pro where X is any amino acid, occurs with the same probability as that of any other gene without such repetition.

mins or metals). Furthermore, we see some areas in the protein chains which contain some repeated patterns, based on which we think that these areas have highly evolved to form these "nonrandom" patterns. But interestingly, in our new view, all such sequences which appear to be highly evolved and nonrandom can simply occur with exactly the same probability as any purely "random" sequence. In other words, if we first generate a purely random sequence and take this as our gene and search for this sequence in a long random sequence, the probability for finding this gene is exactly the same as that for finding a real gene from any living organism with any kind of nonrandom characteristic — as long as the general length characteristics of exons are fulfilled. Again, because of this, we can see that they are probabilistically identical, whether they appear to be highly evolved and nonrandom, or they are random, junk sequences. It is therefore imperative that we should not hereafter be misled by such deceptive appearances.

In general, genes for complex functions such as higher intelligence or for complex organs such as the eye may appear to be highly evolved, because traditionally these functions and organs have been believed to be highly evolved through descent with modification. But, from our discussions we can see that they are not! In our analysis with English sentences earlier in the chapter, did it matter to us if the six-letter word was "heaven" or "enough?" No! As long as the word length was six, all the six-letter words with any meaning occurred in the random sequence. Did it matter to us what the sentences contained? Absolutely not! As long as the longest word was under a certain limit, all the sentences, and any sentence with any meaning whatsoever can occur in the random English sequence. What is the meaning of a gene in biology? It is the biochemical function of the protein it encodes. But no matter what function it harbors, a protein-coding DNA sequence will occur in a random DNA sequence, as long as its longest exon does not exceed a certain limit. This is the crucial theme we shall have to remember.

This is the reason why the genes of any simple or complex organ (such as a little finger or the retina in the eye), or those for a simple or complex organismal function (such as lifting a finger or the termites building their complex termite colonies, or the high human intelligence) can all simply occur in the large universal sequence pool as just a probabilistic outcome. Hereafter, therefore, we shall not be misled by the feeling that proteins giving rise to complex functions or organs should have evolved to reach their complexity; all of them could simply exist in the universal sequence pool as a probabilistic outcome.

THE OCCURRENCE OF ACTUAL GENES IN THE UNIVERSAL SEQUENCE POOL IS INEVITABLE: A SIMPLE COMPUTATIONAL APPROACH

Although we have demonstrated several reasons that stupendously increases the probability of finding a typical eukaryotic gene in the USP, we have not yet explicitly determined its probability and the expected mean length to show that it is well under 10^{30} nucleotides, the size of the USP. We can do this with a fairly simple computation based on our understanding of the extent of codon degeneracy in genes and amino acid degeneracy in proteins.

First, the amino acid degeneracy: In a typical protein sequence, about only one amino acid in every 10–20 amino acids is completely invariant (5–10% invariability). We should also take into account the extent of variability in all the variable positions — since they are not all equally variable. Taking all this into account, we can approximate an overall value of 8% invariance to occur in a typical protein. Thus in a 200-amino-acid protein sequence, on average, 16 amino acids will be completely invariant, and all the other amino acids will be completely variable. Therefore the probability of a 200-amino-acid protein with all its variable and invariant characteristics is simply the probability of 16 invariant amino acids, or P^{16}, where P = probability of each invariant amino acid.

Second, the codon degeneracy: Since 61 codons code for all the 20 amino acids, each amino acid is coded for by an average of 3.05 codons. However, we also know that the frequency of degenerate codons is higher than the nondegenerate codons in a random DNA sequence. This increases the codon degeneracy from 3.05 codons per codon position to about 3.85 codons per codon position (see Figure 7.11). [51]

The average probability of finding any one particular amino acid, with codon degeneracy, is 3.85/64.

The probability for finding a sequence of 16 invariant amino acids is $(3.85/64)^{16}$.

The expected mean length of the random sequence for this 16-amino-acid sequence to occur = 1/probability = 3.4×10^{19} nucleotides.[52]

This 16-amino-acid invariant sequence occurs with the same probability as a 200-amino-acid sequence with typical amino acid degeneracy. And this is equivalent to a 600-nucleotide exon — the maximum length of exons in genes.

This means that for about every 10^{20} nucleotides we search, on average we will find one occurrence of a particular DNA sequence coding for a given 200-amino-acid sequence (with its amino acid degeneracies). Therefore, in 10^{30} nucleotides, there will be 10^{10} occurrences of this exon. Suppose in a gene there is one 600-nucleotide exon and the next shorter exon is 400 nucleotides, the rest of the exons being much shorter. Now we can use our approach to finding the shortest gene, as we did before by the "backward search" method. To find the shortest gene in the USP, we can first locate all the occurrences of the longest exon and then look around each one of them for all the other, shorter exons. If we iterate our search 10^6 times using this approach, the next shorter exon could be found to occur within 1000 nucleotides of the longest exon.[53] At this juncture, it is important to note that most of the exons of even very long genes are much shorter than the 600-nucleotide upper limit — most have 100–150 nucleotides. Therefore, by this process, we can see that even in about 10^{26} nucleotides, any given gene with many exons in it can be found.

In other words, such calculations show that in a sequence of approximately 10^{20} nucleotides, an exon 600 nucleotides long will occur. In a sequence of 10^{26} nucleotides, complete genes with realistic intron lengths as found in genes of living organisms will occur. This is a far smaller amount of DNA compared to what we have computed as available in a primordial pond. In this portion of the USP, therefore, any gene for any biochemical function would occur. Consequently, even a very small part of the DNA possible in a primordial pond would be able to contain almost all the genes needed for the assembly of multitudes of genomes for multitudes of different organisms.

The size of the USP we estimated to be available in a typical primordial pond is 10^{30}–10^{35} nucleotides. In each pond, let the unique sequences occur in multiple copies, say a thousand copies per unique sequence. Taking these into consideration leaves us a unique sequence 10^{29} nucleotides long in each primordial pond.[54] This is 1,000 times greater than the 10^{26} nucleotides needed for all the genes to occur. Thus, in each pond even if only a small fraction of the DNA is available for assembling genomes, it would still be enough. Likewise, even if only one in a large number of ponds gave rise to life, there would have been several ponds from which life would have arisen.

THE PROBABILITY FOR FINDING MILLIONS OF GENES IS THE SAME AS THE PROBABILITY FOR FINDING ONE GENE. THEREFORE A VAST NUMBER OF GENES IS INEVITABLE IN THE PRIMORDIAL POND.

THE PRIMARY QUESTION ANSWERED: WE CAN FIND A MULTITUDE OF GENES IN THE UNIVERSAL SEQUENCE POOL

Let us summarize our discussions of this chapter. In trying to understand the scenario of life on earth, the question we ask is: Can we find DNA sequences for all the genes which code for all the biochemical reactions (that is, all the enzymes, regulatory and structural proteins, and regulatory sequences of all complex organisms) within the USP? We give principal importance here to specific biochemical functions — either enzymatic or structural — rather than to specific DNA or protein sequences.

Our first aim is to find a DNA sequence that codes for a given specific protein: If we are given a protein for a particular function, can we find a DNA sequence that codes for it in a long random DNA sequence? Can we find any one of the immense number of possible DNA sequences that can code for any one of the myriads of variable protein sequences, all of which can specify the very same given biochemical function? We have positively demonstrated that this is highly probable because of many principles. First, because of the high degeneracy of amino acids in proteins, the protein sequence is very forgiving. In other words, many different amino acids can occur at a large number of amino acid positions, and still the protein will specify the same enzymatic function. This means that we do not have to have a strict DNA sequence in order to code for the protein specifying a particular function.

Codon degeneracy is another phenomenon that tremendously increases the number of possible DNA sequences that would code for a protein with a specific function. This phenomenon combined with the amino acid degeneracy enables an enormous number of different DNA sequences to code for the same specific function.

The fact that the eukaryotic gene is split and that only the exon sequences that are considerably shorter than the gene's coding sequence have to be available contiguously, radically increases the probability of the availability of a typical eukaryotic gene in the USP. Most of all, the negative exponential distribution of sequence waiting intervals makes it possible for all the exons and introns of a gene to be present within realistic lengths in the USP.

Because of all these cumulative and synergistic phenomena, the occurrence of a DNA sequence for a gene coding for a protein with a specific function in a random primordial USP is tremendously probable. Indeed the probability is so high that we can find multitudes of genes for a given function. One of the most important principles that we should note here is that if one typical gene could probabilistically occur in the USP, then almost any gene for any particular biochemical function — almost an unlimited supply of distinct genes for multitudes of unique biochemical functions — would occur in the USP. The many principles that we have unearthed and delineated in this chapter make what looked preposterous to become an undeniable reality. This crowning fact that the probability of finding millions of genes in the USP is the same as the probability of finding only one gene makes it more than a realistic probability to find all the genes that would be required for building not only one complex genome, but myriads of different complex genomes based on the principles of independent assembly of genes into genomes. The inevitable consequence of these processes would be the independent birth of myriads of organisms in the primordial pond.

Conclusion

An organism is built and maintained primarily by the actions of proteins coded by genes in the organism's genome. Superficial probabilistic assessments of whether a gene coding for a specific protein could simply occur by chance in the primordial pond have been profoundly discouraging. But these calculations fail to account for several significant characteristics of genes, described in this chapter, that actually make their occurrence highly probable. In fact, these principles of genes cumulatively make it inevitable that a given gene sequence that can code for a specific protein would have been available in the USP. Since the expected mean length of the random sequence is the same for any given gene with typical characteristics, almost any gene coding for almost any protein sequence will occur within this expected mean length of the USP.

We should note that all genes occurring directly in the USP were split into exons and introns — typical of eukaryotic genes. Finally, the notion that the very first cells must have been complex, with nuclei — typical of today's eukaryotic cells — shows that these cells could have been formed directly from the primordial pond. In the next chapter we shall see that these cells could have included genes not only for the propagation of the cells, but also for the development and propagation of whole complex multicellular organisms.

What we are uncovering here amounts to, in essence, an entirely new world view: a realization that the finite quantity of random genetic sequences in a typical primordial pond probably did contain an abundant assortment of complete genes available for assembly into true genomes. We have explored the roles of genes in the construction of organisms, which we define as sets of biochemical structures and functions specified primarily by proteins. Any given protein function can be specified by an immensely large number of different DNA sequences. And most significantly, we have seen that the split structure of the gene, with the length of all its exons under an upper limit, makes absolutely probable the occurrence in the USP of a gene of almost any length. Moreover, within a given length of USP in which one gene is likely to occur, almost any number of other unique genes can also occur.

What is even more interesting is that even the genes and proteins whose internal repetitions of sequences make them appear to be highly evolved and nonrandom, could have occurred in the primordial pond as easily as any junk sequence. Genes that direct the building of complex organs and body parts such as the eye, heart and brain could therefore have simply occurred, ready-made, in the primordial pond. We now understand that it simply boils down to probabilities, and that the likelihood of a full complement of genes occurring abundantly in the primordial pond is much, much higher than previously believed, for all of the many reasons we have described. By all these analyses and considerations, we have now established that multitudes of genes — however complex or simple, and however long they appear in today's living organisms — could have occurred in the USP of the primordial pond. Our task now is to see whether these genes could have been able to conglomerate into myriads of independent genomes, each giving rise to a unique organism.

8

INDEPENDENT BIRTH OF MULTITUDES OF ORGANISMS FROM THE PRIMORDIAL POND

Imagine a lush tropical forest that ripples with life from dawn to dusk and dusk to dawn. What a variety and abundance of creatures live there, plants and animals alike! Millions of organisms of all kinds and sizes and shapes foraging, living, and reproducing. Most of these creatures reproduce by laying eggs, which hatch themselves with little help from the parents. As the parents age and die, their offspring traverse their lives to lay eggs again. For millennia, this cycle of life and death has continued. If these organisms did not originate from one or a few original ancestors by descent with modification, then how did all these innumerable creatures come to be? There was a beginning to the cosmos, a beginning to the Sun and the solar system, a beginning to the earth. There must have been a beginning for life — a point in prehistory when inanimate matter was transformed into living creatures. How did it happen? What is the answer to this "mystery of mysteries"? This is what we shall discuss in this chapter.

In the beginning there was cosmic dust, and in time the dust condensed into multitudes of stars. Somewhere in the immensity of the cosmos there appeared a particular star, our Sun, a mere spec in the vastness of billions of other star systems, floating in one of the billions of galaxies. Around our Sun planets formed, and among these was the earth, planet number three. Physicists and astronomers tell us that the earth, upon its formation, was quite unlike the earth we know today. It was hot and molten, had no atmosphere, had no ocean — not even ponds. It simply hurtled through space and time, around the Sun, for three to four billion years, cooling down ever so slightly from one millenia to the next. While cooling, a unique combination of chemical reactions and physical conditions on this planet led to something beautiful and unique: the formation of life. The hot molten earth facilitated ferocious reactions among many of its hundred or so elements, most of which could react with others. They formed gaseous molecules, including water and many organic molecules. The earth's gravity held these gaseous molecules in bands around the earth — our atmosphere. And as the earth cooled, the water vapor in the atmosphere also cooled, and rain fell onto the molten earth. These conditions were highly conducive for molecules to react with one another, forming more complex chemicals.

Most of these molecules were dissolved in water that formed in hot, boiling ponds of various sizes scattered over the surface of the earth. The chemical reactions continued in these ponds of boiling broth, into which were constantly added a variety of new organic chemicals, all derived from earth's own original, basic elements. The reactions ensued in a continuing cycle of increasing complexity, as small molecules reacted and combined with other small molecules to form larger molecules. Over millions of years, the variety, complexity, and concentrations of these organic molecules increased manyfold. The ponds in which these intense reactions occurred were what we now call the primordial ponds.

Over millions of years, something fantastic happened in the primordial ponds as a result of these processes. The conglomeration of the various molecules produced a new kind of matter graced with a quality we call life. Many kinds of molecules were involved in this process, but two in particular were critical: nucleic acids and proteins. Certainly other molecules were important as well, since molecules like vitamins, for example, give meaning to nucleic acids and proteins. But nucleic acids and proteins were the most important of all because nucleic acids made it possible for organisms to store and transmit coded instructions for their own structural development and functioning, while proteins became the instruments to carry out those coded instructions.

DNA is a linear array of four different types of nucleotide molecules, occurring in various sequences along the DNA strand. The genetic information contained within a strand of DNA corresponds directly to the sequences of the nucleotide molecules, so DNA messages are analogous to written text in a language whose alphabet contains only four letters. As in any language, trillions of "words" and "sentences" can occur simply by chance in extremely long, random DNA sequences. We are presently concerned with the question of whether genes — akin to sentences — were possible in the DNA material that was available in earth's primordial ponds. This is what we explored in the previous chapter, and we concluded that, yes indeed, a very vast number of genes, as found in today's living multicellular organisms, could have occurred in the DNA material in a primordial pond.

From the vast number of genes, one can pick and choose a set that together formed the basis of a living thing. Such a collection of genes is called a genome. Out of a great number of combinations of different kinds of genes, only one genome can produce such a viable organism. It is a random process that assembles different combinations of genes that together could yield a living creature. And the same mechanism could lead to many different combinations of genes that could form many diverse living creatures.[1] That is, the genomes for various creatures could have been assembled with equal ease by this same mechanism, to produce very many different living creatures, independently, directly from the primordial pond. These genomes, although nearly equal in complexity, could have yielded many organisms with distinct anatomical structures, sizes, shapes, and functions. My discovery that the complex unicellular eukaryotes must have been formed directly in the primordial pond from random conglomerations of genes corroborates this possibility that a myriad of creatures arose separately and directly from the primordial pond.

Certainly these assertions are radical now, within the context of an orthodox biology that embraces Darwinian evolution. But take a moment to erase our preconceptions, retrace this line of inquiry, and reconsider all of the evidence we find in the life around us. As we assimilate these new ideas, and observe how they are indeed corroborated by the many details of life on earth, we find that these concepts do carry an intriguing reasonableness that encourages a closer, deeper look.

THE INEVITABILITY OF THE BIRTH OF MULTIPLE, COMPLEX CREATURES IN THE PRIMORDIAL PONDS

When I initially tried to explain my theory to my wife, I said, "All the organisms could have come about just as they are, independently from the primordial pond." Her response was: "Do you really think that an insect or a rat simply came about as it is?" I simply answered "Yes, I do!" She said, "Oh, you say that each of these highly fitting creatures is the outcome of mere chance? How can it be possible?" I know that the response will be similar from anyone who hears about my theory at first. Unless I explain the minute details and the implications involved here, the finesse and the intricacies involved in the new theory may go unnoticed. How is a complex organism inevitable, how are not one organism but multiple organisms inevitable, and how is the perfect fit of the myriad organisms to the physical and ecological environment possible just starting from random genetic sequences in the primordial pond? How does a great order, in fact the highest of orders, arise randomly from total chaos? This is what is precisely explained by the new theory.

Randomness and chance: A new relativistic view in the context of DNA and protein sequences

Is water flowing downhill a chance? No. Likewise, when conditions are right in a solution, the crystallizing of proteins or other chemicals is not a chance; it is inevitable. A physical force (gravity) in the former and a physicochemical force in the latter drive them. It is the same thing with genes and genomes: If an abundant number of genes, far more than that needed to construct an organism, exists in the primordial soup, and if the right conditions and mechanisms exist, then given time, it is inevitable that they come together and construct the genome of the organism. Here also, the physicochemical forces of macromolecular interactions play the central role. However, it is the extremely large amount of DNA sequence, which carries the messages for the enormously large number of proteins, that makes complex life possible. The structure of the DNA double helix, with its power to encode RNA and proteins, and its ability to duplicate itself, is one of the most important and unique ingredients of life. However, it is the immensity of the DNA sequences in the primordial pond that made the occurrence of a vast number of genes and the birth of complex multicellular life possible.

One example of order from chaos is the molecular selection of coding sequences from a random DNA sequence that forms a split gene (with exons and introns) that codes for a protein. The joining of genes to form the genome of a single-celled eukaryote, which is highly complex in itself, is another example. What we are alluding to here is the association or coupling of many more genes to form the genome of a multicellular organism in the primordial pond, where there are million times more genes than needed for constructing one organism. This process is inevitable and is the same as the inevitability of water flowing down hill. Chance becomes a reality in a framework where the total number of events are far greater than required to realize a given event. If the probability of constructing many genomes for multiple organisms is the same as constructing one genome for a single organism, then the construction of many organisms in the primordial pond is also inevitable.

Logically, almost everyone would accept that conducive conditions must have existed in the primordial pond at least for single-celled organisms to have originated, because, without them, absolutely no multicellular organisms would ever have been possible. In fact, chemical evolutionists and other evolutionary biologists unanimously agree that the conditions in the primordial pond must have been conducive for the evolution of the single cells, at least bacterial cells. We have shown that it must have been the unicellular eukaryotes that directly assembled their genomes in the primordial pond, and not the prokaryotes. Therefore, we can logically postulate that there was present such a condition on earth at some time for the formation of the single-celled eukaryotes directly from biochemicals in the primordial pond. If conditions existed for the formation of single cells — as complex as typical eukaryotic cells — then the same conditions would result in equally-complex seed cells of many multicellular organisms leading inevitably to the independent birth of diverse multicellular organisms.

If we take a random trial, the occurrence of a given event depends upon how many times the events are randomly tried. Furthermore, when the number of possible events are fewer, the frequency of a given event increases. There are two possible events in tossing a coin — head or tail. There are six possible events in throwing a die — one through six. If we toss a coin once, the probability that we get a head is one out of two, and there is no certainty that we would have gotten a head, for instance. However, if we toss the coin twenty times, we can be almost certain that we would have obtained a head at least once. At the same time, if we are throwing a die with six sides, the probability that we would obtain a given side is one out of six. Because there are six possibilities, throwing it even twenty times would not ensure that we would certainly obtain a particular side at least once. Instead, if we try one hundred times, the chance of obtaining a particular side becomes almost certain. Thus, the more times we try, the better the chance that we would obtain the given event at least once. The probability by its definition does not change, but the chance of realizing a particular event depends upon the total number of trials. In the context of DNA sequences in the primordial pond, short random sequences is one frame of reference, where it has no meaning biologically. Extremely long random sequences is another, analogous to having many tries at creating life, which have great meaning in that they can contain functional genes.

The probability of a given gene sequence being constant, the chance that it would occur in a random sequence of a billion or even a trillion nucleotides is extremely low. However, as we determined in the previous chapter, in a very long sequence of approximately 10^{26} nucleotides, which is the expected mean length of a typical eukaryotic split gene, trillions and trillions of genes can exist. This length of DNA sequence, this amount of DNA material, is what we determined to be well under what could have really existed in a typical primordial pond on earth, which is approximately 10^{30} nucleotides. Even if one out of a million genes could be biologically useful, there would still be millions of genes available in the primordial pond for the construction of life forms. Under primordial conditions, myriads of genes lead inevitably to numerous distinct genomes for multitudes of organisms. We shall systematically analyze how this is possible, and by the time we finish our analyses, one will clearly see that the independent birth of diverse organisms in the primordial pond is inevitable.

Fundamental principles that enable the formation of independent genomes for multitudes of organisms in the primordial pond

Let us review what we have learned so far in earlier chapters. In addition, we will discuss some new important principles and information required in developing the new theory.

1. Countless different genes existed in the very large universal sequence pool: The universal gene pool

Genes, similar to those in living beings, could have occurred in abundant numbers in the primordial universal sequence pool (USP, the total amount of DNA sequence possible in a typical primordial pond).

  A. The genes that occurred in the primordial pond's random genetic sequences were split into exons and introns. Only the exon sequences were important as the coding portions of proteins. Therefore, only the probability of the individual exon sequences, which are far shorter than the complete coding sequence, needs to be considered while computing the probability of the complete gene.

  B. The degeneracy of the codons in genes and the degeneracy of the amino acids in proteins tremendously increase the probability for the occurrence of a given gene in a random sequence.

  C. The above principles demonstrate that exons similar to those found in living organisms can occur in the USP with an extremely high probability. The negative exponential distribution of sequence waiting intervals in a random sequence enables the multiple exons of a gene to occur as close to each other as in actual genes of living organisms. This makes it possible for complete genes with multiple exons, with lengths just as they are found in actual organisms, to occur randomly in the USP.

D. The primordial ponds could have contained an extremely large amount of DNA sequence, approximately 10^{30} nucleotides. The sequence in which one gene occurs probabilistically will inherently contain a stupendously high number of genes. The total length of the sequence in the primordial pond is far greater than the expected mean length in which a typical eukaryotic gene could occur (10^{26} nucleotides), making certain that myriads of genes could have occurred in the primordial pond.

These discussions indicate that more than all the genes needed for the genomes of all the multitudes of organisms that ever existed on earth would have occurred in the primordial pond's gene pool.[2] What remains to be shown is that, from the vast pool of genes, many different genomes could have been assembled independently through random processes leading to the birth of numerous widely diverse organisms.

2. Unicellular eukaryotes arose directly from the primordial pond

The analysis of the origin of the introns in genes is fundamental to understanding not only which type of genes — split genes or contiguous genes — but also which type of cell — the eukaryote or the prokaryote — came first in the primordial pond. I have demonstrated that the genes which occurred in the primordial pond's genetic sequences (USP) were split into exons and introns, just as the genes of living eukaryotes. The first single-celled eukaryotes were built directly from such split genes, and these cells came with a nucleus.

I may add here that to some of us, it may be difficult to envisage that a eukaryotic cell and its genome could have assembled directly from the primordial pond's genetic sequences and macromolecular machineries. It is important to remind ourselves that the primordial pond had become extremely complex by the time the first cells were constructed. The genetic sequences of the organisms living today contain the most fundamental historical information as to how they could have first originated. Analysis of this information indicates that a high level of complexity must have been there in the primordial pond and that eukaryotic cells must have directly originated in it. It will be beneficial for us to get used to this idea of the complexity in the primordial pond for our further understanding of the complex processes that took place there. In fact, many molecular biologists now agree that it is the eukaryotic cells that came first, and not the prokaryotic cells, corroborating our approach.[3,4]

The possibility that the unicellular eukaryotes could have arisen directly by assembling the split genes in the UGP is perhaps the most crucial phenomenon that enables the independent birth of organisms. It is quite conceivable that the genomes of sexually reproducing multicellular organisms were also assembled as eukaryotic cells — each of which had the ability to grow into a multicellular creature. We call these seed cells, analogous to the fertilized egg of today's living organisms.

3. The genome of a multicellular organism is not too complex compared to that of a unicellular eukaryote

Some of us might say, "My goodness! It was so far difficult to get used to the idea that even single eukaryotic cells originated directly from the primordial pond, but we overcame that difficulty because it is true by all scientific evidence we have seen so far. But how can genomes for the multicellular organisms arise directly in the primordial pond? Are not the multicellular animals and plants far more complex than the unicellular eukaryotes?" Quite interestingly, although there is a world of difference at the level of the organisms, there is not much of a difference between the genomic complexity of unicellular eukaryotes and multicellular organisms.

It is worthwhile to reflect for a moment that the primary difference between the genome of a typical unicellular eukaryote and the genome of a multicellular organism is in the number and variety of genes, not the complexity of genes or genome *per se*. The genetic pathways of the unicellular eukaryote are themselves extremely complex. The genome of the typical unicellular eukaryote consists of approximately 10,000 genes, whereas the genomes of all animal creatures, from worm to human, consist of 10,000-30,000 genes.[5] Therefore, if different sets of genes needed to build different body parts and developmental genetic pathways that could express them at the correct times and places are included in the genome of a eukaryotic cell, theoretically it can differentiate into the various body parts and the whole animal. The basic structure of the cells of the animal body is already defined in the unicellular eukaryote. It is just that the animal body is made up of many different kinds of cells, built by slightly different sets of proteins expressed by slightly varying sets of genes.

There are really no more "intricate" complexities involved in building a "complex" organism than in building a unicellular eukaryote. What is required to build a complex organism compared to a unicellular eukaryote is a number of additional genes, regulatory sequences, and DG pathways, which are individually in no way more complex than the ones needed for building the unicellular eukaryote. Therefore, if the genome of a unicellular eukaryote can be assembled directly from the primordial pond with its large set of genes and the complex genetic circuits for its growth and division, it is almost equally probable that the genomes of many multicellular organisms could have been assembled in a similar manner.

4. BECAUSE THE GENES AND CELLS THAT BUILD THE COMPLEX ORGANS OF AN ANIMAL ARE NO MORE COMPLEX THAN THOSE OF UNICELLULAR EUKARYOTES, THE DIFFERENT SETS OF GENES FOR VARIOUS ORGANS COULD OCCUR WITH EQUAL PROBABILITIES IN THE UNIVERSAL GENE POOL

If we compare the complexity of the individual cells from any animal or plant with those of the eukaryotic single-celled organisms, we can see that there is little difference. The genes and genetic circuits that build the individual cells of different organs and appendages are only different, not more complex, than those that build a unicellular eukaryote. Consequently, the probability of a gene for an organ, occurring in the USP, is the same as that of a gene for the unicellular eukaryote.

For instance, the cells of the liver, bone, or kidney have different sets of proteins and carry out distinct functions, but structurally the proteins from one cell type are not more complex compared to those from another cell type or to a single-celled eukaryote. Therefore, the probability for the occurrence of the genes for the cells of these organs is the same as that for the unicellular eukaryote. Although the vertebrate eye has many substructures, and is a complex organ when taken as a whole, each substructure or tissue of the eye is not more complex than any other tissue that is not part of that complex organ. Similarly, the proteins from each substructure, and likewise their genes, are also of the same complexity. The intricacies of the DG pathways of any body part, whether it be simple like a finger, or complex like the eye, are not truly different. The complexity arises only from the interactions of substructures that culminates in the function of the eye as a whole — the complexity of the organ is not due to the complexity of the set of genes of the eye or the complexity of the gene connections.

5. THE GENOMIC COMPLEXITIES OF ALL ANIMAL CREATURES — WHETHER THEY ARE ANATOMICALLY SIMPLE OR COMPLEX — ARE ALMOST EQUAL

A. The complexity of the genome is not proportional to the size, anatomic complexity, or functional complexity of an organism

When we consider the mammals, the anatomy of a "lower" organism, such as the rat, is no less complex than that of the ape. The rat has a similar number of organs with similar functions, but their sizes are proportionately smaller. A still more "primitive" creature such as the snail, the millipede, or the earthworm also contain many complex anatomical structures. The number of subtissues and organs may be smaller in them. One can, therefore, grade the organisms from simple to complex with respect to their numbers of tissues and organs. A simple animal (such as the worm) may contain approximately 50 different tissues and parts in its body whereas a complex animal (such as the rat or the human) may have 100–300 different tissues and body parts. There are many marine eukaryotic organisms far simpler than the worms that are classified under several phyla, which have fewer cell types, such as the mesozoans, placozoans, myxozoans and poriferans.[6]

Such a wide variation of anatomic complexity in the animal world does not mean that there exists such a variation in their genomic complexities also. It should be understood that the genome of even the most primitive, sexually reproducing, multicellular organism itself is enormously complex — which is indeed not much less complex than those of an anatomically far more complex organisms such as the rat or elephant. The genes and the DG pathways that are expressed to produce each cell (and to construct each tissue) in a worm or in a rat are equally complex. In this sense, humans, elephants, rats, earthworms and even the very small microscopic insects and invertebrates all have nearly the same genomic complexity in terms of the structure and function of the genome. This principle can be established by showing that there exists almost no difference in the structures of genes and genomes among eukaryotes. This is illustrated in Figure 8.1. The genes of all eukaryotes are split into exons and introns, whether they are in a single-celled organism such as a *paramecium*, or in a multicellular animal as big as the blue whale, or as small as the microscopic worm *C. elagans*, or in any of the giant or microscopic plants or trees.[7] The lengths of different genes vary from approximately 1000 nucleotides up to about 1,000,000 nucleotides in all organisms. Research indicates that in the genomes of all animals and plants — whether anatomically simple or complex — long stretches of functionless DNA are present between adjacent genes.[8] The consecutive genes in all these genomes are separated by such useless nongenic sequences, which can be far longer than the genes themselves. Neighboring genes may or may not be expressed in the same tissue or organ. The genes expressed chronologically during development (e.g., homeobox genes) may sometimes be organized in a linear order on the genomic sequence. In all the animals, different sets of genes from the genome are expressed to build different kinds of cells, tissues, and organs based on the same principles of gene regulation. Thus the function of the genome during development and maintenance of a living being is equally complex in all animals.

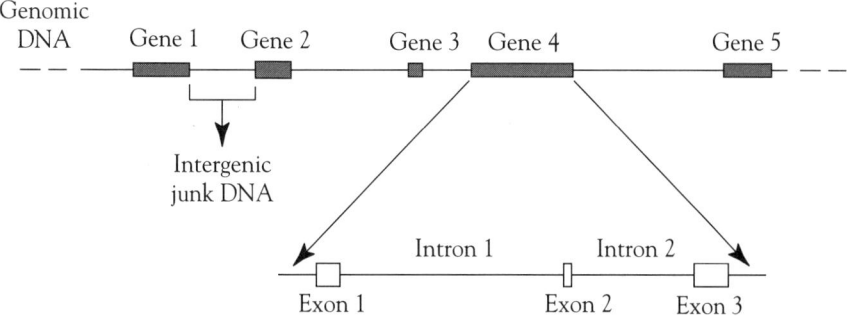

FIGURE 8.1. ALTHOUGH DIFFERENT ANIMALS VARY WIDELY IN ANATOMIC COMPLEXITY, THEIR GENOMIC COMPLEXITIES ARE ALMOST EQUAL. The genomes of all multicellular organisms (from the simplest invertebrate creatures to the most complex vertebrates) are comprised of genes, separated by meaningless nongenic sequences between genes. All the genes present in all the animals, except for rare exceptions, are similarly split into exons and introns. The genomes of widely varying organisms are equally complex in terms of the overall structures and lengths of genes, lengths of intergenic sequences, mechanisms of gene expression, and mechanisms of developing the organism.

We can thus derive a principle that the genomic complexity of even the anatomically "simple" organisms is the same as that of the anatomically "complex" organisms. One can easily derive this by perusing the published literature in molecular biology,[9] which shows that the overall structure of the genes and genomes of organisms — whether primitive or highly complex — is basically the same; and the overall mechanisms of operation of the genome in developing an organism are also the same. In simple terms, all the different genes of all the various multicellular organisms (and unicellular eukaryotes) are of similar structure, and their genetic circuitry is also of similar complexity. This principle clearly illustrates that given that vast numbers of genes, including those of all multicellular organisms, were available in the primordial pond, it is possible for the genomes of the different organisms to have separately assembled with equal ease.

B. *The total number of genes in the different organisms from worm to human is not too different either*

The precise number of genes in the genomes of eukaryotic organisms, including that of a typical unicellular eukaryote, is yet unknown. There are indications that the numbers of genes in all the eukaryotic genomes do not vary widely.

The genome of the bacterium E. Coli has been assessed to contain approximately 5000 genes. It is reasonably expected that the number of genes in a typical unicellular eukaryote is approximately 8,000–10,000.[10] It is interesting to note that the number of all functional genes in the fruitfly *Drosophila* has been estimated to be only about 10,000,[11] which is not different from that of the unicellular eukaryote. The genome of even microscopic organisms such as the worm C. *elagans* appears to contain more than 10,000 genes. Although not precisely estimated, the number of genes in simpler multicellular invertebrates that do not even have defined tissues or organs must be a minimum of 10,000 genes. The number of different RNA transcripts that can be discerned during development of the sea urchin *Strongylocentrotus purpureus* is about 20,000. The gene number in vertebrates, including the human, is supposed to be approximately 20,000–30,000. From these estimates, the number of genes in various multicellular organisms all fall between 10,000 and 30,000 genes. If the average size of the gene is taken to be 10,000 nucleotides, these genes would account for less than 10 percent of the average genome.[12] This leaves the possibility that the distances between consecutive genes in these genomes is ten times longer than the size of the gene itself. Consequently, genes appear to be rarely dispersed islands in very long junk sequences. Does it not show that if the genomic complexities among the multitudes of multicellular organisms on earth are not much different, and if the number of genes are not too variable from each other either, then, in the primordial pond, many distinct genomes could be assembled with equal probabilities?

Although genome complexities do not vary significantly, each organism's genome is unique in terms of the precise number of genes and the amount of DNA.[13] See Figure 8.2. Is this not consistent with the new theory's prediction that if the genomes of different organisms were independently organized in the primordial pond from a common pool of genes, they could have different numbers of genes and different amounts of DNA in each genome?

6. Even the Simplest Life Form Requires a Minimum, but Large Number of Genes

What is life? Living things, as we know of them, have certain characteristics. The simplest living thing that we know of is a bacterium.[14] It reproduces and respires, the two characteristics that we usually ascribe to living things. To make the simplest living entity, one needs a minimum

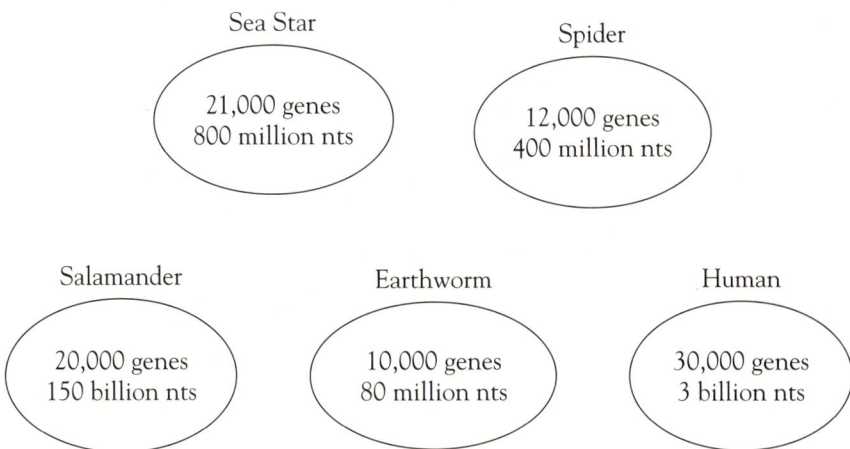

FIGURE 8.2. THE GENOMES OF DIVERSE ORGANISMS ARE UNIQUE IN TERMS OF THE NUMBER OF GENES AND DNA CONTENT. Although the structure and function of genomes are essentially similar (Figure 8.1), they are unique based on the number of genes and the amount of DNA in their genomes. The total number of genes in different organisms are quite variable, although within a range of approximately 10,000 to 30,000 genes. The amount of DNA in their genomes also vary from approximately 50 million (e.g. worms) up to 200 billion characters (in some amphibians), but have no correlation with either the complexity of the organism or its status in the assumed evolutionary tree. The number of genes and the amount of DNA shown are approximate and partly hypothetical. (nts: nucleotides)

set of cellular machineries, enzymes, structural proteins, and genes encoding all these proteins. This set of genes is large — a bacterium's genome contains approximately 5000 genes. Without the minimum number of genes, even the simplest life form cannot exist. Can 10 genes, or even 100 genes, constitute a life form? No. Even 1000 genes do not seem to be capable of that.[15] A minimum number of genes to carry out a minimum set of functions and build a minimum set of structures are needed to construct a basic living form.

Before the formation of any living cell in the primordial pond, there was no selection value for any cellular machineries or genes. Their value had to be tested only in living entities. Thus, unless all the genes and the machineries minimally required for life existed as a prerequisite, life could not have arisen. Therefore, at least this many genes must have existed in the primordial pond.

As I have illustrated, bacteria are only the reductive products of eukaryotic cells. In such a case, the first simplest living cell, being far more complex than the typical bacterium, must have had a higher number of genes — approximately 8000-10,000 genes, the lower estimate for a typical eukaryotic cell. Consequently, the minimum number of genes required for the formation of the first simplest form of life must have been this high.

As a result, if we accept that this many genes did exist in the primordial pond, then in fact millions more were also inherently probable. What we must realize is that when such a large gene pool existed in the primordial pond, all kinds of life forms, from the simplest to the most complex, could be formed. It is only that probabilistically the simpler genetic circuits would be formed first, and these genetic constructs could be used to construct the more complex genetic networks of multicellular organisms in the open primordial pond.[16]

Either the conditions of the primordial pond were such that a very large number of genes existed in it, or the conditions were such that even the minimum number of genes for the simplest unicellular life form did not exist. If simple, unicellular life could be formed, then many complex multicellular organisms could also be formed independently in the primordial pond. In short, it is an all-or-none law — all life, or no life at all. The fact that all the multicellular organisms living today contain approximately 10,000 to 30,000 genes, and that, compared to this, the number of genes in a unicellular eukaryote (~10,000 genes) or even a bacterium (~5000 genes) is not much fewer indeed, is highly corroborative of this.

7. THE EUKARYOTIC CELL PROBABLY ORIGINATED ON EARTH JUST BEFORE THE MULTICELLULAR ORGANISMS

When a unicellular eukaryote could originate from the primordial pond's UGP, it inherently means that the UGP was certainly vast enough to contain genes for almost any protein required to construct multicellular organisms. Therefore, if unicellular eukaryotes could probabilistically arise in the primordial pond, then multicellular creatures could arise. This indicates that multicellular organisms could originate at around the same time or soon after the unicellular eukaryotes came into being. Until recently, it was thought that the first appearance of the eukaryotic cells occurred approximately 600 million years ago. Today, it is thought that the first eukaryotic cells appeared 1,400 million years ago, based on a single fossil finding believed to contain eukaryotic cells.[17] However, the validity of this finding is in doubt[18] (see Chapter 11). If the 600-million-year figure is correct, multicellular organisms that appear approximately 570 million years ago in the fossil record, therefore, came into being soon after the unicellular eukaryotes originated.

THE THEORY OF THE INDEPENDENT BIRTH OF ORGANISMS: OUTLINE AND DESCRIPTION

THE CONCEPT OF THE SEED CELL: THE GENOME OF AN ORGANISM, ASSEMBLED IN A SINGLE CELL, CAN GIVE RISE TO THE DEVELOPMENT OF A MULTICELLULAR INDIVIDUAL IN THE PRIMORDIAL POND

It is not sufficient if all the genes required for the development of an organism is assembled into a genome, and, in a required DG pathway, for it to develop into an individual. First, it must be assembled in a cell. Second, this single cell should be able to express the genome into multicellularity and develop into a multicellular organism. Existing evidence illustrates that this must have been possible in the primordial pond. Before we discuss the new theory in any further detail, let us analyze this evidence and be convinced that if genomes for different multicellular organisms could be assembled in the primordial pond, then there was a mechanism to develop these genomes into the various organisms.

What is a "seed cell"?

In sexually-reproducing animals, development always begins with a single cell called the zygote. In most cases, the male (sperm) and female (egg) sex cells, called gametes, each containing a single (haploid) set of chromosomes, unite to produce the diploid zygote, which develops into the embryo that forms the offspring. This development can occur either within an encapsulated egg outside the body of an animal (oviparous animals) or within a uterus, inside the body of an animal (viviparous). When the embryo is developed in the uterus, the embryo is attached to and supported by the placenta which transfers nourishment from the mother to the developing embryo. In the situation where the embryo grows outside the body in an egg, it is nourished by stored nutrients of the yolk. By whatever mechanism an organism originated, the development of an individual member of the organism starts with a single cell that contains the complete genome.

Genomes and cells were formed together in the primordial pond. Individual genes could code for their protein products even before the first cells were formed. This was first possible because of the primitive transcription and translation machineries available in the primordial pond. Once the proteins from the genes could be decoded, these proteins (along with similarly-formed RNAs) could form the authentic machineries of transcription, splicing, and translation. The assembly of a genome must have happened by the conglomeration of genes in such an environment where the gene-coded authentic machineries were available, and the membranes that surround the cell were also available. It is well established now that the membranes in the primordial soup could self-organize and form a sac-like structure. Thus, once the chance combination of a genome is available in the soup, it could be included in the membranous sac. Out of many chance inclusions, in which the genome, spliceosomes, ribosomes, and other molecules were all included, one right combination would form a viable cell.

In such an environment, if the genome of a multicellular form becomes available in the primordial pond, it could be enclosed as a viable cell. If this has to develop into the multicellular form, additional proteins that would develop the single cell into the embryonic form and then the fully formed organism should also be included. Such additional complexity is small, compared to what is needed for the construction of the single cell. As a result, the probability for this is not too different from the formation of the single cell. Such a single cell that has included the genome for multicellularity and that has the ability to express the genome into the multicellular form, is called here a "seed cell."

Excellent evidence shows that the seed cell concept is valid

The seed cell is analogous to the zygote. It has the ability to grow into an embryo and into an offspring in an appropriate environment such as that of the laid egg of an egg-laying animal, or the uterus of a mammal that supports embryo development. The seed cell, analogous to the developing zygote, could have developed into the offspring in a localized environment that could have existed in the primordial pond with its rich broth of biochemicals: proteins, lipids, carbohydrates, etc.

It is quite possible that such environments provided nutrients for the seed cell analogous to the environments within the eggs of many animals (such as the yolk). In fact, the majority of invertebrates and many vertebrates do not develop fully into the offspring within the egg, because the amount of the stored nutrients within the egg is insufficient. They develop first into a larva, which then must find further nutrients from the environment — and subsequently develops into offspring by metamorphosis.

When we consider the case of the independent birth of mammals, it is reasonable to think that a conglomeration of a large number of cells and biochemicals in the primordial pond could have formed an environment akin to that of the placenta and uterus of mammals. There, a seed cell can differentiate into an embryo and a full-grown offspring.

If the assembly of the genome for an organism from the primordial UGP into a seed cell is quite probable, what we then need is the birth of the first few individuals of an organism to establish a population by mating.

Many observations of living systems support the possibility that seed cells could give rise to individuals in the primordial pond:

1. In all the oviparous animals that produce encapsulated eggs,[19] the eggs contain a single cell, the zygote, from which the whole organism develops.[20] The size of the egg varies, depending upon the size of the stored material. The egg develops by itself into the fully-formed animal completely outside the body of the egg-laying mother. It must be remembered that out of a billion species that ever came on earth, approximately only 5000 species (less than one percent) are mammals,[21] and therefore, more than 99 percent of all animals on earth lay eggs. With few exceptions,[22] only mammals develop their offspring inside the body.
2. In some amphibians and fishes, male and female gametes (sperm and egg) are shed into the water where they unite to produce the zygote; this is a single cell, that directly develops into an embryo completely outside the body of the mother.
3. After egg fertilization in some mammals, the unimplanted zygote (preimplantation embryo) can be flushed out of the uterus[23] and cultured in a test tube (*in vitro*) up to advanced stages.[24,25,26]
4. In some mammals (such as humans) eggs can be fertilized with sperm *in vitro* and can be cultured through several cleavage divisions in the test tube, and then be placed into the uterus for implantation and subsequent development. This is commonly called the "test tube baby."[27]
5. We can clearly see that technological achievements just over the past few decades could achieve the culturing of embryos or growing the baby in a test tube for a considerable length of time. One of the main objectives of embryologists is to prolong the period of *in vitro* culture. In my opinion, a day may come when it will be possible to grow the mammalian zygote to the fully-grown offspring in the test tube. In fact, one of the pioneers in the field of embryology predicts that this should be achievable in the next century.[28] We can then imagine that in the pri-

mordial pond, in geological time, with rich organic and biochemical material and swarming cells, such an environment must have been possible. In essence, a seed cell could have had the ability to grow into an offspring in suitable environments in the primordial pond, even if it requires the nourishment analogous to that of the zygote growing in the uterus of a mammal.

In the new theory, a seed cell can also duplicate and multiply as seed cells. The fact that a zygote can divide into two identical daughter cells under nature, giving rise to identical twins, is supportive of this concept. Furthermore, in the laboratory, each cell of an early-stage embryo can be separated and can give rise to identical offspring. Each cell of the embryo up to this stage has the ability to start as an independent zygote.[29] In some species, an egg can develop without fertilization and give rise to an offspring, which at the adult stage can be reproductively fertile.[30]

In summary, there is no question in the case of egg-laying animals that the fertilized eggs are analogous to the seed cells proposed in the new theory. It is quite conceivable that if their genomes were organized into seed cells in the primordial pond, they could very well develop into fully-formed organisms. In addition, a successful seed cell can divide and multiply into many seed cells and produce many individuals of an organism. (Incidentally, our discussion leads to the conclusion that in the "chicken-or-egg" problem, it is the egg that came first.)

My seed cell concept is applicable to both oviparous (egg-laying) and viviparous (placental) animals. It is convincing — from the details concerning egg-laying animals, as well as the ability of the mammalian zygote to grow in a test tube — that the seed cell could have developed into multicellular offspring in the primordial pond. We must also remember that over a period of geological time, although most such trials would be unsuccessful, rarely some would have been successful, which is sufficient to establish a large number of creatures.

Common bases of life in the primordial pond

There is one more thing that we need to understand before we analyze the theory in detail. There are many common underlying bases for almost all the living organisms, such as the genetic code and basic cellular machineries, which were available in the primordial pond before the first cells were formed. Consequently, any cell or organism that forms in the primordial pond will be able to use them (see Figure 8.3).

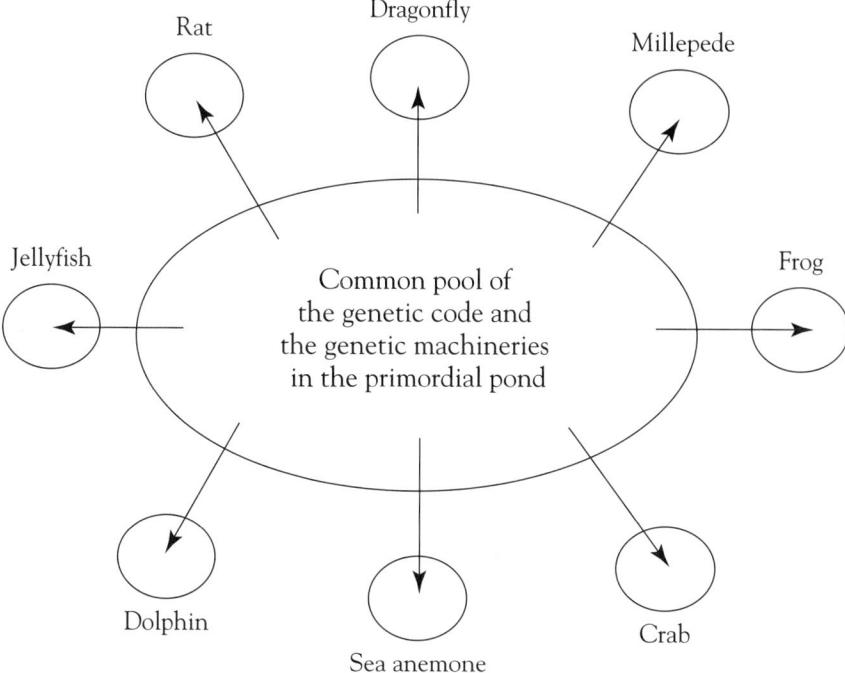

FIGURE 8.3. THE USE OF THE SAME GENETIC CODE AND GENETIC MACHINERIES IN MULTITUDES OF DIFFERENT ORGANISMS INDEPENDENTLY BORN IN THE PRIMORDIAL POND. The genetic code and the genetic machineries such as transcription, splicing, and translation systems had been already established in the primordial pond before any living cells were formed. This pool also contained the DNA-recombination enzymes, such as DNA ligases found in today's living cells, that could recombine different pieces of DNA to help form many genomes. Thus, the primordial pond was a common pool of code, genetic machineries, and genes, from which different organisms could be derived. Consequently, they all used the same code and genetic machineries, but different sets of genes and, more importantly, different DG pathways leading to distinct independently born organisms.

The common bases of life include:

the genetic code;

many enzymes;

cellular genetic machineries;

regulatory proteins and regulatory DNA sequences;

developmental processes and the segregation of sex;

genes for all of the above.

The above are common features of all multicellular organisms, however diverse they are, whether a snail or a raccoon, whether a microscopic plant or a giant oak tree. But they can be derived independently from a common pool of genes, sets of which can lead to these biologically meaningful functions. The genetic code could have become established as the fundamental code for living cells in the common primordial pond so that any living entity stemming from the pond would necessarily use it. It is easily conceivable that the rich broth of the primordial pond could be the *common source* of many such basic functions of life, from which many diverse independently assembling genomes can derive several common necessities for a functional multicellular organism. Once a set of genes for the construction of a basic eukaryotic cell is available as a genome in the primordial pond,[31] the basic genome can be used to assemble and add further complexities, such as the addition of a set of genes for a multicellular organism. Thus, it is not surprising that many basic things about the cells, and the multicellularity of different organisms, have a common underlying genetic basis, although different organisms were independently born from the primordial pond.

Independent and random assembly of genes from the primordial pond into multitudes of genomes led to the independent birth of immutable organisms

So far we have seen that the primordial pond contained a vast number of genes, and that the conditions in the primordial pond became conducive for the formation of genomes and single cells. Then there occurred the formation of many different genomes from the vast pool of genes and from the common source of basic entities for living cells. The vastness of the universal gene pool was such that the number of genes in it must have been several times more than that contained in all the creatures that have ever lived on earth. However, in addition to biologically meaningful genes such as those found in living organisms, the UGP contained a lot of genes for biologically meaningless proteins as well. Random recombinations among the vast number of genes led to many different independent conglomerations of genes. Most of these combinations had no meaning, but extremely rare associations had the right combination of genes organized into the right DG pathways that led to some form of multicellular mass. Again, most of these were meaningless, short-lived entities, but rarely some would have had viable functions. One out of many such living entities were physically fit to earth's conditions. One out of many physically fit organisms were fit ecologically. These were the ultimately fitting and surviving organisms that were perpetuated through geological time.

The new theory of the independent birth of organisms is illustrated in Figure 8.4. Only one out of a very large number of "genomes" assembled into seed cells can become viable. However, because myriads of such genomes can be formed from the vast UGP by random assembly of the genes, effectively a very great number of organisms will eventually become viable in earth's environment. A creature born from a genome should be first physically fit in the environment. If not it dies at birth. If an ecological fit occurs for a physically fit organism, then it survives.

Each independently-born organism has a constant set of genes in its genome, and the DG pathway of each independent creature is rigid. Therefore, the independently-born organism is immutable.

It is clear from the discussions so far that different organisms originated separately in the primordial pond. But can an independently-born organism change into another distinct organism with a new gene or a new body part? The answer is it certainly cannot.

A genome arises as one out of myriads of random combinations. Many such rare successful combinations lead to many different independent organisms in the primordial pond. We must realize that when a genome becomes viable, by the same token, its set of genes becomes fixed. For reasons we discussed in Chapters 3 and 4, the set of genes cannot be changed by any kind of mutation; in other words, for the lifetime of the organism, it is constant.

Likewise, the DG pathway of an organism is fixed and cannot change. The DG pathway of a successful organism is the result of selections of the viable pathways from the myriads of random genetic networks formed in multitudes of genomes in the primordial pond. For the same reason, with whatever DG pathway a genome becomes successful as a viable organism, that DG pathway is fixed for that organism for its lifetime — because it is that DG pathway, by a rare chance, developed a successful organism in the first place. This applies to each organism that was independently born in the primordial pond. Figure 8.5 describes the formation of rigid DG pathways of different organisms. In essence, the constancy of the set of genes and the

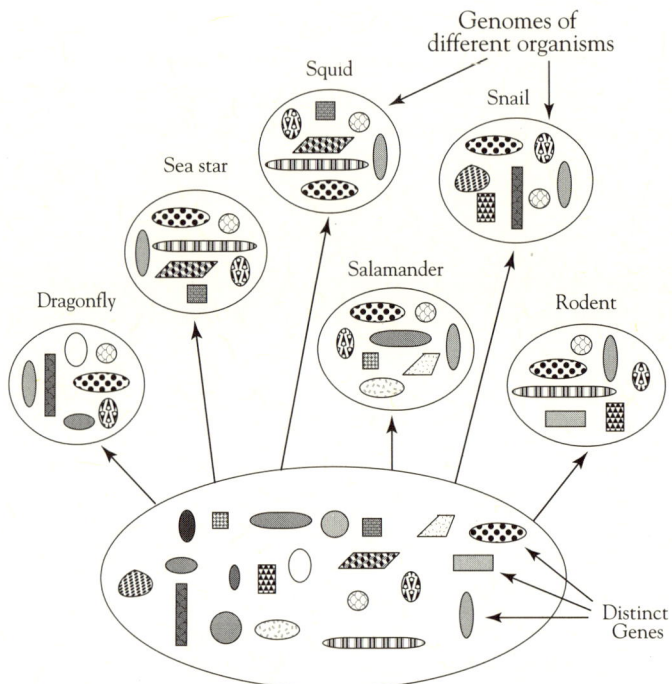

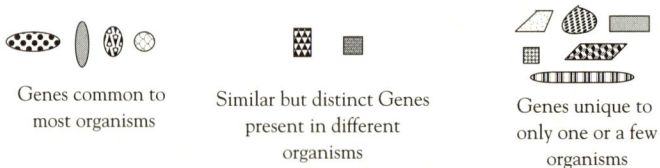

FIGURE 8.4. INDEPENDENT BIRTH OF MULTITUDES OF ORGANISMS BY RANDOM ASSEMBLY OF GENES FROM THE UNIVERSAL GENE POOL (UGP) IN THE PRIMORDIAL POND. The UGP contained myriads of genes for various biochemical functions (shown with different shapes) and multiple copies for each gene (not shown). Random assembly of these genes led to multitudes of independent genomes leading to mostly meaningless multicellular masses, and the rare meaningful organisms. Yet these processes resulted in numerous distinct viable creatures (only some examples are shown). Each successful genome had a unique set of genes and a unique DG pathway. This resulted in the different genomes having a subset of common genes, a subset of similar but distinct genes, and a subset of unique genes. The genomes of various organisms were assembled separately into different zygote-like "seed cells," giving rise to the independent birth of many organisms directly from the primordial pond. Parts of the first successful genomes were included in newer genomes being assembled from the UGP, thereby making it easier to assemble newer genomes resulting in some organismal similarity (see also Figure 8.8).

rigidity of the DG pathway in a genome are responsible for the fixity or immutability of the creature. By the process of independent genome assembly, many different organisms may contain identical or similar genes, because the genes in each genome were derived from a common gene pool, wherein each gene was present in multiple copies. This explains why in reality, living organisms have many common genes. However, by the same process, many organisms can contain a few unique genes. This whole scenario pre-

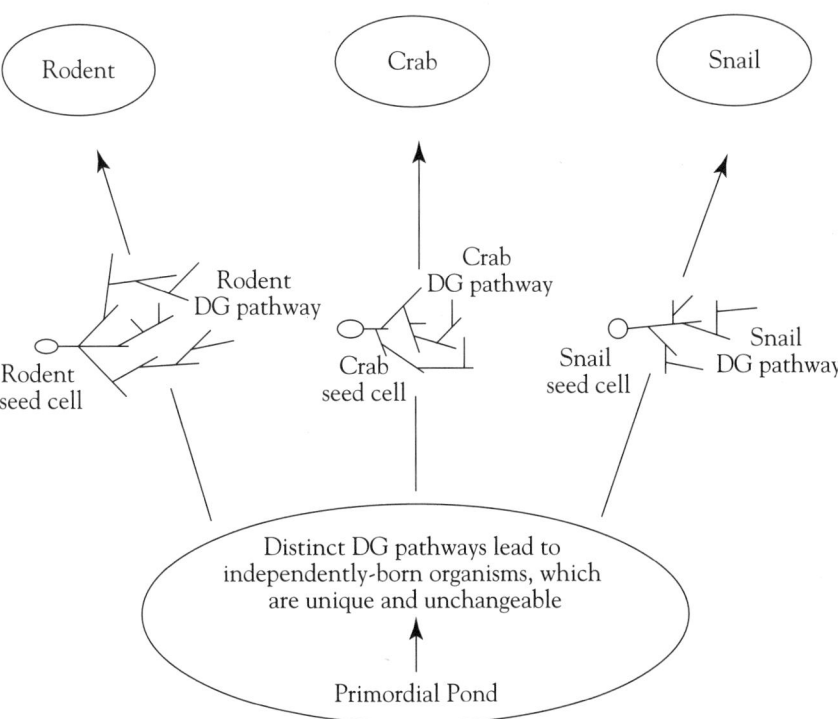

FIGURE 8.5. FORMATION OF A DIFFERENT RIGID DG PATHWAY FOR EVERY INDEPENDENTLY-BORN ORGANISM. From the universal gene pool, different DG pathways for many viable creatures can be formed by random processes. All these organisms can have a set of identical (or at least functionally similar) genes, particularly for building the basic cell, while some sets of genes could be unique. However, the DG pathway of each independently-born organism is distinctly different and rigid. Thus, it is the unique DG pathways that make organisms immutable and unique, although unique genes in many organisms can also contribute to this.

dicted in this theory — the presence of common genes as well as totally unique genes in different organisms — is absolutely consistent with the scenario of organisms living on earth today.

Individual variations of a newly-born organism are produced within its own closed boundary

If different organisms were born independently and if each is immutable (not changeable to another organism) then how were individual variations brought about? Individual variations are caused by sequence variations in the constant set of genes, without affecting the type of any gene, or the rigid DG pathway. Sequence variations are brought about by several kinds of mutation that occur in the genome. However, because the set of genes and the DG pathway of each successful creature is fixed, its boundary of individual variations is also fixed.[32] See Figure 8.6.

At the time of the birth of a creature, there would have been a few male and female individuals of that creature born directly from seed cells in the primordial pond. Once these mate and establish the population, the individual variations can expand through mutations, recombinations and crossing over to fill the constant and defined framework of the immutable organism. They cannot, however, go beyond the closed boundary of the organism. Remember that similar species of an immutable organism could be produced within the closed boundary from individual variations by natural selection or other mechanisms such as mutations in trivial genes (see Figure 8.7). However, they cannot produce a distinctly different organism with new genes or new body parts.

Although each immutable organism thus gives rise to individual variations and similar species within its closed framework, a species — by virtue of its inability to interbreed with another similar species of the same organism — will have its own defined framework of individual variations within the larger framework of the distinct organism.

The limited boundary of individual variations also means that the framework of environmental conditions, in which the individuals of a species of an independently-born organism can be viable, is also fixed. For instance, the temperature at which the individual can survive is a fixed range for every species. The ranges of many such different physical parameters in which an organism is viable can define a multidimensional framework of environmental conditions that is fixed for every species of a distinct organism.[33]

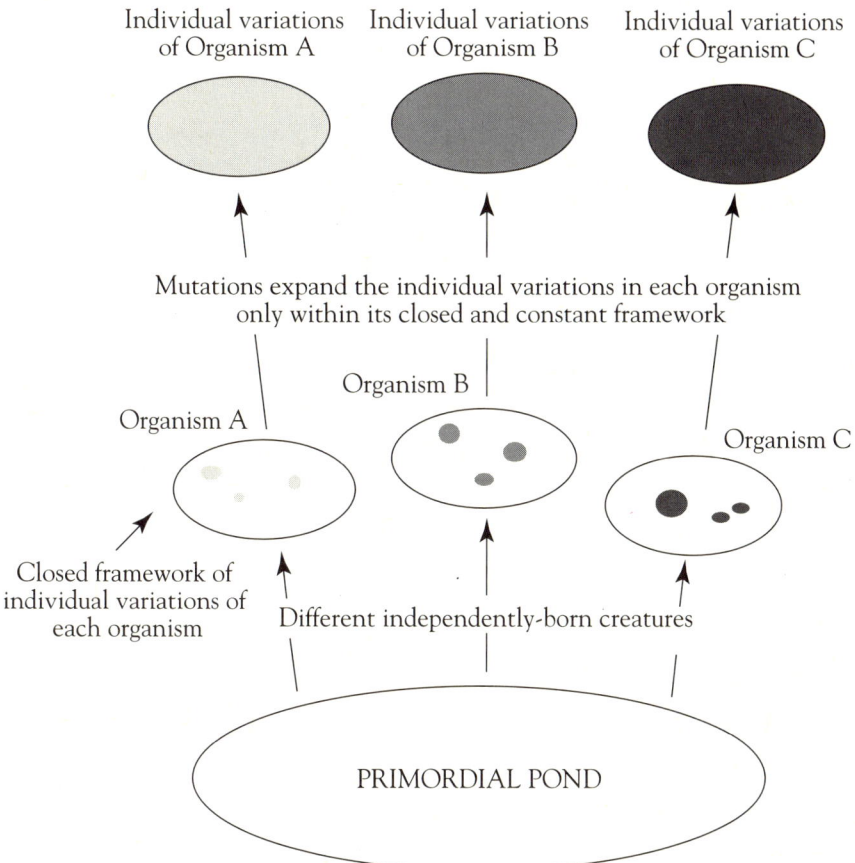

FIGURE 8.6. MUTATIONS CAN EXPAND THE REPERTOIRE OF INDIVIDUAL VARIATIONS OF EACH INDEPENDENTLY-BORN CREATURE ONLY IN ITS CLOSED AND DEFINED FRAMEWORK. Each creature is established by a few male and female individuals born directly from seed cells in the primordial pond. Mutations can expand the repertoire of individual variations within a large, but closed framework. In the figure, the initial framework of a newly-born creature is sparsely populated, and later in time densely populated. This framework is constant however much it is filled and at any length of geological time after its birth.

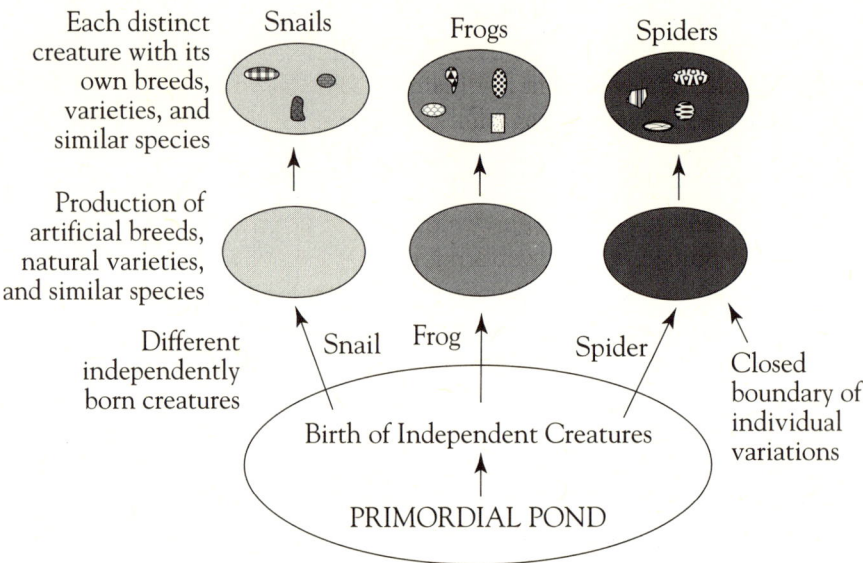

FIGURE 8.7. EACH CREATURE INDEPENDENTLY BORN FROM THE PRIMORDIAL POND CAN GIVE RISE TO ARTIFICIAL BREEDS, NATURAL VARIETIES, AND SIMILAR SPECIES ONLY WITHIN A CLOSED FRAMEWORK. Each independently-born creature is immutable and cannot be changed into another distinct creature by any mechanism. At the same time, a very large population of individual variants can be formed for every creature within a closed boundary characteristic of each creature (Figure 8.6). Within this closed boundary, many varieties and similar species could be formed. The breeds, varieties, and similar species are shown by diagrams with different shapes and patterns within the outline of each distinct creature.

The probability of forming many different organisms is the same as the probability of forming one organism in the primordial pond, because of the equal probabilities for the formation of many different genomes from the open-ended UGP

The probability for assembling the genome of one multicellular organism must be the same or similar to that for assembling another genome of similar complexity. Therefore, given that there existed a vast gene pool with all the genes and regulatory sequences in multiple copies, if one genome could be formed by random assembly of these genes and sequences, then the same process must be capable of forming different genomes for billions of different organisms. In fact, under such circumstances it becomes inevitable. The only reason this is possible is that the UGP is fully open-ended and that the process of forming one genome is the same as that of forming any number of genomes in the common primordial pond. Only the set of genes and the developmental genetic pathways are different in different genomes, despite the fact that they could be overlapping to different extents.

The genomic complexities being relatively equal, different genomes could have separate DG pathways leading to quite distinct organisms with unique organs and appendages. That is, the probability of forming many independent DG pathways is very high. In this situation, there is nothing that can stop this process leading to a large number of very different and unique organisms, with different unique body parts, anatomical complexities, and physiological and functional abilities.

Identical, similar, and unique genes from the common pool of genes in the open-ended primordial pond can be included in different independent genomes. This would lead to different immutable creatures containing the same genes as well as unique genes.

Because the UGP is fully open-ended, it allows the inclusion of the same genes as well as unique genes in different genomes. Consider that there can be multiple copies of a given gene in the primordial pond. Furthermore, there can be many different genes for proteins specifying essentially the same biochemical function. At the same time there can be many unique genes for different unique functions. For example, there can be many copies for the genes involved in glucose metabolism; these can be included in almost every successful genome, thereby having the same gene(s) in all these organisms. On the other hand, there can be many distinct genes for different proteins specifying the same hydroxylase reaction. These could be assembled into different genomes; these genomes would now have different genes for the same function. Another more interesting scenario is that some genes completely unique in both sequence and function could be included in some genomes, e.g., genes that produce a specific metabolic product such as silk, found only in a few organisms, or genes that code for the vertebrate plasma proteins, found only in vertebrate organisms.

What are needed for a genome to express itself into an organism are a set of proteins carrying out particular biochemical reactions and constructing the basic cell, another set of proteins specific to cell types such as liver, eye, and bone, a set of regulatory proteins, and the corresponding regulatory sequences. It does not matter what protein sequence, and thus gene sequence, would lead to the particular enzymatic or structural function. Any of the different proteins that would specify a particular enzymatic or biochemical function is sufficient. Thus the complete assembly of those genes (as functionally defined here) into a genome of a eukaryotic cell can theoretically give rise to an organism. Most organisms use the same metabolic pathways to use nutrients, such as glucose for generating energy, and the same basic unit of life, the cell. Obviously, the genes for these can be essentially the same in many organisms, however widely different those organisms are. In very many cases, organisms do use some special biochemical functions — such as those unique proteins that we described earlier. As we shall see in Chapter 9, many absolutely unique proteins and cell systems are present in numerous distinct creatures. They were also chosen into these genomes, only because it was possible to do this from an open-ended primordial pond. From these several facts, is it not obvious that there is no evolutionary change of a gene into another through organismal evolution?[34,35] As it turns out, what could not be explained by the theory of evolution — the presence of new or unique genes in many different organisms in the living world — is now clearly explainable by the inclusion of the unique genes into the genomes of various organisms directly from the open-ended primordial pond.

The gene pool was immense enough to contain all the billions of genes necessary to give rise to a multitude of creatures. Groups of these organisms have identical gene sequences for some genes, but may contain many organism-specific unique genes. The presence of the same gene in different creatures therefore does not indicate an evolutionary relationship. In this mechanism, similarity at the level of the genome, through inclusion of the same and similar genes from the primordial pond, can be present to varying degrees, resulting in similarity at the organismic level. The commonness of genes can be high enough for many organisms to be classified into groups of similar organisms resulting in a false and misleading scenario as though organisms are related by evolution. In spite of the presence of identical genes and similar genes in diverse organisms, the genomes of organisms are not related by organismal evolution.

EACH NEW CREATURE IS ESTABLISHED FROM ONE OR MORE MALE/FEMALE PAIRS DIRECTLY BORN FROM THE SEED CELLS ASSEMBLED IN THE PRIMORDIAL POND

As alluded to in the new theory, we saw that the genome of a multicellular organism assembled in a single seed cell is capable of developing into a mature individual. The only way to establish a population is for male and female individuals to be born from male and female seed cells, which would mate and start a sexually reproducing population. As we shall see, the probability for the segregation of a seed cell into a male and female seed cells is very high.

Mutations in the genome of a newly-born creature,[36] even at the seed-cell level, contribute to the individual variations of that creature within a finite framework. A small population of each creature (as small as one male/female pair) can expand its repertoire of individual variations through these mutations to fill this finite framework. Hence, one can now see that even one representative genome (male/female pair) defines the boundaries of this framework.

OTHER PRINCIPLES AND PREDICTIONS OF THE NEW THEORY

So far we saw that the independent assembly of genomes in the primordial pond from its common pool of genes leads to independent births of various distinct organisms. This process further leads to many scenarios and principles that are extremely consistent with molecular and organismal aspects of life, as well as the fossil record, absolutely corroborating the new theory. Let us now discuss these principles, predictions, and scenarios.

ORGANISMS BORN FIRST IN THE PRIMORDIAL POND WILL HAVE UNIQUE FEATURES; THOSE BORN LATER WILL INEVITABLY BE CONSTRUCTED WITH SOME BASIC FEATURES OF THE ORGANISMS BORN EARLIER

Pieces of already successful genomes can become part of newly assembled genomes in the primordial pond

We have analyzed the probability of forming many independent DG pathways from the vast number of genes available in the UGP and have come to the general conclusion that it is very high. Thus, at the start of the formation of multicellular life in a primordial pond, there would be many independent genomes whose DG pathways were organized totally independent of other genomes being assembled at the same time — resulting in many unrelated and unique organisms each with unique body parts. Thus there would be multitudes of unique organisms "first-born" in the primordial pond. However, once functional genomes become available, then it becomes unavoidable that these shall be used in pieces in the construction of additional new genomes,[37] although genes from the primordial pond's UGP will also be used. This is because the probability of forming a genome *de novo* from the primordial pond, although considerable and very realistic, is smaller than the probability of forming a genome using already available linked sets of genes as in a genome. Let us not forget that because the UGP was vast, the number of first-born organisms through purely independent assembly of genomes itself could have been innumerable.

What do we mean here by a structurally and functionally linked set of genes? Genes that construct an organism are physically linked into long DNA molecules called chromosomes. This is a reality in today's living organisms. This physical linkage contains genes that are useful or necessary for the construction of the organism. Much of the unwanted or junk DNA could have been removed from the successful genome's chromosomes by random processes. By all these means any large piece of DNA from such a genome, that has a few or more genes, has a better selective value to be used in the construction of a new genome than an equally long random DNA sequence from the primordial pond that had not been used successfully in a genome.

As it turns out, although the genes in a successful genome are already available in the UGP, there is one great difference — the genes in the genome are organized into meaningful sets of genes, with specific timed sequences of developmental switches, that can develop the different body parts, forming a functional organism. In other words, they are not only a physically linked set of genes, but they are also functionally linked, unlike the genes and sequences randomly distributed throughout the vast USP. Because each of these pieces are already successful as parts of viable genomes, despite the fact that they may not have a complete unbroken portion of the DG pathway for a body part,[38] they have a better selective value in the assembly of further genomes — along with other genes and sequences from the USP. Therefore, one or more such DNA pieces from an already successful genome can be included into a new genome that would have otherwise included only individual genes from the UGP by random assortment. Figure 8.8 illustrates this process.

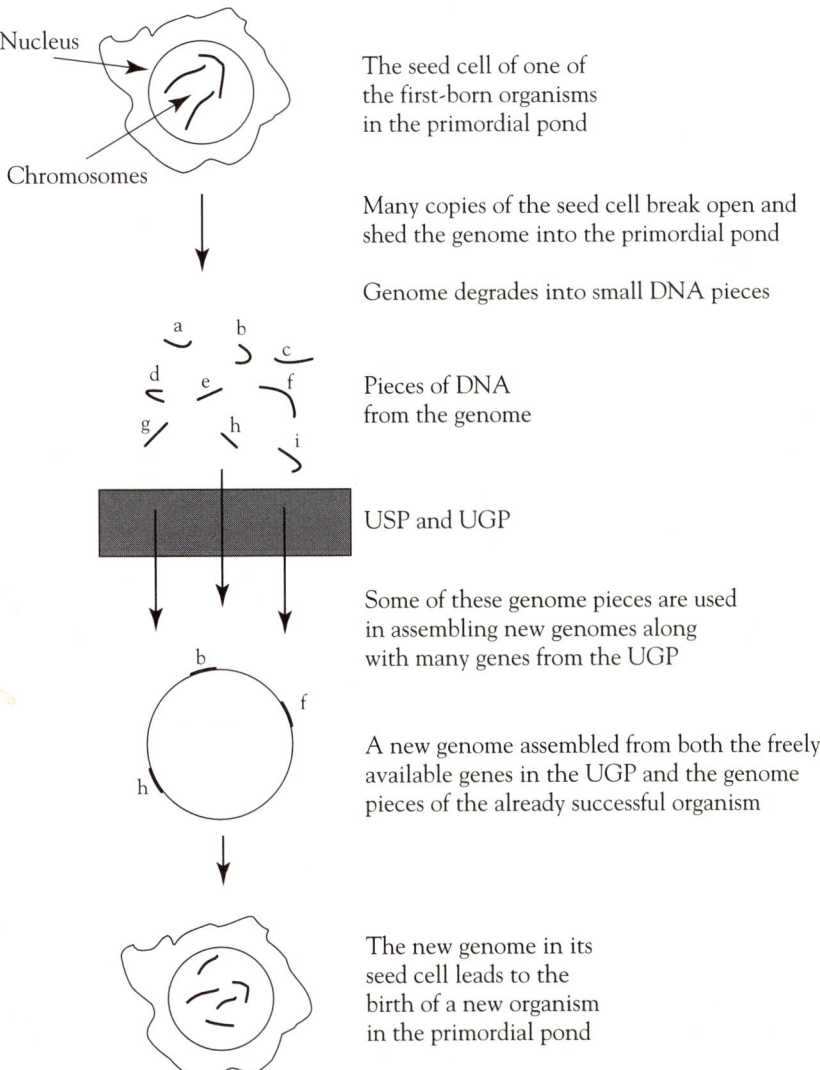

FIGURE 8.8. USING PIECES OF GENOMES FROM FIRST-BORN ORGANISMS TO CONSTRUCT NEW GENOMES OF LATER-BORN ORGANISMS. The different genomes for the first-born organisms were organized independently in the primordial pond resulting in unique organisms. When seed cells (and the individuals made from them) die, they break open, shedding their genomes into the primordial pond. Each of the broken DNA pieces from this genome, because it most likely contained some genes required for the construction of a living organism, had a far greater value than an equally-sized random DNA sequence in forming a genome for a multicellular organism. Therefore, such DNA pieces were bound to be used in the formation of new genomes along with other genes from the primordial pond. This would make some characteristics of the first-born organisms to appear in later independently-born organisms.

Because the use of a linked gene set from an earlier-born organism means the use of a set of biochemical reactions, physiology, or even an organ, the later-born organisms would exhibit these characteristics, not in exact physical structure or biological function, but blended to the structure and function of the newly constructed organism. One ought to expect, by this process, the features of many different first-born organisms can be blended, and new organisms with mixed characteristics can be constructed. That is, if DNA pieces from the genomes of more than one first-born organism were mixed in the formation of a new genome, along with random assortment of independent genes from the primordial pond, the organism from such a genome was bound to have a mixture of features from many first-born organisms. The genome mixing can happen from many genomes that simultaneously became successful in the primordial pond — varying in the amount of DNA from such genomes, and in the number of different genomes from which they are assembled. Thus, these linked gene sets combined with mutation and recombination mechanisms can be used to construct genomes for many disparate organisms with totally unique shapes, sizes, and functions, and with totally different organs and appendages. Let us remember that these new genomes are still assembled independently, although they use pieces of genomes from previously-born organisms. It may be concluded that the later-born organisms are bound to exhibit some of the basic body features of the first-born organisms.

Our observations today are consistent with these principles. When multicellular organisms appeared for the very first time in the fossil record, they did so in an explosion. This happened in the Cambrian period and therefore it is called the *Cambrian explosion*. The organisms in the explosion were too numerous and distinct to have arisen in that short time span by any evolutionary means. When one views this under the new theory, indeed they are truly independent organisms. This is even more true of the Burgess fauna, which I consider to be the result of the origin of life in a separate primordial pond, and each creature was born independent of other creatures in that pond.[39] Many Burgess organisms contained mixtures of characteristics of other organisms in the "Burgess pond." Hence we have reason to believe that the Burgess pond organisms that were born later were produced by using genomes from organisms born earlier in this pond. The basic body plans of the organisms that appear later in the fossil record, starting from the Cambrian period, seem to be built with some body features of those that appeared earlier in the fossil record, especially those that appeared at the beginning of multicellular life in the Cambrian. These facts certainly deserve consideration, and one can see that this scenario is absolutely consistent with what the new theory predicts.

Independent assembly of genomes, and the use of pieces of already successful genomes in constructing later-born organisms, lead to unique body parts and a random mixture of these unique body parts in different organisms. This prediction is true for both living and fossil organisms.

The ability to randomly assemble sets of genes for various organs in the open-ended vast UGP predicts that there should be a random representation of unique organs and appendages in the wide spectrum of living as well as extinct organisms. This is what is precisely seen in the scenario of life on earth. At the same time, the mechanism of genome mixing, that is, using pieces of the genomes of already successful organisms available nakedly in the primordial pond in the construction of further genomes, makes possible the appearance of the same body parts in various combinations in later-born organisms (Figure 8.8). We have wonderful corroborative evidence for this phenomenon.

A strong support from the scenario of living organisms

A large repertoire of organs and appendages of invertebrates excellently illustrate the presence of unique features and body parts as well as their mixtures in multitudes of organisms. When we analyze the invertebrates, they have astonishingly different kinds of body parts, serving unique functions. Let us look at the tube feet of the sea stars or the complex chewing organ, called Aristotle's lantern, in a sea urchin. Look at the mouth parts, stomachs, and digestive systems of different invertebrates, or, for that matter, body parts for any other organismal function such as locomotion. The sea cucumber's mouth parts with its crown of elaborate tentacles, with which it mops the seafloor to obtain organic material, is a unique organ to ponder. Consider the unique filter-feeding structures of totally distinct creatures, the bivalve molluscs and brachiapods, which are grouped under two different phyla (a phylum is a major classification group, such as the molluscs). Let us look at the unique body parts and structures of invertebrate organisms as widely varied as the "walking worm" (e.g., peripatoides, belonging to the phylum Onychophora), shrimp and grasshopper (phylum arthropoda), water bears (phylum Tardigrada), "peanut worms" (phylum Sipuncula), leech or earthworm (phylum Annelida), sea gooseberry (phylum Ctenophora), sea pens, jelly fish and sea anemons (phylum Cnidaria), and snails, sea slugs and squids (phylum Molluscs). We can go on endlessly here.[40] Again, within the large grouping of each phylum, each distinct organism is significantly unique; they only share some common features and therefore are lumped together by taxonomists, but otherwise they are quite disparate from each other. We can conclude, based on the new theory, that these are independent organisms — and indeed they are placed in groups (such as classes and phyla) which are in fact evolutionarily unconnectable.[41] It is important to keep in mind that almost every invertebrate organism (that is genuinely not a variety or a similar species of a distinct creature) is built by a unique DG pathway. Such innumerable and widely disparate unique organisms — almost all of which appeared at the very bottom of the Cambrian period in a geologically miniscule amount of time when multicellular life first appeared on earth — indeed profoundly attest to the independent birth of creatures in the primordial pond.

There are at least a hundred distinct types of eyes in living animals as determined by evolutionists Ernst Mayr and Salvini Plawan. Because these are evolutionarily unconnectable, Mayr and Plawan had tried to show that eyes could have evolved in these many distinct independent lines of organisms from totally blind organisms. Even within every eye lineage that they tried to connect by evolution they faced great problems, and if all their buttressing were avoided, there should be hundreds of distinct eyes in the animal world. They studied only representative organisms, and if all the organisms were studied, we can be sure that there would be thousands of distinctive kinds of eyes that are evolutionarily unconnectable. In looking at nature, one can see that this fact corroborates the independent birth of creatures; each organism with a different kind of eye was independently born in the primordial pond. Just based on one organ, the eye, the living world is a clear proof that an organ for light perception can be constructed in thousands of uniquely different ways. Certainly, we can see that organs and body parts for every organismal function, such as locomotion, sensory perception, sexual reproduction, predation, digestion, and so on, can all be constructed in myriads of ways — just by specific gene connections originating independently in the primordial pond's universal gene pool — thereby constructing myriads of organisms.

The organs and appendages of all organisms, living and extinct, seem to have been randomly assembled from a stock or pool of a variety of unique organs and body parts. This is indeed seen from the many random combinations of these in many organisms. For instance, *Peripatus* is an invertebrate that has the soft body of an annelid and the jointed limbs of an arthropod.[42] *Lophophorates* are invertebrate organisms belonging to a set of three distinct phyla, which have many mixed features among them, but otherwise unconnectable by evolution.[43] The squid is an invertebrate, but its eyes are similar, and in fact superior, to those of vertebrates. The duck-billed platypus is a mammal because it suckles its young, but it lays eggs like a reptile and has a duck-bill. The marsupial characteristic, developing the prematurely-born offspring in a pouch while the offspring is continually suckling, is an odd character in animals with otherwise regular mammalian properties. The extinct *archeopteryx* and the living hoatzin seem to be fully developed birds, but with some reptilian characteristics. Snakes are supposed to be reptiles without legs. But they have poison glands that other reptiles and amphibians lack. Spiders are invertebrates but they have many single-lens eyes,[44] which are unlike the compound eyes of many invertebrates. The structures of marine mammals such as dolphins and whales are another very good example. They have many characteristics of mammals. But they do not have limbs like land mammals. Instead, they have limbs similar to fish — front flippers and tail flippers. Other characteristics of their body are highly suited to aquatic living.

Because of mixed characteristics present in animals, many organisms are considered to be intermediates between assumed predecessor and successor organisms. But truly they are not. They are really independent organisms with a mixture of characteristics. This is corroborated by the fact that even evolutionary biologists agree that the duck-billed platypus does not have the characteristics to be an intermediate between reptiles and mammals;[45] similarly, marsupials cannot be the intermediate between them. In fact, almost all purported intermediates actually are not intermediates at all. They are independent organisms with mixed characteristics that are misleading as intermediate characteristics. The inability to connect such organisms through evolution indeed shows that these organisms must have originated independently in the primordial pond.

Imagine a bag containing all kinds of limbs and organs. Let us now suppose that the bag contains multiple copies of each of these, and you are asked to put your hand in the bag and randomly choose pieces and construct an organism, while discarding those pieces which do not seem to fit. Using your imagination you can create hundreds of differently shaped and sized organisms. This is similar to children playing with Legos; if we assume each Lego block to be a distinct organ or body part, we can build bizarre organisms — like the four-eyed fish, millipedes with very many legs, spiders with eight legs, fishes with many kinds of fins, the dragonfly with its unique body structures, a dinosaur, and a squid with large vertebrate-like eyes. According to the new theory, this is what has essentially happened in the primordial pond. Those random combinations of organs and appendages that would lead to an organism having body parts with the right shape, size, and function would, as a rule, be viable and successful.

Consider the genes and genomes required for the assortment of a mixture of body parts in the repertoire of organisms. Imagine that different sets of genes representing various body parts are put into some bottles and stored on a laboratory shelf. They are marked as to which body parts they represent. There would be millions of such bottles on the shelf. Suppose we mix the contents of various bottles randomly. Most of the mixings will lead to meaningless products, but one out of many will lead to a viable organism. We may also produce organisms that are not viable and therefore will die. Only rarely will a good organism be made by such a random process. However, if only there is an extreme supply of these bottles, then by these random processes, many viable organisms with different body parts and shapes and sizes will accumulate. This is what has happened in the primordial pond.

Fossils of extinct organisms exemplify the random mixtures of body parts

The phenomenon that random assembly of genes from the primordial pond would lead to multitudes of organisms with unique body parts must have been especially true when the USP and UGP were fresh and no organism had yet been formed.[46] However, soon after that, the use of genome pieces from these organisms would lead to a random mixture of these unique body parts in later-born organisms. This is what is precisely seen among the organisms that were formed at the start of multicellular life according to the fossil record — the Cambrian explosion and the Burgess Shale fauna. The multitudes of creatures found among these fauna had unique organs and appendages. This is why they have been classified into a considerable number of distinct groups (phyla and other higher taxonomic groups), which are evolutionarily unconnectable. If we forget about these classifications for a moment and perceive them as individual creatures, we cannot but discern that they are unique creatures with unique body parts, born independently as predicted in the new theory. Furthermore, many of these organisms also contain a random mixture of these unique body parts, illustrating the predictions of genome mixing.

This phenomenon is especially striking in the Burgess fauna. In my opinion, Burgess Shale organisms were born in a separate pond. These organisms were extremely diverse. The organs and limbs of most organisms were unique; those of others are found to be a truly random mixture of these unique body parts. Stephen J. Gould, noted paleontologist and evolutionist, is puzzled about such a scenario because he looks at them with an evolutionist's view. He is confounded how these unique body parts and their random mixtures can be brought about in these organisms by evolution. This is illustrated in his elegantly written book, *Wonderful Life*:

> What order could possibly be found among the Burgess arthropods? Each one seemed to be built from a grabbag of characters — as though the Burgess architect owned a sack of all possible arthropod structures, and reached in at random to pick one variation upon each necessary part whenever he wanted to build a new creature.[47]
>
> ... The Burgess had been an amazing time of experimentation, an era of such evolutionary flexibility, such potential for juggling and recruitment of characters from the arthropod grabbag, that almost any potential arrangement might be essayed (and assayed).[48]

... Now who ever dreamed about a connection between the rear end of a shrimp, the feeding appendage of *Sidneyia*, a squashed sea cucumber, and a jellyfish with a hole in the center? Of course, no one did. The amalgamation of these four objects into *Anomalocaris* came as an entirely unanticipated shock.[49]

... Whittington and Briggs concluded that *Anamalocaris* "was a metameric animal, and had one pair of jointed appendages and a unique circlet of jaw plates. We do not consider it an arthropod, but the representative of a hitherto unknown phylum."[50]

... I suspect that such a strange phenomenon prevails in Burgess time, and that we have had so little success in reconstructing Burgess genealogies because each species arose by a process not too different from constructing a meal from a gigantic old-style Chinese menu — one from column A, two from B, with many columns and long lists in each column.[51]

... I think that Dereck Briggs had a model like this in mind when he wrote of the difficulty in classifying Burgess arthropods: "Each species has unique characteristics, while those shared tend to be generalized and common to many arthropods. Relationships between these contemporaneous species are, therefore, far from obvious, and possible ancestral forms are unknown." I also think that the model of the grabbag might be extended to all Burgess animals taken together, not only to the arthropods separately. What are we to make of the feeding appendages on *Anomalocaris*? They do seem to be fashioned on an arthropod plan, but the rest of the body suggests no affinity with this great phylum.

... The jaws of *Wiwaxia* (recalling the molluscan radula) and the feeding organ of *Odontogriphus* (recalling the lophophore of several phyla) come to mind as other possible features from the mega-grabbag.

Is it not obvious that the organisms in the Burgess pond with unique features must have been born independently, and that is why they are evolutionarily unconnectable? Does not the whole scenario fit precisely with what is predicted from genome mixing in the primordial pond? On the whole, the evidence is so striking and remarkable that the predictions of the new theory are abundantly borne out in the scenario of the living as well as extinct worlds.

Evidence for the use of basic body plans from earlier-born organisms in the later-born organisms by using prior genomic pieces

Because pieces of genomes of earlier born organisms were used in the later-born ones, the organisms which are born later in the primordial pond can also have the basic body plans of those that were born earlier. This is true in living and extinct organisms at many different levels of body plans. For instance, the *coelom* (the internal body cavity that houses the major organs) is supposed to be a general body plan. There are coelomate animals and acoelomate animals. A substantial fraction of the animals are built on the basic coelemate structure. Most animals have either a radial symmetry (starfish) or a bilateral symmetry (human). There are chordates (with a backbone-like structure),[52] which is another basic body plan among coelomate animals. Some invertebrates and all vertebrates are chordates in the sense that they follow this definition. Varied groups of invertebrates, each of which appear to adhere to a certain basic body plan, are grouped under different phyla. For instance, echinoderms, molluscs, and arthropods each follow a different unique body structure.[53] But in all these animals defined under a basic body structure, most organisms are evolutionarily unconnectable. Most coelomate organisms are evolutionarily unconnectable. Likewise, most chordate classes, orders, and most arthropods are again evolutionarily unconnectable.[54] The only way this could have come about is if the basic body plans were used in an open-ended primordial pond as DNA pieces, but the organisms' genomes were assembled separately and the organisms were born independently. All these indeed clearly corroborate the predictions of the new theory!

Other implications of genome mixing in the primordial pond

The use of linked sets of genes from the genomes of earlier-born organisms in constructing the genomes of later-born organisms has many further implications regarding the independent birth of creatures.

1. *It increases the probability of assembling newer genomes at the start of multicellular life leading to a burst of organisms.* Once one organism was successful in the primordial pond, the process of using small or large pieces of an already successful genome would tremendously increase the probability of assembling newer genomes leading to a burst of organisms. By this process, additional genomes can construct body parts and features over and above the already constructed body features because the probability for assembling gene connections using existing templates is obviously more than if these have to be organized only from randomly distributed genes.

2. *Same structures and functions, from subcellular to whole body parts, can be used in many unique organisms.* Once a linked set of genes for a given function, for instance a metabolic cycle, is available in the primordial soup in multiple copies, it can be used in a variety of subsequently assembled genomes. By this process, gene sets can be used commonly in many separately-born organisms. Certain basic genomic mechanisms, such as gene regulation, can be thus employed. It is conceivable that the use of linked gene sets would continue in the primordial pond for a considerable geological time.

3. *Genome mixing will still lead to unique organisms.* Although many sets of linked genes from different organisms could be used in assembling many new genomes, they would be necessarily organized into unique DG pathways which lead to absolutely distinct, independently born, immutable organisms. In all these, the genes in the UGP are also used to different extents and all the genomes are still independently constructed. Therefore, the resulting organisms are still independently born, although they may have some body parts common to previously-born organisms.
4. *Genomic repatterning without mixing from other genomes.* It is also possible that the genome of an organism available freely in the primordial pond could change drastically by random processes without involving other genomes. In such a case it can lead to a distinct creature, however still resembling the organism that the genome originally represented. This explains why in some of the closely resembling organisms in the living world, there are drastic differences in their genomes and in their anatomy that could not have happened through organismal evolution; they could only have occurred in the open-ended primordial pond.[55]
5. *Slowing the formation of unique and mixed genomes in the primordial pond.* Although entirely new body plans and structures can come in later geological time, there is one thing that would limit this — the gradual depletion of the USP and the UGP in the primordial pond. In spite of the fact that the chemical activities of the primordial pond would be replenishing and adding to the resources of the USP and UGP, the biological activities, i.e., the formation of biological cells and organisms, will slowly deplete them. Firstly, the uniqueness of the USP and UGP will be exhausted and slowly converted to more and more linked gene sets of successful organisms. Secondly, the DNA material itself will be drained from the primordial pond by the formation of the organisms' genomes. Thus, eventually the primordial pond will be depleted and the birth of creatures will cease. However, by then a large repertoire of creatures would be stably living in all the nooks and corners of the earth.

It may be inferred from the above arguments that at the beginning of multicellular life in the primordial pond, organisms born will be purely unique for a certain amount of geological time — a prediction that is supported well in the fossil record. Soon, the use of linked gene sets for the basic metabolic pathways, tissues, organs, and body parts from the first-born organisms in the newly constructed genomes will lead to later-born organisms that have these features in them — another prediction that is proved to be true both in the fossil record and in living organisms. In time, more and more linked gene sets would be available for a wide variety of body features, functions, biochemicals, and metabolic pathways. For instance, many organs and body parts such as the eye, bone, liver, kidney, heart, lung, wing, feather, hair, and so on, are used as basic organs in a variety of different evolutionarily unconnectable organisms. The mechanism of genome mixing also provides a thoroughly faultless justification for the random mixture of characteristics in many organisms such as the squid, the *peripatus*, and the duck-billed platypus. Furthermore, it explains the use of basic regulatory mechanisms such as the homeobox genes and their general pattern of deployment during development in organisms as diverse as the worm to the human.[56]

"RANDOM PERFECTION:" THE NEW CONCEPT THAT EXPLAINS THE PERFECT FIT OF ORGANISMS TO VARIOUS ENVIRONMENTS

When one hears that an organism such as the lobster or rat originated as complex as it is in the primordial pond, it would be indeed difficult for anyone to believe it. The initial reaction would undoubtedly be shock and total disbelief. Although it seems impossible and improbable, when we consider this in the context of the random processes by which organisms were born in the primordial pond, it would become clear that what is impossible and preposterous in one context is inevitable in another.[57] Just as the occurrence of a given side in the throwing of dice is almost absolutely certain when we throw it one hundred times, the occurrence of a meaningful living entity among meaningless multicellular masses would be certain when the random events leading to these masses are vast enough. In other words, when the number of random events are large enough, the unbelievable will certainly happen. This is what made possible the birth of the beautiful organisms among the vast number of meaningless ones in the primordial pond.

Organisms are well-suited to the earth's physical, chemical, and ecological conditions

The organs and appendages of various kinds of organisms are well suited to their environments.[58] Snails, earthworms, squids, sea-stars, insects, fishes, and birds are all very distinct organisms living in dissimilar environments, but their body structures and appendages are almost perfectly suited to their respective physical and ecological environments. This has been explained in evolutionary theory to be the result of the adaptation of organisms to the ever changing environments while they also transform from one organism to another. But we shall see that this is not true. We shall demonstrate that this is the result of a process in which among the vast repertoire of organisms born independently in the primordial pond, only those organisms suited to an environment survives while others perish. Because it is the result of statistical and probabilistic occurrence, I have called this mechanism "random perfection" or "stochastic perfection" of immutable organisms. We will use the simpler term, random perfection, from here on.

In this mechanism, among a vast repertoire of organisms with grotesque and bizarre ingrowths and outgrowths, a whole spectrum of organisms with all imaginable organs and appendages will be born without concern to the earth's environment — for instance, organisms totally unfit for an oxygen atmosphere, but may be fit in an environment without oxygen. Because a vast number of different kinds of organisms is made possible by the very mechanism of the random assortment of genes into a vast number of genomes in the primordial pond, leading to a wide spectrum of internal and external organs and body parts, the probability of the occurrence of organisms that fit the physical and ecological environment is a realistic one. Although all creatures were born without concern to the environment, multitudes of these would fit the earth's various environments purely by coincidence.

All kinds of multicellular masses, mostly meaningless, are possible by the differentiation of independent genomes assembled by random processes in the primordial pond. Organisms that had organs and appendages fitting in the earth's physical and chemical conditions must have occurred purely by chance among these myriads of multicellular masses.

Between the single-celled organism and the perfect multicellular organism, all kinds of life forms, some meaningful and mostly meaningless, must have been possible. These must have been born by the differentiation of the great number of randomly assembled genomes. Most forms may not have been able to use any kind of food and have any meaningful function in any environment on earth, because they did not possess any useful, well-defined organs and limbs. But one out of these many multicellular masses would be able to use some food, reproduce sexually or asexually, and live on earth.[59]

Only if a multicellular mass produced by the combination of genes within a seed cell can exhibit some of the locomotive functions, food consumption and excretion of wastes, and some self-defense mechanisms, then it can be defined as an organism fit to live on earth. Only if it is capable of asexual or sexual reproduction then it can self-perpetuate. Only such organisms that fulfill these minimum requirements will become "viable" and "live." Therefore some form of energy-generating system and respiration are the first conditions for a mass to be alive. The energy transfer in one case seemed to have required organisms being fixed to the earth, deriving their energy directly from sunlight, air and earth's chemicals — what we call plants. But another set of organisms, which could move on the earth's different media (water, air, or land) could find its energy in the form of "food," which are other living forms — these we call animals.

One out of a very large number of randomly-assembled genomes can produce a multicellular mass, which would have some parts and structures meaningful as organs and appendages that can carry out some basic "animal" functions. The power of the primordial pond's molecular mechanisms was such that they could produce, by the independent assembly of genomes, a wide repertoire of organisms from the anatomically simplest organisms to the anatomically complex invertebrates and vertebrates; from those with almost no intelligence to those that are highly intelligent — all among a far larger repertoire of mostly meaningless multicellular masses, with and without grotesque internal and external protrusions, among which coherent body parts occurred. In short, the organisms that are perfectly fitting on earth and those that seem meaningful and beautiful are truly a statistical result of the occurrence of a great number of organisms in the primordial pond that were mostly bizarre and grotesque.

Only those organisms that were suited to the fundamental physical and chemical environments on earth — gravity, temperature, land, water, air, light — could be fit to live. Therefore, for every organism that emerged as a viable life form, there were many, many multicellular masses that perished.

"Random perfection" might also be termed "random selection by environments," but we shall not use this term because of a possible misunderstanding and misinterpretation with Darwin's term of "natural selection," which in fact has absolutely no connection to the new concept of random perfection.

As a rule, an independently born creature can become viable only if it is fit in the earth's physical environments. Therefore, the corollary is that if a creature becomes viable, then it would contain the organs and appendages for all the minimally essential complex functions such as locomotion, finding food, sexual reproduction, etc. Hence it is conceivable that the organisms that became viable on earth, (and those that lived for a considerable time and became extinct), must have had perfectly fitting organs, limbs, and appendages at the time of their birth from the primordial pond — with many differing shapes and structures, obeying the only rule that they should fulfill the basic requirements of a "living organism." However astonishing the fit of organisms to their environment, whether physical or ecological, all of them can be accounted for through random perfection.

Now let us turn to the DG pathway of an organism. Given the universal set of genes that occur in all organisms, a vast number of DG pathways can be formed in myriads of permutations and combinations. Furthermore, using pieces from already successful genomes in the construction of genomes for new organisms — giving it an added power for the simultaneous burst of independent creatures — must also be taken into account in this process. These processes would bring forth an almost unlimited number and variety of creatures from among which those that are physically and ecologically fit would survive. In essence, only because of the vast number and the variety of organisms possible, all the perfectly fitting organisms living on earth and those that became extinct could arise. This is still an extremely small subset of the organisms born in the primordial pond. In this process, there is no connection between the genome and the environment. The genome contains only a set of genes organized into a series of genetic networks resulting in a multicellular mass. It is the environment that determines which multicellular masses are suitable. Thus according to the new theory, the power of random processes, starting from a vast gene pool, could bring forth the millions of creatures that ever came on earth independent of each other. It must be remembered that there is no adaptation, discussed in Darwin's theory, in these processes.

The sieving effect of random perfection

Consider the physical sieving of some materials such as a crude collection of pebbles from the beach with random sizes and shapes (including grotesque shapes) in order to obtain pebbles with the desired size and shape. When a sizable quantity of the pebbles are sieved, through a series of specifically sized and shaped sieves, it will result in the elimination of most of the randomly shaped pebbles. Only a few pebbles out of all the pebbles would make it through all the sieves, and consequently would have the specific shape and size for which the sieves were designed. This process enables one to get the pebbles with desired shapes and sizes from the starting collection of randomly shaped and sized pebbles. What comes out of the seives would appear to be highly designed, although we started with a crude, random collection of pebbles.

This is the approach we have to take to comprehend random perfection of organisms. Here the random sieve that selects the organisms to be fit on earth are natural forces, defined by the physical and chemical conditions of the earth first and then the ecological conditions. Remember that this is not "natural selection," where starting with a population of a species with individual variations, the subpopulation with favoured variations in an environment are preserved while the unfit ones are culled, allowing one species to change into another. On the contrary, what we are dealing with here is an abundant number of independently-born immutable organisms. They are automatically sieved through the physical and ecological environmental conditions resulting in the best fit ones to the environment, while eliminating those which do not fit at all or as well as others. There is no "adaptation" occurring here.

The random sieve therefore results in highly varied creatures that would appear to be perfectly designed to the physical, chemical, and ecological environments. However, though the selected organisms are immutable, slight refinement of every organism is possible within the closed framework of the immutable organism (as discussed below), which would have aided further in the perfect fit of organisms to the environment. But, it is imperative to keep in mind that the refinements are strictly confined only to those organs and appendages with which an immutable organism is born in the primordial pond, because no new body part can ever be evolved.

Perfect ecological fit is due to the selection out of diverse immutable creatures, not individuals within a species. "Fine tuning" of different subpopulations of an immutable creature to different environments can happen by natural selection, but only within the closed framework of the creature.

What is fine tuning? The individual variations in the population of a particular creature may allow the fine tuning of already existing body parts in an independently-born organism by the principles of natural selection. This principle can operate only here, and cannot lead to the evolution of distinctly new organisms with new body parts. For example, the proboscis (the feeding tube) of the butterfly can be fine tuned in length to suit the flowers, but the proboscis itself could not have evolved from an organism that lacked it. It must have occurred, at least as a primitive organ, in a butterfly born in the primordial pond. Its fine tuning could have happened only within a closed framework. The acacia tree and the ant *Pseudomyrmex* are perfectly fit to the earth's physical environment as well as to each other[60] (see also Chapter 10) because of the random occurrence of both of them among the myriad organisms that were born independently and immutably in the primordial pond, with each of their special characteristics at least as primitive as possible. But the fine tuning of their characteristics, with which they were already born, to fit to each other better would have happened by the process of natural selection.

The independently-born creature that is not fit to the physical environment will not survive and cannot perpetuate. If a physically fit organism had a conspicuous color, shape, or size, it would make easy prey. For instance, a small insect that is delicious for predators, and that could be easily spotted by its color and shape, would have certainly been eaten away soon and become extinct. What is left will be perfectly fit physically and ecologically.

All these random perfections happen just as surely as liquid flows down hill due to gravity. If one pours water into a container with many interconnected tubes of different shapes and sizes, the water takes these different shapes automatically. In the random perfection of the perfectly fitting creatures on the earth out of the myriads of independently-born creatures, such a mechanism was occurring automatically and smoothly.

Natural selection and adaptation can occur only within the closed bounds of the independently-born immutable creatures and can lead only to varieties and similar species of every creature

Although an independently-born organism is unchangeable to another distinct organism, it can have a large repertoire of individual variations leading to many varieties and similar species, all of which will fall into its characteristically constant, closed framework. Thus, the new theory should be understood at two levels: 1) Different creatures are born independently and are immutable. 2) Each independently-born creature has an innate ability to change within a finite boundary based on individual variations and mutations, producing artificial breeds, natural varieties and similar species characteristic of the distinct organism. See Figure 8.7.

Because there is considerable room for swaying within the defined limits of a creature, the creature can adjust to physical and ecological environments as the environment changes. This may be one reason for the perfect fit of organisms to one another and to the environment. All these refinements, however, can happen only within the boundary of an immutable organism. By these refinements, an earthworm cannot transform into a snail or a centipede even through a large number of intermediates, but one earthworm variety can change into another earthworm variety or one snail variety into another snail variety. An invertebrate could not have changed into a vertebrate such as a fish with its new body parts of bones and the complex single-lens eyes, but the fish could have only originated independently at least as a primitive fish with its bones and eyes and could have refined into some other fishes.

It is of paramount importance for us to understand the distinction between the immutability of a creature and the flexibility of a creature to sway within its closed framework of individual variations. The powers of Darwin's mechanisms — natural selection and adaptation — are now reduced to be operative only within the confines of each immutable organism, capable of producing only varieties and similar species.[61]

THE INDEPENDENT ASSEMBLY OF GENOMES LEADS TO DISTINCT GAPS AMONG CREATURES: THEREFORE GAPS AMONG CREATURES ARE REAL AND MISSING LINKS ARE IMAGINARY

The gaps between organisms and the missing link problem

What are gaps between organisms? Let us take the crab and the human. Obviously they are totally different from each other. There is a definite gap between them. But this is not the gap that we are now concerned with. We are interested in the gap that exists between the supposedly related organisms — between one creature and another that has supposedly evolved from the first. Such a gap might be between an earthworm and another worm-like creature such as a millipede, centipede, or velvet worm. The gaps between creatures from two different classes within a phylum, or between the supposed invertebrate ancestor of the vertebrates and its supposed vertebrate descendent, are examples of gaps in evolution. These gaps are clear and distinct. If we look in evolutionary biology textbooks, we can see that in the "evolutionary trees" these gaps are connected by dotted lines, meaning that the intermediates, or missing links, have never been found.

We are not concerned with gaps between similar species of a distinct organism, usually described under evolutionary theory as gaps between species. The similar species of the cat — the tiger, ocelot and panther, each with its own confined framework of individual variations — have gaps among them. These are not the gaps that we are trying to explain by the independent birth of organisms. These are explainable by either natural selection or trivial-gene mutations within the closed confines of an immutable organism. It is the distinct gaps among distinct creatures that are supposed to be evolutionarily related — but cannot be explained by any evolutionary mechanisms — that we are explaining here by the independent birth of organisms.

Independent assembly of genes into unique genomes leads to gaps among distinct, independently-born creatures

When one understands clearly what is an independently-born organism, one can see that there ought to be gaps among such organisms. If individual creatures originated directly in the primordial pond as independent entities, there ought to be gaps between any two such creatures. This is what we precisely see in the living world. Gaps among creatures is a genuine phenomenon; it is not an artifact resulting from extinction of assumed intermediate organisms. It is the process by which the different genomes were separately organized and the creatures were independently born in the primordial pond that is primarily responsible for the gaps among them.

Independently-born organisms are unique in many distinguishing characteristics (Figure 8.4). Because the genes and the DG pathway of a creature's genome are fixed and unchangeable (Figure 8.5), the independently-born creatures are also immutable. Hence, we may conclude that the gaps between any two distinct creatures originated when they were born in the primordial pond, and these gaps are permanent.

DIFFERENTIAL EXTINCTION OF INDEPENDENTLY-BORN CREATURES: THE MECHANISM OF EXTINCTION IN THE NEW THEORY

If distinct organisms were independently born, why should there be extinctions, and why should certain organisms become extinct in a changing environment, while others survive? The range of physical and ecological conditions in which the individuals of an independently-born creature can be viable is finite and fixed, and is characteristic of the creature. When environmental conditions exceed and exclude this range, the creature would become extinct. This finite range is different for different creatures, although they could overlap. Therefore, in one extreme set of conditions, some organisms would become extinct, and some would survive.

With all its physical flexibility of sequence changes and mutations, the genome of an independently-born organism is functionally fixed. This means that the ranges of many environmental parameters within which the individuals of an organism can live are indeed fixed. Environmental conditions can include many different parameters, such as temperature, pressure, and chemical balance (e.g., oxygen, carbon dioxide, and chemical poisons). The minimal amount of food required by an individual is another parameter because the availability of food can vary in different environments. One must note that the genome fixes the range of these parameters as inherent abilities of the organism. With all the sequence variability of the genome, which offers a tremendous range of possible variation of the organism, we can see that still the range of environmental parameters in which the individuals of a given organism can be viable is fixed. Once we understand this principle then we can straightaway see that if the environment goes out of this specific range crucial for the survival of an organism, then the organism will become extinct.

A creature cannot change endlessly according to the change in the environment producing new distinct creatures, as believed in Darwin's theory of evolution, because the set of environmental conditions within which the population of a creature can be viable is finite. The population of a creature living at a given time and in a given set of environmental conditions need not represent all the individual variations of a creature possible within its finite bounds. It may represent only a small subset or portion of the whole framework. A changing environment may vary the population within the closed circumference of the creature, so that only that subset population fit in the current environment will survive (Figure 8.9). In this sense the population of the creature follows the environment, but it must be kept in mind that this occurs within the closed framework characteristic of the creature. However, if and when the environment returns such that it is well within the creature's complete framework of allowed environmental conditions, the small surviving subpopulation is sufficient to expand within the constant framework of individual variations, because the subpopulation contains the intrinsically constant set of genes and the rigid DG pathway of that creature.[62,63]

FIGURE 8.9. EXTINCTION IN THE NEW THEORY. The genome of a distinct creature allows myriads of varied individuals in its population, which can survive in slightly different environmental conditions. The survivability of these individuals, however, are confined to a set of environments within a fixed range (which corresponds to the constant boundary of individual variations of the creature). This constant framework does not change no matter how much the environment itself changes. When the set of environmental conditions shifts or constricts (Situation 1), the range of viable individual variations changes according to the changing environment. However, if and when the environmental conditions change beyond the boundary of the creature so as to exclude all the conditions in which any of the individuals of the creature can survive (Situation 2), then no individual of the creature survives. This is also true for each of the similar species of a distinct organism.

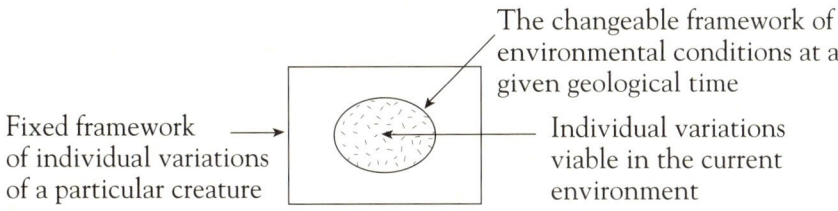

Situation 1: When the environmental framework shifts, the range of viable population moves accordingly. When the environment moves considerably, it constricts the living population within the constant boundary of the creature.

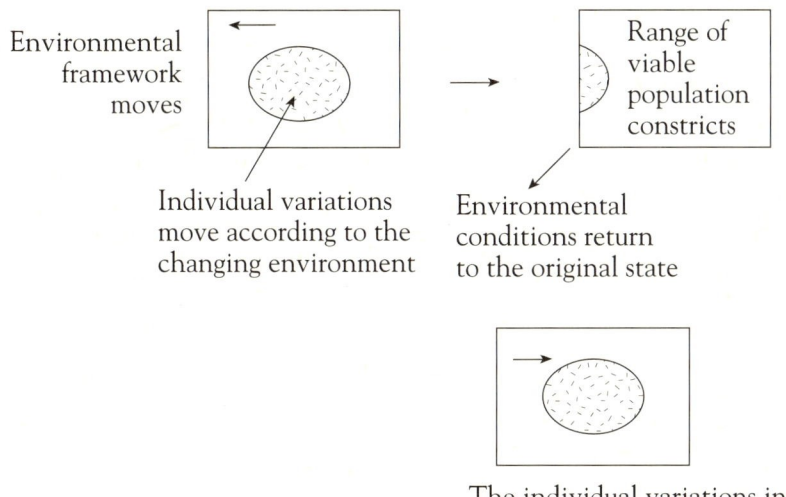

Situation 2: When the environmental conditions go completely out of the range of the fixed framework of the organism, no individual survives and the organism becomes extinct

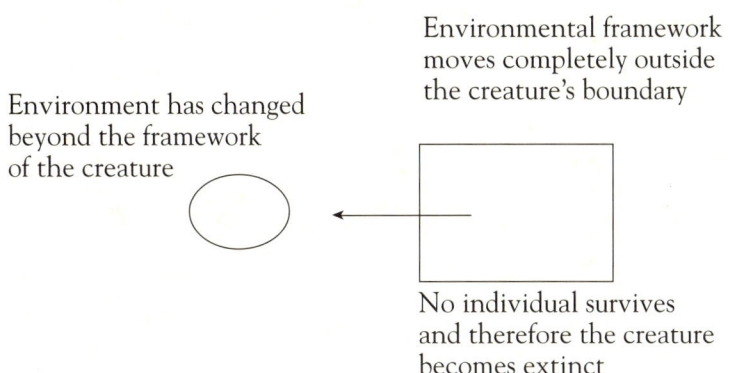

341

Now let us turn to the many great extinctions on earth as revealed by the fossil record. Extinction occurs throughout the fossil record, and sometimes extinction becomes widespread. Swaying environments are the reason for the regular extinction of creatures. Periodic drastic changes in environmental conditions are responsible for the simultaneous extinctions of many creatures. The changing environment can affect different creatures variably, because the allowed environmental boundaries for different creatures are distinct. Hence it can be expected that, in one extreme set of conditions, some organisms would become extinct while others survive.

It is said under Darwin's theory that in a changing environment an organism alters accordingly, producing new creatures; but in a rapidly changing environment, a creature does not have enough time to adjust and therefore dies. On the contrary, my theory proposes that the independently-born organism does not change into another distinct organism. It adjusts only within its defined bounds according to the changing environment, as long as the change is within the bounds. Furthermore, the slowness or the rapidity with which the environment changes does not matter as much as how drastic the environmental change is. When environmental conditions change beyond the defined limits specific to a creature, extinction becomes inevitable.

Reasons for the overall similarities common to all independently-born organisms

An overall view of the living world would show that there are many similarities among the various organisms — albeit they are independently born, and contain entirely different body parts and features. Consider the functional similarities in millipede legs and human legs that share absolutely no structural similarities. During the separate birth of diverse organisms in the primordial pond, as we shall discuss, these basic similarities must have been brought about by three major unifying principles:

1. Functional constraints of organisms.
2. Environmental constraints on earth.
3. Use of the same or similar genes from the prior genomes in the assembly of newer genomes.

Similarity in independently-born organisms due to the minimum functional requirements for basic living

Every living organism must have a minimum set of structures and functions. In the case of a multicellular animal, it has to be able to move, eat, digest, excrete, and reproduce. It must have some sensory organ by which it can perceive its environment. There can be very simple to very complex organisms in the wide repertoire of organisms that were independently born. One can see that each animal born will have some body parts to fulfill each of the above requirements to live on earth, but each body part that fulfills one particular function can be quite varied in different organisms with respect to its structure and the mechanism by which it functions. For instance, if we consider all animals that have some mechanism of movement on land, the organs and/or mechanisms used for movement can be very different in each — such as the hundreds of legs of millipedes, the "leg" or "foot" of snails, the legs of crabs, rats and elephants, and the mechanism of locomotion of worms. If one views the variety of body parts in multitudes of organisms with the new theory in mind, it is clear that the predictions of the theory are well borne out. In other words, the scenario of body parts of all animals that live on earth and that became extinct — a wide variety of organs and appendages to fulfill a small, basic set of "animal" functions — can be precisely explained by the independent birth of organisms from the primordial pond.[64]

Similarity imposed by the earth's constraining physical and chemical conditions

Most organisms, although born independently, have overall similarities in terms of their functions, because the earth's physical and chemical conditions — gravity, water, air, land, light, temperature — require them to do so. The scenario of an apparent overall similarity of organisms on earth is still the result of random perfection from myriads of organisms that were made without concern for the environment in which they were born. But because of the narrowly defined constraining physical conditions on earth, the many creatures surviving would tend to have an overall similarity in their body parts because many were subject to similar conditions. For instance, the limbs of all organisms on earth are fit so beautifully to work in a gravitational force. However, suppose that the earth did not have gravity. The kinds of animals that would have survived in that environment would have been very different. Without gravity, perhaps, only those organisms with limbs that had hooks to hold on to the ground, instead of limbs to work against gravity, would have been able to survive. Although this is a rather crude example, it illustrates that many kinds of organisms with organs and appendages unsuitable to the set of earth's physical conditions were also born, but those with appendages suitable in that set of conditions only survived. A constraining set of physical and chemical forces therefore unifies the general nature of body plans with overall similarities in limbs to walk, fly, or swim, sensory organs to hear (in air and water) and see (with light), and a body trunk to hold all these things together.

Can we wonder then, within the small domain of earth's physical and chemical environments, there can be an extremely large set of diverse body parts in different organisms that would carry out essentially the same or similar set of functions and be fit equally? Indeed a vast spectrum of body parts are theoretically possible from entirely different DG pathways, which would carry out essentially the same overall function. Due to this unifying effect of the earth's environments, the multitude of unique organs and appendages of the millions of organisms that were born independently in the primordial pond can carry out similar functions.[65]

Organismal similarity is also due to the use of prior genomic pieces in the assembly of new independent genomes in the primordial pond

Because different living things were born in a grossly similar or same earthly environment, all of them can be expected to use some common genes, genetic pathways for general metabolism, as well as some basic structures of cells and tissues. Furthermore, as we saw before, the use of parts of a genome from one organism in constructing the genome of another organism in the primordial pond can also contribute to such similarity among organisms.

How can the precise DG pathways for many different multicellular organisms be assembled from the vast UGP?

One might say that although by all analyses the new theory is quite acceptable, there is one question that comes to mind concerning the independent birth of creatures: the DG pathways of many independently-born organisms must be unique. Even given that there exists a vast number of genes, how can these DG pathways come about in the primordial pond? Among the random gene connections, how can meaningful DG pathways simply occur as probabilistic byproducts?

If an animal genome consists of 10^4–10^5 genes, and if the UGP consists of 10^{10} unique genes, the probability of assembling a given genome with a given sequence of on-off switches is miniscule. But there are several considerations here that make this probability very great, mainly because the UGP is fully open for the random assortment of genes in any possible arrangement.

There are other factors to be considered: 1) Millions of genes coding for the same protein sequence, but with quite different DNA sequences and exon-intron structures can occur in the UGP. 2) Multiple copies of the same gene brought about by DNA-copying enzymes, to the extent of millions, can occur in the primordial pond. 3) Many different genes with entirely different DNA sequences coding for different proteins, but all of which can carry out the same biochemical function, can also occur in the primordial pond. 4) The physical organization of genes in a genome does not matter; it is only the sequence of on-off switches — the genetic networks and DG pathways

— that are important except perhaps in rare occasions. 5) A very large number of different combinations of genes would have been equally successful as genomes. Thus even if one in a billion or trillion combinations is successful as a genome, still trillions of successful genomes are possible. 6) The availability of already successful genomes as linked sets of genes in single DNA pieces greatly increases the probability of the assembly of newer genomes. 7) The genetic combinations need not occur in a short time. The primordial pond could have been productive for a very long geological time, which increases the probability of such occurrences. 8) DNA pieces containing only genes are not needed. The assembly of large DNA pieces which contained one or a few genes and mostly junk DNA is sufficient to lead to a viable genome.

A large number of different rigid developmental genetic programs are simultaneously and independently possible from the open-ended USP. This is what is found in all the living organisms, strongly supporting the theory of the independent birth of organisms.
We analyzed the probabilities of DG pathways and concluded that the unique series of on/off switches of genes constructing an organ can never be evolved from an organism lacking it.

Based on the new theory, there is no need to change the developmental program of one organism into that of another to explain the scenario of diverse creatures on earth. However complex a DG pathway is, it is possible to assemble the series of developmental switches from the vast repertoire of "regulatory" sequences in the USP, and the "regulatory" proteins that can bind them. Let us discuss the reasons why this should be possible. A DG pathway is the functional connection of different genes through the physical combinations of different pairs of regulatory DNA sequences and regulatory proteins. A regulatory switch pair is any protein and DNA sequence that can specifically bind to each other reversibly in response to a specific chemical signal. Such pairs can occur in enormous numbers in the USP and UGP randomly. DNA-binding proteins with specific DNA-binding properties can, like any protein, occur in numerous forms in the UGP. Therefore the probability for many different proteins that can bind DNA sequences is very great. Furthermore, the typical DNA sequence that binds a protein is usually not long, and can have many variations.[66] The developmental genetic pathways of body parts and whole organisms use such DNA sequences and their corresponding binding proteins as on-off switches for their genes. Let us also remember that almost any switch pair (DNA sequence and its binding protein) can be used to control a specific gene, making it possible in the primordial pond for a specific DG pathway to originate with great probability.

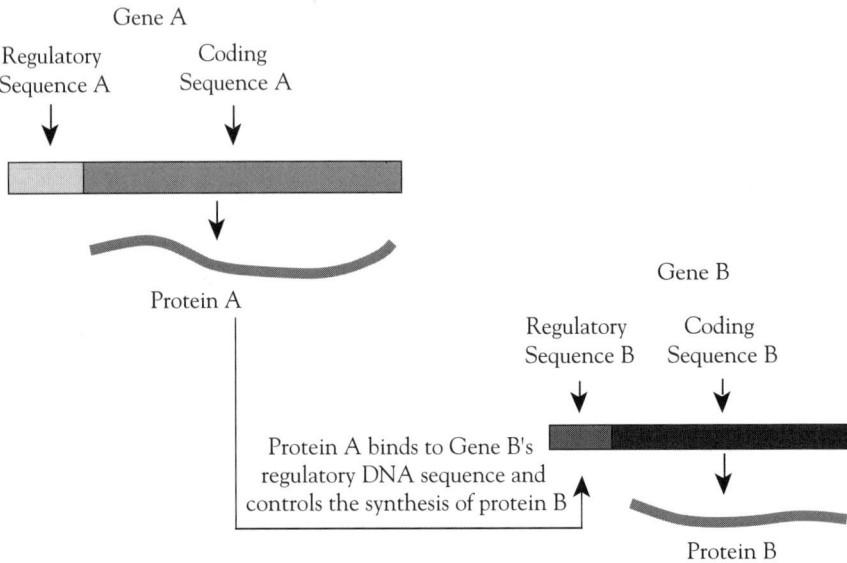

FIGURE 8.10. THE MEANING OF REGULATORY PROTEINS AND REGULATORY DNA SEQUENCES AND THEIR HIGH PROBABILITY IN THE PRIMORDIAL POND. In regulating the expression of one gene by another gene, the protein product of one gene binds to another gene's regulatory sequence and controls the expression of its coding sequence. The regulatory DNA sequence is situated usually close to a gene's coding sequence. The probability for the occurrence of many such regulatory DNA sequences and the proteins that can bind them in the vast USP is extremely high. The probability for the DNA-binding proteins to occur as a domain in many other functional proteins is also high, enabling proteins with other biochemical functions to also be regulatory proteins.

Consider a regulatory protein, coded from gene A, binding to a sequence next to gene B regulating its expression (as depicted in Figure 8.10). The product of gene A becomes the regulatory protein for gene B, by binding with the regulatory sequence of gene B.[67] But essentially what is happening here is the placement of a DNA sequence that could bind specifically to another protein in front of a gene. It is highly probable that several millions of such pairs could have existed in the UGP, which, let us not forget, contained trillions of genes. When we think of such a vast random gene pool, it is easily discernible that there can be many DNA sequences that can bind many different proteins, and vice versa. In the inducible regulation of a gene, such as the induction of the genes coding for enzymes that degrade lactose (see Genetics Primer), a repressor protein normally bound to the regulatory sequence is released by the binding of a biochemical such as lactose. Among the myriads of proteins possible in the UGP, an immense number could have had the properties of reversible binding to one or more biochemicals as well as to DNA, as seen in the living systems.[68] Thus, the number of regulatory systems available in the UGP could have been very high.

When such a vast repertoire of regulatory switch pairs are available in the UGP, they only have to be picked up to form the right combinations of genes that would assemble into the right DG pathway of one organism. This is, however, a difficult process and cannot happen over a short period. But because of the great variety and multiplicity of these materials, random recombination mechanisms in geologic time must have inevitably occurred thereby leading to the right DG pathways. The essential theme is that the set of regulatory switch pairs of one creature is so small when compared to the universal set of switch pairs available in the UGP that the probability to assemble one set in the right order is certainly very realistic and high.

This principle can be extended to show that the probability of assembling the DG pathways of multitudes of different creatures is also the same as that for one creature, because the complexities of the DG pathways of all organisms are similar. Consider the DG pathways of a few widely distinct invertebrate organisms such as the sea star, crab, dragonfly, octopus, and earthworm. The same thing holds true for the Cambrian or the Burgess organisms which were invertebrates. This applies as well when we include the vertebrate organisms, such as the raccoon, fish, opossum, duck-billed platypus, bat, and eagle. We can perceive that the DG pathways for these creatures are different but equally complex.

We must also remember that the DG pathways of different organisms can be mixed in the primordial pond by mixing the genomes of already born organisms leading to new creatures with similar characteristics from one organism or with mixed characteristics from different organisms. This is an extremely important phenomenon that makes possible the building of myriads of DG pathways using DNA pieces from the already assembled genomes of first-born organisms; these pieces contained sets of genes and some portions of DG pathways for some body parts. This process could have been used variously in the constructions of multitudes of newer genomes in the primordial pond.

Inclusion of largely the same set of genes but different DG pathways in different genomes from the primordial pond can lead to entirely different creatures

Theoretically, a given set of genes can be organized in two different DG pathways, i.e., two different blueprints for constructing two entirely different organisms. In fact, the genes for a large number of enzymatic proteins (and perhaps for many structural proteins) are the same or very similar in many different organisms, but their DG pathways are quite different. This is analogous to using exactly the same kind of building blocks to build two buildings entirely different in size and shape. This is precisely the basic difference between two different creatures. The building blocks, the genes, can be largely the same, but the blueprints can be entirely different. According to the new theory, different sets of largely the same genes could be assembled into entirely different DG pathways leading to unique organisms that are immutable. When this is the case, the fact that numerous unique genes and proteins are present in many creatures only strengthens the new theory even further.

We can clearly see that the widely different invertebrate organisms are really not different in terms of their structural and functional complexity. Is it not obvious that their DG pathways, although distinctly unique from each other, are not at all different in terms of complexity, and therefore could have been separately assembled directly from the primordial pond with equal ease using the common pool of genes? We can also see that vertebrate organisms are different from, but not more complex than, the invertebrates — for example, in genes responsible for bones and plasma proteins. Bones may form larger vertebrate bodies, but they are truly not far more complex in terms of structure and function. Among vertebrates, again, there is not much of a difference in anatomical complexity, and therefore their DG pathways — different as they are — are equally complex, and each could have been assembled with equal ease using its own set of proteins and genes.

Independent assembly of genomes in the open-ended primordial pond leads easily to highly complex organs in organisms: the explanation for the origin of highly complex organs

We saw that Darwin's theory is unable to explain the evolution of highly complex organs such as the eye. In fact we have demonstrated that it is highly improbable to evolve an eye from an organism lacking it by any evolutionary mechanism. On the contrary, the new theory provides an explanation for the origin of highly complex organs in organisms born separately in the primordial pond as equally well as it explains the origin of simple body parts. This is because, under the new theory, it is as probable to bring about the set of genes and the specific DG pathway of an extremely complex organ as those of a simple body part, through independent assembly of genomes directly from the open-ended primordial pond.

The set of genes for a highly complex organ is no more complex than the set of genes for a simple organ
The genes expressed in a simple organ are comprised of the same basic structure, namely the exons, introns, and regulatory sequences as those

expressed in a highly complex organ. The genes expressed in both types of organs — the complex eye and the simple finger — synthesize proteins, which either function as enzymes, bind regulatory sequences, or function as the building blocks of cellular structures in these organs. There is absolutely no difference whatsoever among the general structures and functions of the genes expressed in these organs at extreme ends of complexity. Then why is the eye perceived as far more complex than the finger? The answer lies in the manner in which these organs function — not in the complexity of the set of genes or the DG pathways that construct them. The specificity of the DG pathways for the various organs are different, but the complexities and probabilities of the different DG pathways are the same.

In regard to the eye, the iris cells are comparable to those in any other organ except that these cells carry out the specific function of contracting and relaxing thereby adjusting the diameter of the pupil. The proteins that form the cells of each subtissue carry out uniquely specific functions, which harmoniously fit with the function of the whole organ. The cells that form the cornea, the retina, and the lens are likewise no different than those forming any other simple tissues of the body. Thus, each tissue of the eye is constructed by cells that are no more complex than any other types of cells that are present in the body. When correctly organized into appropriate substructures, these cells do the function of "seeing." As discussed before, specific genetic networks expressed from the same genome construct the different subtissues of an organ. These differential expressions of genes result from a programmed, timed sequence of various sets of genes, constructing the cells and suborgans in the required three dimensional shape and size of the eye. When we reflect upon a simple organ, it is also built in a similar fashion — through complex, timed gene networks and positioning of the specific cells and tissues at specific three-dimensional coordinates. It is just that one set of gene networks goes to construct an organ, which happens to be extremely complex at the level of the organism, and another set of gene pathways goes to construct another organ supposedly simple, but one which is no less complex at the level of the genome. Thus the genes and the DG pathways that construct complex organs are no more complex than those that construct much simpler organs. Therefore, it is appropriate to conclude that if the genetic network for a simple organ can be assembled into a genome from the primordial pond, the genetic network for a complex organ can also be assembled with equal ease. In short, when we look at the complexity of an organ — not in terms of its anatomical and functional complexity, but in terms of its genomic complexity — there is no real disparity between a complex organ and a simple organ. Consequently, what we are looking at here are nearly equal probabilities of genetic networks for both a simple organ or a highly complex organ.

Biologists tend to think that the genes of highly perfect organs might be highly complex, and, therefore, how could these genes simply occur in the UGP? But consider that in the open-ended UGP, the complexity of an organ has no meaning. The organ is complex only at its gross anatomical level, but not at the gene level (or even at the cellular level).

This argument can be extended for the series of DG pathways through which the organ is built. A cell that starts to build the eye during embryogenesis will switch different sets of genes to be expressed in the succeeding cells, which are arranged in a particular three dimensional pattern. But the process of switching the various sets of genes in successive cells at different locations of the eye, as the eye is being built, is no more complex than the similar process of building the small bones of a little finger or a particular tooth or even an ear lobe.

Consider the steps of constructing a building. There are certain building blocks and certain procedures to lay these blocks. One can construct a simple and ordinary-looking building or a very complex-looking building with intricate shapes. However, the process involved in constructing the two buildings is nearly the same. Both require the same basic operations and materials. The important difference is in the design or the blueprint. The construction process of the organs of the body is similar to that of a building. The basic building blocks (the genes and cells) and the process (the developmental process) being the same, the different blueprints (DG pathways) bring forth a complex organ and a simple organ.

Biochemically, the function of our little finger is no less complex than that of our eye. We can contract the muscles of the little finger voluntarily, whereas the iris can contract its muscles through an involuntary action upon receiving information from the brain in response to varying amounts of light that enters the eye. Both the signals are delivered by the brain, and for the organ this makes no difference. The function of the bone, the muscles, and the nerve cells of the little finger are no less complex than the function of the retina and the nerve cells attached to it and the muscles of the eye. The overall function of the eye appears to be much more complex than the overall function of the little finger, but biochemically, physiologically, and genomically, there is absolutely no difference in complexity. This illustrates that the probability of the assembly of the set of genes for both the organs from the primordial pond is the same. The paramount implication of these considerations is that if a little finger can come into being from the UGP, so can an eye.

Random perfection of a highly complex organ

The different parts of the eye work together culminating in the intricate function of seeing. This function involves the processing of the biochemical information from the retina through the nerve cells and the brain and the feedback response from the brain that causes the iris to contract or relax.[69] All this was made possible by the random assembly of eye parts and their simple biochemistries through their genes, which have no meaning individually, but have great meaning when they are assembled together. (It is analogous to building a car. Different subparts are built by different groups of experts, with each part being useless on its own. But when the subparts are put together following the specific blueprint, the whole unit works.) The idea here is that such an intricate function was not evolved through organismal evolution. It is the result of the building of the DG pathways for the suborgans by purely random processes in the primordial pond. The organs and appendages of the first-born organisms in the primordial pond occurred, purely as statistical byproducts, among mostly meaningless and grotesque outgrowths. Only those organisms with useful body parts survived — but these were immutable. Under these circumstances, the value or the complexity of an organ has a meaning only at the level of the whole organism, but not at the level of the genome, especially when we consider that the assembly of the genes and the DG pathways for different organs from the primordial pond's UGP is equally probable — and that the structural and functional complexities of the genes and proteins that build the various organs are not at all different!

A complex eye may be constructed by more genes and a larger genetic network in the DG pathway than those that construct an optic nerve. Therefore the prior availability of a linked gene set for the suborgans, such as an optic nerve, can aid this process of building an eye with more genes. Once a primitive organism with a primitive light-sensing mechanism had originated in the primordial pond, the set of genes for this mechanism and its genetic network would become available freely in the primordial pond. Thus, these gene sets could function as reagents with which further construction and "improvement" of the eye design could happen. But we must remember that all these processes occurred purely randomly in the primordial pond. As more and more sets of genes for tissues and suborgans from already successful organisms became part of the UGP, the probability of assembling an organ with further complexity using these reagents increased steadily. But effectively, the random perfection was the process by which all the organs and appendages have been perfected. Indeed, the fact that there occurs a repertoire of different kinds of eyes in the living world, which are absolutely not connectable by organismal evolution, indicates that the only way they must have originated was by the independent births of the organisms with their unique eyes and not by organismal evolution. The scenario of the living world thus profoundly corroborates the independent birth of creatures.

"Complexity" of an organ has no meaning in the theory of independent birth of organisms, whereas it has great meaning in Darwin's theory

If we have an extremely large number of genes in the open-ended UGP, then assembling a set of genes for a "highly complex" organ has the same degree of difficulty as assembling the set of genes for a "simple" organ. At this juncture, I would like to highlight the distinction between the ease of achieving a highly complex organ based on my theory and the improbability of achieving this through the mechanisms of evolution.

In reference to the extreme difficulty of evolving a highly perfect organ such as the eye, Darwin has stated in his *Origin of Species*,

> To suppose that the eye, with all its inimitable contrivances for adjusting the focus to different distances, for admitting different amounts of light, and for the correction of spherical and chromatic aberration, could have been formed by natural selection seems, I freely confess, absurd in the highest possible degree.[70]

But Darwin thought that the presence of a repertoire of eyes in the invertebrate world that could be assorted into a gradation from simple photoreceptors to complex eyes showed that eyes had evolved through organismal descent with modification. However, in his later correspondence, he appears to be even less confident. He wrote to Asa Gray in February 1860,[71]

> The eye to this day gives me a cold shudder, but when I think of the fine known gradations, my reason tells me I ought to conquer the cold shudder.

In Chapter 3 we demonstrated that the presence of fine gradations of eyes in the animal world notwithstanding, the different types of eyes could not be connected evolutionarily. In fact the scenario is absolutely corroborative of the new theory of the independent birth of organisms. Independently-born organisms can have quite distinct kinds of eyes, without any organismal connection. As we discussed in Chapters 3 and 4, it is virtually impossible by Darwin's mechanisms of descent with modification to "evolve" an organ with any special function such as the eye. Because of the mechanisms by which an organ is supposed to evolve, the complexity of an organ has its full extent of meaning in Darwin's theory — the more complex the organ, the longer and more difficult the path of natural selection of individual variations.

What this means is that, we must not forget, the unique genes and the unique DG pathway that develops the eye should be evolved through mutations in the genome of an organism, and, as we have demonstrated in Chapters 3 and 4, this is highly improbable. While complexity of an organ has a great meaning in Darwin's theory of evolution, it can be clearly seen that there is no more complexity in bringing forth a complex organ compared to a simple organ by the theory of the independent birth of organisms. The genes and the DG pathways are organized in the open-ended primordial pond, just as for any other simple organ. The combination of genes for the eye has nearly equal probability as the combination for any other organ. The number of gene connections in the genetic pathway or the network developing the eye may be somewhat more than that of another organ such as the tongue, but not widely different, and the process of assembly must be the same, indicating that the eye and the tongue are nearly equally probable.

The origin of sex, sex organs, and instincts in the new theory

People usually think that sex is unique and that the segregation of multicellular organisms into two "opposite" sexes, the male and female, is something incredible. What we are going to see here is how this sexual segregation originated in the multitudes of organisms each born directly and independently from the primordial pond.[72] What we shall see is that the sex organs and instincts are not really different from any other organ system in the animal. We have seen that any other organ or body part is just the result of certain specific gene circuits originating from the primordial pond's vast genetic sequences. This is precisely true for the genes and genetic circuits forming sex segregation, sexual organs, and associated instincts. Furthermore, there are a variety of sex organs and systems in the animal world that are unique to many organisms and therefore unconnectable by evolution. The wonderful entity called sex is not really unique in the usual sense and it is no different from any other simple organ system in terms of its origin and function.

The availability of the vast number of genes and the power of random permutations and combinations of these genes giving rise to various DG pathways can lead to the formation of a sex organ, instinct, or intelligence, similar to the formation of any other complex organ such as the eye.

Origin of sex: The origin of differences in the male and the female, including sex organs

Instead of addressing the advantage of having sex differences in the male and the female, we shall ask, given their advantages, how these dissimilarities could have originated according to the theory of the independent birth of organisms. Remember that although only very rare combinations of genes can give rise to organismal structures and functionalities, the extremely high number of random combinations possible in the primordial pond in geological time can lead to many structural and functional complexities.

After the basic unicellular eukaryotic genomes were successful in the primordial pond, most initial genome combinations that could result in multicellularity would have lead to multicellular organisms without any sexual segregation. But a rare combination of genes leading to the male and female entity is not difficult to envisage. It requires a few sets of genes, even fewer than those of most organ systems such as the heart, eye, or brain. Even if once, in the primordial pond, such sexual segregation into male and female had occurred, this gene combination becomes available as a general mechanism in the open-ended primordial pond.

In connection with the present subject, one ought not to look at the process of sex as sexually attractive features of a male and female that bring them together. Let us here look at it as the ability to produce sperm and egg, and the ability to produce certain hormonal proteins. The rest are organs and tube systems to place the egg and the sperm together into a container, and the sex instincts and sexual act to bring them together. These can be built in a variety of ways, by many different independent combinations of genes. Furthermore, if the basic genetic mechanism of sperm and egg formation is established by random gene combinations once in the primordial pond, and the segregation of an organism into a male and female entity is successful, then this basic mechanism can be used in many different ways, in independently assembling genomes to produce many diverse sex organs and systems, as different as those of worms, insects, fishes, and mammals. In other words, one should see no difficulty in concluding that once the basic process of sperm and egg production is built into gene connections in the primordial pond, the gene connections, available in multiple copies, can be used in building a variety of sex systems and organs with several diverse shapes, sizes, and instincts.[73]

A wide variety of unique sex organs and systems exists in the living world which cannot be connected evolutionarily

In looking at the living invertebrate world, one cannot but wonder at the numerous kinds of sex organs and reproductive systems. It is astonishing to see so many unique types of sex body parts and the manner in which they copulate. The reproductive system and copulation in earthworm are unique.[74] Earthworms are hermaphrodites, meaning each individual has both male and female sex organs. During copulation, each inseminates the other. Sperm travel along grooves to receptacles on each earthworm, while a slime tube holds them together.

Some leeches use a penis to ejaculate sperm directly into the female. Other leeches intertwine and deposit a packet of sperm (a *spermatophore*) against the body of the other. The spermatophore penetrates the body wall, and the sperm is released, which then migrate to the vagina to fertilize the eggs. The eggs are deposited in a cocoon where the offspring develop.

In some fish, such as wrasses, individuals are one sex early in life and the other sex later, in a process called sex reversal.

Millipedes have their sex organs close to the base of the second pair of legs.[75] During copulation, the male uses his seventh leg to collect sperm from his sex gland and then moves to deposit it in her sexual pouch.

The male damselfly holds the female behind her head with special clasping organs. The female inserts the tip of her abdomen into a pouch on the male to take up sperm that had been deposited there earlier.

Male squids produce a spermatophore which is carried by the penis into a mantle cavity; the penis is never inserted into the female. During copulation, one of the male's arms, the *hectocotylus*, picks up spermatophores from the mantle cavity and moves them into the female's mantle cavity. When properly stimulated, the spermatophores release their sperm. The female uses her arms to pull fertilized eggs from her body and attach them to the sea floor, forming clusters called "dead man's fingers".[76] These are but a few examples. In looking at the different kinds of sex organs and the methods of copulation, fertilization, and the growth of the young, each one seems to be a marvelous unique invention.

In vertebrates also, there are many different types of sex and reproductive systems. Males of many vertebrate creatures do not have an external organ for copulation. Instead, they have a *cloacal vent*, an opening that also serves the urinary and digestive systems. Some fish and amphibians simply release gametes into the water where fertilization occurs. Internal fertilization uses more complex organs, such as those present in mammals.

When we analyze the methods of fertilization and bearing young, almost all possible manners in which this can be achieved that one can think of are represented throughout the animal world in a random fashion. Most invertebrates and vertebrates are oviparous. Oviparous female creatures deposit eggs into the environment, where they hatch.[77] In some animals, eggs hatch fully-developed offspring, while in others, a hatched larva must undergo metamorphosis to the fully-developed form. Viviparous animals, such as mammals, develop the young completely inside the body. Aside from mammals, some fish, amphibians, lizards, snakes, and invertebrates are also viviparous. Some of these animals hold their eggs in the reproductive tract until they hatch and the embryos must derive most of their nourishment from the egg yolk. This birthing process, called *ovoviviparity*, occurs in sharks, reptiles, and many insects. In sea horses, for example, males carry the eggs in a pouch, becoming "pregnant" with developing eggs. Some frogs store their eggs beneath their skin: Surinam toad females incubate their eggs in specialized pockets on their backs.[78] In viviparous lizards, such as the European skink *Chalcides chalcides*, the female has a placenta that delivers nutrients to the embryo.[79] Some sharks hatch their eggs inside the female, and the early embryos feed from a placenta in the uterus.

Let us not be misled by such definitions of oviparity, viviparity, or ovoviviparity. They are only large general groupings to categorize the methods of bearing young. When one thinks of it, in fact these are the only ways in which the process of bearing the young can be categorized, although there are numerous sex and reproductive systems unique to most organisms in the living world.

Methods of copulation also vary widely. In some amphibians, such as plethodontid salamanders, a male stimulates the neck and head of a female, then moves under the female so that her throat is above his back legs. While rubbing the female's throat, the male releases a spermatophore onto the ground for the female to retrieve.[80] In frogs, the male holds the female tightly, applying pressure that helps the female release her eggs. As the eggs appear, he sprays them with sperm fertilizing them externally. Mammals have conspicuous sex organs, a penis that enters the vagina and deposits the sperm inside. When one thinks of sex organs and systems, what usually comes to mind is the mammalian penis and vagina. But let us remember that it is but one of thousands of kinds of sex systems when we look at the whole living world. In short, like the many different eyes in the animal world, these distinct reproductive systems are evolutionarily unconnectable, and could not have originated through organismal evolution. Each is brought about by unique genetic circuit and developmental genetic pathway.

The unique sex organs and systems could have originated only by independent birth of creatures in the primordial pond

After the foregoing discussion, one can see that all these unique sex systems could have originated as sets of genes available in the UGP, and randomly selected from the primordial pond in the separately assembled genomes of different organisms.[81] Thus, the creatures developed from various seed cells could be distinct, with unique sex organs. Furthermore, there could be much similarity of sex organs in many diverse creatures because of the inclusion of some duplicate sets of genes in different independent seed cells.

It may be asked how the extremely complex and precise genes for primary and secondary sexual organs and reproductive cycles could have been available in the primordial pond and selected into genomes. My answer to this is the same as that for the complex organs such as eyes. As we have alluded to in Chapter 7, the sets of genes for these highly complex organs and organ-systems can occur with a high probability in the UGP. What remains to bring forth sex differences is to separate the sets of opposite-sex genes in two individual seed cells. When the set of genes for totally unrelated organisms such as the crab and rat can occur in the UGP and be separated into two different genomes, according to the new theory, the set of sex-specific genes for the different sex organ systems can also exist in the UGP, and can be enclosed in copies of the same genome, making them male- and female-specific genomes. The sets of genes for sexual reproduction have only the same complexity as those of the highly complex organs. Furthermore, with basic gene sets as reagents, subsequent genomes could construct a variety of differently-shaped sex organs with dissimilar sex instincts, because the open-ended gene pool and sequence pool of the primordial pond is also available to them. It can be thus inferred that many additional genes and regulatory sequences can be derived from the UGP and USP, while using the pieces of DNA from the first successful sexual organisms.

As we discussed in Chapters 3 and 4, it is virtually impossible to evolve whole sets of genes for a highly complex organ and the corresponding specific DG pathway, by Darwin's mechanisms of organismal descent with modification, starting from an assumed primitive "ancestral" creature that lacked that complex organ. In the same manner, we can see that the evolution of the highly complex sexual reproduction systems is also improbable to be arrived at by Darwin's mechanisms. Darwin did not have a mechanism to explain how life was "breathed" into the first one or a few creatures. Thus, he could not explain the appearance of sex and reproductive systems in the first one or a few creatures. Even if we take it for granted that in the first original creature some form of primitive sex system existed, the "evolution" of highly complex reproductive systems, such as the placenta of mammals, is unexplainable by the theory of evolution. In contrast, this is easily solved by the genome theory proposed here by the random selection of sets of genes for different organs and systems separately from the vast open-ended UGP in the primordial pond.

Male and female seed cells lead to male and female individuals

A genome is included in a seed cell. The seed cell is capable of multiplying as a single cell. At the same time, it is also capable of developing into a whole multicellular organism (analogous to the zygote of living organisms). Both male and female seed cells can be assembled, which are capable of developing into the male and female members of a population.

Male and female individuals of a creature are born from different seed cells containing genomes that differ only in the set of sex genes. Although this may seem unreasonable, there are strong reasons for its high probability. Reproductive systems are only as complex as any other organ system. The only important consideration is that the sex-specific genes for male and female have to occur in two separate seed cell genomes, all else being equal. In fact, it is known in many instances that the same genome can give rise to male and female individuals under different environmental conditions. The same turtle egg has the potential to give rise to a male or a female offspring depending upon the temperature of incubation. The same thing happens in crocodiles. Furthermore, several animals change sex during their lifetime. There are many hermaphrodite organisms in the animal world (having both sexes in the same individual), as seen before, which indicates that the same genome produces the male and female characteristics in the same individual. All these facts signify that making a male or female is easily done from the same genome, which can be induced by simple parameters such as temperature. This clearly shows that the DG pathway that develops an individual into a male or a female is just like the DG pathway that develops a specific organ. One can infer that it is not difficult to segregate the genes for a male or a female into a specific chromosome and in two different sex cells.

Some multicellular structures in various multicellular life forms produced from myriads of seed cells could be gonads (testis/ovary); some would be secondary sex organs (sex-attracting organs, such as plumage in birds or breasts in the human, sex hormones called pheromones, and colors and patterns in invertebrates). Perhaps among the organisms produced in the primordial pond, some had only secondary sex organs, but no genital organ to copulate; whereas other organisms would have had the latter but not the former. Both the above situations may or may not have had the reproductive cycles of sperm/egg production. There could have been many seed cells producing individuals, with wrong combinations of male and female sex organs and secondary sex characteristics. Rarely, some seed cells will possess all the three sets of genes for all these three functions — attractiveness by secondary sex features, copulation by genital organs, and reproduction by sperm/egg cycles. This is analogous to many seed cells giving rise to individuals with improper or incomplete organs, which will not survive. Only those individuals with the absolutely right organs will survive. Therefore, only one out of myriads of seed cells may form a viable organism. This may explain why it would have taken geological time for seed cells to be formed with genomes capable of producing viable organisms.

Origin of sexual instincts

Sexual instincts are no different than any organ or body part. They are induced by certain proteins or some small molecules synthesized by proteins, which function as sex attractants, and some protein "receptors" that bind them. In other words, they are induced by a set of genes, which would have been available in the primordial pond's UGP, just as the genes for the proteins in any other organs.

ORIGIN OF INSTINCTS AND INTELLIGENCE

All instincts and intelligence of different organisms, from the primitive intelligence of some simple animals to the highest human intelligence, all are brought about by specific genetic circuits. They could have originated in organisms from the gene connections in the primordial pond, when the organisms were born independently in it. We shall analyze this with an example of the social systems and intricate behavior patterns in some insects.

Social hierarchy with caste systems and intricate structural and behavioral patterns in some insects

One cannot but wonder at the behavior patterns of insects — termites, ants, wasps, and bees — because they are so mind-bogglingly unique and complex. These insects live in colonies organized in a social hierarchy, with different *castes* of individuals. Members of each caste have different body parts and perform different tasks. Males, fertile females (queen), and infertile female workers are typical castes. Workers produce *royal jelly*, a special nutrient that is fed to an immature female to produce a queen. The mature queen secretes a *queen substance*, which the workers ingest and in turn feed to the developing larvae. Larvae fed with queen substance develop into sterile female workers rather then queens. When the queen grows old, the supply of queen substance dwindles. This stimulates the workers to build royal cells, where they feed eggs with royal jelly to produce a new queen.

The presence or absence of queen substance directs the genome to produce a worker or a queen. Workers will not build royal cells while the queen is producing queen substance. Workers continually share the contents of their stomachs with other workers, spreading the substance throughout the colony. This process ensures that there is only one queen at any given time. When there is not enough queen substance to go around, such as when the colony grows too large, the colony may split in two and generate a new queen. This complex behavior arises from specific nerve circuits determined genetically and developmentally.

An insect society consists of a great number of different kinds of individuals existing as a superorganism. They are the offspring of a single pair of adults and they are incomplete as individuals, incapable of an independent life. In termites, the workers are blind and sterile. There exists a support system and any individual would die for the lack of it. Soldiers guard and defend the colonies. Their jaws are so large that they cannot use them to gather food, so they have to be fed by the workers. The queen is helpless and trapped at the center of the colony; her large body cannot even fit through the passages that lead out. Her body is almost wholly devoted to producing 30,000 eggs per day. The workers must deliver food to her constantly to keep her alive and productive, and they must tend to all her eggs. The only sexually active male is a wasp-sized king who stays with the queen, and also depends upon the workers for survival.

As with bees, chemical signals develop the colony. The workers continually collect pheromones from the queen and distribute them throughout the colony. Although eggs can become either sex, the queen's pheromones inhibit sexual development, resulting in offspring that are sterile, wingless, and blind. A full complement of soldiers also produce a pheromone that circulates in the colony and prevents any larvae from developing into soldiers.

Origin of social systems and hierarchies of an insect superorganism in a single insect genome by genome assembly in the primordial pond

We can see that in these social insects, several social classes of the same organism are produced from a single genome. That is, the same egg can hatch into several classes of social insects, under the influence of various chemicals. It is certainly not possible to explain the origin of such a complexity from an organism lacking such a system based on the theory of evolution. But we can certainly explain the origin of such a thing under the new theory based on the independent assembly of genomes. The set of genes that produce the various classes of insects are subsets of the same genome, and the specific DG pathways that achieve this also are part of the same genome. Only the different sets of genes and the different sub-DG pathways are triggered under the influence of the specific biochemicals. These gene sets must have been assembled into a genome of the insect only by random processes in the primordial pond. The probability involved with this is the same as the probability with any other organ or organism. All the complexities involved in the production of the hierarchy of social insects stem from the differential expression of distinct sets of genes from a given genome. It is my view that there is no difference in the assembly of these sets of genes in the primordial pond compared to those for the other kinds of animals lacking this social system.

If the DG pathways for all the organs in an organism are simply specific genetic circuits that start to operate in specific space-time coordinates in the genetic pathway of a zygote, the same mechanism can be applied to produce the various castes of individuals starting from the same genome. One pheromone could induce certain points in the DG pathway of the insect and inhibit others, thus leading to one type of individual, such as the worker. Another hormone would induce the specific DG pathway of another kind of insect, such as the soldier. When no hormone is used, the male and female develop into the king and the queen. All these DG pathways are subsets of a common DG pathway for the insect. What we must note here is that the mechanism for production of one kind of insect individual is just the same as the mechanism for producing an organ; and producing another kind of individual is like developing another organ.

The unifying theme in the new theory is that different genes, no matter what function they encode, are not different in terms of their probability, and the various gene connections, no matter what organ, appendage, or instinct they specify, are only a mathematical probability — all of which can occur in the primordial pond leading to independent births of distinct creatures.

FROM MEANINGLESS GENE CONNECTIONS IN THE PRIMORDIAL POND, COME THE MEANINGFUL STEREOTYPED INSTINCTIVE BEHAVIORS OF INSECTS — EVEN LEADING TO THE INTRICATELY COMPLEX TERMITE COLONIES

Termites build large-scale buildings in which the superorganism lives and reproduces. These buildings are ingeniously designed and efficient with intricate heating and ventilation systems. Such a building is the result of thousands of blind workers, each instinctively carrying tiny pellets of mud. Neither the termites design it, nor the workers know what they are doing ultimately. How is this possible? It is truly the result of the myriads of random gene connections that occurred in the primordial pond, producing myriads of meaningless things, out of which a specific gene connection produced a meaningful colony of insects. Otherwise, let us be clear, this astonishing construction is truly as bizarre, grotesque and meaningless as anything that could be produced by any other bizarre gene connection, but is meaningful only in a certain context — both for the society of the insects and for the human mind that comprehends it. What we can see is that each termite worker is totally unaware of what she is building. She is simply following an instinct to carry the pellet of mud in a specific direction and place it as her instinct directs. But the result of the combinations of activities of all the insects is the beautifully-constructed, astonishingly-designed termite colony. All this can only result by the random processes from among myriads of meaningless gene connections that occurred in the open-ended primordial pond from independent genome assembly, and not by organismal descent with modification from a "lower" organism that lacked such social hierarchy and the abilities to build such complex buildings.

An analogy: a robot is operated by a set of programmed instructions. This set of instructions is equivalent to the neuronal instructions of the stereotyped termite castes, which are in turn genetically determined. Even if one programmed instruction is wrong or absent, the robot will not work after that instruction. It is possible to program a set of robots to build a building as complex as the termite's house. This is the same as the gene connections in the termite's genome. In other words, the set of actions of the termites in their construction of the intricate building is simply a set of gene connections, which are as probable as the gene connections for building an organ.

The whole thing is similar to the construction and functioning of the eye. The substructures of the eye do not mean anything by themselves — it is only when all of them are put together that the whole eye can be structurally meaningful and perform its function. Now consider that each eye substructure is purely the outcome of gene connections. Thus, it is the totality of all the circuitry of genes for all the substructures that makes the structure of the whole eye possible as well as its astonishingly complex function. Truly, the genetic pathway for a given substructure by itself is a meaningless entity. But when it is put into operation in the context of a functional genome in a developing organism, all the meaning comes to it. In the case of the termites also, their social hierarchy and their stereotyped actions are the result of a genetic circuit that was organized in the primordial pond purely by chance. Only in the context of the termite's society, and only when all the pieces of the stereotyped behaviors of the different sets of individuals of the superorganism come together, will the genetic circuit express a beautiful meaning. A gene connection simply occurs arbitrarily among the myriads of genetic circuits in the primordial pond; the selection of some of those with contextual meanings therefore occurs randomly. But again, remember the help from the earlier basic constructs of a living organism that would have come to help the more complex genome constructions in the primordial pond. This is what must have happened in the independent birth of the termites and other insects with their unique and stereotyped social behavior.

The formation of organisms was possible because of the vast number of genes in the primordial pond. Out of the myriads of genes, some sets were useful for body structures, some for sexual behavior, some for other instincts, and some for intelligence. At the molecular level, these sets of genes were qualitatively indistinguishable and of equal complexity. All these genes were selected and organized into very many genetic networks leading to low-level intelligence of a worm, social instincts in insects, sexual characteristics, and even the highest level of human intelligence.

One final thing to note here: We have seen that each of the various castes of a given superorganism is developed by a subset of a single DG pathway. If this is the case, did the superorganism originate in the primordial pond as the superorganism itself, or did a regular male and female pair first originate, and in time, the sub-DG pathways of the various castes arise? Even if the latter seems probable, we should note that there is no new gene or DG pathway that has evolved to produce the castes. It was the reduction or retrogradation of an already existing DG pathway, independently assembled in the primordial pond, that led to the sub-DG pathways for the various castes. This in no way affects the new theory, rather, it strengthens it. Again, looking at the need for many specific hormones that would initiate or stop the DG pathways at various points in the DG pathway of the superorganism to produce the various castes, it seems highly probable that the superorganism was born in the primordial pond as a superorganism.

The new theory unifies all living organisms born independently on Earth through their common origin in the primordial pond

We can see that the new theory connects all the organisms in the living world — although they were all separately born in the primordial pond — through a fundamental unifying theme. Their genomes were all assembled independently from the same pool of genes by the same basic principles. All the organisms are constructed with essentially the same kind of building blocks leading to a great diversity of organisms — from microscopic worms to the elaborately constructed invertebrates and vertebrates. Because the fundamental requirements for life on earth are so narrowly defined, all the organisms must employ a considerable number of genes common to all of them — while using many unique genes.

The second principle in the new theory that unifies the different organisms concerns their genetic circuits. As we know, organs and organisms, as well as instincts and intelligence, are built by specific developmental connections of genes. All these specific gene connections of roughly equal complexity originated among the myriads of random networks of genes available through biochemical processes in a common pool of genes. Thus, numerous distinct DG pathways could be constructed using gene sets from a common pool. Moreover, if the genetic networks for simpler systems, tissues, or even body parts could be used to construct other systems, more complex or not, they can be included in future genomes constructed in the primordial pond. Many basic genetic processes in all the independently-born multicellular organisms can thus have a common origin.

In essence, any complexity, whether it is a complex organ such as the eye, sexual segregation, instincts and high level intelligence, or social hierarchy in insects and their ability to build intricate houses, all of these existed as genes and genetic networks in the primordial pond. Our central theme is that once genes existed in the primordial pond to the extent that unicellular eukaryotes could be formed directly from it, then it becomes inevitable that many different multicellular organisms will be formed with all these characteristics independently. Thus the new theory unites all organisms through the commonness of genes, biochemistry and physiology — based on essentially the same genome-construction process from the common pool of building blocks in the primordial pond.

Retrogression of an independently-born organism by the loss of some gene functions is possible

We have said consistently that an independently-born organism is immutable. However, it is possible that an organism may loose some characteristics and still be viable. For instance, the legs of an animal can be lost due to some mutations in the genes or regulatory sequences responsible for the development of legs. This loss of legs may not be detrimental to the organism, and the organism may still be able to live and reproduce. Under such circumstances, a new organism may be formed. It is known that some salamanders loose legs. It may simply be the cutting off of that particular branch of genetic circuitry that develops the legs in the complete DG pathway of the organism.

It has also been established that in the salamander, metamorphosis from the larval stage to the adult stage requires the hormone thyroxine. If this hormone is deficient, the salamander can continue to live as a larva all its life and even reproduce.[82] In fact this larval form is categorized into a different species from the salamander. However, we must note that this is a special case wherein an independently-born creature changes into a degraded form by the loss of some developmental function by mutation. Here there is no evolution of new genes nor new DG pathways.

The eye of an animal has a specific developmental genetic pathway. In an eyeless mutant of the salamander, the optic cup fails to develop because it is incapable of responding to an inducer — a genetic defect. Likewise, in some cave populations of the Mexican characid fish *Astynax mexicanus*, the eyes are degenerate because of a genetically diminished ability of the optic cup to induce lens development.[83] Thus in these cases, the loss of structures and functions is only due to the loss of one or more gene functions. There is no evolution of new gene nor a new DG pathway in any instance of loss of structure or function.[84] In short, in the new theory, it is possible for an independently-born creature to degrade from its original state of birth and produce an animal in which some structure or function is lost.

One might say, by the same token, that snakes could be derived from a reptile by loosing legs. On the other hand, there are reasons to indicate that this may not be so. They have poison glands, fangs, and poison proteins that are unique to them and not to any other known reptiles and therefore cannot be evolved from other reptiles. Looking at the innumerably different kinds of organisms, there is no reason organisms cannot be born in the primordial pond without legs if they can survive without them. It is quite possible that from the genome of an animal such as a reptile, the loss of legs and the gain of other organs (such as the fang and the poison gland) with other genomic modifications could have happened in the open-ended primordial pond. This might even explain the presence of residual hind legs in some snakes.

THE PRECISE GENOMIC ARCHITECTURES PREDICTED BY THE THEORY OF INDEPENDENT BIRTH OF ORGANISMS IS PRESENT IN TODAY'S LIVING ORGANISMS

If the new theory of the independent birth of organisms is correct, then what it predicts regarding the genomes of different independently-born organisms should be observed today in living organisms. We shall examine the predictions of the new theory, and see how the scenarios of the genome structures in living organisms precisely fit these predictions. Note that the very same genomic architectures of living organisms are unexplainable by any theory of evolution.

A. The new theory predicts islands of genes in long random DNA sequences in the genome of each organism: The origin of intergenic "junk" DNA and the genomic architecture

If a large DNA piece in which a gene is only a small part (the rest being useless) is combined with another DNA piece containing another gene as its small component, what would happen? Obviously, it would lead to meaningless DNA regions between genes. Figure 8.11 depicts this process. This is what must have happened in the assembly of genes leading to genomes in the primordial pond, and that is precisely why we have what we call intergenic "junk" DNA in all the genomes of animals and plants.

In the random mechanism of the assembly of genes in the primordial pond, it is not possible to precisely connect only genes together. In fact, only random recombinations between DNA segments of the USP could occur in the primordial pond irrespective of whether they contained genes. Although all the minimum necessary genes for a viable organism must be present in a successful genome, long, nongenic, meaningless sequences would occur between genes. Under such circumstances, much of the genome can be nongenic junk DNA. This is found to be absolutely true from the data on the genomes of many animals and plants. Computing from the content of DNA in a genome and the possible number of genes in it, it appears that a vast majority of the genome is indeed nongenic DNA occurring between consecutive genes. Approximately 80-99 percent of the genomic DNA of most organisms seems to be junk (see Genetics Primer). This scenario is indeed plainly corroborative with the predictions of the new theory.

One might say that the lesser the junk DNA, the better the genome in its efficiency; but it may not have been possible above a certain limit for the random processes to eliminate the unnecessary DNA. Therefore, the junk DNA has persisted to this day. Some people call the intergenic region "selfish" DNA.[85] We can see that it is only a metaphorical name, and selfishness is not the reason for its origin. The origin of the junk DNA is due to the random manner in which the genome was directly assembled in the primordial pond.

B. According to the theory of independent genome assembly in the primordial pond, the genomes of different organisms should contain different amounts of DNA — this precisely explains the so far enigmatic "C-value paradox" in living organisms

The amount of DNA in a genome is termed the C value. As we saw, this amount is far more than that represented by genes. The lack of correspondence between C values and the amount of genetic information in the

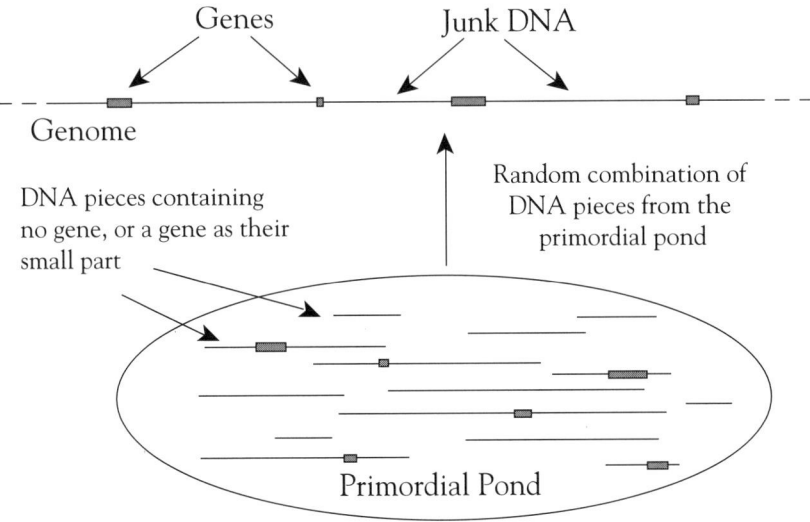

Figure 8.11. Assembly of a genome by random combinations of DNA sequences in the primordial pond leads to long "junk" DNA between islands of genes. The DNA pieces in the primordial pond contained random sequences. These can be short (containing only a few nucleotides) or long (up to several millions of nucleotides); each large piece may or may not contain a gene as its part; in fact, only rarely a piece may contain a gene especially useful in a living system. Random combinations of these DNA pieces occurred in the primordial pond to form a genome. Because the joined DNA pieces contained either no gene, or a gene only as their small part, the genome would be mostly random DNA sequence with only small "islands" of genes scattered in an ocean of meaningless DNA. Such an architecture actually exists in the genomes of all living multicellular organisms, with the intergenic sequences termed "junk DNA."

genome is known as the C-value paradox. The C values of different organisms do not correspond to supposed "evolutionary trees" (see Table 9.1).[86] They widely vary among organisms without any evolutionary correlation. For instance, the C value of some amphibians, such as the salamander or conger eel, are 50 times larger (150 - 200 billion nucleotide characters) than that of the human (three billion characters). Moreover, the genome sizes of even supposedly related organisms vary widely. These observations cannot be explained by the evolutionary theory. If organisms had evolved from an original ancestral creature as claimed by the evolutionary theory, then the genome size should have a reasonable gradation, either generally increasing, decreasing, or remaining constant, along the evolutionary scale. On the contrary, it is erratic.

The new theory provides a perfect solution to this paradox: the independent assembly of genomes in the primordial pond should lead to distinct C values in different organisms. When the genomes were separately organized from the primordial pond's universal gene pool, they were assembled by random recombination of the DNA material. The only underlying requirement for a successful genome is the presence of the minimum necessary genes organized into a specific DG pathway. Under such circumstances, there is no control over the amount of nongenic material that will be introduced into each genome. Thus, the DNA content of every genome should be unique and random. This is what we precisely witness in organisms today (see Table 9.1). The genomic DNA content along the supposed evolutionary tree from fish to human does not increase or decrease, but in fact the genomic sizes are randomly distributed. Even among organisms classified within the same taxonomic group, such as amphibians, there are no C value correlations among different families or species.

Even when a new genome is organized from the pieces of already successful genomes, this principle can be followed. Thus it will be reflected in the genomes of somewhat similar organisms. This is consistent with the observed C values of different organisms. For instance, two amphibian species whose overall morphologies are similar may have a 10-fold difference in their relative amounts of DNA.[87]

In short, what is paradoxical to the evolutionary theory about the haphazardness of the DNA content of organisms is not a paradox to the new theory of the independent birth of creatures. It is, in fact, a strong corroboration.

SIMULTANEOUS BIRTH OF MULTITUDES OF UNIQUE AND INDEPENDENT CREATURES AT THE START OF MULTICELLULAR LIFE AND CONTINUATION OF THE BIRTH OF NEW CREATURES IN LATER GEOLOGICAL TIME: PREDICTIONS OF THE NEW THEORY UNDENIABLY CORROBORATED BY THE FOSSIL RECORD

When the fossils of multicellular life first appear in the fossil record, they do so in a burst called the "Cambrian explosion." The scenario found in the fossil record thus perfectly fits with the new theory of the independent birth of organisms.

The Cambrian explosion occurred within about 10–20 million years, with a burst of widely diverse and unique organisms. Interestingly, recent research shows that the Cambrian explosion is restricted to only 5–10 million years,[88] about 540 million years ago. Geologically speaking, 5–10 million years is a very short time. Compared to the age of the earth, which is 4600 million years, even 20 million years is but a tiny fraction. If the earth's life history is condensed into a one-hour movie, the Cambrian explosion would play for only 10 seconds. Even if the movie starts at the appearance of the multicellular eukaryotes, the Cambrian explosion would still occupy only the first minute. When it began, it all started in a spurt. Does it not plainly illustrate that the multitude of organisms in the explosion must have been independently and simultaneously born in the primordial pond? What more evidence do we want that would so strongly and clearly corroborate the independent birth of creatures than the fact that all the diverse organisms appeared simultaneously on earth at the very start of multicellular life? It is clear that however long it took for the earth to cool and the primordial ponds to reach their biochemical complexity to the extent that unicellular eukaryotes could be formed, all kinds of unicellular and multicellular organisms were inevitably formed in a spurt. When we view such clear and strong corroborations for the new theory from the Cambrian explosion, combined with the fact that the same Cambrian explosion is in total contradiction to Darwin's theory, and has been one of the greatest difficulties and puzzles for it, it becomes even more evident that the predictions of the new theory are what must be correct.

The abrupt start of life with its full complexity has been puzzling to evolutionists: Why do the anatomically complex organisms in the Cambrian explosion have no direct, simpler precursors in the fossil record of Precambrian times? The answer is that they do not need to have simpler ancestors. If the biochemical complexity of the primordial soup reached a sufficient level so that a living entity as complex as a unicellular eukaryote can come about, then, by the same mechanisms, multitudes of highly complex multicellular organisms must also inevitably appear — all independently and simultaneously. Otherwise not one living thing can originate! This all-or-none law is superbly corroborated by the fossil record.

We will discuss this and other fossil evidence in more detail in Chapter 11.

The emergence of extreme order out of pure randomness

From all our foregoing discussions, we can see that extreme order emerged from chaos on primitive earth. From random chemical reactions emerged biologically meaningful macromolecules such as DNA and proteins, among other meaningless ones. Still, in the DNA and protein sequences, only random sequences existed. From these purposeless sequences and the randomly-generated macromolecules emerged meaningful living entities — a great number of organisms, both unicellular and multicellular. The genomes of organisms born first in the primordial pond were formed from pure randomness. Except for the aid from pieces of the genomes of the first-born organisms, the genomes of the later-born organisms were assembled independently in the primordial pond. Every uniquely beautiful creature on earth is thus the outcome of this emergence of order out of pure chaos.

Just as the beautiful crystallization of salts in solutions is the outcome of random and chaotic interactions of salt molecules, and the formation of stars and planets, mountains and oceans are the result of the interactions of cosmic dust, the formation of life also was the result of random interactions of molecules in the primordial pond. Such interactions led to myriads of distinct independent life forms through the same process of constructing genomes.

Conclusion

The origin of organismal diversity and disparity on earth has been among the most interesting and enduring quests of biological science. Although Darwin's theory seemed to resolve the matter in many respects, the theory fails to accommodate many crucial aspects of life on earth. Even after a century of scrutiny and refinements, Darwin's original theory and all of its descendents remain plagued by glaring inconsistencies and contradictions in our observations of the natural world. While recent iterations of classic evolution theory appear to have succeeded in smoothing out some of the wrinkles, all evolution-based explanations are ultimately weighed down by many crucial observations. But virtually all of our observations and evidence are easily accommodated by our new explanation for organismal diversity: that genes were in fact abundantly available in the primordial pond, that by purely random processes they could have conglomerated to produce multitudes of complete, viable genomes, and that these genomes could have independently, simultaneously, spawned a spectacular array of different life forms.

We have seen how the biochemical complexity of the primordial pond increased over geological time, to ultimately yield a large amount of extremely complex macromolecules such as DNA and proteins. Then we learned that multitudes of complete genes, with precisely the same architectures as those present in today's multicellular creatures, could simply occur in the long DNA molecules in the primordial pond. The abundance of genes combined with the complexity of biochemical reactions made it inevitable that various genes would recombine, and that many of these combinations would constitute viable genomes for both unicellular and multicellular organisms. Like the zygotes of today's multicellular organisms, these genomes, enclosed in individual eukaryotic cells called seed cells, would independently develop into individuals of different organisms. This is what we have demonstrated in this chapter, by illustrating the following principles:

> If one particular gene is likely to exist in the primordial pond's universal sequence pool, then an almost unlimited number of unique genes are probable in the same amount of random DNA sequence.

> If the complexity of the primordial biochemistry and the number of genes required for the construction of a unicellular eukaryote could exist in the primordial pond, then it was absolutely inevitable that multicellular life would be formed.

> If the genome for even the simplest multicellular organism could be formed in the primordial pond by random processes, there is no question that genomes for innumerable unique multicellular life forms with widely varying anatomical and functional complexities could be also formed — all independently and simultaneously — by the very same processes.

> Finally, under such circumstances in an open-ended primordial pond, it is inevitable that pieces of already successful genomes could be mixed in the construction of other new genomes, leading again to many new and varied organisms containing features similar to earlier-born organisms.

The scenario of life on earth, both past and present, actually corroborates this theory. What better evidence could we hope to find for the simultaneous assembly of many separate genomes than the almost identical levels of genomic complexity between the simplest and the most advanced multicellular life forms on earth today? What could be more convincing than the simultaneous appearance of innumerable unique organisms in the Cambrian Explosion, at the very start of multicellular life? Moreover, the lack of evolutionary connections among living organisms at all levels of taxonomic hierarchy, as verified by the consistent absence of transitional forms in the fossil record, absolutely corroborates the predictions of the new theory.

The new theory explains the origin of highly complex organs such as the eye as well as it explains simple organs. Instead of regarding organs and body parts in terms of anatomical complexity, our new theory explains them in terms of genomic complexity. It then becomes apparent that there is truly no difference in genomic complexity between a simple organ and a complex organ. The probabilities for the genes and genetic circuits for any organ to originate in the open-ended primordial pond are relatively equal. This ability to explain complex organs as well as simple ones can be extended to also explain the origin of sexual segregation, and even instincts and higher level intelligence, since genetically these characteristics are no different than complex organs. The new theory's explanation for the origin of every unique organ, appendage, instinct and intellect in terms of gene sets and genetic circuitry is further corroborated by the inability of these features to evolve by organismal descent with modification. And finally, all of our observations of life on earth are consistent with our new theory's ability to unify, in terms of gene sets and circuitry, the origins of anatomically diverse organs and appendages.

While the separate assembly of genomes was responsible for the independent origins of multitudes of creatures in the primordial pond, each organism thus born is unchangeable to another with a new body part. This is entirely consistent with our primary finding in Chapter 3: that the genome of any organism is both physically and evolutionarily closed. Again, according to the new theory each organism, although unchangeable to another organism, could diversify within a limited range to produce a vast repertoire of individual variations as well as many similar species through mutation and genetic recombination. But these artificial breeds, natural varieties and similar species are inherently confined within the closed bounds defined for each distinct creature. Beyond limited variation, mutations can produce only developmental defects and diseases; they cannot change the constancy of the gene set or the DG pathway.

We have conclusively shown that this principle works clearly with all living organisms at all levels of complexity. For instance, an invertebrate such as a snail can change into many different snail varieties but an invertebrate can never be converted into a vertebrate. A squid with its unique vertebrate-like eyes and other unique body structures and functions could only have originated independently in the primordial pond, and its descendents could include only other varieties of squid. The characteristics of an organism in terms of its basic fundamental structures as well as its gene set and fixed DG pathway will never change. In consideration of all of these principles and all of their corroborating evidence, it is clear that all organisms were born independently, and were — and are — also immutable. This is the secret of life, and of its origin and history. This will be the truth of life for as long as life survives on this planet.

Let us remember that the history of science has always shown that open-minded and fair hearings of new, scientifically valid concepts are the bottom-line prerequisite for scientific progress — however startling, shocking, or heretical the new concept may at first appear. Our new theory of the independent birth of organisms rests on absolutely sound scientific principles and corroborating facts, and is a perfect fit with all of what we know about life on earth, both past and present.

9

THE MOLECULAR SCENARIO OF LIFE: EVIDENCE FOR THE NEW THEORY

The concepts developed in the previous chapter demonstrated that various organisms were independently born from a common pool of genes in the common biochemical environment of the primordial pond. Numerous genomes were constructed independently, and fragmentation of genomes and recombination of their segments with the primordial pond's gene-pool produced still more genomes. If this were the case, what could we expect — at the molecular level — of the scenario of organisms living today?

First we must ask if we have sufficient molecular and cellular information about living organisms to be able to test our predictions. Fortunately, the answer is yes. Over the past few decades researchers have accumulated much detailed information concerning biochemistry, genes, immune systems, bone marrow systems, and protein systems. We know how genes code for proteins — the actual working machineries of cells and organisms. What we need for our analysis is DNA and protein sequence information, as well as an understanding of the workings of the genome. Starting in the 1980s, considerable amounts of sequence information became available thanks to the concerted efforts of molecular biologists, and today we can use this information to determine whether organisms were born independently in the primordial pond.

Does our understanding of biochemistry, genes and genomes truly fit the concept of the independent birth of organisms, or does it fit the theory of evolution? We will show that the molecular scenario is indeed consistent with multitudes of distinct creatures born independently of one another by the independent construction of their genomes from a common pool of genes. The presence of unique genes, protein systems, and cell systems in numerous organisms prove that the organisms could not have evolved from a common ancestor, and that they could have originated only independently from the primordial pond. We will also show that the commonality of many biochemical features and of many genes in various organisms is indeed due to their independent origins from the common pool of genes in the same primordial pond. Molecular details show conclusively that similarities among distinct organisms cannot be attributed to evolution of organisms from a common original ancestor by descent with modification.

What kind of molecular scenario do we expect in organisms living today if distinct creatures were all born independently in the common primordial pond eons ago?

We saw that the primordial pond contained a vast number of genes and had the conditions required to organize these genes into various genomes. The construction of various genomes capable of being born into various organisms widely differing in their anatomical structures and biochemical functions gave rise to numerous distinct creatures. Because these genomes were organized from a large common pool of genes, they could have many of the same genes common to many or all genomes, and many genes that were unique to only one or a few genomes. Therefore, we should expect that although common genes could be present in the genomes of today's organisms, many of them should contain absolutely unique genes. Do we in fact find such a scenario? The answer is a definite yes, which unequivocally proves that numerous genomes were indeed constructed independently from a common pool of genes. While the presence of common genes in organisms can be argued as a support for the theory of evolution, the presence of unique genes can support only the theory of independent birth of organisms and indeed argues totally against organismal evolution having occurred in these creatures.

Additionally, what kind of biochemical scenario should we expect in the living organisms if our predictions of independent assembly of genomes is correct? The independent genome-assembly process should have led to the production of the same biochemicals in those organisms with essentially the same genes, and unique biochemicals would be present in those organisms with unique genes. Again, do we find such a scenario in organisms living today, which would prove that the organisms were born independently? The answer is indeed yes!

Looking at the many other details in a similar manner, we can say that in general, if the new theory of independent birth of organisms is correct, life on this planet should show the following molecular evidence:

1. The genome of each organism should be unique with a unique developmental genetic pathway, leading to distinct, unrelatable creatures.

2. Because genes present in the random DNA sequences in the primordial pond were randomly assembled into various genomes, genomes should contain meaningful genes as well as significant amounts of junk DNA material.

3. Again, because the genomes were assembled independently from the primordial pond's gene pool, the sizes of the genomes of various organisms should also be random without any order or relationship with each other.

4. Because each genome was assembled randomly from the primordial pond's gene pool, different genomes may have unique genes.

5. Because multiple copies of genes were available in the primordial pond, different genomes should also have many identical genes, which could change only into the gene variants, with allowed sequence variations.

6. The independent birth of organisms allowed the inclusion into genomes of many distinct, unrelated genes for proteins with similar biochemical functions. This process would require today that various living creatures can have evolutionarily unrelatable genes for similar biochemical functions, but with a minimal protein sequence and protein structural similarity.

7. Because all the genomes and seed cells were assembled from a common pool of genes and a common biochemical environment, wherein the basic biochemical processes of cellular life had already been established, they would all be based on the same biochemical features, and the same genetic code and genetic machineries.

8. If multicellular eukaryotic genomes were assembled directly in the open primordial pond, there should be molecular evidence for the genes and genomes of at least the single-celled eukaryotes to have directly originated from the primordial pond's random genetic sequences.
9. If any special feature is present in the genome of an organism, then it should be present uniquely in one or a few organisms, without evolutionary relationship with other organisms.

We have touched upon some of these aspects in Chapter 8 to show that the new theory is consistent with, and is corroborated by, the actual scenario of life on earth. In this chapter, we shall systematically analyze the molecular details of living organisms more elaborately and thoroughly, and illustrate that all of them indeed corroborate the independent birth of creatures. If we can show that the picture of life on earth at the molecular level is absolutely consistent with what is predicted by the independent birth of organisms, and also demonstrate that many of these details are improbable by the evolution of one organism into a distinct creature by descent with modification, then we would have essentially proved that the new theory is correct. In fact we shall show that while Darwin's theory seems to explain a few aspects of the molecular scenario of organisms, it cannot really explain many crucial details of the whole scenario. The new theory, however, is able to explain almost all the details precisely and convincingly.

We shall see that the new theory unifies all living organisms on the basis of the common universal gene pool of the primordial pond from which all distinct organisms were independently born. The genomes of these distinct organisms were constructed taking common genes, unique genes, and similar genes from the rich pool of all these entities. Consequently, they produced a wide variety of organisms with some similarity as well as distinct uniqueness at the level of genes and at the level of the organism. At this juncture, we should remember that we are not saying that creatures with the same body parts or the same set of genes in their genomes are born independently. As we discussed in previous chapters, these are the varieties and similar species of the same distinct organism. We are alluding to only those creatures that have unique body structures and/or unique genes.

This chapter contains many technical terms, used to illustrate the distinguishing unique characteristics of distinct organisms. We shall discuss these terms as we go along without undue digression. Let us remember that it is not necessary to fully comprehend the jargon in order to understand the concepts herein.

Unique and unchangeable developmental genetic pathways are found in the genomes of various organisms. Such a phenomenon can arise only by the independent assembly of genomes in the primordial pond.

We have established that the developmental genetic pathways of different creatures are unique and rigid — one of the main themes of the theory of the independent birth of organisms. This concept states that the developmental pathway of an organism is so unique that it cannot be changed into that of another through organismal evolution. Can we show this to be a fact by analyzing the developmental genetic pathways of various living organisms? The answer is yes!

If the genomes of various organisms were independently assembled in the primordial pond, one should expect that necessarily the genetic pathway of development of each successful genome was also organized independently. If this were the case, obviously we shall expect that the DG pathway of each organism living today should be distinctly different. Fortunately, we have sufficient molecular details in this field from the past 20 years, which we shall analyze objectively and show that this is indeed true. We can find ample evidence from modern research in embryology that the DG pathway of each organism is unique and distinct from those of other organisms. If we first establish this as a fact here, then we can be sure from our discussions in Chapter 3 that the unique DG pathways are also unchangeable and rigid. This conclusion will show beyond doubt that the only way unique and unchangeable pathways of organismal development could have originated was by their independent organization in the primordial pond.

On the contrary, evolutionists believe that early development of all organisms must be an evolutionarily "conserved" process; this belief stems from early embryos of many widely different creatures that "look" superficially similar. Therefore, people have traditionally believed that the DG pathways of various organisms are similar and are derived from that of the original common ancestral animal. But we can show that this concept, that "ontogeny recapitulates phylogeny" (*i.e.*, an organism repeats the ancestral evolutionary stages in its development), is absolutely wrong. What I am trying to show here is that we have not looked at these details so far with an objective and open mind from completely outside the domain of evolutionary theory. When we do this, we can certainly see that the DG pathways of various organisms are unique.

Embryonic cellular developmental maps of different organisms are distinctly unique

The timed mapping of the growth and development of embryonic cells in three dimensions, starting from the zygote, can be equated to the developmental genetic pathway of an organism. This is because the cellular developmental programs leading to the different types of cells in the body, that gives a unique size, shape, and structure to the different body parts, is the same as the developmental program of the whole animal. This cellular program can be equated to the developmental genetic program of the organism, because the pathway taken by all the different cells of the organism indeed reflects the genetic program that is operating in the organism. Consequently, if the cellular developmental programs of different organisms are completely different from each other, we can conclude that the developmental genetic programs of various organisms are also completely different.

There are several organisms in which the total number of cells in the fully developed animal is limited to a few thousand. In these organisms, it has been possible to map the developmental pathway traversed by each cell of the animal. That is, from the zygote, the timing and positioning of each new cell in the geography of the animal's body has been determined. For example, the organism *Aplysia* contains about 1000 cells in its body. The exact times of differentiations and three-dimensional paths of all these cells, starting from the zygote, have been determined. Likewise, the developmental pathways of all the cells in the microscopic worm, *C. elagans*, that contains about 1000 cells, is known. This has also been worked out in other organisms such as *sea urchin*. In fact, one can see that the patterns traversed by the cells of the embryo are absolutely different in each organism. The published literature[1] shows that the "cell division maps" of the three organisms, *C. elagans, Aplysia,* and *sea urchin,* are quite distinct.

The few animals in which such a cell growth map has been worked out illustrate this phenomenon of the distinctness of the DG pathway. The entirely different pathway that the zygote of each organism traverses to build the fully formed animal illustrates that indeed it is the DG pathway of each organism that is unique. The unique pathways that bring forth the structural and functional specificity of the fully-formed embryo provides perhaps one of the best molecular clues that the developmental genetic pathway of every organism is highly specific and unique.

Modern embryological research provides ample evidence that the DG pathways of organisms are absolutely unique. They show that even at the very early stage of embryogenesis, distinct patterns of embryogenesis appear in different organisms. For instance, Eric H Davidson[2], a pioneer in embryology and eukaryotic molecular biology, clearly shows by his research and those of others that the cellular developmental pathways traversed by various organisms are indeed quite distinct. Although he is a staunch evolutionist, his remarks beautifully underscore that our concept is certainly correct. He writes in a recent review article (italics mine),

> Classical authors, however, and those of their successors who have attempted to deal with more than one embryonic form, have been struck by the amazing variety in the modes of embryonic development that exist in the various phylogenetic reaches of the Animal Kingdom. All embryos do indeed achieve the imposition of spatial patterns of differential gene expression, and yet some begin this process by intercellular interaction, and others even before there are any cells that could carry out such interactions; some rely on lineages that are autonomously committed to given functions from the moment they appear, others deal wholly in plastic, malleable cell fate assignments; some utilize eggs that before fertilization are cytoskeletally organized in both axes, some in one axis only, some apparently in neither; ... *For the differences among taxa in their modes of embryonic development are anything but trivial and superficial (certain hopeful reductionist delusions of recent years to the contrary).*[3]
>
> ... Table 1 considers the development of representative animals of eight different taxa for which there is considerable information available. *It is evident at a glance that these organisms utilize profoundly different strategies to achieve development.*[4]
>
> ... If by *regulatory architecture* we mean the specific designs of the mechanisms utilized to establish differentially functioning spatial territories in various embryos, the plan of the regulatory interactions required, the levels of hierarchy amongst regulatory genes and so forth, *then it is clear that the architectures underlying cell specification in the various embryonic forms are themselves very diverse.*[5]
>
> ... Given that the naive belief that *there is some general regulatory architecture that will explain all forms of embryonic development is unsupportable,* it is the *particular* regulatory architecture underlying the embryonic process in each animal taxon that is the key to understanding how its embryos achieve development.

Davidson concludes his review article saying:[6]

> ... So how do embryos work? Or, the question to which we have come, what regulatory structures are required to explain the appearance of differential spatial patterns of histospecific gene expression in embryos of diverse taxa? *The answer is clearly different for different embryos,* but there are certain sets of strategies that groups of phylogenetically unrelated animals share.[7] ... *It is clear that a casualty of these arguments is the 19th century concept that early development must be an evolutionarily conserved process.* We see that during evolution

the regulatory genetic elements controlling embryogenesis have been reassembled in many different combinations. [8]

From Davidson's writings on the comparative embryonic development of different creatures (belonging to various taxa), one can see that the developmental processes among different organisms appear to be similar only at a superficial level. At a deeper level, however, the differences underlying the developmental processes, in fact, far exceed the apparent similarity among them. I believe that the superficial similarity of these developmental processes is akin to the similar processes of basic cellular genetics and cellular machineries in all the different organisms, such as the DNA replication, transcription, splicing, cell division, etc. As Chapters 3, 4 and 8 demonstrate, these similarities and common processes do not mean an organismal evolutionary connection, and the same is true for the developmental processes. It only means that all the different organisms derived all these basic mechanisms in their genomes from a common pool of genes.

It is interesting to note that Davidson, who is a staunch follower of evolutionary theory, shows by his most recent and advanced research that the concept of "the evolutionary conservation of early embryological development among various animals" is absolutely incorrect and that the embryonic pathway of development for each organism is indeed distinctly different. The establishment of these facts is undoubtedly a strong corroboration to the new theory of the independent birth of organisms.

Besides looking at the uniqueness of the actual cell division pathway of embryonic development, one can also derive the above conclusion by a systematic analysis of genetic programs that construct the different organs and appendages of various organisms. Based on the logic that creatures with unique organs must follow unique developmental programs for that organ, we saw (Chapter 3) that the developmental program of a creature with a unique organ or body part is quite different from that of an organism lacking in that body part. Even a minute body part with a specific structure and function requires a large number of genes organized into a specific developmental genetic program to construct that body part; often it requires the action of unique genes not present in other creatures, supposedly in the lower steps of the assumed evolutionary ladder. While this is so, the unique shapes and sizes of multitudes of organisms with countless unique organs and appendages occurring in the living world demonstrate our concepts clearly. We have seen earlier that only because numerous creatures are highly unique in their structure and function, they are classified into several distinct higher taxonomic groupings such as phyla, classes, and orders. Now consider the fact that the structure, shape, and function of the unique body parts in many organisms, even those belonging to the successive steps in the supposed evolutionary ladder, are quite distinct. Furthermore, many creatures have unique types of cells (obvious examples are eye cells, bone cells, etc.). These facts corroborate the observations that the cellular developmental genetic programs of different organisms are distinct.

From all our analyses here and in Chapter 3, we can conclude that the DG pathways of different organisms are distinctly unique — despite the fact that various organisms may use some similar developmental strategies due to the commonality of the primordial pond and environmental constraints.

The molecular evidence for the uniqueness of the DG pathways of various organisms and their rigidity is also corroborated well from the fossil record. As we have discussed in Chapter 3, the sudden appearance of each organism in the record in its full form, and its remaining virtually unchanged throughout the length of the geological record, precisely illustrate and corroborate our principle that the DG pathway of an organism is unique and cannot change through organismal descent with modification. While this is a total enigma and a major blow to any theory of evolution, it is an absolute corroboration to the new theory.

In summary, we have established that the DG pathways of diverse organisms living today on earth are indeed quite distinct. Thus the predictions of the new theory that the various genomes with distinct DG pathways assembled independently in the common primordial pond is proven to be true.

Molecular evidence that the genes and genomes of animals and plants could have directly originated from the primordial pond's random genetic sequences

One of the most important requirements of the new theory is that there should be evidence that the genes of the typical eukaryote could directly occur in the primordial pond's random genetic sequences. Indeed we have the best evidence for this. I have shown through my research that it is only the split genes, typical of eukaryotes, which could occur directly in the random sequences in the primordial pond, and not the contiguous genes of the prokaryotes. By this, I have shown that it is the unicellular eukaryotes that should have originated directly from the primordial pond, and not the prokaryotes. Since we discussed this elaborately in Chapter 7, we shall not go into details here.

Molecular biologists now accept that it is the unicellular eukaryotes that should have originated first. However, they say that the first cells were "urkaryotes," meaning that they did not have a nucleus, although their genes had introns, and that after the prokaryotes evolved from them by losing introns in their genes, the prokaryotes moved inside the urkaryotes as nuclei, and made them eukaryotes. They seem to say this in order to be consistent with a previous theory called the endosymbiotic theory, in which a prokaryote engulfed another prokaryote that became a nucleus.

However, in my theory of the origin of introns and the origin of cells, I have shown that the cells that could directly come from the primordial pond must have been unicellular eukaryotes with the nucleus (Chapter 7). Furthermore, the genomes and cells of typical multicellular eukaryotes are not any different from that of a unicellular eukaryote. Thus, this provides direct evidence that the genomes of the multicellular eukaryotes could likewise assemble in the primordial pond.

THE SIMULTANEOUS PRESENCE OF IDENTICAL GENES, UNIQUE GENES, AS WELL AS FUNCTIONALLY SIMILAR BUT STRUCTURALLY UNRELATED GENES IN VARIOUS ORGANISMS: EXPLAINABLE ONLY BY THE THEORY OF THE INDEPENDENT BIRTH OF ORGANISMS

The gene pool in the primordial pond contained an immense number of unique genes, but each of the genes was present in multiple copies to various extents. When many distinct genomes were assembled from this pool, the genomes would have had the following characteristics:

1. Unique genes should be present in the genomes of one or a few organisms but absent in most organisms.

2. There should be evolutionarily-unrelated genes coding for essentially the same overall biochemical functions. The proteins may have low or high sequence and structural similarities.

3. Unrelated genes will be present in many different organisms, each of which would code for a protein with a distinct biochemical function.

However, the proteins may have similar subfunctions such as binding the same cofactor, and may have sequence similarity over such functionally-similar regions. Yet, these proteins (and their genes) will have no sequence similarity over the functionally dissimilar regions, which can constitute a considerable portion of the protein.
4. Many essentially identical genes should be present in the various independently-assembled genomes.

We shall analyze the details of the molecular scenario in living organisms and show that these predictions are precisely found to be true. We shall demonstrate that such a scenario can arise only if the organisms were independently born. This scenario cannot be explained by any theory of evolution.

Presence of unique genes in numerous creatures can be explained only by the independent birth of organisms

Consider a hypothetical situation where we are given a pool of genes, and each gene exists in multiple copies to various extents. If we are asked to randomly assort these into various genomes, what sort of distribution of genes will be found in the resulting genomes? Some genomes would have more than one copy of certain genes, only one copy of a few other genes, and no copies of yet other genes. Under this situation, many genomes would include one or a few of the same genes. At the same time, there would be present some unique genes in one or a few genomes that would be totally absent in other genomes. The unique genes and the proteins they encode are the true indicators of what had actually happened in the primordial pond.

If, as claimed by the evolutionists, various organisms originated from an original common ancestor through evolution, then the genes of every organism living on earth should be related to those of all other organisms, and unique genes — unrelated to any other genes in any other organisms — can never be present in any organisms, as we have unequivocally demonstrated in Chapters 3 and 4. However, the presence of unique genes is extensively exemplified in creatures on earth. The identical genes in the genomes of various creatures can only indicate that these genomes were derived from the common pool of genes in the primordial pond by including copies of the same genes in distinct creatures — especially in light of the fact that the presence of unique genes can be explained only by the independent organization of genomes in the primordial pond.

The unique genes can specify unique functions. Groups of unique genes can specify unique systems of function. Similarly, unique cellular systems can work in distinct organisms for specific cellular functions. If this theory is correct, examples should abound throughout the living world. When we analyze the available molecular details, this is indeed found to be absolutely true, strongly supporting the possibility that creatures were independently born in the primordial pond. It is true at various levels of taxonomic categories, from the highest to the lowest. It is obvious that at the highest level, organisms belonging to various phyla or classes do have many unique proteins and genes. It is also equally true that there do exist many unique proteins in organisms classified into orders and families, showing without a doubt that multitudes of distinct creatures must have been born independently in the primordial pond.

Unique proteins in various organisms: Explainable only by the theory of the independent birth of organisms

We do not know about all the proteins of even one organism, let alone from all the organisms on earth. However, we have sufficient knowledge and information from many organisms concerning their proteins, genes, and cell systems, from which we can determine that there exist many unique genes in various organisms, some of which we discussed in Chapters 3 and 8. Indeed, if we know all the proteins of all creatures, we can be sure that unique genes and proteins will be found to be the norm of organisms rather than the exception. Although we briefly noted this principle in Chapter 8 when we delineated the theory of the independent birth of organisms, let us discuss this more elaborately here.

We can either take a look at the molecular scenario from the bottom-up or from the top-down in their assumed evolutionary taxonomic hierarchy. Either way we shall see the phenomenon of the presence of unique proteins clearly. Let us take a look from the top-down. There are proteins in the primates that are not present in other nonprimate mammals. Proteins such as the *chorionic gonadotropin* (a placenta-specific protein hormone) are only present in the primates[9] (with the exception of horse) and not present in the reptiles or other animals. Similarly, proteins in mammals are not present in the nonmammalian vertebrates. Examples are the placenta-specific proteins, and many hormones. Vertebrates have numerous proteins which are now established to be totally absent in the invertebrates. Remember that all the vertebrates are part of only one phylum, whereas all the organisms on earth are classified into 34 entirely distinct phyla. This means that the various invertebrate creatures are even more distinctly different among themselves than the various vertebrates differ among themselves. Within the invertebrates, there are numerous examples of animals with unique proteins and protein systems. Groups of invertebrate organisms have special proteins that are absent in all other organisms. Numerous proteins such as silk proteins, poison proteins, hormones, and antifreeze proteins are examples. Thus, we can illustrate that unique proteins and genes are present in creatures at all levels of organismal hierarchy such as the family, order, class, and phylum. This often may not be true at the level of genus, and even at the level of family, because the creatures within these taxonomic categories may be derived from a single, independently-born creature — *i.e.*, they may be the varieties and similar species of a distinct organism.

Protein systems and cell systems that are absolutely unique to various creatures: possible only if the organisms were independently born in the primordial pond

We saw in Chapters 3 and 4 that there are approximately 600 proteins in the blood plasma of the vertebrates, which are not present in any of the invertebrates. A search for the presence of any protein similar to these proteins in the invertebrates, from which the plasma proteins of the vertebrate blood are believed to have evolved, has been totally futile from the start of this search about 20 years ago. Such studies have actually revealed the distinctness of cellular systems, immune systems and other protein systems in vertebrates. Very interestingly, these studies have also revealed unique cellular systems and proteins present in particular invertebrates, which are absent in other invertebrate and vertebrate creatures.

All the proteins of the vertebrate blood plasma, such as *albumin, fibrinogen, fibronectin* and *transferrin*, are totally absent in the invertebrates. These are proteins each with many repetitions of a basic domain. The search for even the basic domain for each of these proteins in the invertebrates has been totally unsuccessful. Even in the case of fibrinogen, the only case wherein a protein in an invertebrate has been found with some similarity in sequence, it is in a form from which the vertebrate fibrinogen could not have evolved — even by the account of molecular evolutionist Russell Doolittle, who is a pioneer in the field of plasma proteins.[10]

In fact, we can categorize the biological functions of an organism such as the circulatory system, immune system, and reproductive system and show that each of these systems in different organisms are indeed unique in terms of their genes, proteins, and cells. We can very well see that these functions have to necessarily occur in one form or another in any living creature, even when they are all independently born. And so, the presence of these functionally similar systems in various creatures are in no way reflective of organismal relationship with one another. At the same time, indeed, the absolute distinctness or uniqueness of these systems show clearly that these are totally independent genes, proteins, and systems not at all related by organismal kinship.

Unique immune systems in many different creatures

Vertebrate immune systems consist of many cell systems and protein systems (the immunoglobulins). The vertebrate immune system is completely absent in the invertebrates. The invertebrate immune systems are distinctly different, with absolutely no evolutionary relationship with those of the vertebrates. The invertebrates do not have the white blood cells common to vertebrates such as humans. They do not have the T cells that all vertebrates have. These are the cells that attack and destroy any invading organisms, viruses or bacteria. The invertebrates have their own unique circulating "blood" cells to fight invaders. These cells are generically called *hemocytes*. Furthermore, there may be different kinds of cells for this purpose in the various invertebrates.[11] For instance, the body fluid of the earthworm contains cells called *chloragocytes*, apparently present only in these animals.[12]

The immune response of the mammals is based on their capability to have a memory to a previous immune reaction. For instance, once we get an infection such as chicken pox virus early in life, we are immune to the same infection for the rest of our lives. This is because vertebrates have B-cells, which produce antibodies (a kind of protein molecule) specific to the virus (the antigen), and this memory is registered into the animal system for its life. In contrast, this kind of immune memory is absent in invertebrates, and their immune response is nonspecific.

Many invertebrates, particularly marine forms, live in areas rich in organic nutrients with an abundant microflora that, apart from providing food, also act as a potential source of infection. Although invertebrates lack the immunoglobulins or other immune-recognition molecules, they still manage to keep their internal body fluids sterile, because they have their own effective means to distinguish between self and nonself.[13] The invertebrates have a totally different kind of system. Some of these creatures have antibacterial proteins and genes specific to them. Insects, for example, have molecules called *lectins* for recognizing invading foreign bodies. The lectins are nonimmunoglobulin proteins with a capability to bind sugars nonspecifically. These have absolutely no relationship to vertebrate immunoglobulins. Furthermore, lectins are a diverse class of proteins with little or no structural or functional relationship among the different kinds of lectins. Lectins do not remember a previous infection. There seem to be many other proteins in various invertebrate creatures for recognizing foreign elements and destroying them.[14] For instance, there exists a large spectrum of proteins in the body fluid of the earthworm, allowing earthworms to neutralize invading bacteria.

With the understanding that different protein molecules such as the lectins are grouped under a particular generic name based solely on their functional similarity, we can see that various invertebrate and vertebrate organisms have a random pattern of having and lacking these kinds of molecules. For instance, the C-type lectins are present in insects, crustaceans, echinoderms, tunicates, birds and mammals, but absent in centipedes, millipedes, arachnids, annelids, mollusks and many other invertebrate groups, and is also absent in reptiles and amphibians.[15] The *major histocompatibility* molecules are present in mammals, birds and amphibians but absent in reptiles and all invertebrates. A *C-reactive* protein is present in mammals and absent in birds, reptiles, and amphibians. Among invertebrates, it is present only in tunicates and arachnids. Cockroaches, giant silkmoths, tobacco hornworms, and other insects also rely on at least one other mechanism to combat infection: a variety of defensive proteins such as the antibacterial *ceropins*.

Uniqueness of the respiratory systems, cells, and proteins in distinct creatures

It is interesting to note that the circulatory systems of insects do not have anything to do with breathing. They do not have red blood cells to carry oxygen to the various tissues and body parts. Their circulatory system is used only for transporting nutrients, water balance, and immunity. Insects and most invertebrates breath through their *tracheal* system (a system of tubes which is totally distinct from the vertebrate lungs). Many invertebrates breath through their skins simply by diffusion of the air into the body fluids from which the cells exchange oxygen and carbon dioxide (this happens even with some vertebrates such as the salamander). The aquatic invertebrates such as the lobster and crayfish have gills, from which oxygen is directly diffused into the body fluids. A few insects have circulatory systems and oxygen-carrying cells. These organisms are unique and live in low-oxygen environments.[16]

In the vertebrates, the *hematopoietic* cells, which circulate in the blood, originate from cells in the bone marrow. However, in invertebrates, the origin of the cells in their blood is unknown. Obviously there is no bone marrow in the invertebrates. Note that we use the term "blood" here loosely to denote the circulating fluid in invertebrates. There is nothing in common between the blood of vertebrates and invertebrates. The blood proteins of the invertebrates are totally distinct from those of the vertebrates.

Also, there are apparently different blood proteins among the various invertebrate creatures, as we can predict if different invertebrates had originated independently in the primordial pond. In light of the theory of the independent birth of organisms, it is not surprising that there are many differences and uniquenesses in these systems among the different invertebrates, which can never be attained by any evolutionary mechanism.

Different invertebrates use unique oxygen carriers. Hemoglobin is the well-known protein that carries oxygen in our red blood cells from the lungs to the various body parts and brings carbon dioxide from them back to the lungs to be exhaled. This protein is present in all the vertebrates. In addition, it is also present in many invertebrates such as some worms, molluscs, insects, and a few others.[17] Very dissimilar hemoglobins are present in a few invertebrate organisms. Only functionally (*i.e.*, oxygen-binding function), these are similar to the typical hemoglobin. Also, there are numerous invertebrates that do not contain hemoglobin.

Hemocyanin is a large protein that contains copper and binds oxygen reversibly. This protein is the oxygen carrier in some invertebrates. For instance, molluscs such as *Helix* and *Octopus* have hemocyanin. Bivalves, which are also classified as molluscs do not have hemocyanin. *Limulus* (classified under Chelicerate) and *Palinurus* and *Homarus* (classified under Crustacean) also have hemocyanin. It is possible that the various hemocyanins are structurally variable, although functionally similar. Hemocyanin protein has absolutely no relationship with hemoglobin or other oxygen carriers.

Chlorocruorin (a protein containing iron) is present in some worms (four families of polychaetes, classified under annelid). Apparently this protein is absent in other invertebrate or vertebrate animals so far tested.

Heamerythrin, another protein with iron, is present in a few invertebrates but not in others. This protein is found in some annelids, sipunculans, and priapulids, which are supposed to be fairly similar to each other, but is also present in another totally distinct group of organisms, the brachiopods, thought to be evolutionarily unrelated to the other three. It is also very interesting to note that the single class Polychaeta exhibits within itself three of the four respiratory pigment molecules, in apparently random patterns amongst the families.

The process of "coagulation" and proteins used for it are distinct in different creatures

When we have an injury, such as a cut on the skin, very soon the blood "clots" at the site of the injury, sealing the hole. A large repertoire of proteins called "blood clotting factors" participate in this process. In addition, a type of blood cell called the platelet also participates in these reactions. These proteins and cells form a kind of mesh, producing a sealent around the site of injury. This kind of blood clotting is unique to the vertebrates.

The invertebrates do not have this kind of blood clotting. The system of blood coagulation and the proteins associated with it in the vertebrates are totally absent in invertebrates. Not only that, there are some other kinds of systems and proteins in some invertebrates that prevent the leaking of the body fluids. These systems and proteins are totally different from those present in the vertebrates — indicating the uniqueness of these systems and proteins to these organisms. Some of the invertebrate creatures have a jelling substance called *coagulon* in their body fluids, which gels at the site of injury.[18] We can see that the proteins and other molecules used here have to be entirely unique to them as are the corresponding genes. It should also be noted that different invertebrates may have different systems of preventing the leakage of body fluids. The details of most of these systems are yet to be worked out. We can be very sure that as these details become known in the near future, they will strengthen the theory of the independent birth of organisms.

We can go on and on talking about such unique proteins and systems in different creatures, invertebrates or vertebrates. There can be numerous hormones and proteins involved in sensory perception, reproduction, digestion, etc., which may be unique in various creatures. While we can continue to elaborate and produce a whole list of unique proteins, genes and cell systems, it is not our aim here to produce a compendium of such things. Our aim is to illustrate the concept of uniqueness by giving sufficient examples. Certainly, the examples we have discussed give us a flavor of the uniqueness of these things in living creatures. When we realize that what we have seen is only the tip of the iceberg, a substantial portion of which will be revealed in the near future by the genome studies currently undertaken, we can understand the magnitude of the number of proteins, genes, and other systems uniquely present in distinct creatures living today.

We should also note that we are often speaking about vertebrates versus invertebrate differences as a whole. It is important to see that there are innumerable differences among the various invertebrates. In fact, most of the details are not yet well-defined, for in general the research interests of the scientists have been so far focused on the vertebrate organisms. We need to have no doubt at all that the differences among the various invertebrates are numerous, which can originate only if these creatures had been born independently in the primordial pond.

The presence of unique lens crystallin proteins in various organisms

Before we conclude our discussion of the uniqueness of genes and biomolecules in distinct organisms, let us return to the eye. The lens of lens-containing eyes is made up of proteins with transparent properties. These proteins are called *crystallins* (if a protein is present very abundantly in the lens tissue, it is defined as a crystallin).[19] Crystallins form a transparent matrix. No other known function for these proteins exist in the lens. Different creatures exhibit different crystallins. Many of these crystallin proteins are unique to groups of organisms and are therefore called taxon-specific crystallins. Some appear to be the same or similar in many organisms. Interestingly, some of these crystallins function as enzymes in some tissues other than the lens.

Because some of these crystallins are found to have enzymatic functions in the metabolism of biochemicals such as amino acids, sugars, and nucleotides, molecular evolutionists believe that during the evolution of the lens from an organism lacking a lens, existing genes in that organism were "recruited" to form the lens. When we analyze the whole scenario from our perspective, it will be seen that such a belief is totally incorrect; it will show that there are many proteins unique to the lens and most likely absent in the supposedly "lower" organisms with lens-less eyes or with no eyes at all. Knowing that entirely new genes cannot be evolved in the genome of an organism even in trillions of years, such an analysis will illustrate that the only way the genes for the various crystallin proteins could have originated was by the independent assembly of these genes from the open primordial pond.

If the vertebrate eye had first evolved in a primitive fish, then the pattern of proteins expressed in all further descendents of the fish (that is, all other vertebrates) must be the same or similar. But there are taxon-specific crystallins which are absent in other taxa even among vertebrates. Invertebrate creatures with lens eyes also have unique crystallins. Ducks have ε-crystallin, chickens have δ-crystallin, turtles have τ-crystallin, squids have SIII-crystallin, and frogs have ρ-crystallin as taxon-specific crystallins. These have sequence homology to different enzymes, although we do not know if they actually have these enzymatic functions. Our point is that once a lens had evolved in the primitive fish, then there is no reason to remove it and put another protein in its place. Even if it had happened for reasons that the other protein is better evolutionarily, then we must see an evolutionarily-connectable pattern. But what we see in the taxon-specific proteins is a random pattern. This pattern fits only the random assortment of genes from the universal gene pool.

The delta (δ) crystallin is the major lens protein in birds and reptiles but apparently is not present in mammals. There are two delta crystallin genes: δ1 and δ2 in the chicken genome. The δ1 gene is actively expressed in the embryonic lens and encodes the major structural protein, whereas the δ2 gene appears to be weakly expressed in various tissues. The δ2 protein has been found to have high homology to the rat *arginosuccinate lyase* enzyme. However, there is a discrepancy: the δ1 is present in reptiles and birds, but is absent in mammals. To counter this discrepancy, evolutionists propose that the δ1 and δ2 were evolved by gene duplication of an ancestral gene in the ancestor to reptiles, birds, and mammals, and that the δ1 gene is lost only in mammals. We must remember from our discussions in Chapters 3 and 4 that such precise losses are highly improbable. Except for the homology of the δ2 crystallin of the chicken with the arginosuccinate lyase of the rat, it must be noted that this protein is not a crystallin in the rat eye lens. Furthermore, the δ1, which is a crystallin in the chicken eye lens, is not even present in the rat genome. These facts make it clear that the chicken crystallins are entirely different from those of the rat.

If the various crystallins in the various lens-containing organisms are not related, and if they had originated independently from the primordial pond's gene pool, why do some of these proteins exhibit enzymatic functions or at least have homology with enzyme proteins? It is possible to use one gene for more than one function in organisms born in the primordial pond. Such multifunctional proteins could be put under different regulatory controls and function differently in two different tissues. There is no reason such compartmentalization cannot occur if the two different functions do not interfere with each other. We can be sure that this is what has happened here — not what molecular evolutionists have called the evolutionary recruitment of old genes for new functions.

Let us consider how the eye lens proteins could have been selected from the primordial pond. The universal gene pool contained a large number of genes for structural proteins and enzymatic functions. An enzyme carries out its function only in the presence of its *substrate*, the molecule that the enzyme acts on. Without the substrate the enzyme is only an inert protein. One of many such proteins in general will have the property of being transparent. Some have the property of being nicely packed in a cell and form a transparent matrix with a refractive index useful in a lens. If the substrate is not provided in these cells, then the enzyme, which now functions as a transparent matrix, cannot function as an enzyme. Using this principle, any gene for proteins from the universal gene pool can be selected to build the lens as long as they have these transparent properties, no matter what functions they have in the other tissues and cells of the body of a living system. This is what might have happened in the selection of genes for eye lens in the primordial pond.

The interesting point is that there probably exist many crystallins in the different taxons and creatures that are possibly unique to many of them. Many of these crystallins may not have any similarity to any other known genes in the genome, and may not have any enzymatic activity (*i.e.*, they may be proteins with only the passive transparent matrix property). Let us read what Joram Piatigorsky, the noted authority on lens crystallins, and his associates have to say:

> Thus, despite the structural similarities between the cubomedusan (jellyfish), squid and vertebrate lenses, their crystallins appear very different.[20]
> ... The lenses of frogs of the genus *Rana* contain a major protein distinct from other known crystallins. ... it has no relationship with the protein found in birds and crocodiles, and has been renamed ρ-crystallin. ... As yet no enzyme activity has been identified for ρ-crystallin.[21]

The β and γ crystallins are purported to have some weak similarity at the tertiary structure level to the bacterial spore coat protein S. But this is an erroneous expectation in the hope that only then the scenario fits with Darwin's theory. In fact many scientists do not believe they have any similarity to the bacterial proteins. Thus these can be said to be the unique crystallins of the lens, which support the new theory. The lens is only one tissue of the eye. We can be confident that there will be many proteins that will be found in the suborgans of the eye which will not be expressed in any other tissues. Further, we can be sure that many genes used in these suborgans of the eye are not present in their supposed ancestors.

PRESENCE OF UNIQUE METABOLIC PRODUCTS IN VARIOUS ORGANISMS IS MORE EVIDENCE FOR THE PRESENCE OF UNIQUE GENES

We have so far shown that there exists numerous genes encoding unique proteins in the genomes of various organisms. Because these genes could not be evolved by descent with modification, we demonstrated that the organisms having such unique genes could not have evolved from those that lacked such genes. There is yet more molecular evidence existing in organisms that goes to prove the same phenomenon. A number of unique metabolic products are present in different creatures; these are not proteins but small molecules which are the result of enzyme actions on other small molecules. Usually, a metabolic product is the result of a set of enzymes that act sequentially on small molecules to synthesize the final metabolic product. Many color-forming pigments are examples of metabolic products. Furthermore, in many organisms, there exist unique metabolic pathways yielding unique metabolic products. These genes, metabolic pathways, and products are unique to one or a few creatures.

For example, several chemicals are synthesized by some moths and butterflies that would make them unpalatable to predators. The end-products of metabolism excreted from the body are different in birds, mammals, reptiles, amphibians, and fish. *Chitin* is a chemical that is used to build several structures in the invertebrates. Several poisons are produced by snakes, frogs, spiders and other creatures. Many more examples can be given.

The available information is more than sufficient to illustrate the phenomenon of the wide distribution of unique genes in various organisms. I am confident that future research will uncover more examples illustrating the uniqueness of metabolic products in organisms. We should also note that the data we have provided here are from ongoing research. Thus, even if some of the proteins and genes we have discussed here are found not to be unique in the future, the presence of unique proteins in multitudes of organisms will still remain true.

We have amply demonstrated the distinctness and uniqueness of proteins, genes, and functional systems in various creatures living today. Even if creatures were independently born in a primordial pond, all the creatures on earth have to be based on tissues, cells, proteins, and genes. Each organ and tissue has to be built by proteins and they have to necessarily work based on genes and gene mechanisms, because these are the only ways on which living entities could be based and could originate on earth. They should necessarily work based on metabolism, which on earth usually needs oxygen; they should also necessarily have a system for protecting themselves from invading organisms, thus needing an immune system. But when the different creatures originated in the same primordial pond from different subsets of genes of a large universal gene pool based on the same fundamental gene mechanisms, they can easily contain entirely distinct genes, proteins, functional systems, cells, tissues and organs.

Our aim here is to show that unique proteins and genes are abundant in living creatures. Molecular evolutionists believe that all these multitudes of unique protein genes have evolved from a single common ancestor — even a single-celled ancestor. But as we have shown in Chapters 3 and 4, they could never have evolved from a common ancestor over any length of geological time. Molecular evolutionists simply say this without any evidence, only because they have to work within the domain of evolution. But we now know for certain from our demonstration in Chapter 7 that genes for all these numerous unique proteins must have simply occurred in the universal gene pool in the primordial pond, and must have been independently selected into the genomes of these multitudes of creatures that were born independently. Our illustration of this phenomenon here indeed shows us that even similar genes present in various creatures could have only originated independently from the primordial pond.

THE NEW THEORY PREDICTS THE PRESENCE OF UNRELATED GENES THAT CODE FOR ESSENTIALLY THE SAME BIOCHEMICAL FUNCTION IN DIFFERENT ORGANISMS

More than one particular protein structure can carry out a given biochemical function. Distinct protein structures, with varying lengths and amino acid sequences can carry out the same biochemical function.[22] Note that here we are not talking about the same protein structure (*i.e.*, essentially the same protein) with variations in its amino acid sequence. We are talking about entirely distinct proteins with identical functions.

In the primordial pond there could have been several distinct genes coding for distinct proteins each of which can carry out the same biochemical function. Under these circumstances, it should be possible to include different genes for the same biochemical function in different genomes. What should we look for in today's living organisms to verify this prediction? We should look for proteins that perform the same biochemical reaction in different organisms, and whose genes are so unrelated in their structure that they cannot be connected by organismal evolution. Do we see them? The answer is certainly yes!

We shall look at a few examples, remembering that there are only a limited number of proteins and genes known from several organisms, which enable us to make such a comparison. We shall then extend and generalize our conclusions.

Bacterial protein subtilisin and mammalian serine proteinase

A well-established case wherein distinct proteins can specify exactly the same basic biochemical function is that of the bacterial proteinase subtilisin and the mammalian serine proteinases such as chymotrypsin. In these proteins, the catalytic mechanisms are essentially the same but there is no sequence similarity beyond a use of the same kinds of amino acids in the active site.[23] It is clear that these genes must have originated in the primordial pond independently.

Distinct collagen genes in different organisms

A good example for the presence of structurally distinct genes for essentially the same biochemical function is the protein *collagen* that we find in animal cartilage. It has a repeating sequence and three-dimensional structure. The protein has a repetition of the triplet amino acid sequence Glycine-X-Y, where X is any amino acid and Y is hydroxyproline (a modified form of the amino acid proline). Any variations of this repeated amino acid sequence essentially functions as a collagen. We have evidence from the literature showing that there are many genes unrelatable by evolution that code essentially for collagen. The structures of many of these genes in the various animals are different. The structures of the collagen genes from the nematode, fruitfly and the human are distinct, as judged from the number of introns and their positions.[24] From these structures we can confidently say that these genes have not evolved from a common ancestor. As we have demonstrated in Chapter 7, numerous distinct genes can occur in the primordial pond, all of which code for the repeated Gly-X-Y collagen sequence.

Examples from the prokaryotic world

Excellent examples in the prokaryotic world are the restriction enzymes. A restriction enzyme recognizes a specific DNA sequence in the DNA double helix and cuts the DNA precisely at this sequence. For example, one restriction enzyme called *Asu I* recognizes the sequence GGNCC (N is any nucleotide) and cuts at that sequence wherever it occurs in a DNA molecule. This enzyme is found in the bacterium *Anabaena subcylindrica*. Another restriction enzyme called *Sau96 I*, from another bacterium *Staphylococcus aureus* PS96, also recognizes and cuts the DNA at exactly the same sequence. Biochemically, the two different enzymes carry out the same function. However, the proteins, and thus their genes, are quite distinct. These are entirely different proteins, unrelated by evolution. The enzyme called *BstZ I* from the bacterium, *Bacillus stearothermophilus*, and the enzyme called *Eco52 I* from an entirely different bacterium, *Escherichia coli* RFL52, cuts DNA at the same sequence, CGGCCG. Thus we establish here an important phenomenon that totally different proteins can specify the same biochemical function, and that such proteins exist in the living world.

In the above examples, we have shown that multiple genes for a given biochemical function can occur in the living world, whether they occur within the same organism or in different organisms. However, not all the possible biochemical functions in the living world can be catalyzed by more than one unique protein structure. We should also keep in mind that not all the possible proteins for a given biochemical function may have existed in the primordial genetic sequences (USP) as genes. However, it is possible that for some biochemical functions, more than one distinct gene could have existed in the USP. Out of these, a few could have been included within the same or different genomes. This is what we have shown evidence for in the living world with the examples above.

Can we demonstrate this phenomenon of the occurrence of distinct genes for the same biochemical function in the primordial pond (*i.e.*, a random DNA sequence) using computer simulation? Theoretically it is possible. However, practically it is not possible to simulate this phenomenon because we do not know the general parameters of distinct proteins that can specify the same biochemical function[25] (such as the subtilisin and the chymotrypsin). It is, however, possible to show that the genes for the protein sequences for subtilisin and chymotrypsin can occur independently in a long random DNA sequence (split into exons and introns,[26] see Chapter 7). As we can see conceptually, no matter what protein sequence we take, two distinct protein sequences, whether they specify the same biochemical function or distinct biochemical functions, will certainly occur in primordial sequences.

We can also prove that given a protein, its gene can occur with many different gene structures (exon-intron structures, lengths, and sequences) in entirely different random DNA sequences. Our computer simulation experiments in Chapter 7 corroborate this phenomenon. Given a prototypic protein sequence (such as that of the protein collagen), it is clearly possible to find many distinctly unrelated genes in the primordial genetic sequences that would essentially code for the same protein. In fact, when we do not specify at what positions the sequence is broken into exons, many different occurrences of the collagen coding gene, each of which can have different exon lengths, exon numbers, and exon positions in different genes (and entirely different intron sequences and lengths), are found to occur.

In the situation of selecting genes for living organisms from the primordial pond, nature imposes certain constraints. The structure of collagen is such a constraint. The triple-helical structure of collagen, required for its function, puts an absolute constraint that the repeating Gly-X-Y sequence be maintained throughout the sequence. Thus, when we simulate the search for such a gene in the computer, we can impose this repeating structure, and we do obtain many distinct genes which can code for this collagen sequence. We can therefore say that nature imposed such a constraint and selected very many distinct collagen genes from primordial genetic sequences into the genomes of various organisms.

Ovalbumin and α_1-antitrypsin.

Another good example of gene similarity occurring among distinct genes is that of *ovalbumin* and *α_1-antitrypsin*. Ovalbumin is the principal protein in chicken egg white. Its function is not known. α_1-antitrypsin is a human plasma protein involved in the control of elastase; individuals deficient in it have a high risk of lung disease.[27] They are 24% similar in their amino acid sequences. However, their gene structures (*i.e.*, the position of the exons, introns, and their sequences) are so distinct that they could not have evolved from a common ancestral gene. The chicken ovalbumin gene contains seven introns, all of them located in the 5' half of the mRNA, whereas the three introns in the human α_1-antitrypsin gene are all in the 3' half of the mRNA. The introns in question do not show significant sequence similarity. All this makes it clear that these two genes must have originated independently in the primordial pond, with their sequence similarity. Although the underlying reasons for their similarity does not seem to be obvious because we do not know the exact function of the albumin, we can see that there should be a similar functional constraint, at least in portions of the two proteins, which would impose a similar sequence constraint.

It is interesting to see that molecular evolutionists struggle to show that somehow these genes can be evolutionarily related — by some roundabout mechanisms of intron losses and introductions through descent with modifications of organisms.[28] Although our discussions concerning similar genes can by themselves explain the independent origin of similar genes, evolutionists have interpreted the similarity among genes as the strongest proof for evolution having occurred. Therefore, it is important for us to show how the interpretation is flawed. Having proved that genes which code for proteins with similar functions can occur independently, we shall analyze later in this chapter why evolutionists mistakenly think that similar genes are related by evolution.

THE NEW THEORY PREDICTS THAT FUNCTIONALLY-DISTINCT PROTEINS CAN HAVE SIMILAR SUBFUNCTIONS AND SIMILAR AMINO ACID SUBSEQUENCES

According to the new theory, evolutionarily-unrelated genes should be present in many different organisms, each of which would code for a protein with distinct overall biochemical functions. However, the proteins may have similar subfunctions such as binding the same cofactor, and may have sequence similarity over such functionally-similar regions. Yet these proteins (and their genes) will have no sequence similarity at all over the functionally dissimilar regions, which can constitute a considerable portion of the proteins. Based on the new theory, such genes could occur independently in the primordial pond. In fact, it is theoretically and practically possible to demonstrate this phenomenon by computer simulation.

We demonstrated in Chapter 7 that almost any gene coding for any protein sequence can occur in the DNA sequence pool of the primordial pond. Let us now consider if there can exist different genes in the primordial pond coding for distinct biochemical functions, but which may use some similar subfunctions. Probabilistically speaking, the answer is a definite yes. In reality too, there are many distinct biochemical functions that use similar subfunctions. As an example, the sugars such as sucrose, maltose, and lactose are similar in their structures, and the biochemical reactions which break them down into smaller molecules are also similar. Thus, the enzymes that break down the three different sugars can be similar to some extent in their structure, and also in amino acid sequence. The genes for these enzymes can certainly occur independently in the primordial genetic sequences.

Almost all enzymes in the living world bind small molecules, called *cofactors* or *coenzymes*, and use them in their biochemical reactions. These enzymes bind the cofactors only as part of their biochemical function, and have one or more other protein regions or domains that carry out other subfunctions such as breaking or making a chemical bond. There are some coenzymes and cofactors (such as vitamins and metals) in the living world that are used by almost all of these thousands of enzymes. It is obvious that the portions of the different enzymes which bind the same coenzyme (for example, the coenzyme called NAD, nicotinamide adenine dinucleotide) can be similar in structure. The same structure for this portion, or domain, of the proteins may require a specific amino acid sequence (albeit with sequence variations). Thus, all these various enzymes, though distinct in their overall structure and biochemical functions, should have structural and sequence similarity in that domain of the protein which binds the same coenzyme, NAD. Figure 9.1 describes such a scenario wherein distinct enzymes have functionally the same domain as well as functionally distinct domains. Several proteins in the living world also have multiple domains that carry out different subfunctions, for instance, binding a cofactor, binding a DNA sequence, and breaking a specific biomolecule. In these cases, the domains in various multifunctional proteins with the same subfunction will be essentially the same and will have essentially similar protein structures and amino

THE MOLECULAR SCENARIO OF LIFE: EVIDENCE FOR THE NEW THEORY 401

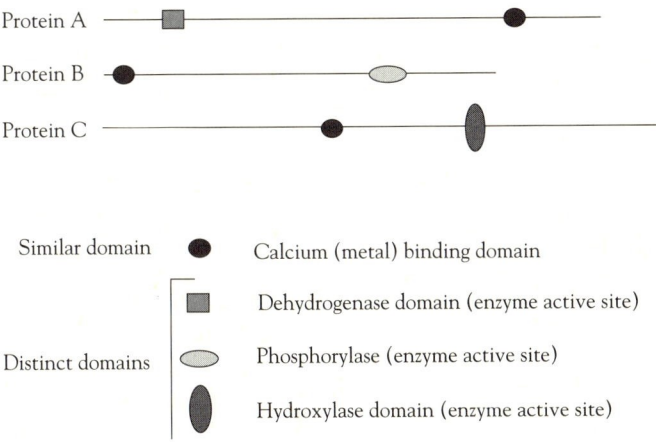

FIGURE 9.1. THE OCCURRENCE OF SIMILAR PROTEIN DOMAINS IN TOTALLY INDEPENDENT PROTEINS. Distinct proteins with quite different overall biochemical functions often require similar subfunctions such as binding a metal, cofactor, DNA, etc. Because of the functional constraints of these similar protein domains, they could have similar structures and similar amino acid sequences. Thus the DNA sequences that code for the similar portions of these proteins could be similar — even though the genes for these proteins occurred independently in random primordial sequences. Also, the rest of the protein portions can be of varying length and sequence and can have unique, unrelated subfunctions.

acid sequences, and the other domains with distinct subfunctions will have no structural or sequence similarity.

Now our question is, can distinct genes for such distinct proteins occur independently in the primordial pond with similarity in only the small regions of the proteins binding the same coenzyme, and have different amino acid sequences in other parts of the proteins? The answer is certainly yes (Figure 9.2). We have essentially shown this in Chapter 7. We can conduct simulation experiments in the computer, wherein we can demonstrate that distinct genes coding for distinct proteins with similarity in one or a few portions can occur easily in the primordial pond, while the other regions of the proteins are dissimilar.

Our next question is, can we find actual examples to corroborate our arguments in the living world? The answer again is a definite yes. In fact, as DNA sequences of more and more genes are worked out in laboratories around the world, this phenomenon is being unraveled in an unprece-

A   Distinct, independent proteins with some similar coding sequence and the rest unique coding sequence:

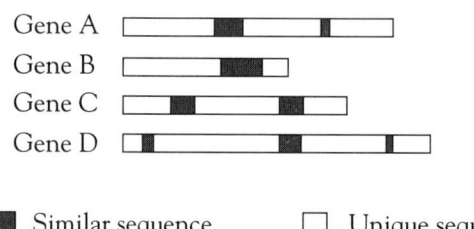

B   Independent inclusion of distinct genes with sequence similarity in various creatures in the primordial pond:

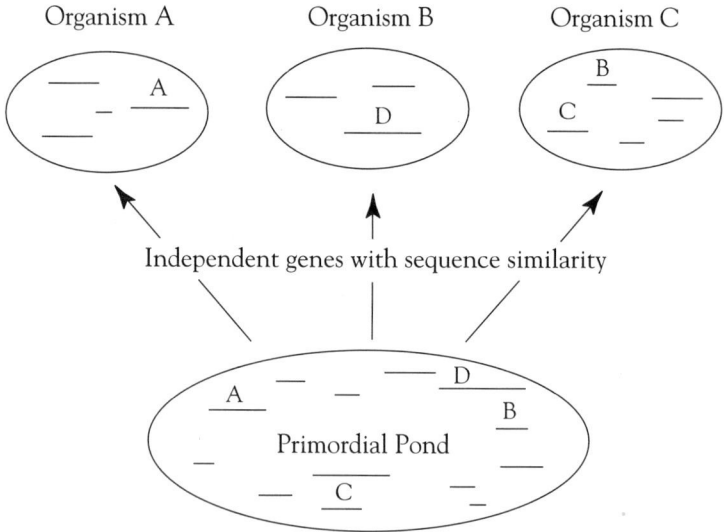

FIGURE 9.2. DISTINCT PROTEINS WITH SEQUENCE SIMILARITY CAN ARISE INDEPENDENTLY FROM THE PRIMORDIAL POND. (A) Distinct proteins with similar subfunctions will have structural and sequence similarity over these domains, as described in Figure 9.1. These proteins often have similarity only in small portions, the rest of the sequence being unique. The sequence for a similar domain can fall in two or three different regions in a protein. (B) The coding sequences for such proteins can occur independently in the primordial pond in distinct random DNA sequences (i.e. distinct genes) and could have been assembled into independent genomes. Thus distinct proteins from independent genes could have similar subfunctions, substructures and subsequences.

dented manner. The only thing is that so far their interpretation has been through evolutionary eyes. We now have to unearth the underlying truth in these observations based on the new theory of the independent birth of organisms by exploring and analyzing the details with our new view.

Evolutionists focus on the short portions of the proteins (and genes) that are similar, even in totally distinct, independently-originating proteins — that in the first place mislead them to think that these are evolutionarily related. We focus on the dissimilar portions — often the major portions of genes and proteins — which cannot be in any way explained by evolution, but that can be very well explained by the independent birth of creatures. Figure 9.1 shows that only one domain is similar among the distinct proteins while all other domains are distinct. It is possible to extrapolate this concept and show that more than one similar domain could occur in independent proteins, while other regions could have domains with totally distinct functions, structures, and sequences.

EXAMPLE 1: MANY DISTINCT PROTEINS WITH A COMMON ATPASE DOMAIN

Here we shall talk about three proteins which have distinct biological function but have some common "motifs" or domains. We shall analyze the results of a recently published paper.[29] Our concepts become clear when we see what the authors say about these proteins and how they attempt to connect them evolutionarily. Let us read some of the authors' discussions (italics mine):

> In spite of their different biological function, actin, Hsc70, and hexokinase contain similar three-dimensional structures. No overall sequence similarity between these three protein families can be detected with standard pairwise sequence alignment algorithms, so the structural similarity came as a surprise. All the three bind and hydrolyze ATP. The ATPase activity of actin is involved in the control of polymerization; that of Hsc 70 (a member of the hsp70 family of heat shock proteins) is involved in a variety of "chaperonin" functions, such as keeping protein chains translocation competent, preventing aggregation, or aiding in their refolding from aggregated states; and that of hexokinase is involved in phosphorylation of glucose at the entry to the glycolytic pathway.
>
> The common structural feature is that of two domains of similar fold on either side of a large cleft with an ATP binding site at the bottom of the cleft. ... These three structures are not only similar in three-dimensional fold but probably also in some aspects of mechanism: their ATPase active sites are lined with identical or similar residues ... *Thus, the structural similarity of the three proteins is correlat-*

ed with partial similarity of function. A common evolutionary origin of this ATPase fold is therefore likely. Here, we exploit the similarities of the three remotely, but clearly, related three-dimensional structures and the information contained in the multiple sequence alignment within each of the three families and define a common sequence pattern.

The sequence similarity among the three biologically different proteins that the authors allude to is over five short conserved regions, each of approximately 20 amino acids (in total ~95 residues), whereas the total protein lengths vary from ~350 to 500 amino acids. The rest of the protein regions (regions other than the above-mentioned conserved regions) should be involved in the specific biological activity of the proteins. For instance, a long region between two of these conserved stretches in the hexokinase is most probably involved in sugar binding. This region has no counterpart in actin or Hsc70. It also becomes clear from the authors' research and analysis that the invariant sites within these conserved regions are only the active sites for the ATPase function of the three proteins. The authors themselves agree that sequence similarity between any two of the three known structures (actin, hsc70, and hexokinase) is very low, although they have the same fold (or three dimensional structure) in subdomains Ia and IIa (meaning only in the subdomains involved in ATPase activity).

All these analyses and findings are in fact precisely fitting our predictions. Clearly, distinct proteins with different lengths and sequences and with distinct biological functions have similar subdomains for similar subfunctions. Even in these similar regions, the sequence similarity is restricted only to the few active-site amino acids. There is no other sequence similarity in all the rest of the regions. These details are a fact, but the authors interpret them through incorrect evolutionary assumptions.

In fact, the authors also find, by searching for similar sequence patterns, several other proteins which may have similar subdomains and three-dimensional structures. These include several sugar kinases — fucokinase, glycerokinase, gluconokinase, xylulokinase, and ribulokinase. Remember that one of the three original proteins that the authors had in their family of proteins is hexokinase, which is a sugar kinase.[30] The authors say that these sugar kinases have other conserved regions that are not present in actin or hsp 70. It means that they have their own distinct domains for their sugar-specific functions (note that the sugars that these distinct proteins attack are different) in addition to the similar ATPase domains.

The authors also uncover yet other proteins with totally different biological functions which may have an ATPase domain. These include prokaryotic cell-cycle proteins, a protein involved in cell division, and another protein which may be involved in DNA stability.

In fact the authors say:

> Many examples of proteins with essentially the same fold and very low sequence similarity are known,

which strongly supports our concepts — presence of similar subdomains in entirely distinct proteins that are totally unrelated by evolution. However, it is the evolutionary thinking that misleads the authors, as evident from their following conclusion of their analysis in the same article:

> Conceivably, an ancestral protein of ~150 residues acquired the capacity to dimerize and bind ATP in an active site between the two subunits. ... Later, gene duplication and fusion led to the common ATPase fold of ~300 or more residues characterized here. One major branch evolved into the ancestors of actin, the eukaryotic cell cycle proteins, and hsp70, in part by insertion of subdomains. The other major branch evolved into sugar kinases. Since members of the Hsc70 family are present in eukaryotes and prokaryotes, the divergence between sugar kinases, actin, and Hsc70 has to be anterior to prokaryote-eukaryote branching. ... In summary, a dimeric ancestral ATPase appears to have evolved into what is now a diverse set of prokaryotic and eukaryotic enzymes with functions as different as muscle action, construction of the cytoskeleton, protein refolding, metabolic phosphorylation of sugars, and, possibly (indirect evidence only), control of some aspects of bacterial cell division.

Indeed, we know based on the new theory that these are distinct proteins with similar subdomains for similar subfunctions which are required as a prerequisite for their overall biological function. Because the authors assume that evolution is an established fact, they say that these distinct proteins evolved from one another or a common ancestral protein. In essence, they say that an ancestral ATP binding domain evolved into many different proteins such as the hsp70, Hsc70, all the sugar kinases, and proteins functioning for muscle action, construction of cytoskeleton, and so on. In the light of the new theory, we can see that their interpretations are incorrect. The details and arguments we have uncovered here only far too well illustrate our predictions that totally distinct, unrelated proteins with similar subdomains can independently originate.

Example 2: The story of the G-proteins and the G-protein-linked receptor proteins

In a multicellular creature, the cells in its various organs and tissues receive signals from outside the cell through the interaction of other molecules circulating in the blood (secreted by other cells located at remote sites in the body). There are protein molecules on the cell membrane, called receptor proteins, that transmit this molecular information from outside the cell to the intracellular biochemicals. The extracellular portion of the receptor can interact with specific signaling molecules to change the conformation of the receptor protein. This change is transmitted biochemically to the receptor portion inside the cell, which can now recognize another protein, bind with it, and cause structural and biochemical changes in it. From such changes, the various biochemical activities of a particular cell are coordinated in an organism. This kind of transmission of a biochemical signal from the exterior of the cell to the interior of the cell is called *signal transduction*.

There are many receptor proteins known in living systems. Each one has a similar overall structure — a seven-domain structure that spans the cell membrane. It has extracellular domains that protrude outside the cell membrane that bind with the signal molecule and another domain inside the cell that interacts with cell proteins. These domains can be distinct on different receptor proteins. In a moment we shall see that while some portions of the various receptor proteins can be similar, other portions are so distinct that these receptor proteins must be unrelated, independent proteins.

Just as there are different receptor proteins, there are various intracellular proteins that bind with the intracellular portion of the receptor protein. A class of such proteins is called G-proteins, for they also bind with the nucleotide GTP and hydrolyze it as part of their signal-transducing mechanism. Just as the intracellular portion of the receptor protein is distinct in different receptor proteins, the domains that bind the receptors in the various G-proteins are distinct. However, in these proteins, the portion that binds and hydrolyzes the GTP nucleotide is similar. Again, we are in a situation that exemplifies our concept — that totally distinct proteins can have similar subfunctional domains, in this case the GTP binding and hydrolyzing domain.

About 20 different G-proteins have been found so far to be involved in various signal transductions. Based on the fact that these proteins all have the common similar domain of binding GTP, molecular evolutionists have proposed that all these proteins have evolved from a common ancestor. But, when we analyze the whole scenario, it will become clear that these are distinct, unrelated proteins.

It is very important to understand that these proteins have been arbitrarily named GTP-binding proteins because they all bind GTP. One could have called them by totally different names based on their binding with different receptors. Thus what we have to drive home is the point that distinct proteins having different domains for various biochemical activities *also* have a similar domain for one particular subfunction.

The case of the receptors for GTP-linked proteins. They all have the same membrane-spanning region, but different effector regions and different receptor regions for various GTP-linked proteins.

Several different G-protein-linked receptor proteins all have a certain amount of sequence similarity and all have a similar subfunction — although they also have many distinct subfunctions. From such similarity of subfunction and similarity of some sequence, molecular evolutionists believe that these proteins are related by organismal evolution — and have classified them into one family of related proteins.

We shall illustrate that these proteins originated independently in the primordial pond — and are thus unrelated by evolution — and that they were selected for similar structures for similar subfunctions in different situations from the primordial pond. The need for essentially the same subfunction imposes similarity of protein structure, which in turn imposes similar amino acid sequences in the distinctly independent proteins. We shall also demonstrate that the amount of similarity found among these proteins, which are claimed by the evolutionists to be high, is actually possible to be obtained in independent proteins. If we simulate the independent selection of distinct proteins with some amino acid sequence constraints representing the structural constraints, we can show that far higher identity or similarity of amino acid sequence can be obtained in independent proteins.

By these analyses we will finally show that the evolutionary interpretation of similarity which is the crucial and perhaps the only molecular foundation on which evolutionary theory stands can be shattered and that a new and perfectly valid interpretation can be given for similarity of genes and proteins in various organisms based on the theory of the independent birth of organisms.

The serotonin 1c receptor (called the 5HT1c receptor) shares some sequence and structural properties with the family of receptor molecules that has been predicted to span the lipid bilayer seven times.[31] This family includes rhodopsin and some other opsins, the α– and β–adrenergic receptors, the muscarinic cholinergic receptors, the substance K neuropeptide receptor, the yeast mating factor receptors, and the oncogene c-*mas*. Each of these receptors is thought to transduce extracellular signals by interacting with G-proteins.

We should note, based on the essential properties of a "transmembrane" protein, that any membrane-bound receptor protein will necessarily have the minimum biochemical features — seven domains which span the cell membrane. Each of the domains has to have some specific structural and sequence properties to be in a transmembrane region, such as being hydrophobic. Between each consecutive transmembrane domain, there is a protein portion alternating outside the cell and inside the cell.

The transmembrane portions have to have certain specific amino acids for their properties of "transducing" the signals. Given these basic structural features, one cannot escape noting that even if several proteins having the transduction properties arise independently, they all should share some fundamental sequence properties.

We have amply demonstrated in Chapter 7 that the DNA codes for many distinct protein sequences can independently occur in the primordial universal genetic sequence pool. What we want to show here is not much different. We need only to show that some of the domains among the many distinct proteins are similar in independent proteins. As we have reiterated in Chapter 7, the probabilities for the occurrence of distinct proteins is essentially the same whether the proteins have absolutely no similarity or have some specific similarity in one or a few regions among them. If we ask the computer to search for distinct protein sequences with similarity in only one or a few portions, and all other regions of the various proteins being totally distinct in sequence, sure enough the computer will find genes that would code for these proteins in random DNA sequences. In other words, such proteins can occur independently in the primordial pond, and absolutely need not be related by evolution.

Let us take the comparison of the 5HT1c protein with the other proteins in the family. The 5HT1c receptor shares 25-percent sequence identity with the β2-adrenergic receptor, and 20-percent identity with the muscarinic and substance K receptors. Compare this with what we can get in our computer simulations — 40- to 80-percent identity between independent genes for the same protein occurring in a random DNA sequence — with only allowed sequence variations as found in natural proteins. It shows that DNA codes for several distinct proteins with sequence similarity can occur in random primordial sequences.

The differences in these proteins far outweigh the similarities among them. Although one can argue that the similarities can be explained by evolution, we can see that the differences can never arise by any evolutionary mechanism. Let us look at these details. The third cytoplasmic loop, thought to interact with the different G-proteins, is of widely varying length in different receptors. No sequence similarities within this domain are apparent among these receptors. In the 5HT1c receptor, this loop consists of 77 amino acids. The hamster β2-adrenergic receptor has 25 amino acids in this loop, the bovine substance K receptor has none, the human α2-adrenergic receptor has 125 amino acids, the rat muscarinic M3 receptor has 207 amino acids, and lastly, the bovine rhodopsin has none. How can a protein with only the transmembrane structure come to evolve into various proteins by changing only the third cytoplasmic loop to be able to bind various specific G-proteins? It is absolutely incorrect to even expect that such a thing is possible by random mutations and recombinations. It is simply impossible to come up with the necessary variations of length and amino acid sequence of this domain by means of random mutations and selections.

The problem of the evolution of a given receptor – G-protein system is made all the more complex because, for a specific effector (signal molecule) outside the cell, there should be a specific receptor, for which there should be a specific G-protein, in turn for which there should be another effector inside the cytoplasm. If from one ancestral receptor all the receptors had evolved, and similarly if from one ancestral G-protein all the G-proteins had evolved, then how did all the complementary binding domains on the receptors and G-proteins evolve in a coordinated manner? And how did this evolution happen for the dozens of receptor proteins and G-proteins that are specific for various effectors, receptors, and G-proteins, all ultimately from a unicellular eukaryote? So why should a prototype transmembrane protein exist at all and how did it originate?

While there is no answer to any of these questions by any evolutionary means, in contrast, it is easy for us to see that the DNA codes for all these various proteins with all their functional activities, structures, and amino acid sequences could have easily occurred in the primordial pond genetic sequences independently.[32]

One important point we should note here in all these analyses is that molecular biologists tend to pursue studies on proteins and genes similar to one another for several reasons. For one thing, once they have a protein or a gene, they want to study the "related" proteins and genes because it is easy to identify and isolate genes which have some sequence similarity to this original protein or gene. In this fashion they have something to study in a straightforward manner. The second thing is that there is no straightforward method to identify unique, unknown proteins or genes. Thirdly, scientists tend to study what is in vogue, and what is currently in vogue is the study of similar proteins and genes. Thus, we should not forget the reason why we see so many analyses in today's scientific journals pertaining to similar genes.

EXAMPLE 3. THE CASE OF MANY DISTINCT PROTEINS WITH DIFFERENT BIOLOGICAL FUNCTIONS ALL OF WHICH CONTAIN A SIMILAR PROTEIN TYROSINE PHOSPHATASE DOMAIN[33]

It has been found that phosphorylation (addition of a phosphate group) of tyrosine may play an essential role in the regulation of diverse cell activities. To date, 15 human proteins containing protein tyrosine phosphatase (PTPase) domains have been reported. Six of the cloned human PTPases are cytoplasmic enzymes (located inside the cells), whereas the nine other known human PTPases are transmembrane proteins with extracellular receptor-like regions connected to cytoplasmic PTPase domains. Four transmembrane PTPases have also been identified in *Drosophila*. Again, just as in the case of the GTP-binding proteins, we can see that only the PTPase domain is similar in all of these proteins. It appears clearly that these have been named PTPases only because all of them have the functionally similar PTPase domain, and because the scientists who isolate them do so primarily to study the PTPase function. Furthermore, there is an underlying evolutionary assumption that all these have evolved from a single ancestral protein with PTPase activity. However, except for the similar PTPase domains, the other regions of these different proteins are indeed distinct. As we shall see, these are indeed quite distinct proteins, unrelated by evolution, and that all could have occurred independently in the primordial pond and could have been selected for their particular biological functions in living cells. Let us first see, in the words of the authors, how these proteins have several domains and regions, and how they differ from each other.

> In contrast to the highly homologous PTPase domains, the extracellular receptor-like regions of the transmembrane PTPases are distinct in both size and structure. For example, the extracellular regions of LAR, PTPδ, PTPμ, DLAR, and DPTP contain varying numbers of immunoglobulin (Ig)-like domains and fibronectin type III (FN-III) domains. ... In contrast, the extracellular regions of PTPβ, DPTP99A, and DPTP10D are composed only of FN-III domains. LCA, PTPα, and PTPε have extracellular regions of various sizes with no obvious similarity to any known proteins. ... PTPζ and PTPγ comprise another subfamily ... we ... describe a type of receptor structure that is homologous to carbonic anhydrases.

As the authors have found, the regions other than the PTPase domain are quite unique to the proteins. This is absolutely consistent with our arguments that these proteins are indeed unique, despite the fact that they have the PTPase domain similar in all of them. What we see here is a concept in exact opposition to that of the molecular evolutionists: these proteins are not related just because they have a common PTPase domain; they are independent and unique proteins with a *functionally*-similar domain for a particular subfunction. As we have reiterated in the other examples, these proteins can occur independently in the random primordial sequences with all their characteristics completely, each of which could have been selected independently for its different biological function in living systems.

EXAMPLE 4: THE CASE OF THE HOMEOBOX GENES: PRESENCE OF THE COMMON MOTIF REQUIRED IN ALMOST ALL THE DEVELOPMENT-REGULATING GENES, WHICH ARE OTHERWISE QUITE DISTINCT AND UNIQUE

The homeobox is a 183-base-pair sequence encoding a trihelical DNA binding domain. This domain has been found in several genes in a wide variety of eukaryotic organisms. The protein products of some of the homeobox genes act as ubiquitous transcription factors, whereas most are involved in the control of embryonic development. In the fruit fly Drosophila as well as in the mouse, the segment identity along the axis of the embryo is specified by homeotic selector genes belonging to the Antennapedia-type of homeobox genes. Both in insects (HOM genes) and vertebrates (HOX genes), these genes appear in clusters. The beetle Tribolium has its homeotic genes arranged in one complex (HOM-C), whereas Drosophila has them split into two complexes [Bithorax complex (BX-C) and Antennapedia complex (ANT-C)]. In mammals, there are four clusters (Hox-1 to Hox-4).

To gain insight into the evolution of developmental control mechanisms, several studies of the evolution of the Antennapedia-type homeobox genes were done by comparative sequence analyses (for detailed references see the article by Schubert[34]). In the words of Schubert et al.,

> While these investigations could reveal some aspects of homeobox gene evolution, many of the evolutionary steps which led to the linear order of the genes in the ancestral cluster of insects and vertebrates and, more recently, to the final organization of the homeobox gene clusters still remain obscure.

The studies by Schubert et al reiterate what the previous studies have revealed (italics mine):

> Most of the homeobox genes examined are part of the regulatory network controlling embryonic development in metazoans. It is tempting to link the radiation of the major types of these homeobox genes to the origin of metazoans. ... From our analysis of the Drosophila and human homeobox gene clusters we propose the following model for the evolution of these genes. Starting from a single ancestral gene, the three precursors of the 3', central, and 5' classes of the Antennapedia-type genes are derived. The insect and vertebrate derivatives of these three classes correspond to the anterior (3' class), central (central class), and posterior (5' class) regions

of the embryo. The division into three classes fits nicely with the proposed function of the Antennapedia-type genes to determine the head, trunk, and tail in the ancestral arthropod. The next duplication in the ancestral cluster, dividing the 3' class into the *pb*- and *lab*-like genes, preceded the separation of diploblasts and triploblasts, possibly more than 1 billion years ago. Before the divergence of vertebrates and arthropods, another duplication separated the *Antp* and *Dfd* precursor, the common ancestor of higher metazoans in the Precambrian already contained a cluster of at least five genes. Independently during the evolution of insects and vertebrates, further duplications generated the 11 genes found in *Drosophila* and the 13 genes proposed for the vertebrate ancestral cluster. *The number of genes was increased in vertebrates by duplications of the whole cluster. Drosophila, on the other hand, used different strategies, such as multiple promoters and alternative splicing, to increase the complexity and the coding potential of its anterior-posterior differentiation control system.*

What we derive from the evolutionists' discussion is that they propose that somehow from the one original gene, all other complexes and clusters of the homeobox-containing genes have evolved. They even go to the extent that the Drosophila use different strategies such as multiple promoters and alternative splicing to increase the complexity and the coding potential of its differentiation control system. One of the crucial things to note here is the fact that in the homeobox proteins, only the homeobox "motif" is common, and that the other region(s) of these proteins by which they control various genes are entirely distinct.

From our determinations in Chapters 3 and 4, such an evolution of these systems from preexisting genes is virtually impossible. What all these scenarios really show is that these systems independently originated from the common pool of homeobox-containing genes in the open primordial pond. As our computer simulations on the availability of genes prove without a doubt, several genes with different functions can contain a common homeobox domain in them very easily (*i.e.*, probabilistically). These multitudes of genes containing the homeobox domain can be assorted and organized in various ways to give rise to distinct systems of development control, producing many distinct organisms. It is because of the fundamental *biochemical* processes by which development of organisms is constrained — meaning the development is controlled by genes in only a few possible biochemical ways — that the many different independently-born organisms use the same homeobox domain and somewhat similar biochemical mechanisms for development control. We should note that DNA binding of proteins is an essential part of development control. Again, once the basic process is defined in the genome of even one metazoan (the basic multicellular organism), this can be used in the genomes built after that from the pieces of the first genome, although, we should remember, the later genomes were all constructed independently in the open primordial pond.

We should note, therefore, that similar "master control" genes, that control the development of various organs (whether structurally or functionally similar or distinct organs), could be independently included in entirely distinct unrelated creatures from the common pool of primordial pond genes. Thus the presence of similar (or even the same) developmental master control genes in distinct creatures — triggering the development of the distinct unrelated eyes, or the fruitfly or butterfly wings, or a vertebrate bone — does not mean that these creatures are related by organismal evolution. Our concept is corroborated by the fact that there are numerous totally unrelated genes in these distinct organisms, say invertebrates and vertebrates, for their "blood," "immune" systems, "coagulation" systems, and "breathing" systems, which simply disproves that these creatures are related by descent with modification.

An important thing that we should be aware of is that the presumed evolution of proteins with many subfunctions from preexisting proteins requires preexisting genes or sequences for the various domains for all biochemical functions in the genome, such as those for binding the various effectors and receptors, only on which the genetic recombination can happen in the first place. Why should these domains be present in the genome, if they were not functioning in the cell before the evolution of such proteins with these subfunctions? Even if we take it for granted that they were present to start with, the number of random recombinations required to bring them together to evolve a useful multifunctional protein are too large to occur within the 600 million years since the start of multicellular life.

It is crucial to understand that it is not sufficient even if the genes for these multifunctional proteins can evolve in the genome. They should be precisely organized in the DG pathway so that they are switched on in the right place in the developing embryo and are expressed precisely in the right cells and tissues. This is most crucial, for without such a precise position in the cell and the organism, there is no use for these proteins. The receptors for the GTP-binding proteins are transmembrane proteins, meaning that they are situated inside the cell membrane, with a definite orientation such that the correct portion of the protein is hanging outside the cell and the portion that binds GTP is situated inside the cell. The effector protein molecules should be synthesized in particular cells and not in other cells. Such specificity comes from the DG pathway. Even if we assume that all these genes for all these proteins could be evolved in the genomes of organisms, the organization of these genes in the precise DG pathway is highly improbable. We can see that by both these considerations it is simply incorrect to say that such genes evolve within the genomes of organisms.

"Consensus" sequences are very short and their probability in the USP is very high. Therefore, different proteins with the same consensus sequence can occur in the USP with a high probability.

So far we discussed the improbability of the evolution of proteins with some similarity among them, while major portions of the proteins are unique, and, at the same time, the high probability for the DNA messages for such proteins to occur in the random DNA sequences in the primordial pond. We shall now briefly describe reasons for another phenomenon, the presence of consensus sequences among the functional domains of enzymes.

There are consensus sequences for some functional domains of enzymes and other proteins. However, such sequences are short and their probability in the USP in the context of the assembly of genomes is very high. The biological meaning and implications of this is that various genes can easily occur in the USP with all these features and would have been selected for their functions. We can verify this prediction by the analyses of the consensus sequences in today's genes.

The consensus sequences for splice junctions is a good example. In my analyses of these sequences I found that they do occur in a random manner in a random sequence. The splice junctions are short and also they occur with considerable variations in them. In actual genes, not all the occurrences of the consensus sequence were real splice junctions. Some occurrences happened to be real splice junctions used in genes and many others were not. Sometimes, the real splice junction does not have the consensus sequence.

Similarly, such a phenomenon can be shown to occur in protein sequences. When Russell Doolittle looked for the consensus sequences for the nucleotide-binding domain of proteins in a protein sequence database, he found that they occurred nonspecifically.[35] That is, the consensus sequence was not found only in the genuine nucleotide-binding proteins, but rather frequently in proteins that had nothing to do with nucleotide binding. Furthermore, a good number of nucleotide-binding proteins did not have anything close to the consensus sequence.

Such consensus sequences are easy to occur purely probabilistically in the USP, and therefore, in the genes that occur in the random sequences. The occurrence of consensus sequences for a particular domain specifying a particular biochemical function such as nucleotide binding in many different proteins does not in any way indicate that these proteins are related by common ancestry. Different proteins with the same consensus sequences could have been encoded by different genes which occurred independently in the USP and be selected for their specific function into the various genomes of organisms.

Conclusion: The scenario of similar genes in various creatures cannot be due to organismal evolution. It can be only due to the independent birth of creatures in a common primordial pond.

When we thoroughly analyze the scenario of gene similarity, what do we really see? Do we see the evidence for organismal evolution having occurred? Certainly not! On the contrary, what we see is actually evidence against it — and what we see indeed absolutely corroborates the independent assembly of genomes in the primordial pond and independent birth of organisms.

What we see when similarities among many genes are shown by "aligning" their sequences (meaning that they match in certain positions of amino acid sequences) is that there are only few places at which there is match of amino acids. We see many many "gaps" that must be inserted in a gene sequence in order for it to align with the sequence of another gene. These gaps are assumed to be deletions in one gene or insertions in the other gene. These gaps may run up to several tens or hundreds of amino acids.

For the evolution of one gene into a distinct gene, by changes of point, deletion, and insertion mutations, the gene has to change from one function into another function (such as from the β-adrenergic receptor to the rhodopsin or *vice versa*) in the genome of an evolving organism. Furthermore, it has to get integrated into the DG pathway of the organ (rhodopsin into the eye DG pathway) and so on.

We have very well seen that the evolution of a gene within a genome is highly improbable. However, it is clear that such a scenario is possible by the independent selection of separately-occurring, similar genes from the vast random sequences in the primordial pond, and their independent assembly into distinct genomes. Given that a specific biochemical function has to be accomplished, we can find very many distinct proteins for that function in the random sequences. And we have shown that these genes which have to obey certain similar biochemical parameters will have at least some structural and sequence similarity. This amount of similarity is precisely what we happen to see among the similar genes with similar functions in nature. Truly, there is nothing more to it!

We have shown by computer simulations that many distinct genes can be found in random sequences, each of which can specify the same or similar biochemical function and have high sequence similarity. In fact, when we constrain the invariant amino acids and specify the variable amino acids in a computer simulation experiment (as we did in Chapter 7 in searching the multiple copies of a given gene), we obtain protein sequences of distinct genes with 30 to 80 percent similarity.[36] It would be even more easy to find such similarity if we allow gaps of amino acids as we find in genes in nature.

It is imperative that we are not deluded by gene similarities. If we look at these similarities with an absolutely open mind and consider that such genes can independently occur in the primordial pond and give rise to independent births of organisms, then the fog will clear and we will certainly begin to see the truth. It is time to free ourselves from the shackles of evolution that say similar genes are evolutionarily related. When we do so we can see that there is nothing left to validate the theory of evolution! Truly, let us ask ourselves what is there to support the theory of evolution? The only thing that keeps alive the concept of evolution is the similarity of genes tying organisms together. If this scenario can be shown to be not due to evolution and to be only due to independent birth of organisms, there is nothing more to support evolutionary theory. This is indeed where we are now!

The literature almost always tries to connect even the remotest similarity of genes and proteins by organismal evolution. Therefore, it is a jumble of misinterpretations through which we must sort and sift to find the right kind of evidence that belongs to each of the categories predicted in the new theory. We are demonstrating a few examples for three categories — the presence of the same genes, unique genes, and similar but unrelated genes in various organisms. Because the literature is just starting to accumulate information that can be used for our analysis, we can only use complete information, which is unequivocal. In one or two situations in the literature, therefore, there may be errors and gaps in the data or faulty interpretations, and therefore our interpretation may be wrong. However, I am confident that the majority of our analyses and conclusions, and thus our overall concepts, will remain intact.

IDENTICAL GENES IN WIDELY DISTINCT ORGANISMS CAN BE ONLY DUE TO THE INDEPENDENT ASSEMBLY OF GENOMES FROM THE COMMON POOL OF GENES IN THE PRIMORDIAL POND

When many genomes are assembled from a common pool of genes in the primordial pond, it is inevitable that not only unique and similar genes, but even identical genes will be found scattered among distinct, independently-born organisms. This concept is very consistent with all our observations so far. The presence of essentially the same gene in distinct genomes does not constitute organismal evolution, rather, it can be very well explained by the independent birth of creatures from a common pool of genes in the primordial pond. Not only that! In light of the fact that the presence of unique genes and similar but unrelated genes in distinct creatures cannot be explained by evolution and can be explained only by their independent births, the presence of essentially the same genes in them can be explained only by the independent births of creatures (Figure 9.3).

Let us now see some examples of the same or similar genes in various creatures.

Which theory explains the combined scenario of genes in organisms:
The New Theory of the Independent Birth of Organisms
or the Theory of Evolution?

Type of gene scenario	Evolution		Independent Births	
	Individual Cases	All Cases Combined	Individual Cases	All Cases Combined
Unique genes in distinct organisms	No	No	Yes	Yes
Similar but unrelated genes in distinct organisms	No	No	Yes	Yes
Essentially the same genes and similar relatable genes	Yes	No	Yes	Yes

FIGURE 9.3

Example 1. Intron positions in the actin gene are conserved in some creatures but not in others

Because actin is a protein required for the separation of chromosomes into dividing cells, which is a fundamental process in all living cells, actin genes are present in almost all creatures, whether unicellular or multicellular. The typical actin gene has a nontranslated leader (less than 100 nucleotides), a coding region (~1200 nucleotides) and a trailer (~200 nucleotides).[37] Most actin genes contain one or more introns. The positions of the introns can be aligned with regard to the coding sequence in most of these genes. When we do this, we find that almost every actin gene is different in its pattern of intron interruptions. When we take all the genes together, introns are found to occur at only 12 different sites, although no individual gene has more than six introns. Rat and chicken have six interruptions in common. More notably, rat, chicken and sea urchin have four interruptions at the same locations.

This situation could have arisen by the presence of many more introns in the original gene in the primordial pond, and the loss of some introns in the genes that were included in various creatures within their genomes over geological time. However, even if only a few introns are common to creatures such as the rat, chicken, and sea urchin, this fact tells us that the genes are related, or they are the same gene with variations in the number of introns. This is an excellent example wherein we find essentially the same gene in vertebrates and in an invertebrate. While this fact by itself may induce us to think that invertebrates and vertebrates are related by organismal evolution, the equally important fact that vertebrates have more than 600 proteins in their blood plasma that are totally absent in any invertebrate, and the fact that the invertebrates have many proteins and cell systems not present in vertebrates, authentically tell us that this scenario cannot be due to organismal evolution. Instead, this situation can be explained only by the independent birth of creatures in the common primordial pond, wherein, in addition to including unique genes in distinctly assembled genomes, essentially the same genes could have also been included. Looking at the whole scenario, it should not be difficult for anyone to accept that this is the only correct explanation.

Example 2. Positions of introns in the triosephosphate isomerase gene are well conserved between maize and chicken

The organization of the gene for triosephosphate isomerase from a fungus, plant, and bird is interesting for our discussion. The chicken gene has six introns, of which five are at identical positions to the introns in maize (corn). Two introns are at common positions between maize and the *Aspergillus* fungus.[38]

While this is a good example for the presence of the same gene in distinct organisms, in this case an animal and a plant, it is also revealing in other respects. Molecular biologists interpret this data to say that introns were present in eukaryotes before plants and animals diverged. In other words, their implication is that all organisms, plants and animals, originated from a single-celled eukaryotic ancestor. Now, when we consider this argument against the fact of the presence of numerous unique genes and proteins in multicellular organisms that cannot be evolved by organismal evolution, it becomes obvious to us that this argument is incorrect. Again, the only way to explain this scenario is by the independent birth of organisms.

Example 3. Globin genes in different organisms have essentially the same exon-intron structure

Introns occur at homologous positions relative to the coding sequence in all known active globin genes, including those of many mammals, birds, and a frog. There are three exons in all these genes, and the two introns are located at constant positions relative to the coding sequence. It is clear that this is a common gene included in these organisms.

It is interesting that the globin gene is also present in leguminous plants, with three introns, two of which are in homologous positions to those in vertebrates. Again, the same kind of argument we elaborated to the above examples holds good here.[39]

Globin is the protein part of the hemoglobin in vertebrates, the iron-containing protein that carries oxygen. Most invertebrates do not have hemoglobin. They have heamocyanin instead. Thus, the variable presence and absence of globin in various organisms indicate that this scenario is consistent with the new theory.

Although we have not made a systematic study of the presence of essentially identical genes, we have provided a few well-known examples that illustrate the point. Certainly, many more examples may be found by perusing the research literature.

We stated in the new theory that an already successful genome could be used, in part or full, in the construction of the genomes of later-born organisms. This process offers another mechanism for the presence of identical genes in different organisms. For example, the same set of basic-cell genes for a unicellular eukaryote can be used by many different multicellular organisms. This is an important phenomenon, by which the first genes that became part of single-celled organisms will appear in the later-born organisms, whether unicellular or multicellular, just by sheer chance that they were part of the first living cells. It does not mean that they were the best possible proteins to carry out the particular functions. It just means that they were selected purely by chance, as part of the first cells to enable the life of the cell. Thus, once some genes became part of the first successful genomes, they could become available in multitudes of copies in the open primordial pond. As a consequence, they would then be available easily in the primordial pond, and, for this reason, would have become part of the genomes which were independently assembled later.

Thus, in considering the origin of organisms in the primordial pond, we can see that even if the genomes of various organisms were independently constructed, and even if such organisms were entirely different and distinct in their body structures and functions, they can contain many essentially identical genes.

What is the number of possible proteins and biochemical functions?

When we think of life on earth as we know it, it is important to consider the universe of all possible biochemical functions. It is largely true that biochemists have already studied the majority of biochemicals present in living things. By going through a biochemistry book, one can see that there can be no more than a few million biochemicals. Let us then assume that the number of possible biochemicals is in the millions. Consider how many biochemical reactions there can be. Since each biochemical can react with a few other biochemicals, there can be on the order of trillions of biochemical reactions possible in living systems. We have here considered only the kind of reactions that break or make a chemical bond. But even those interactions between macromolecules such as protein-protein or protein-DNA, which do not make or break bonds, are also biochemical reactions.[40] Even when we consider all these kinds of reactions, we can roughly say there are no more than one thousand trillion (10^{15}) biochemical reactions in living things on earth. This is a very liberal estimate.

Let us now compute how many possible protein sequences there can be, given a maximum length of 3000 amino acids for a protein. This is truly immense — 20^{3000}. Out of these, we can be sure, only a tiny fraction will have any enzymatic or other biochemical or protein function. (Although the protein sizes in living systems can vary from 100 to 3000 amino acids, we can indulge in such rough estimates for our overall understanding of the problems concerned here.) Even though many amino acid variations of a protein specify the same biochemical function, still this general statement is true. Although it is simply impossible to compute such a number, we can see that it can be definitely far higher than the number of all the biochemical functions present in all living systems on earth. What this tells us is that many distinct proteins could be capable of carrying out one particular biochemical function. Any one of these might possibly be useful to carry out that particular biochemical function in a living system.

How does this computation help us in our understanding of the origin of life on earth? It helps us in understanding that the number of functional proteins possible from the primordial pond's genetic sequences are far more than what is required to construct all living forms. As we saw in Chapter 7, many fundamental phenomena make it possible for almost any protein specifying a particular biochemical function to have occurred as a gene in the finite amount of genetic material in the primordial pond. Any one of the multitudes of proteins that can carry out a specific biochemical function will do for that function. No matter how long the protein chain, no matter what its overall structure or sequence, that protein should be able to function in a living organism for that particular function. A genome is successful because it codes for needed functions. Therefore, any organism that came to life by the independent assembly of genomes in the primordial pond should include certain basic functions in it, immaterial of what genes it included. Under these circumstances, most of the organisms that came on earth independently should have included many functionally-similar proteins, which were unrelated. Protein chemists and molecular biologists certainly agree that functionally-similar proteins, even if they are not related, will usually have structural and sequence similarity.[41] This would in turn reflect in the DNA sequence similarity of the genes, although the genes are not at all related.

We must also realize that the complete set of functional proteins and genes that we find in all organisms today, and that existed in all organisms that were born in the primordial pond, is still far smaller than the set of all theoretically possible functional proteins and genes. We must not think that the functional set of proteins in living things are the only or unique set that is capable of forming life. In other words, what we find in all living organisms is an extremely small subset of what can be formed theoretically, and, most likely, of what could exist in the primordial pond. Even if the entire USP contained a very small subset of all theoretically possible biochemical functions (*i.e.*, genes for functional proteins), it is still a vast number of genes that would be sufficient to give rise to multitudes of independent organisms.

The genes and proteins in all living organisms have to necessarily belong to a limited set of biochemical functions, the members of each set being similar among themselves — even when all organisms had originated independently: Explanation of "gene families"

We can see that the biochemical reactions in a living cell are categorizable in major groups, such as the metabolism of sugars, amino acids, nucleic acids, fats, and carbohydrates. When we take one type of biochemical such as the sugar, many sugars and their metabolic products are very similar in structure. This means that the enzymes that work on the sugars are functionally similar and will have structural and amino acid sequence similarity. Similarly, the enzymes that synthesize these sugars from their precursor molecules will also be similar to varying degrees among themselves. Thus, these enzymes will be categorized in one or a few groups of similar enzymes — even if they originated independently. More importantly, their genes will be categorized as similar genes, and will be classified into families of genes by molecular evolutionists.

All the biochemicals in all organisms can be classified into a few major types. As a consequence, in living organisms there are only a few fundamental types of biochemicals and biochemical reactions, and there can be only that many families of genes. Only rarely will enzymes and proteins not fall into these few categories. This must be true, even if billions of organisms were born independently by the same biochemical mechanisms of DNA and proteins. Indeed what we see in the living world is a reflection of this phenomenon.

The presence of enzymes and proteins categorizable into a limited number of families does not mean that each family of proteins has evolved from a common ancestral protein through organismal evolution. Even when numerous organisms were born independently of each other from the common pool of genes in a primordial pond, we shall necessarily see the scenario of families of genes, each family consisting of many members which have functional, structural, and sequence similarity.

The presence of many functionally-similar biomolecular reactions, even in the simplest living entity

What we want to show here is that a set of genes that are functionally similar should have actually occurred independently in the primordial pond, and that they are absolutely required for even the simplest living creature to come into existence. In the above, we set out to demonstrate that the biomolecular reactions in the living world can actually be categorized into several sets of similar biochemical reactions, each of which is generically called a gene family.[42] Here we shall show that such families of similar genes are required even for the simplest living entity. This will illustrate that the new theory of the independent birth of organisms is perfectly valid — even on the face of the presence of scores of similar genes and proteins in the living world in totally distinct organisms. This will also illustrate that the interpretation of evolutionists — that a set of similar genes is the result of evolution from a common ancestral gene — is indeed incorrect.

SIMILAR GENES, NOW GROUPED UNDER A "FAMILY," ARE NOT RELATED TO ONE ANOTHER BY EVOLUTION. EACH ORIGINATED INDEPENDENTLY IN THE PRIMORDIAL POND.

Contrary to the belief of evolutionary biologists that a set of similar proteins that they call a "family" had to be evolved from an ancestral "root-stock" gene in the original organism by organismal evolution, we can demonstrate that many proteins, independently originated from the universal gene pool, can be categorized into families of proteins based on structural and functional similarities. They need not have any evolutionary relationship at all. The new theory gives a unifying theme about redundancies of proteins and genes. Large numbers of biochemical reactions operate on similar biochemicals. Therefore, the functions of enzymes that operate on similar biochemicals are redundant. This *functional* redundancy in proteins requires a possible *structural* redundancy and in turn a redundancy in protein and gene *sequence*.

Based on the independent assembly of genomes in the primordial pond, totally independent proteins can be grouped into families of proteins based only on their similar structure, which can be imposed by similar functions — not based on evolutionary relationship. The nature of the independent assembly of genomes imposes that a large minimal set of essential functions, required for an organism's development and function, must be included in a genome for it to become viable as a creature. This would result in the inclusion of many similar proteins for similar biochemical functions within one genome. However, we see only the end result of all these processes, and not the means by which these originated initially in the different organisms. When viewed with evolutionary eyes, the end result can appear as if due to evolution, but we can see that it is not evolution that led to the families of proteins, but rather it is the functional similarities of proteins that independently originated in the primordial pond's universal gene pool.

EXAMPLES OF FUNCTIONAL SIMILARITY AND STRUCTURAL
SIMILARITY IN KNOWN PROTEINS THAT ARE MINIMALLY
REQUIRED FOR EVEN THE SIMPLEST LIVING ENTITY

There are many biochemical reactions within a cell that are similar. There are many enzymes that catalyze similar reactions — many different hydroxylases, transaminases, peptidases, carboxylases, kinases, phosphorylases, dehydrogenases, and esterases are examples. Each set of enzymes catalyze similar biochemical reactions on different biomolecules. There are many different enzymes, each of which transaminates a different biochemical, for instance. All these proteins, although different from each other, have functional similarity and therefore tend to have structural and sequence similarity. Many of the proteins that bind similar metals, cofactors, or coenzymes are similar in those functions, such as the iron-binding proteins, calcium-binding proteins, or proteins that bind the cofactors NAD (nicotinamide adenine dinucleotide) and FAD (flavin adenine dinucleotide). A set of proteins that bind a given metal or cofactor may have sequence similarity over the domain that binds the cofactor. Evolutionary biologists may say that one iron-binding protein could have evolved from another protein that binds iron. But as we have now determined, two different enzymes for similar functions could have originated from the universal gene pool as distinct genes.

As an example of the potential similarity of functions of the metabolic enzymes, let us consider the oxidation of glucose. Several enzymes sequentially act on the successively degraded product of glucose, in several steps, completely converting glucose into CO_2 (carbon dioxide) and H_2O (water). Although different enzymes carry out the steps, the steps share many similar functions. We may expect that at least some of the enzymes could share structural and functional similarity — even when these enzymes had originated independently.

As another example, food contains the sugars sucrose, maltose, and lactose. These are called disaccharides, because they are made up of two smaller sugar units such as glucose, fructose, or galactose. Before the disaccharides can be used in the body of an animal, they must first be enzymatically hydrolyzed (broken apart),[43] to yield their smaller sugar units as follows:

$$\text{maltose} + H_2O \xrightarrow{\text{maltase}} \text{glucose} + \text{glucose}$$

$$\text{lactose} + H_2O \xrightarrow{\text{lactase}} \text{galactose} + \text{glucose}$$

$$\text{sucrose} + H_2O \xrightarrow{\text{sucrase}} \text{fructose} + \text{glucose}$$

The enzymes *maltase*, *lactase*, and *sucrase* break down the sugars maltose, lactose, and sucrose to their corresponding smaller units. In these cases, we can see that the three enzymes work on similar sugars, and that at least one of the reactants (H_2O) and one of the products (glucose) are exactly the same in all three cases. Therefore, all three enzymes can have structural similarity.

Another set of enzymes with similar functions is involved in the biosynthesis and degradation of amino acids. The α-amino groups of the 20 amino acids commonly found in proteins are removed at some stage in their degradation.[44] These amino groups are collected and ultimately converted into a single excretory end product, which in human beings and most other terrestrial vertebrates is *urea*. The removal of the α-amino groups from most of the amino acids is performed by enzymes called *transaminases*. In these reactions, called transaminations, the α-amino group is transferred from the amino acid to the α carbon atom of α-ketoglutarate, leaving behind the corresponding α-keto acid analog of the incoming amino acid, and causing the amination of the α-ketoglutarate to form glutamate:

α-amino acid + α-ketoglutarate ⟷ α-keto acid + glutamate

Some of the transaminases, which are named for the amino-group donor, are designated by the following reactions:

alanine transaminase
alanine + α-ketoglutarate ⟶ pyruvate + glutamate

aspartate transaminase
aspartate + α-ketoglutarate ⟶ oxaloacetate + glutamate

leucine transaminase
leucine + α-ketoglutarate ⟶ α-ketoisocaproate + glutamate

tyrosine transaminase
tyrosine + α-ketoglutarate ⟶ p-hydroxyphenylpyruvate + glutamate

It is clear that the four transaminases, even if they are different enzymes, carry out similar biochemical reactions and act on similar biochemicals. This is true with a large number of biochemical reactions — in fact, the examples we have seen here illustrating this concept is just the tip of the iceberg. Although distinct enzymes bind specifically to different substrates, if the three-dimensional structure of the substrates are similar, the enzymes can have structural and sequence similarity. If the reactions they catalyze are similar — e.g., transferring amino groups from amino acids — those parts of the different enzymes that carry out similar functions may have a structural similarity. Furthermore, the molecules they bind during such reactions may have considerable structural similarity, and the binding function of the transaminase enzymes also can have similarity. As we saw above, although the many transaminases are distinct enzymes, the different amino acids they bind have similarity in major portions of the amino acid molecule, and in all the different transamination reactions we saw above, one of the reactants (α-keto glutarate) and one of the products (glutamate) are the same.

The very same set of enzymes we have discussed here may or may not have similarity of sequence,[45] but it is logical that such *functionally* similar enzymes can be expected to have some *structural* as well as *amino acid sequence* similarity. These examples are brought here to emphasize that many enzymes involved in the metabolism of a given biochemical can have similar binding domains and reactive domains. This principle can be extended to the peptidases, peptidase inhibitors, GTP-linked binding proteins, DNA-binding proteins, and receptors of a given molecule, all of which are grouped as belonging to various families of proteins by evolutionary biologists.[46,47]

The above examples illustrate that the successive metabolic products of a biomolecule would necessarily have some structural similarities. Therefore there is an inherent redundancy in functions of many enzymes within one organism, even within a unicellular eukaryote. This is because all the functions of a metabolic cycle are minimally and crucially required within any living system, even a bacterium. It is not difficult for anyone to see that the degradation of a sugar such as glucose, or the transamination of many amino acids is required in every living cell. Thus even in a unicellular eukaryote, a very high degree of functional and structural similarity of various proteins is expected.

When diverse organisms could independently arise from the universal gene pool as proposed in the theory of the independent birth of organisms — based on the occurrence of genes for all the biochemical functions in the UGP — the membership in each family can inherently be expected to be very large, although the total number of families of proteins could be restricted to a fairly small number. In fact, we have shown by computer simulation experiments that when one specific protein sequence can exist with an *a priori* probability in the USP of the primordial pond, then myriads of proteins with sequence similarity are also bound to occur in it — either with similar functions or entirely different functions.

Based on these considerations, one can see that it is unreasonable to say that the many enzymes constituting a particular metabolic pathway (many of which may fall under a family of proteins) have all evolved from a common ancestral gene through descent with modification — because these sets of proteins are minimally required for even the simplest living cell to come into existence, and because the genes for these proteins can occur independently in the primordial pond as we have demonstrated in Chapter 7. How can even the simplest cell arise on earth unless it has the enzymes to metabolize a sugar like glucose and is capable of utilizing energy? How can a cell exist without the ability to synthesize the amino acids required for constructing its numerous proteins? Unless and until all these minimum kinds of genes are available, a genome cannot at all lead to the development of a viable cell. When even the simplest cell comes into existence in the primordial pond, it will necessarily have many proteins and genes categorizable into distinct families based on similarity of structure, function, and sequence.

Numerous metabolic enzymes are required for the structures and functions of even the simplest living cell, because these are required for the metabolism of the basic chemicals which are required for the life of any living system. At least the set of enzymes catalyzing similar types of biochemical reactions can be expected to have some structural similarity and therefore sequence similarity among them. Likewise, even in the simplest living cell, there must be DNA polymerase, RNA polymerase, ribosomal proteins, many DNA-binding proteins that will serve as gene-regulatory proteins, and so on, which can be categorized into several sets of proteins with similar biochemical functions. Thus we can prove that even in the simplest living entity, there must exist several families of proteins simply based on the similarity of their functions. As we have reiterated, genes for all these enzymes must have been available in the primordial pond's genetic sequences — and only then it was possible for even the simplest living system to come into existence. By the same token, if similar genes for similar biochemical functions could exist in the USP simply based on probability, why could not the similar proteins that are found in the various organisms come into existence in the primordial USP by the same mechanisms?

Evolutionary theory presumes that there was a root stock of proteins in the original organism, the origin of which is not clearly discussed under Darwin's theory even by modern molecular evolutionists. Through descent with modification when the original species speciated, diverged, and diversified, the root-stock genes for proteins in the genome of the original creature had duplicated, mutated, and modified to become diversified into proteins with similar as well as new functions (which they call evolutionary "innovations") categorizable into families of proteins. Because each family started with a root-stock protein, the members of a group would have structural similarity and sequence similarity. Note that, based on this concept, many protease inhibitors are grouped into one family, a vast number of protein kinases are grouped under a protein kinase family, protein phosphatases into one family, and similarly the G-protein-linked receptor proteins into another family.[48] Our foregoing analyses and considerations illustrate that this concept arose due to a false assumption that organismal evolution was an established fact, and that the evolutionists' interpretation of similar genes and proteins is fundamentally incorrect.

Proteins that have similar function are defined as "homofunctional" proteins. Some "heterofunctional" protein families (proteins that have different functions, but with some similarity of structure or similarity at the level of the amino acid sequence) were also apparently uncovered by database searches.[49] However, these also could be due to similar domains in these proteins originating independently in the primordial pond USP, as we discussed previously. Our foregoing analyses and discussions clearly illustrate that the molecular scenario that exists in the living world can be thoroughly explained by the theory of independent birth of creatures.

Molecular evolutionists say that redundancy in proteins and genes demonstrate Darwin's theory of descent with modification. But as we demonstrate here, redundancy is the property of proteins and genes even in the simplest creature, and in the multitudes of creatures that originated independently in a common primordial pond. Redundancy in various genes and proteins is an absolutely unavoidable phenomenon even when numerous creatures originate from the common pool of genes, based on the general mechanisms of genes and proteins. Again, molecular evolutionists say that only a few hundred or a few thousand families are possible in the living world, which, according to them, illustrate evolutionary theory. But we can see that even when numerous creatures originated totally independently in a primordial pond, the biomolecular reactions in them would be such that they all would be classifiable only into a few hundred or a few thousand similar biochemical reactions; and their proteins and genes would be categorizable into only that many families.

Redundancy is inherent in any kind of information — and more so in genetic information

Let us now consider the fact that redundancy is inherent in any information — which will exemplify that the vast amount of genetic and biological information in living organisms will inherently have even more redundancy by its very nature. In any language, for instance English, there is a great deal of redundancy between sets of sentences in a long passage. Each sentence is formed by similar rules, and contains a subject, object, and a connecting verb. Furthermore, they all have many prepositions. Because of this we use similar words in different sentences. The verbs and the prepositions are highly common to many sentences. For example, words such as: the, for, of, to, from, where, when, why, is, are, etc. are very common. But the sentences can convey entirely different meanings. The manner in which the sentences are constructed follow certain basic rules. Therefore, many sentences can have high similarity by having similar words. The larger the number of sentences in a passage, the more sentences with similarities we would find. But the sentences are different and have different information.[50]

This basic similarity is built into the nature of all sentences. The similarity between two sentences will increase when two different sentences have the same meaning. That is, if two people are asked to construct a sentence with that meaning, they both can be expected to use synonymous and identical words in the two sentences they constructed independently. This process would force and direct the two sentences to be far more similar than if the sentences do not have the same or similar meaning.

This principle can be applied in the context of the proteins. Protein folding follows certain basic rules. The presence of hydrophobic amino acids in the buried regions of a protein and their general absence on the protein surface is one of them. Turns, or "elbows," in proteins use particular constellations of amino acids, which occur commonly in proteins.[51] A natural rhythm of nonpolar residues in alpha-helices is another rule. These basic rules may impose a certain amount of similarity recognizable in some totally different proteins. Therefore, although the messages in different proteins are different and have independent origins, they can still have similarities to varying extents. Sets of domains that bind different metals, cofactors, DNA, and nucleotides such as GTP, all can be distributed in sets of different proteins that would contribute to similarities, although the final specific overall biochemical function of each protein may be different. These domains can be equated to the prepositions in our English sentence example we discussed above.

When stringency is imposed on two different sentences to have the same meaning, then the general construction of the two sentences will also be similar. Functional and structural similarity in certain groups of proteins such as those exhibiting certain kinds of biochemical or enzymatic functions, may accentuate the similarity already imposed by the inherent redundancy in proteins by the general nature of folding of all proteins. But the important fact is that the information for all these proteins could originate totally independently of each other in the primordial pond from the universal sequence pool. Without a knowledge of this, if we look at the genomic scenario of organisms, one has to agree that it is quite misleading in that it presents a picture as though many of the proteins are evolutionarily related.

Considering genetic similarities, indeed there is nothing to say that organisms are related, or that these proteins are related by evolution from an ancestral gene, except for the fact that these proteins have sequence similarity. Only when there is identity between two given genes in terms of exon-intron structure (*i.e.*, the same exon lengths and positions in the gene) the genes could be said to be the same genes.

Our foregoing analyses very clearly pictures what must have gone on in the primordial pond to give rise to numerous independent genomes which in turn gave rise to the independent birth of multitudes of organisms. Our analyses and discussions also clearly show the reasons why the similarity of sequences in genes and proteins are misconstrued to be the result of evolution. Is it not clear that the absence of a totally and radically different theory involving absolutely no organismal evolution can easily mislead one to see things differently, that the scenario is due to organismal evolution? Is it not obvious that this whole argument of root-stock genes in the original ancestral organism giving rise to gene families is the result of an evolutionary premise for which no other molecular proof exists except for the gene similarity, which itself is explainable clearly without involving evolution? Combined with all other molecular scenarios and other details of life on earth that clearly corroborate the theory of independent birth of organisms, one can see that our considerations here are well founded and perfectly valid. We can thus confidently conclude that numerous genes with functional, structural, and sequence similarity must have occurred independently in the primordial pond, and many of them must have been selected for similar functions in the genomes of all creatures.

COMPUTER SIMULATIONS INDICATE THAT SIMILAR GENES FOR FUNCTIONALLY SIMILAR PROTEINS COULD HAVE EXISTED INDEPENDENTLY IN THE PRIMORDIAL POND

We have demonstrated in Chapter 7 that genes split into exons and introns for almost any given protein function could occur in the random DNA sequences available in a primordial pond. We have shown that genes with internal sequence repetitions could occur as easily as genes with no such repetitions. Here we want to show that several genes with sequence similarity, coding for distinct proteins either with functional or subfunctional similarity, could exist in the primordial sequences independent of each other.

In fact we can see that all the proteins that have sequence similarity in all living systems fall into one of the following categories: 1) variants of the same gene; 2) distinct genes encoding similar functions; or 3) genes with no overall functional similarity but which have similarity over regions of similar subfunctions. All three categories of genes can be shown to be possible in the USP by computer simulation.

Two different independently-originating proteins, each with many domains, may need one or two similar domains for their functions. Based on the principles we discussed in Chapter 7, we can see that the probability of occurrence of such part similarity in two different proteins can be very high. For instance, the probability of the occurrence of a given exon in multiple locations in the universal sequence pool is extremely high. When I tested this in simulations of searching for hypothetical genes, each of the exons of the gene is found to occur at numerous locations in the hypothetical universal sequence pool. In fact, when I searched for two different protein sequences with part similarity in computer simulations, they were found in the same random sequence. Consider one gene with 10 exons and another with seven exons. Let four exons be common to both genes. If we look for each of these genes in a random sequence using the computer, they will be found in different locations of the random sequence — in fact, each of the genes occurs many times with sequence variations. The essence of the theme is that part similarity in two different genes can occur independently of each other, with no evolutionary connection. In fact, in the scenario we are considering — the primordial pond with the USP in which myriads of genes can occur — such a thing is absolutely inevitable.

In addition to this principle, there could be many other possibilities by which such part similarity can occur in two proteins whose genes originated in the universal sequence pool. Multiplication of a random sequence and recombination of these in the USP when the DNA sequences were forming is one way this could have happened (refer to Chapter 6 for explanations of these prebiotic mechanisms). Consider the situation where copies of the same DNA sequence could have recombined with two different unique sequences. If two distinct genes occurred in these two recombined sequences encoding entirely distinct biochemical functions, then these genes would have part similarity. These two genes may be selected in a given genome or two different genomes because they contain the same protein module with a functional characteristic required in a genome.

Why do evolutionists believe that similar genes are related by organismal evolution? The problems of evolutionary arguments

We have demonstrated in various ways that independent births of organisms in the common pool of genes in the primordial pond will necessarily lead to the presence of many genes in distinct organisms with functional, structural and sequence similarity. But this is not enough, because the whole fields of biology and genetics are filled with problems concerning similarity of genes due to the constant attempts to connect these genes by organismal evolution. Therefore, it becomes necessary for us to separately discuss how and why the evolutionists' beliefs are incorrect.

There are two types of errors on the part of molecular evolutionists: 1) conceptual errors in interpreting genuine similarity, and 2) practical errors in database searching of similar sequences. We shall briefly discuss these in this section.

Molecular geneticists and evolutionists proclaim that similarity of genes is due to organismal evolution. They have grouped the similar genes found in all living organisms into various families of similar genes, and say that each gene family evolved from a single ancestral gene. In fact, based mainly on the similarity of genes, they proclaim that the theory of evolution is absolutely proved. It is imperative for us in light of the new theory to discuss why this belief persists and why it is incorrect. If the basis of gene similarity can be explained by the theory of independent birth of organisms, and if it can be shown that none of the evolutionary theories is required to explain the similarity of genes, then essentially we can show that the theory of the independent birth of organisms is at least as proven as any evolutionary theory.

Why do evolutionists think that genes, similar or not, have evolved from common ancestral genes?

What makes molecular evolutionists believe that similar genes had evolved from a common ancestral gene present in a common ancestral organism through organismal evolution? Why do they continue their beliefs?

- They believe that all organisms are evolved from a common ancestral organism, based on Darwin's theory of evolution. This requires that the genes of all these organisms originate from the starting original organism. This is the first totally unfounded assumption.

- There exist many similar genes in the same or different organisms, which is a misleading support for Darwin's theory.
- Because they work from an evolutionary premise, biologists try to connect these similar genes by organismal evolution, without realizing that many such similar genes ought to have been present even in the very first, even if most primitive, original organism that came on earth, no matter how it originated.[52]
- Evolutionists fail to note that many similar genes do occur even in the simplest organisms living today, including single-celled eukaryotes and even prokaryotes[53] — and, therefore, it is unnecessary to explain the presence of similar genes for similar functions in various animals by the evolution of a family of genes from a common ancestral animal. How could the various similar genes come into being in the prokaryote by evolution if it could not even survive without these genes?
- When evolutionists try to find evidence for the theory of evolution by relating similar genes through organismal evolution, they do not satisfactorily address the presence of entirely new or unique genes in various distinct creatures, and do not have an answer as to how unique genes can come about in these creatures. They simply say, without any evidence, that these unique genes could have evolved from preexisting genes — only to be consistent with the theory of evolution.
- Computer simulation research demonstrating the presence of complete genes in the primordial pond have neither been possible nor been carried out until now. Only here we give such an unequivocal proof that complete genes could have simply occurred in the primordial pond. Only here we show that numerous genes with sequence similarity could occur independently in available random DNA sequences. Thus, there was no other way of explaining the presence of similar genes until now than by invoking evolutionary connection.

When we dig through the molecular scenario more deeply and ask whether many of the details fit the evolutionary theory or the theory of the independent birth of organisms — such as the presence of unique genes that we saw above — it becomes obvious that the evolutionary theory is indeed incorrect and that they illustrate the independent birth of creatures in the primordial pond. It becomes clear that evolutionists have been playing with the same details with only one thing in mind, the theory of evolution — without even looking at these details through any other mechanism outside the theory of evolution.

It is crucial for us to realize that there is absolutely no direct proof for evolution having occurred, and the molecular evolutionists' claim is based only on an interpretation of circumstantial evidence from today's living organisms. We can show that their interpretation is wrong, and that there is a better interpretation by a theory that is absolutely outside of the domain of evolutionary theory.

An inherent problem in the approach of an evolutionary geneticist in seeking similarity of genes

Evolutionary geneticists deal with an inherent problem when they analyze protein similarities looking for assumed evolutionary relationships. They start with a prior, strongly-rooted notion of evolution. Therefore, according to them, those proteins with functional similarities have evolved from one another. Consequently, they expect the proteins to have structural similarity and sequence similarity. So if they find sequence similarity between two functionally-similar proteins or genes, they believe that it is a direct proof for Darwin's theory of evolution.

Because evolutionists expect two proteins which are functionally similar to be evolutionarily related, they look for sequence similarity even before one knows whether these proteins have sequence similarity. When a sequence similarity is found — which is expected simply because of the functional similarity even without evolutionary connection — they confidently provide it as evidence for evolution having occurred. On the other hand, if there is little or less significant sequence similarity, they try to bend the methods of aligning or searching for similarity of sequences in order to "improve" the similarity. We now know that because of the strongly-rooted evolutionary ideas, evolutionists do not consider that their observations may stem from a nonevolutionary process.

The main theme of evolutionary arguments for sequence similarity

According to Darwin's theory, all organisms had evolved from some original organism. The modern synthesis of Darwin's theory, and the modern evolutionary geneticists, do not discuss how this very first organism originated. However, granted that it had somehow come into being, it is assumed that natural selection produced new organisms with astonishingly unique structures and body parts. Moreover, under evolutionary theory there is no concrete or well-founded concept as to how the genes of the original organism could have originated. Evolutionists simply take it for granted that somehow the root-stock genes constituting the genome of the very first organism appeared. Each root-stock gene gave rise to many new genes with similar or new functions — as the original creature evolved into many new creatures. Based on this assumption, evolutionists categorize similar sequences in the living world into gene families,[54] each of which descended from one root-stock gene present in the original organism.

Their thesis is that because common ancestry is the essence of evolution, the similarity of protein and gene sequences that exist in the living world therefore apparently supports Darwin's notion of descent with modification. Let us not forget that such categorical conclusions are purely based on their interpretations of the molecular scenario existing in organisms living on earth today, and that there is otherwise no direct proof for evolution having occurred. Therefore, these details are open to entirely other interpretations if such are possible.

The belief of molecular evolutionists, based on sequence similarities, can be illustrated by the following quote from Russell Doolittle, published in 1989:[55]

> All living organisms must trace back to a common ancestor, and it is reasonable to think that some very early ancestor had a relatively small genome coding for a relatively small inventory of prototypic proteins. Most contemporary gene products are the result of past gene duplications and subsequent divergence resulting from gradual amino acid replacement. As a result, many proteins have already been grouped into families. There are sequences for scores of each of four major protease families and correspondingly large numbers of protease inhibitors. There are vast numbers of protein kinases, all apparently descended from a common ancestor, and we can anticipate a similar multitude of protein phosphatases.
>
> Hindsight not withstanding, some of the unanticipated sequence resemblances are nothing less than awe-inspiring. Consider, for example, the remarkable family of G protein-linked receptor proteins. When the β_2-adrenergic receptor was cloned and sequenced, it was found to be approximately 25% identical to bovine rhodopsin. This is a degree of resemblance that leaves little doubt of common ancestry, even ignoring the fact that both proteins are transmembrane signaling devices that interact intracellularly with G proteins (transducins).
>
> Since that initial finding, more than a dozen other receptor proteins have been found to have related sequences, including a wide assortment of receptors for biological amines, biologically active peptides, and, in the slime mold, cyclic AMP!
>
> ... The buzz-word that is emerging more and more in the sequence comparison field is 'redundancy'. This is a happy situation in that redundancy is what allows the reconstruction of past events. Common ancestry is the essence of evolution, and nowhere is Darwin's notion of 'descent with modification' more apparent than in the amino acid sequences unraveling before us.

Russell Doolittle stated similar thoughts in a publication in 1979 as follows.[56] In it we can see how the functionally-similar proteins that are similar in sequence are connected by evolutionary argument. We can also see an expectation that functionally-similar proteins should be related by evolution — even before knowing that they have sequence similarity (italics mine):

> One of the foremost objectives in the study of protein evolution is establishing the origins of individual proteins. Given the general rule that "new" proteins originate from other existing proteins, present day proteins can be categorized according to their heritage just as well as according to traditional classifications which depend on function or source. Many important metabolic enzymes are ubiquitous to animals, plants, and microbes, the lines of descent from eukaryotic systems tracing directly back to prokaryotic ancestral types. For the most part, these enzymes are quite conservative (cf. phosphoglyceraldehyde dehydrogenase), and their relatives will probably be recognizable from their amino acid sequences without much difficulty. *Also it will not prove surprising if many of these enzymes, particularly those employing the same cofactors, turn out to be related to each other. There is already some evidence that phosphoglyceraldehyde dehydrogenase and glutamic dehydrogenase, both of which employ pyridine nucleotide coenzymes, have homologous primary structures. Also, several other dehydrogenase enzymes have been found to have similar sequences around their "essential thiol" groups. X-ray diffraction studies have shown that several dehydrogenases requiring pyridine nucleotide cofactors have identical three-dimensional structures in the cofactor binding region.* At this point no one will be overwhelmed if many enzymes catalyzing similar reactions in different situations turn out to be evolutionarily related. In many other instances, however, we simply do not have the faintest clue as to where the existing proteins originated.

As we have discussed elaborately in Chapters 3 and 4, we saw that the whole field of connecting the proteins by evolutionary relationships, such as the vertebrate plasma proteins with those of invertebrates, is full of problems and improbabilities. We saw that the assumed ancestral proteins are indeed totally absent in the assumed ancestral organisms. One who proposes the evolutionary origin of these proteins should also be able to connect these through evolution. It is not sufficient to show that an arbitrary set of proteins in organisms have functional similarity, structural similarity, and sequence similarity among them, and then to simply say that these are evolutionarily related by gene duplication that started in an ancestral organism. An overall analysis with a comprehensive perspective will unequivocally show that the belief of evolutionists is due to incorrect and prior assumptions seen through the mist of evolutionary theory, and that all the facts point to the independent assembly of genomes in the primordial pond.

TECHNICAL PROBLEMS THAT EVOLUTIONARY GENETICISTS FACE IN SEARCHING FOR AND INTERPRETING SIMILAR GENES IN SEQUENCE DATABASES

The misinterpretations of evolutionists concerning similarity of genes are due to fundamental conceptual errors. These errors lead them to misinterpret the actual biomolecular scenario of similar functions, structures, and sequences of genes and proteins to be the result of organismal evolution. In addition, evolutionists also face several technical problems in searching for similar sequences in gene and protein sequence databases, misleading them to interpret false or chance similarity to be genuine similarity.[57] The concept of evolution is strongly rooted in the minds of almost all scientists who begin their careers in the biological and biomedical sciences. The reason is that they are taught in school to believe that Darwin's theory is an established fact, and that only creationists oppose it for religious reasons. The minds of students are programmed to think that, scientifically speaking, evolutionary theory is correct. Because of this, they always look only for "evidence" for the evolutionary theory and ignore anything that is against the theory. Scientists practicing molecular biology are no exception. In their field, what they look for as evidence of evolution is similarity of proteins and genes in various organisms. To them, even new and unique genes evolve by evolutionary mechanisms of mutations. They assume that proteins and genes with different (*i.e.*, unrelated) functions will have structural and sequence similarity. Therefore, when they find sequence similarity among proteins and genes that have unrelated functions, they group such proteins into a family of "heterofunctional proteins," implying that these evolved from a common ancestral gene from a common ancestral organism.

Today, sequence databases for proteins and genes are available. To find proteins or genes with similar sequences, a scientist "searches" a database for such proteins or genes. Besides our main aim to demonstrate that their concept of evolution itself is fundamentally incorrect, we can show here the many errors in database search analyses that are made because of this incorrect concept of evolution. For instance, scientists misconstrue the seeming similarities between unrelated sequences that arise due to methodological and judgmental errors as genuine sequence similarities; they then use these wrong results for evolutionary interpretations.

Whenever a new DNA sequence or a protein sequence is worked out in a laboratory, the first thing that the scientists do is to search the database for DNA and protein to find if there are any similar known sequences. If they find a similar sequence, it may or may not be *genuinely* similar. Usually, the scientists go by certain arbitrary assumptions about what is a genuine similarity, because there are no concrete rules to set forth which is genuine similarity and which is false similarity.

Although we can elaborate the many kinds of technical and judgmental errors in database searching for similar sequences,[58] we shall not go into the details for want of space and because it is too involving. It suffices to say that there are many kinds of technical and practical problems in searching for and interpreting similar sequences. For instance, because of what is called the "gap" problem (gaps allowed within a sequence, creating artificial insertions or deletions), even sequences that do not have genuine similarity will show themselves to be similar. Such problems are discussed elegantly by the noted molecular evolutionist Russell Doolittle.

> If we move another step closer to reality and permit the existence of internal deletions and insertions (gaps), the situation becomes more complicated. It is obvious that gaps increase the matching of unrelated sequences as well as related ones, and if unlimited gaps are allowed, two unrelated sequences that are very long can be arranged in a fashion that achieves virtual identity over their aligned portions.

THE THIRD BASE OF THE CODON CAN BE HIGHLY VARIABLE IN GENES WHILE THE FIRST AND SECOND BASE ARE "CONSERVED:" WHAT IS THE NEW THEORY'S EXPLANATION?

In analyzing the coding sequence of a given gene found in many organisms, there exists a phenomenon concerning the variations of codons. If we take one gene and analyze its coding sequence in many different organisms, we naturally find sequence variations. We saw that this is because the amino acids at many positions in a protein can vary with a set of allowed amino acids. Likewise, the codons at almost all the positions of a coding sequence are degenerate. Usually there are three or four codons, with the same first two bases but different third bases, that code for the same amino acid. As a result, if we analyze the frequency of the nucleotide differences at the three possible codon positions in the sequence of a gene from many different organisms, they vary most at the third codon position, less at the second and the first. Based on the principles we have so far discussed, we can show clearly that this phenomenon can arise when organisms were independently born — by mutational changes of the same gene in each organism without altering the basic function of the protein. This phenomenon can also arise even if two gene sequences coding for functionally the same protein arise independently of each other (Figure 9.4). But evolutionists believe that this phenomenon is due to the evolution of organisms from one another.[59,60]

Evolutionists believe that the same gene in widely different organisms has in fact evolved by descent with modification, simply based on the

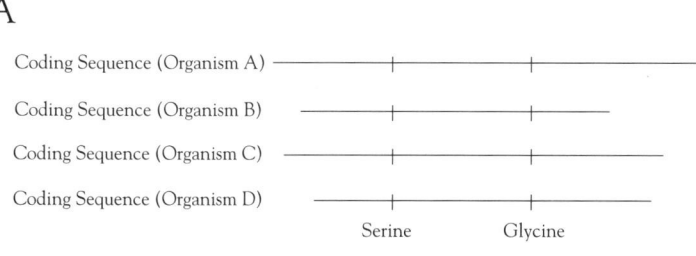

FIGURE 9.4. THE THIRD BASE OF THE CODON IS MORE VARIABLE THAN THE SECOND AND FIRST IN SIMILAR CODING SEQUENCES IN DISTINCT ORGANISMS — EXPLAINABLE BY THE INDEPENDENT OCCURRENCES OF GENES IN THE PRIMORDIAL POND. (A) The sequences of similar genes in various organisms can be aligned with each other to match the similar sequences. (B) When the frequency of the nucleotides at one particular aligned codon position (e.g. at the serine position) is computed, it is seen that the third base of the codon has the highest variability. This is expected due to the codon degeneracy even if the codes for the similar proteins occur in totally independent and unrelated DNA sequences. This is also expected even if essentially the same protein in distinct independent organisms undergo mutations independently leading to normal variations.

fact that there are nucleotide differences in the gene's coding sequence (and amino acid differences in its protein) in various organisms — which they assume are due to mutational changes during evolution. Molecular biologists compare the coding sequences of the same gene present in various organisms and show that the differences are most common at the third base position of the codons. This is precisely expected under our new theory: if the same gene were included independently in different organisms, it will undergo such mutations that either do not affect the basic function of the protein, or lead only to a defective protein. That is, organisms that are independently born can have the same gene, and in each organism the gene undergoes random mutations producing normal variants of that gene independently of any other organism, while all organisms remain immutable. Thus, the phenomenon of the presence of the same gene in various organisms and the presence of sequence variations in them — which is claimed to be the strongest molecular evidence for the evolutionary theory — can be explained by the theory of the independent birth of organisms. This is also true for the similar portions of entirely distinct genes.

The same gene included in two entirely distinct genomes from the primordial pond can be changed in the two creatures, although these creatures originated independently. These changes can occur in those amino acid positions that can tolerate substitutions. Where amino acid substitutions are not tolerated, often changes in the third, most degenerate codon position are allowed. This is variation in an immutable genome within its confined genomic framework.

OTHER MOLECULAR SCENARIOS IN CREATURES LIVING TODAY REFLECT WHAT HAPPENED IN THE PRIMORDIAL POND WHEN VARIOUS GENOMES WERE INDEPENDENTLY ORGANIZED

THE SET OF GENES OF ALL GENOMES ARE "ROOTED" IN THE OPEN-ENDED PRIMORDIAL POND — NOT IN THE INCORRECTLY ASSUMED "ROOT STOCK" OF GENES IN THE GENOME OF THE FIRST ORGANISM

Evolutionary geneticists assume that all genes evolved from the root-stock genome of the original creature. Clearly this is an incorrect assumption. The reality is that the stock of genes is the universal gene pool in the primordial pond, from which the genes were variably assorted into different genomes.

A number of known details in the scientific literature strongly support this theme. Evolutionary biologists compute the supposed time of divergence of different organisms based on phylogenetic studies. They find the number of nucleotide differences between the different sequences of a given gene in various organisms and use certain algorithms to compute their "evolutionary distances." Then they use the data from the fossil record to assign the geological time of divergence of different creatures from common ancestors. In doing so, they also assign the time of the invention of the earliest original protein or that of the earliest duplication event of a gene from which a new gene is assumed to have evolved. In almost all the cases, this time happens to be quite large such as one billion years, which is far before the earliest multicellular organism (the original creature) was supposed to have come into existence. In other cases it is approximately 600-800 million years ago, which goes back to the time of appearance of multicellular life. Time and again, in almost all their calculations molecular evolutionists come up with a time of at least 600 million years ago and up to and more than one billion years ago, for the origin of the genes of even vertebrates.[61] Remember that according to the fossil record, this is the time (~570 million years ago) when multitudes of multicellular creatures originated almost simultaneously in the Cambrian explosion.

While we understand that the evolutionary basis for these computations are not correct, we are bringing in this computation to show that the time of "invention" of new proteins, or duplications of some genes, by the account of evolutionary biologists themselves is at around the time of the appearance of multicellular life. The theme of the new theory is that all the proteins would have become available in the primordial pond around the time when multicellular life appeared on earth, and that a majority of life forms were independently born simultaneously from the primordial pond. The fact that these proteins seem to have been available at or much before the Cambrian explosion strongly supports our concepts.

Mosaic scenario in genomes of today's organisms show evidence that genomes were assembled in the primordial pond

The scenario presented by the genomes of living creatures is what we would expect from the independent organization of genomes as proposed in the theory of independent birth of creatures. Order from chaos could arise from the random sequences in the primordial pond only by random processes.[62] What we see in the multitudes of genomes is evidence of what happened in the primordial soup's universal sequence pool during the independent organization of the genomes. In the assortment of genes into a genome from such a pool, it is possible to avoid a large amount of unwanted sequence material through molecular evolutionary processes. But at the same time, it is not possible to avoid it entirely, and a lot of junk DNA would be included in the genome. As described below, this is what we see. Similarly, in various genomes, different amounts of junk DNA should be included. This again is found to be true. During this process, in addition to good genes, some defective genes could be included, which are also found in the genomes. Unique genes should be included in various genomes, which we have elaborately shown to be true.[63]

Such different kinds of unique characteristics in each genome creates a mosaic molecular scenario. Only the mechanisms of the independent organization of genomes from the primordial universal sequence pool could be responsible for this. It could not arise due to any activities within the genomes of supposedly evolving organisms.

The mosaic molecular scenario — the presence of unique genes, identical genes, different patterns of repetitive sequences, unique transposons, highly varying C values — can be explained only on the mechanism of independent assortment of genomes in the primordial pond, as we described in Chapter 8. There must have been thousands of genetic mechanisms in the primordial soup at the time of the birth of creatures. When the normal mechanisms were organizing a genome into a viable seed cell and a creature, one or a few such mechanisms could get into the genome as a parasite. Organisms bear many parasites, good, bad, and neutral. The genome is no exception. Transposons could be one such parasite that got into the genomes from the primordial soup.

Chemical evolution must have led to many precellular molecular machineries in the primordial soup. At the time of the birth of many creatures simultaneously, there must have been many activities and changes by these molecular machineries that led to the organization of a genome. In doing so, many errors could have crept into the sequences. For example, DNA replication is a normal mechanism needed in cell division; DNA recombination is another normal mechanism. Such normal mechanisms can sometimes create errors and produce pseudogenes. Under these situations, we could expect the seed-cell genome to contain many nonfunctional gene duplicates with errors. On the other hand, after a genome becomes established in an independent creature, some genes in the genome could be duplicated and mutated to become defective genes, which we see as pseudogenes. But the pseudogene cannot lead to a new gene, as molecular evolutionists believe. Such errors in an established genome can only lead to genetic wastes, not new genes.

The junk or garbage or genetic waste cannot be "expelled" from the genome — just because the probability of the specific destruction of a pseudogene or an unwanted sequence in the genome of a multicellular organism is too small. So they persist, giving the false impression that they are on their way towards forming new genes. But truly, this does not mean that a pseudogene is *en route* to forming a new gene.

We do not yet completely understand the complexity of the genome — both its organization and function. However, we can be confident that our predictions are theoretically correct. There may be several aspects of the genome that we may not be able to explain based on the new theory at the moment. But the majority of the crucial and important aspects of the genome can be explained by the new theory. We can be sure that when the details of the genomes of many organisms become complete, we can more clearly explain the history of all of the details of the genomes based on the new theory.

MANY UNIQUE FEATURES EXIST IN THE GENOMES OF NUMEROUS CREATURES WITHOUT EVOLUTIONARY RELATIONSHIP WITH ANY OTHER CREATURES: A DEMONSTRATION THAT CREATURES MUST HAVE INDEPENDENTLY ORIGINATED

THE ORIGIN OF JUNK DNA AND THE C-VALUE PARADOX ARE SOLVED BY THE THEORY OF THE INDEPENDENT BIRTH OF ORGANISMS

The amount of DNA in the haploid[64] genome, such as that in the sperm nucleus, is called the C *value*. C denotes the "constancy" of the size of the haploid genome within any one species. While the C values of each organism remain constant, they vary widely among eukaryotes (see Table 9.1). We saw in Chapter 8 that there is considerable discrepancy in the DNA contents of the genomes of living organisms. There is a lack of correlation between C values and the presumed amount of genetic information required to construct an organism (called the C-value paradox[65]), and there is absolutely no evolutionary correlation of C values of different organisms. As one goes up the supposed evolutionary ladder of organisms, there is no correlation of the genome size with the supposed evolutionary sequence. If all reason tells us that such a scenario cannot arise by evolutionary descent with modification, how then did such a wide and haphazard variation of genome size come into existence?

While this is a great stumbling block for the evolutionary theory, it is clearly explained by the new theory. In the new theory, the genomes are assembled independently from the universal sequence pool of the primordial pond. The amount of DNA included as genomes in the different seed cells can vary widely because each genome includes a different number of genes, and, more importantly, a different amount of nongenic, random sequence. Organisms independently born in the primordial pond need not have similar amounts of DNA in their genomes. The intergenic "junk" DNA included in the different genomes could vary even far more significantly because of the manner in which each genome is separately assembled from the primordial genetic sequences. All these could contribute to the different C values (see also Chapter 8).

The junk DNA originated in the primordial pond but could not be eliminated from the genomes of multicellular organisms for several biological reasons. One such reason is that the seed cell and the cells of the organism were built with the size of the genome that was organized originally, and has become biochemically, physiologically and mechanistically tied to that

Taxon	Range in genome size (kb)	Ratio (highest/lowest)
Protists	23,500 – 686,000,000	29,191
Euglenozoa	98,000 – 2,350,000	24
Ciliophora	23,500 – 8,620,000	367
Sarcodina	35,300 – 686,000,000	19,433
Fungi	8,800 – 1,470,000	167
Animals	49,000 – 139,000,000	2,837
Sponges	49,000 – 53,900	1
Annelids	882,000 – 5,190,000	6
Mollusks	421,000 – 5,290,000	13
Crustaceans	686,000 – 22,100,000	32
Insects	98,000 – 7,350,000	75
Echinoderms	529,000 – 3,230,000	6
Agnathes	637,000 – 2,790,000	4
Sharks and rays	1,470,000 – 15,800,000	11
Bony fishes	382,000 – 139,000,000	364
Amphibians	931,000 – 84,300,000	91
Reptiles	1,230,000 – 5,340,000	4
Birds	1,670,000 – 2,250,000	1
Mammals	1,420,000 – 5,680,000	4
Plants	50,000 – 307,000,000	6,140
Algae	80,000 – 30,000,000	375
Pteridophytes	98,000 – 307,000,000	3,133
Gymnosperms	4,120,000 – 76,900,000	17
Angiosperms	50,000 – 125,000,000	2,500

TABLE 9.1. RANGE OF C VALUES IN VARIOUS EUKARYOTIC GROUPS OF ORGANISMS. The DNA content of the genomes of various organisms vary widely, and in an evolutionarily unrelatable manner (kb: kilobase). [Adapted from : Li, W-H. and Graur, D., 1991, *Fundamentals of Molecular Evolution*, Sinauer Associates, Sunderland, Massachusetts. With permission from Sinauer Associates.]

size of the genome. For instance, the size of the nucleus, and the size, organization, and number of chromosomes may be some of the possible reasons. Once this had happened, then the junk DNA became an integral part of the genome and the cell, and started to have size-related functions. Therefore, it persisted in the genome to this day.

THE EXISTENCE OF DIFFERENT TRANSPOSONS IN VARIOUS GENOMES CORROBORATES THE INDEPENDENT BIRTH OF CREATURES

In Chapter 4 we demonstrated that transposons are incapable of bringing about a new gene or a new developmental genetic pathway. There are many transposons in the living world, and the transposons in various organisms are different. How is such a scenario possible if the multitudes of organisms all had evolved from an original ancestor? In fact, for the evolutionary theory to be correct, the same transposons should be present in numerous creatures, or at least they should be the modified versions of a prototype, which would indicate an evolutionary relationship.

It is highly improbable that only the tips of the evolutionary branches have developed unique transposons and sequences associated with them. If it is true that creatures originated by evolution from one another, then it is also improbable that the ancestral organisms lacked the transposons and that only the present organisms had evolved them in a variable manner. If the ancestral organisms had some transposons then it is not possible for the descendent organisms to have abolished them and evolved new ones.

Although we are not making a systematic analysis of transposons in various organisms, we can see that distinct organisms have different transposons. Some organisms may not have a transposon, while others have many transposon systems. Bacteria have several transposon systems (*Tn3*, *Tn5*, *Tn10*, etc.). Yeast has a heterogeneous class of transposons called *Ty* elements. The fruit fly *Drosophila* seems to have several of its own transposons (*e.g.*, *P* elements, *I* elements, *copia*, and a copia-like element called *Gypsy*).[66] Maize has its own set of transposons (*e.g.*, *En/Spm*, and *Ac/Ds*).[67]

It appears that transposons are named generically. A system of genes for a few enzymes (usually 2-5) and some repetitive sequences (usually one or two) constitute a transposon system that could transpose itself and other sequences around a genome. Because it involves a few different enzymes for recognizing a repetitive DNA sequence, and cutting and recombining DNA, one can theoretically conceive of many distinct, unrelated transposon systems. This is what we find in living organisms.

As we discussed in Chapter 4, a transposon is a fairly simple system requiring only two or three enzymes and one or two short sequence repeats. It is quite probable for the genes of these few enzymes to have occurred in the vast UGP. Therefore it should have been possible to assemble such a fairly simple system from the primordial pond. In fact, when the gene sets for complex cells and organisms were randomly assembled, it is highly probable for a variety of distinct transposons to have also assembled separately into different genomes. We can be sure that the scenario of distinct transposons in the genomes of various organisms is the result of their differential inclusion in the different genomes while they were being assembled in the primordial pond.

It can be seen that the transposon is just a "selfish" genetic system that simply moves around without the real effect on evolution that evolutionary geneticists propose they have. It is well known in the field of virology that many virus genomes contain similar terminal repeats. A virus genome, deleted for all but the terminal repeats can be replicated independently if the required replicating enzymes are synthesized by another virus in which only the terminal repeats have been removed.[68] The enzymes that replicate the complete virus genome also replicate just the two terminal repeats joined end-to-end because the repeats contain the necessary structure for binding the replicating enzyme and for the replication mechanism to operate. The terminal repeats by themselves are packaged into virus particles known as "defective" particles. These particles are a waste produced during the production of the normal virus, just as wastes in any chemical or other processing industry. There can be many such instances of "genetic waste" in the biological world. One can clearly see the similarity between the virus system and the transposon *Ac/Ds* system that we discussed in Chapter 4.

The duplication of an animal genome involves many replication, transcription and repair enzyme systems. In the origin of such a large genome from the primordial soup, it is reasonable to expect that some parasitic mechanisms involving one or a few such enzymes will creep into the genome, as long as it is harmless to the propagation of the genome. It may cause cancer, congenital abnormalities, or genetic diseases, but as long as it is not detrimental to the genome as a whole, it will persist.

Therefore we can see that selfish genetic systems such as the transposon could have originated from the primordial soup itself. There could have existed many gene systems that could take away sequences from one DNA and insert into another. A transposon would persist as a parasite, replicating itself along with the large genome, and carrying out whatever functions its sequence specifies. It may be that one or more of its genes are used in the normal functioning of the genome. The new theory does not preclude a normal gene to be part of a transposable system.[69] It may be that it was impossible to avoid the inclusion of such parasitic systems while organizing such a large genome with several larger and more complex genetic replication and recombination mechanisms in the primordial pond. In other words, the parasitic inclusion of such simpler mechanisms in various creatures may be inevitable.

Different patterns of repetitive sequences in various genomes: A support to the theory of the independent birth of creatures

The eukaryotic genome is characterized by the repetition of certain sequences. Some nucleotide sequences of various lengths and compositions occur several times in a genome either in tandem or in a dispersed fashion. These are called *repetitive sequences*. DNA sequences that are not repeated in the genome are termed *single-copy* or *unique* sequences. The proportion of the genome constituted by repetitive sequences varies widely among organism. Greater than 50% of the genome in many cases can be repetitive DNA.

Why should there be such repetitive sequences in the genome of a eukaryote? Is there a function for such repetitive sequences? Research over the last two decades could not reveal any function. Regardless of function, if the genome of Darwin's original ancestor had some repetitive sequences in it, the genomes of organisms that descended from the original ancestor should exhibit the same or at least a very similar pattern of repetitive sequences. It appears, however, that the genomes of different creatures exhibit distinct patterns of sequence repetition.[70] Such a pattern is unlikely to result through descent with modification from an ancestral genome.

Such a varying pattern of sequence repetition can originate by the independent assembly of genomes in the primordial pond. While assembling genes into different seed cells, the different genomes could have developed different patterns of repetitive sequences. The repetitive sequences may have been needed for a smooth organization of the chromosomes or some such molecular or cellular physiological process. It is possible that only those seed cells which could achieve sequence repetition could become viable. In this process, different seed cells could have formed different patterns of repetitive sequences. Alternatively, if a genome were assembled without repetitive sequences, and included a transposon parasitic system, the transposon could generate the repetitive sequences in the genome. And different transposons in different genomes would lead to distinct patterns of repetitive sequences.

If a unique mechanism of repeating the sequences existed in the primordial pond, and if the repetitive sequences of the USP were unique, then it would have been possible that all the genomes assembling in it could have made use of these common repetitive sequences. However, from what we see in the genomes of living creatures it appears that such a thing did not happen, and it seems quite likely that each independently-assembling genome had independently developed a unique pattern of sequence repetition at the time of assembly — perhaps for some biological reasons such as chromosomal stabilization with DNA-binding proteins. What is discernible here is that while it is improbable to arrive at the various patterns in various organisms from a unique pattern of sequence repetition in the genome of one assumed ancestral organism (or in which no sequence repetition was present), it is extremely probable for each independently-assembling genome to arrive at a distinct pattern of sequence repetition.

It is absolutely clear from all our analyses and discussions that the genomes of various organisms were independently organized in the common pool of genes in the primordial pond, and therefore the various organisms had originated independently as separate units of living entities that were never organismally and evolutionarily connected to each other. It is of utmost importance for us to realize this fact, for its implications and repercussions are enormous.

Conclusion

Although Darwin's theory seems to be validated by some molecular aspects of cells and organisms, many crucial details of molecules and cells fundamentally contradict evolution theory. Meanwhile, our new theory of the independent birth of organisms predicts a certain molecular picture in creatures living today, and almost all of these predictions are verified in genomes and cells. Our new theory easily accommodates all of what we know at the molecular level: both the evidence that supports Darwin, and the evidence that contradicts him.

The independent assembly of genomes from a common pool of genes in the primordial pond allows for the same genes to be present in different organisms, which we know to be the case. Genes that are unique to only one or a few organisms, however, should also be present in numerous creatures, which we also know to be the case. If such unique genes cannot be evolved within the genomes of any organisms, even in trillions of years, as we learned in Chapters 3 and 4, how else can they originate except by independent assembly into different genomes? Moreover, how can the genomes of various organisms contain enormous amounts of "junk" DNA, and how can the quantity of this useless DNA vary so wildly from one organism to another, unless a varying quantity of meaningless primordial DNA was independently included in each genome from the beginning? In addition, the total lengths of the genomes of various organisms are equally variable and uncorrelatable.

Remember that the commonness of the genetic code and the presence of sets of the same genes in various organisms is due to the assembly of all the genomes from the same pool of genes, and the same genetic machineries, genetic codes and basic genetic principles. There simply is no other reasonable explanation for the occurrence of common genes in different genomes that also contain absolutely unique genes, a mosaic scenario that could not have occurred by organismal evolution.

Biochemical and cellular similarities also come from the common pool of the primordial pond. It is the genes that are fundamentally responsible for these biochemical and cellular characteristics. When the genes of various independently forming genomes can come from a common pool of genes, the biochemistry and cellular structures of all distinctly originating organisms can also come from the same common pool. Thus, multitudes of organisms can originate each independently in the primordial pond, yet with the same basic biochemistry and cellular structures and functions.

Under our new theory, families of genes — genes that encode functionally similar proteins and therefore are similar in structure and possibly also in sequence — should be found even in the unicellular eukaryote or prokaryote, whatever the simplest living entity may be. This is because these genes encode enzymes that catalyze metabolic pathways — fundamentally and minimally required for the simplest living entity — and because these pathways should necessarily include numerous similar biochemical reactions. How can the simplest living entity come into existence without the set of genes responsible for the metabolism of at least one sugar for energy, and the set that facilitates the metabolism of amino acids, which are the basic components of proteins? Even this minimal set of genes fundamentally required for the existence of the simplest living entity can be categorized into many families of genes. Furthermore, such genes, as we have clearly demonstrated by computer simulation experiments, should have been abundantly present in the primordial genetic sequences, and should have been included in the independent genomes. Remember that the concept of the gene family — that a particular family of genes has evolved through organismal evolution from one single ancestral gene — has been articulated only because some genes encode functionally similar proteins that also have structural and some sequence similarity. But it is incorrect to say that this phenomenon is the result of evolution.

We have demonstrated that genes that are unique in most regions, but similar only in regions wherein they specify subfunctions such as binding a cofactor, could occur independently in random sequences in the primordial pond and could be separately included in the genomes of distinct creatures. Thus, it is unnecessary and incorrect to say that such distinct genes with partial similarity can evolve within the genomes of organisms from an ancestral gene and be responsible for the evolution of new organisms.

While the presence of similar or essentially the same genes in various organisms is seemingly explained by Darwin's theory, the presence of unique genes and unique DG pathways is absolutely unexplainable by evolution. All these details are, however, explained by the theory of the independent birth of organisms. Darwin's theory is supported by a statistical argument of variations in basically the same genes, without fundamentally realizing that the numerous unique genes present in many organisms cannot be evolved from any preexisting genes by any known mechanisms of genetic mutation.

The presence of absolutely unique genes in numerous creatures is something known only in recent years. Evolutionists almost always assume that unique genes could be evolved from preexisting genes, however distinct they may be. At least some evolutionists (e.g., Russel Doolittle), however, are puzzled by the complete absence of these genes in presumed evolutionary ancestors. It is particularly puzzling without the new concept that such genes could simply occur in the primordial pond's genetic sequences, and that these unique genes could be assorted into distinct genomes independently. Computer simulations can now demonstrate that genes for almost any given biochemical function could have occurred in the primordial pond's genetic sequences. We can also demonstrate unequivocally that no mechanism of genetic mutation and rearrangement could have been responsible for the occurrence of unique genes in numerous creatures. It's no wonder, then, that the evolutionists have been debating among themselves about changes in genes and genomes, and the evolution of creatures either gradually or rapidly, and so forth. They are simply asking the wrong questions.

The theory of evolution was formulated before science was able to examine and consider molecular details. When the first few molecular details became known, most observers regarded the new evidence as confirmation of Darwin's basic theory, simply because some of the new evidence (such as similar genes) coexisted with it. But our ever-expanding knowledge of molecular genetics has also posed mountains of new problems for evolution theory. Indeed, the more we learn about genetics at the molecular level, the more contradictions we find between molecular realities and evolution theory. But in the absence of a fundamentally different competing theory, scientists have simply tried to put aside such problems, however substantial they are.

We can see that no theory of evolution can explain the presence of unique genes in multitudes of organisms. No theory of evolution can account for the junk DNA in all multicellular organisms. No theory of evolution can account for the vast variations in the quantity of junk DNA among almost all organisms. Evolution cannot explain the origin of unique DG pathways in the genomes of unique organisms by any or all kinds of mutations. It cannot give any reason for the constancy of genomes if evolution is an ongoing mechanism. It cannot show why only neutral variations of genes with respect to their function exist in the genomes of all organisms. It cannot show how the various patterns of repetitive sequences as well as distinct transposons can occur in the genomes of various organisms.

But all of these problems are solved by our new theory of the independent birth of organisms. Our detailed analyses and considerations in these first nine chapters all point to the same conclusion: that the independent birth of organisms is the only reasonable way to explain the mosaic molecular scenario in the living world. The theory unifies all living organisms on the basis of the common universal gene pool of the primordial pond, from which all organisms were independently born by the assembly of common genes, similar genes and unique genes from the pool, thereby producing a wide variety of organisms with similarities as well as differences at the genetic, genome and organismal levels.

We have been in a rut, spinning our wheels, for the past 130 years. Darwin's theory of evolution has been so universally accepted that we have begun all subsequent inquiries with the uncritical assumption that evolution is true, and with the primary objective of reconciling new evidence to the old theory. Within this context, it was only reasonable to interpret the strong similarities of basic genetic machineries across different organisms as evidence of evolution. But we must remember that evolution remains, after all these years, just a theory — unproven and, as we are now finding, unprovable.

The theory of evolution only seemed to explain a few of the molecular details of life on earth, but no more.[71] The new theory of the independent birth of organisms better explains — by far — even the most fundamental molecular and genomic scenarios. Today we are blessed with sufficient information and the technologies to be able to test this new theory — to subject it to the deepest level of scrutiny by analyzing the genome data. Our analyses and arguments so far give us the confidence that the new theory will endure any level of scrutiny and analysis.

10

MOST ORGANISMS ARE DISTINCT AND EVOLUTIONARILY UNRELATABLE

Having shown in the previous chapter that the biomolecular scenario in creatures living on earth indeed corroborates the new theory, and contradicts Darwin's theory, we are now challenged to demonstrate that the observed characteristics of creatures fit only the theory of the independent birth of organisms, and not the theory of evolution. Independent birth predicts that creatures will be distinct and evolutionarily unrelatable. The new theory also predicts a random distribution of structural and functional characteristics among organisms. That is, there should be no hierarchical gradation or relationship of structural characteristics among animals, as is predicted in evolutionary theory.

In this chapter we will find that our new theory, when attempting to explain the scenario of life, is virtually immune to the problems faced by the theory of evolution. Our new theory requires that organisms originating directly from primordial genomes will be unique and unrelatable by organismal evolution, and we can confirm this prediction by analyzing the creatures living today. An inventory of the creatures around us today is also consistent with our notion that many new genomes could have been constructed by recombining or mixing the first-formed unique genomes and the primordial pond's gene-pool, and that these new genomes could have produced creatures with mixed resemblances to the first creatures.

Evolutionary theory misinterprets the relationships among strikingly similar animals (for instance, the many similar species of a snail), which are actually varieties of the same distinct organism. Evolutionists have believed that the mechanisms that produce minor variations among similar organisms are responsible for all of the diversity of life on earth. Snails and crabs, for example, are presumed to share a single common ancestor from which snails and crabs divergently evolved. But we have already seen that all genomes are closed to evolutionary changes of a scale that could produce two such different creatures. Only minor variations are possible by that mechanism. Evolutionists may believe that millipedes evolved from a shrimp-like crustacean.[1] But while we know of many similar species of shrimp, and many similar species of millipedes, the shrimp and millipede are separate organisms, unrelated by evolution. The shrimp and millipede therefore must have originated independently.

This invalid extrapolation, integral to evolution theory, is the primary source of a hoard of problems, notably in explaining phenomena such as the evolutionary origin of higher taxa. This again is due to the lack of any alternative theory that scientifically explains the origins of organisms without requiring the evolution of all organisms from a single original ancestor.

The fossil record reveals that the major taxonomic groupings into which we now group all creatures, living and extinct — all the phyla, classes, and orders — originated all at once, when multicellular life first appeared. If we take a fresh look at the scenario of life on earth, forgetting for a moment the traditional view of evolution with its assumed hierarchical groups, it becomes clear that multitudes of distinct creatures originated on earth simultaneously, and only a few slight variations of each original organism appeared gradually over the intervening millennia. It means that these original creatures have been living forever essentially unchanged.

If evolution's doctrine of connectability were not so deeply ingrained in our thinking, we would not likely see the "relatedness" among all species that we now mistakenly take for granted. Our new theory does not dispute the occasional connections among different families within an order, or the more common connections among genera within a family or species within a genus. But even a casual look at the variation of life on earth reveals clear boundaries between distinct organisms, and our more careful investigations have shown that creatures belonging to different phyla — or to different classes within a phylum, or to different orders within a class — are unconnectable.

In this chapter, we shall take another look at life on earth, from a new point of view that discards the preconceptions of evolution theory. What we will find is an abundance of mostly unique creatures — without any evolutionary connection — as well as creatures with a random mixture of unique characteristics.

WITHIN THE SAME CONSTRAINING DOMAIN OF PHYSICAL AND CHEMICAL FORCES ON EARTH, NUMEROUS LIFE FORMS WERE POSSIBLE FROM MULTIPLE, INDEPENDENTLY-ASSEMBLED GENOMES

Can all the creatures on earth, whether invertebrates or vertebrates, plants or animals, single-celled or multicelled, all have similar basic units of cells, even if numerous creatures were independently born? We answered this question showing how numerous creatures could originate independently based on the same biochemical and molecular biological principles, and from a common pool of genes in the same primordial pond. We should constantly remind ourselves that if we accept that fundamentally distinct creatures could be produced in the primordial pond independently, then slightly changed creatures could also be produced in the primordial pond by mechanisms of genome mixing and genome alteration or restructuring.

Because genomes were assembled randomly from the universal gene pool, with some unique genes, but more importantly, unique developmental genetic pathways, there is no continuity among the various genomes assembled in a primordial pond — *i.e.*, each independently-assembled genome is a distinct entity giving rise to a creature that is also distinct and unrelated to others. Therefore there would be gaps among independently-born organisms. However, each organism gave rise to many varieties or similar species based on individual variations through natural selection, and sometimes by mutation. Once we understand this main principle, then it is easy to extend this theme and to show that the independently assembled genomes mixed among themselves to produce new mixed genomes in the primordial pond. This led to independent creatures with mixed structural and functional characteristics.

At a gross level, an overall similarity of limbs and appendages were imposed in distinct, independently-born animals by gravity and other physical forces at the time of the birth of creatures. For instance, all creatures on earth have to either walk on land, swim in water, or fly in air — because land, water, and air are the only three media available on the surface of the earth. Therefore, any organism that walks on land — whether it is a snail or crab, a millipede or a cat — has to have some form of legs. Any organism that swims in water has to have some appendages that would specifically aid in swimming — fins, paddles or flippers. And any creature that flies necessarily has to have some form of wing — as in a dragonfly or a hummingbird, a mosquito or a bat. In a similar manner, the limited set of chemical forces imposed a certain basic functional similarity in metabolisms, cells, tissues, and organs in every creature born on earth. But within this domain of limited or constraining physical and chemical forces, the mechanism of the independent assembly of genomes (and independent birth of organisms) would lead to numerous distinct creatures with organs, limbs, appendages, systems, and techniques unique to each of them, and in a sense peculiar. However, they may exhibit varying degrees of similarity due to overlapping sets of genes and developmental genetic pathways which could be included in the different genomes from the same primordial pond.

In essence, if creatures were actually born independent of each other in a common primordial pond, then they should have an underlying commonality of sets of genes, biochemistry, and cellular structure, and an overall *functional* similarity of organs, limbs, and appendages. But the creatures should be widely different from each other in their structures and functions and be extremely unique in a manner unconnectable by evolution.[2] In fact, when one peruses the scenario of living organisms, one can certainly see that life on earth exemplifies what is predicted in the new theory, and disputes what is predicted in evolutionary theory.

We shall primarily deal with animals here and not plants. However, our concepts and discussions on animals will apply equally to plants also.

Unique structures, organs, and limbs in organisms exemplify the new theory

An important principle of the new theory is that the genomes of organisms arose from random combinations of genes from the universal gene pool that would lead to a living organism as long as it was viable and capable of reproduction. It is immaterial what the organism looked like (within physical limitations of size and shape). It does not matter what organs and limbs the creature had. The basic requirement is that a mobile organism (animal) should have organs that would make its body viable (such as for digesting food, transporting biochemicals throughout the body, excretion of the unwanted materials, respiration, and reproduction) and have limbs and appendages suitable for finding its food, protecting itself from predators, and for finding its mate. The presence of a large number of widely varying combinations of organs and limbs in living creatures, which precisely fit these requirements, would show that the new theory must be correct.

We shall first briefly address the uniqueness of invertebrate creatures illustrating the concepts of the new theory. We shall then describe the mixed structural characteristics of many organisms showing that the other predictions of the new theory are borne out. We shall then discuss the uniqueness in the vertebrate organisms. Vertebrates are grouped into one half of one phylum, whereas all the organisms living on earth today are grouped under 35 distinct phyla. Extinct creatures are grouped under another 30 or so distinct phyla. The reason for classifying the invertebrate creatures into so many phyla and numerous classes and orders is that each distinct creature is highly unique. Even evolutionists agree that the structures of these organisms classified into these higher taxa are so unique and distinct that they are unable to connect them by any evolutionary means. This problem is traditionally known in the field of evolutionary biology as the problem of the "origin of higher taxa." The reason for classifying all vertebrates as part of one phylum is that they all have bone as the underlying common tissue or organ. Even with this underlying commonality, we shall prove that there are a great number of unique vertebrate organisms, which are evolutionarily unconnectable among each other and which could have only originated based on the principles of independent birth of organisms.

It should be noted that some gene commonalties are found to exist between all organisms — in organisms as widely differing as bacteria, mammals, and plants. However, such gene commonalties are explainable by the independent birth of organisms in the common primordial pond, while the unique genes and structures of various creatures are also explainable by the independent birth of organisms. In contrast, while the evolutionary theory can explain the gene commonalities among organisms, it cannot explain the presence of unique and unrelatable genes in organisms. This is a crucial distinction.

It is clear that beyond the similarity of the organisms grouped within the small sets such as the genus and family, organisms belonging to the various higher groupings cannot be connected evolutionarily. In the new theory, a set of similar species may represent the varieties of an independent, distinct creature. Beyond this, there is no evolutionary connection among organisms. On the other hand, a set of similar creatures may represent the ones produced from the same seed cell by its modification in the primordial pond.

Let us discuss a few examples of the unique characteristics of animals. These examples exemplify the typical distribution of organisms throughout the living world. If we were to cover all the unique organisms that are known, it would take a whole book by itself. But if we were to read a zoology book such as that of Mitchell and associates,[3] or *Life on Earth* by David

Attenborough,[4] with our new outlook in mind, then we would clearly see that all the details of life on earth fit the new theory. We shall touch here on some representative examples. One can, of course, find many more from the world of organisms.

UNIQUENESS AMONG INVERTEBRATES

One might say that the millipede evolved from a worm like the earthworm, because they look somewhat similar. This is improbable because millipedes have legs and antennae and other structures that the earthworm lacks. On the other hand, some evolutionists say that the millipede evolved from a crustacean when it climbed onto the land from the water. But the uniqueness of millipedes clearly show that they are independent creatures and could never have evolved their unique structures from a crustacean. Crustaceans breathe through feathery gills that grow out from their legs.[5] In contrast, millipedes use a system of breathing tubes, the trachea. From openings in the shell, the tubes branch internally into a fine network that delivers air to all the tissues and cells. The reproductive systems also show that the crustaceans and millipedes have absolutely unique systems. The crustaceans rely on external fertilization in water. In millipedes the system is totally different — by sexual contact the male transfers sperm to the female, using a form of internal fertilization. It is clear that when one tries to explain the origin of the organisms based on the theory of evolution, one almost always tries to find a most reasonable ancestral creature to a given creature, and tries to connect them through evolution. This approach is superficially appealing, but in reality it simply does not work.

Starting from the simplest of the multicellular creatures, which are supposed to be primitive, evolutionists have found it difficult and indeed impossible to connect various invertebrate creatures. Take the case of the evolution of the simplest multicellular organism from a unicellular eukaryote. It is accepted by adherents of evolutionary theory that the origin of multicellular animal life from single-celled ancestors is the most enigmatic of all phylogenetic problems.[6]

When we peruse the literature and look at the phylogenetic separations of creatures and their structures and functions in biology textbooks, the distinctness of these creatures become plainly evident. In fact, we do not even have to take much effort to prove our concepts. Biologists, zoologists, and evolutionists themselves struggle to connect the creatures based on evolution. We need only to simply point these out to show that they fit very well with our concepts of distinctness and uniqueness. It becomes plainly clear in the words of biologists and zoologists that there is only much confusion in trying to collect and connect various creatures into evolutionary groupings. These groupings, as we can see, are truly artificial. However, in these discussions we should not forget that we are also dealing with many creatures that are varieties (often termed species) of a distinct independently-born creature. These are perhaps now grouped under a genus or family. Only these are organically related. But when we go beyond this level, there is no relationship. It is because of this confusion that the problem of the inability to connect creatures by evolutionary mechanisms arises.

Figure 10.1 is an evolutionary tree showing how all creatures have supposedly evolved from an original ancestor into the various taxonomic groupings. According to this scheme, the higher taxonomic categories should all be connectable. Figure 10.2 shows that the organismal relationships purported by such an evolutionary tree do not actually exist. Figure 10.3 shows what kind of scheme is predicted by the new theory, and how it precisely matches what we see in nature. If we dismiss the arbitrary, artificial groupings of creatures into taxonomic categories such as the phylum, class, order, and so on, all we are left with in reality is a random collection of numerous distinct unrelated creatures.

Even when we consider the numerous unicellular eukaryotes, we have evidence that they must have originated independently in the primordial pond, although they could have used common genes and common genetic machineries from the common pool of genes. This is proved from the fact that most of the unicellular eukaryotes are very distinct from each other and show no evolutionary connection. The 35,000 or so unicellular living forms have been classified into four or five phyla and into numerous classes and orders. There are marked differences among the unicellular eukaryotes belonging to different taxa, and the evolutionists find it hard to connect them.[7] They are so distinct that evolutionists think that the single-celled eukaryotic organisms had lived for over 1000 million years separately, and back at one billion years ago, they could have originated separately from the prokaryotic stock. As we demonstrated in Chapter 7, prokaryotic ⟶ eukaryotic evolution is highly improbable.

A single marine species, *Trichoplax adhaerens*, has been classified into a separate phylum (Placozoa) because of its unique body organization. This unusual creature has only two distinct cell regions. It moves its flat body with flagella on its surface, and reproduces by budding or by splitting in two. This creature is unconnectable to any other known creature by means of evolution, and that is why it is classified into a separate phylum.

The dicyemids have a unique structure and reproductive systems and are classified into the phylum Mesozoa. Another distinct group of invertebrates, the orthonectids, has also been classified into this phylum. Similarly,

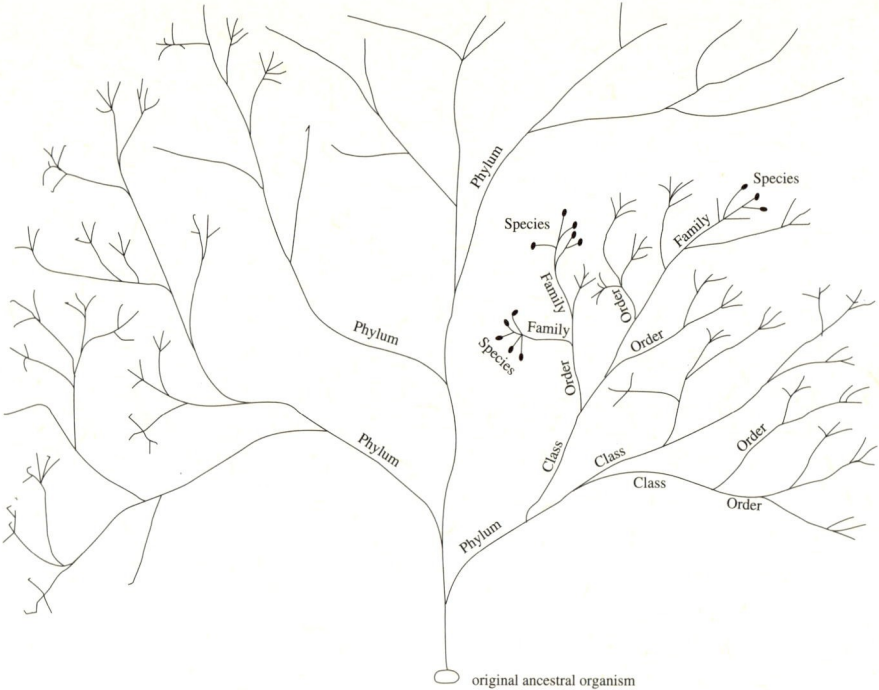

FIGURE 10.1. WHAT DOES AN ASSUMED EVOLUTIONARY TREE PREDICT? The original ancestral organism gave rise to a few related creatures, each of which in turn gave rise to a few more new related creatures. This process led, in geological time, to all the organisms on Earth, living and extinct. This would require — even if all the transitional forms had become extinct without leaving fossils — that the evolutionary lineages could be traceable. It means that various families should be connected within an order, various orders within a class, various classes within a phylum, and various phyla should be connectable through an original, even hypothetical ancestor.

MOST ORGANISMS ARE DISTINCT AND EVOLUTIONARILY UNRELATABLE 461

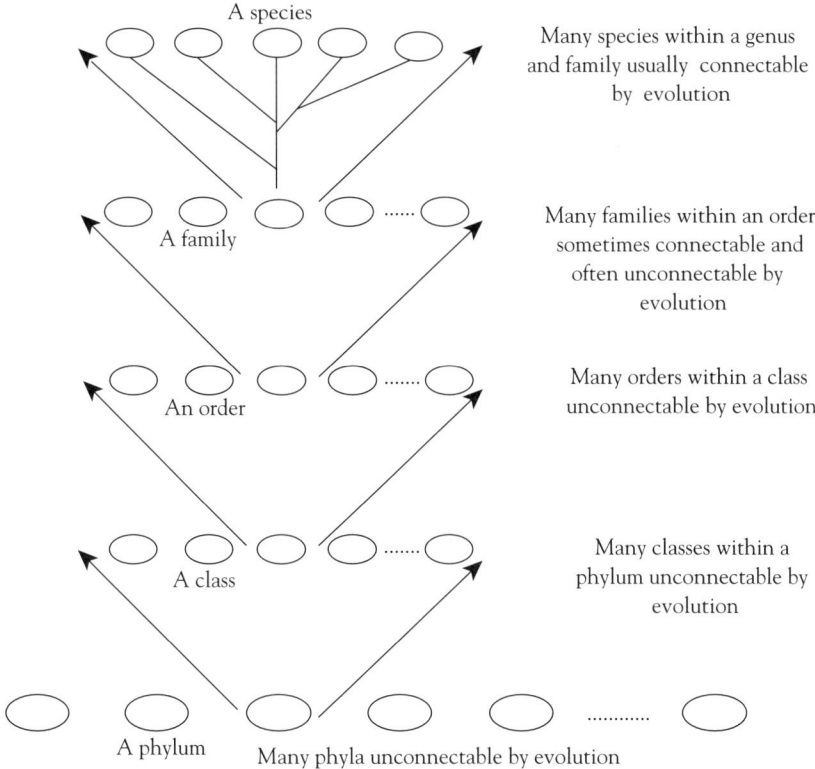

FIGURE 10.2. WHAT ARE THE ACTUAL OBSERVED RELATIONSHIPS AMONG ORGANISMS? Organisms are classified into approximately 35 living and 25 extinct phyla, based on assumed evolutionary connections. However, in reality the different phyla are unconnectable. The different classes within a phylum and orders within a class are also unconnectable. Sometimes the families within an order and often genera within a family and species within a genus are connectable by evolution. Thus, if we dismiss these purely artificial groupings of organisms, the "evolutionary tree" becomes nothing more than a random collection of numerous distinct creatures, with each giving rise to only a few similar species of its own.

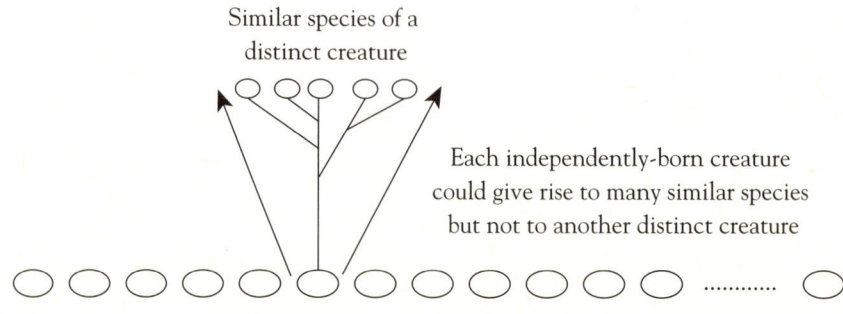

FIGURE 10.3. THE SCENARIO OF ORGANISMAL CHANGE: PREDICTED BY THE NEW THEORY. The new theory predicts that numerous distinct creatures were born independently in the primordial pond and live forever unchanged — except that each distinct creature could give rise to many of its own very similar species. It requires that all organisms, living and extinct, should present a random collection of distinct unrelated creatures, and that each distinct creature should be represented by many similar species. When we objectively analyze the scenario of life on Earth, we see that it actually presents exactly the same picture as predicted by the new theory.

several classes of sponges, *e.g.*, calcarea, hyalospongiae, sclerospongiae and demospongiae, are quite unique and are classified into a separate phylum called Porifera.

When we start looking at the more complex forms of invertebrates, we find their uniqueness to be even more striking. This can be seen throughout the invertebrates. When we discuss the uniqueness of these creatures, we shall describe the major groupings such as the phyla or the classes to illustrate our concepts of uniqueness and unconnectability to other organisms. However, we should note that even when we go further down into the orders and families, the details of the creatures will reveal that this concept is fundamentally true almost down to the genus and sometimes species level.

The creatures classified into the phylum Cnidarians, containing the classes hydrozoa (*e.g.*, obelia, hydra, and marine jellyfish), scyphozoa (sea nettles), anthozoa (sea anemones, organ-pipe coral, sea pen) are all indeed unique in their structures and functions.[8] These are so distinct, that even some evolutionists think that these could have evolved from unicellular eukaryotes independently of other invertebrate animals.

The sea gooseberry classified under the phylum Ctenophora is a unique creature. It has rows of ciliated plates that provide propulsion in the water. There is a special balancing organ that coordinates the beat of the cilia. It has tentacles with adhesive cells (collocytes) for capturing food. The animal feeds by forming its tentacles into a sticky screen that filters small prey from the water. There is no specialized circulation or breathing system.[9]

Ctenophorans have unique attributes that distinguish them from Cnidarians. They have a mouth, pharynx, stomach, digestive canals, and anal pores. Their collocytes are developmentally and anatomically distinct from similarly-functioning cnidarian cnidocytes. Also their life cycles at the larval stage are completely different.

These structures are highly unique and they should have corresponding genes and unique developmental genetic pathways in the creature's genome. When we analyze these unique structures — especially asking questions about the unique DG pathways and their unique genes — one can clearly see that they cannot evolve from creatures that lack these structures.

Worm-like "bilateral" animals (whose bodies consist of two mirror-image halves, as in the human) grouped under three distinct phyla are called acoelomates, because they lack a body cavity. These are: Platyhelminthes (flatworms), Gnathostomulida (gnathostomulids), and Nemertea (ribbon worms). An examination of the creatures grouped under these large categories would show that most of these are unique in their structure and function. Flatworms are themselves unique. Take for example the monogenetic flukes which have a muscular organ called the *opisthaptor*, which they use to attach to their hosts. Tapeworms have a unique front attachment organ, the scolex, attached to segments called proglottids. These creatures are again unique. The creatures classified under the phylum Gnathostomulida are also unique, which must make zoologists feel uncomfortable. Authors of the book *Zoology* say the following:[10]

> The phylogenetic relationships of gnathostomulids are enigmatic.

The ribbon worms are also unique. They have a distinctive long hollow tube, the proboscis, held in a fluid-filled cavity. When muscles contract around the cavity, the proboscis is hydraulically thrust out. Often the proboscis has sharp stylets for hunting.

The pseudocoelomates are a grouping of nine completely distinct phyla. Zoologists themselves say that these may not be related to one another although they are grouped together.[11] This is a common situation, whereby animals are grouped based on traits that are not evolutionarily relatable traits, a problem that zoologists and evolutionists themselves point out:[12]

> Nine phyla are grouped together as the pseudocoelomates, a diverse assemblage of aquatic eumetazoans. ... Phylogenetic affinities among the nine phyla and with other phyla are obscure.

When we study the creatures of these phyla, it becomes clear that each creature is unique and contains structures and functions that are totally distinct. For instance, the body wall of a nematode (phylum nematoda) consists of a unique multilayered cuticle, an underlying epidermis, and longitudinal muscles.[13] The male reproductive system include testes, sperm ducts, and seminal vesicles, which open into the rear end of the intestine. The male's back end has specialized organs that hold the female genital pore open during copulation. The female has ovaries, oviducts, and uteri, as well as a vagina that forms the genital pore. These creatures are highly unique and cannot be connected to any other creatures by any evolutionary mechanism. Their unconnectability to other creatures is described in the words of Mitchell et al.:[14]

> The affinities of nematodes with other phyla are vague. There is no other living group, including the other pseudocoelomate phyla, believed to be closely related to these worms.

The creatures such as the hairworms (phylum Nematomorpha) and spiny-headed worms called acanthocephala (phylum Acanthocephala) have unique body structures and life styles. Rotifers (phylum Rotifera) have several distinctive features.[15] For instance, most rotifers have a crown of rotating cilia for feeding and swimming. Nearby is a jaw-like structure called the mastax. Some rotifers have a foot with movable toes that secrete an adhesive. All these creatures are doubtlessly unique and distinct from each other.

Zoologists are uncertain about the phylogenetic origin of the rotifers. The state of affairs is clear when we read Mitchell:[16]

> The affinities of rotifers with other phyla are obscure, but these pseudocoelomates have several features in common with certain acoelomates. The rotifer mastax resembles the jaws of gnathostomulids, but it is not known whether these structures are homologous. ... Rotifers probably had their origins in the earliest acoelomates, or acoelomates and rotifers may have had a common ancestor among the earliest bilateral metazoans.

The creatures of the phyla Gastrotricha and Kinorhyncha are unique in their body structures. For instance, Kinorhynchs are unique among pseudocoelomates by having body segmentation, both externally and internally. If we look at their other structures, we can see clearly that these creatures are indeed unique. Zoologists feel that[17]

> ... without more knowledge of kinorhynchs, it is difficult to speculate about their phylogeny.

Some marine sand-dwelling creatures discovered in 1983 could not be placed in any known phylum because they were not found to be similar to any known organism. These creatures were classified into a new phylum called Loriciferans. They are named for the *lorica*, a tough, armored case that covers them. The discovery of these distinct creatures adds another excellent example to the uniqueness of creatures on earth and their unconnectability to other organisms.

Other examples of creatures with absolute structural uniqueness are found in the phylum Priapulida. These creatures have a mouth at the end of a proboscis that is surrounded by prey-catching spines. A muscular pharynx with curved teeth gathers food. A thin flexible shell covers its body and is molted periodically. There are no specialized organs for circulation or breathing. Some species have amoeboid, oxygen-carrying cells that contain hemerythrin. The creature has a nerve ring around the pharynx and a single nerve cord similar to those in annelids.[18] Again, there is no connection of these creatures to any others on earth. This fact is simply exemplified by what zoologists Mitchell and associates have to say:[19]

> There are more questions than answers about priapulids. Debate over their phylogenetic position has waxed and waned since their discovery during the Linnaean era, and they have been classified as coelomates some of the time and as pseudocoelomates at other times.

How various creatures totally unrelated by evolution are grouped into a category such as the pseudocoelomates is exemplified by looking at the structures of the creatures belonging to the phylum Entoprocta. Their structures are highly unique. Their uniqueness can be understood just by reading what the zoologists have to say about their phylogenetic origins:[20]

> The phylogenetic position of the entoprocts is controversial. We have included them in this chapter mainly because they have a pseudocoel.

Again, the fact that many distinct creatures are lumped together into one phylum can also be seen by analyzing the various creatures in one phylum. While creatures grouped into different phyla are totally distinct and unrelatable by evolution, we have been seeing that creatures grouped into classes and orders within a phylum are also distinct and truly not related by evolution, except, in general, below the level of the family and genus. One of the best examples of such a phenomenon can be seen in the phylum Mollusca. Based on superficial similarities, animals as widely different as neoplina, sea slugs (nudibranchs), sea butterflies (also called pteropods), sea hares (aplysia), terrestrial snails, land slugs, squids, nautiluses, cuttlefishes, octopuses, bivalves, and cones have been grouped under the single phylum of Mollusca (although in different classes and orders). A close look at these creatures shows that they possess distinct structures that are evolutionarily unconnectable.

For instance, animals of the class Monoplacophora have changed little in body structure for hundreds of millions of years. Monoplacophorans have a single shell, and a muscular foot. Unlike other molluscs, several body parts are repeated. The foot stems from the bottom of the shell and has eight pairs of retractor muscles. The creature has up to six gill pairs, two pairs of atria in the heart, two pairs of gonads, and six pairs of excretory organs. The foot has ten pairs of nerves.

The aplacophorans (in a separate class called Aplacophora) are worm-like, marine bottom dwellers. The polyplacophora is a class composed of creatures called chitons. These include fossil creatures dating from the Cambrian (over 500 million years ago). Chitons have a unique shell composed of eight overlapping plates.

Snails, slugs, limpets, abalones and many other creatures are in the class Gastropoda within the phylum Mollusca. Gastropods are classified into three subgroups based on the arrangement of their breathing organs. Most *prosobranchs* have gills in the front, and some have lungs. Some are herbivorous (abalones and limpets) and some carnivorous (cowries and whelks). The *opisthobranchs*, another subgroup of gastropods, have gills at the rear. Opisthobranchs include bubble shells, sea slugs, sea butterflies, and sea hares among about 3000 species. Bubble shells have a large head with sensitive tentacles. Other species have special projections on top of the body for gas exchange. Sea butterflies have unusual oar-like extensions that they use for swimming.[21] Sea hares have earlike tentacles and resemble rabbits in the way they hold and munch on algae. The *pulmonate* subgroup has about 7000 species. These creatures have an internal lung for breathing air. Snails and land slugs are grouped under this subgroup.

We can thus see that creatures as widely different as neopilina, sea slugs, sea hares, sea butterflies, and snails are grouped into one class, the gastropods.

Another class within the phylum Mollusca is the *bivalves*. It is a distinctive group of about 30,000 living species. These include creatures such as clams, oysters, mussels, and scallops. Bivalves live in a shell with two hinged valves. They have a muscular digging foot that can protrude from between the valves, and one or two pairs of large gills on either side of the visceral hump. The gills are covered with cilia for filter feeding, but they're also used for breathing. Special siphons conduct water across the gills. Unlike other molluscs, these creatures do not have a head or a radula (a food-gathering organ with tooth-like projections). The unique body structures of the bivalves illustrate clearly that these are distinct creatures lumped within the larger group of the molluscs based on superficial traits upon which these organism groups are defined. The obvious reasons for this forced lumping is the underlying belief that all organisms must be related by evolution, and that by some means of commonality, they should be shown to be related. When we analyze the subgroups of these creatures, we can be sure to find unique organisms there as well.

Another class called Scaphopoda groups approximately 350 living species, collectively called tusk shells. They have a cone-shaped shell with holes at both ends. From one end, sticky tentacles and a digging foot protrude. Tusk shells burrow in the mud, trapping microbes with their tentacles. A toothed radula brings food inside. They breathe through their body, as they lack gills. By moving the foot in and out, water is drawn across the body. Scaphopods resemble bivalves only in their digging method, a poorly developed head, and developmental patterns.

Cephalopoda is another distinct class of animals grouped within the phylum of Mollusca. The nautiluses, squids, cuttlefishes, and octopuses are members of this class, along with about 600 living species and about 9000 fossil forms. These creatures possess unique body structures and organs. Many have a ventral muscular funnel that forces water from the body, providing jet propulsion. They have beak-like jaws, a radula, and long, suction-cup-bearing tentacles or arms around the mouth. Cuttlefishes and squids have eight arms and two tentacles. Octopuses have eight arms and no tentacles. The nautilus has up to 90 tentacles, and no suction cups. One of the most striking features of some of these creatures is that they possess lens-based eyes that are different and more advanced than human eyes.[22] The presence of creatures with lens-based eyes within one class, which is grouped into a phylum with other classes of creatures with other kinds of eyes or with no eyes,[23] is one of the best examples of how entirely unconnectable creatures have been lumped together into one phylum.

Such widely distinct creatures are forcibly lumped together into the phylum Mollusca and are believed to have evolved from a common ancestor. The basis of lumping together is a set of common features among these creatures: foot, visceral hump, head, mantle, mantle cavity, and shell. To evolutionists, just the presence of a mantle, a mantle cavity and some type of muscular foot is sufficient to be "diagnostic" of a mollusc.

Molluscs date back about 600 million years to the Cambrian period, when multicellular creatures first originated. This fact also supports the new theory that many distinct creatures lumped together as Molluscs must have originated independently of each other right in the primordial pond when all the other distinct creatures were being born. When millions of creatures originate independently in a small, constrained environment, with only a few major parameters such as gravity, water, earth, and air, their locomotive, eating, and breathing organs in their bodies can have some similarities at least within small groups of creatures. This is precisely what happened when the creatures were independently born in the primordial pond.

We have covered about 15 phyla so far. The story of the rest of the living phyla is the same: each phylum is distinct from the others; classes within phyla, and orders within classes, are also mostly distinct. For want of space, we shall only briefly touch upon the details of the rest of the phyla.

About 15,000 creatures are grouped into the phylum Annelida, divided among four classes: Polychaeta, Oligochaeta, Hirudinea (the leeches) and Branchiobdellida. When we look at the body structures and organs of the creatures within the phylum, they are distinct from the creatures in other phyla. The creatures within the various classes of this phylum are also distinct from each other. For instance, clam worms have a head with sensory organs, five pairs of tentacles, and two pairs of eyes. An earthworm (class Oligochaeta) has many structures and organs unique to itself. It is clear that the earthworm consists of numerous organs and structures with specialized functions (see Figure 10.4). Leeches (class Hirudinea) also have unique specialized structures and organs. Some annelids have a clear blood plasma, tinted red by an oxygen-carrying protein. Some others have a green blood plasma whose color is due to the oxygen-carrying pigment *chlorocruorin*.[24] Some breathe through gills. Earthworms and other annelids without gills breathe directly though their whole body surface. In tube and burrow dwellers, body movement circulates water over the body surface. In earthworms, surface pores secrete mucus onto the body from special glands, aiding in gas exchange. The esophagus in some annelids also includes a grinding and mixing organ called the gizzard.[25]

Different annelids use different organs and different methods to seek, detect, capture, and devour their prey. For instance, the Glycera, a burrow-dwelling polychaete, has pressure receptors to detect a small invertebrate such as a crustacean when it passes by its network of tunnels. Glycera then rapidly extends its pharynx, seizes the prey, and paralyzes it with poison. Earthworms do not have eyes; leeches have up to ten eyes, but without lenses to focus images. Clam worms have eyes with a lens that seems to concentrate the light. The earthworm cocoon formation is a unique process, involving unique structures. The *clitellum* secretes a mucous "slime tube" around the front third of the body. Then inside the slime tube, the clitellum surrounds the whole body in a nutrient-rich cocoon. The cocoon slips off of the worm, picking up sperm and eggs as it passes over the sex glands. There are numerous other details of unique structures and organs of the various creatures

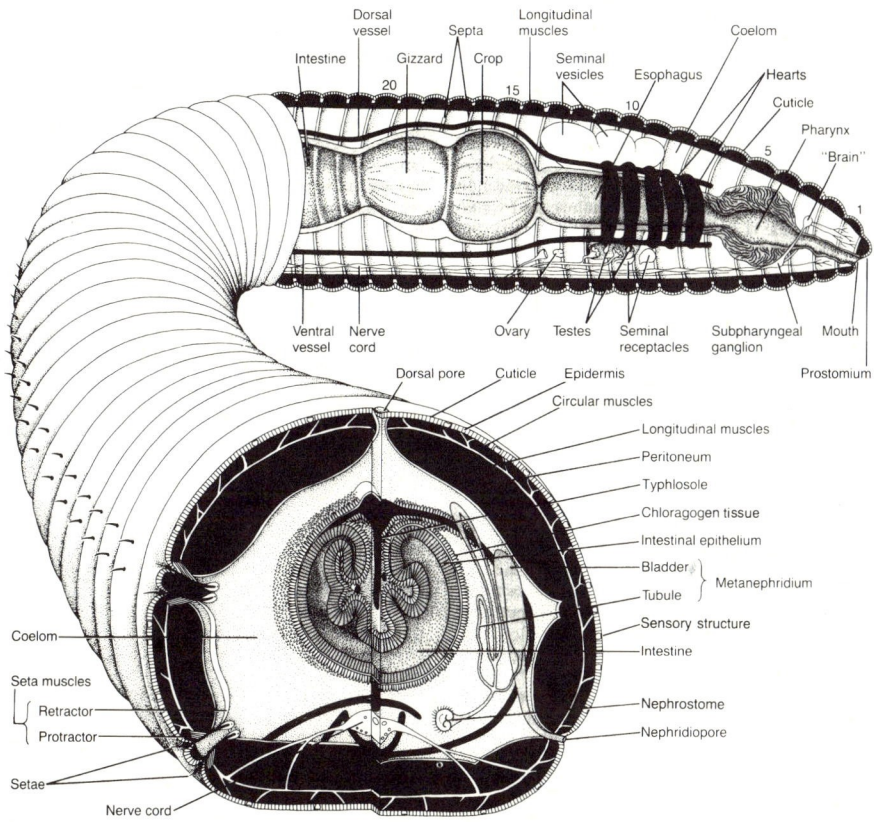

FIGURE 10.4. ANATOMY OF AN EARTHWORM. Even an earthworm, commonly thought to be a primitive organism, contains many complex organs and body parts. (From:Zoology by L. Mitchell, J. Mutchmor, and W. Dolphin. Copyright © 1988 by The Benjamin/Cummings Publishing Company. Adapted by permission.)

grouped within the phylum annelid, whose details we cannot be going into for want of space.

Let us look at the unique structures and organs in the Sipuncula phylum. The body has an unsegmented trunk and a mouth-bearing proboscis, called the *introvert*. The introvert is extended hydraulically. It has spines and a ring of cilia-covered tentacles that gather food. Sipunculans have no specialized breathing or circulatory structures; only internal body fluid and coelomocyte cells with an oxygen-carrying pigment called hemerythrin. Amoebocyte cells collect waste that is excreted by a pair of kidney-like organs, which also release the gametes.

Although organisms classified into the phylum Echiura appear to resemble the sipunculans, they are distinct creatures with unique organs. They have a flat proboscis, distinct from the introvert type of sipunculans. The proboscis has mucus-coated cilia and a gutter that guides food into the mouth. They may have several pairs of *metanephridia* (kidney-like organs). Most echiurans have a closed circulatory system similar to that of earthworms. Some also have coelomocytes with a type of hemoglobin. Except for speculations, the phylogeny of the Echiurans is unknown, that is, they cannot be connected with any other creatures by means of evolution.

Members of the phylum Pogonophora (beard worms) have a "beard" of pinkish tentacles. They are distinguished among nonparasites by a lack of a digestive tract. They may feed by trapping food particles on their tentacles or by absorbing dissolved nutrients from sea water. The tentacles have inward folds called pinnules to increase surface area. Cilia appears to propel water through a central canal formed by the tentacles. The body has three distinct regions. A forward part bears from one to about 200,000 coiled tentacles, depending on the species. The central part, a long trunk, has cilia and surface projections. The rear part is segmented like an earthworm. Each body region has its own distinctive internal compartments. There is a closed blood vessel system with a type of hemoglobin. Capillaries may provide gas exchange, excretion, and even feeding in some species. There is no question that the pogonophorans are distinct creatures.

The phylum Arthropoda is the largest and most diverse phylum in the animal kingdom. Although the arthropods are speculated to have arisen from annelid-like ancestors based on some annelid-like features, these creatures are indeed distinct from annelids. The animals within this phylum are grouped into four subphyla: Trilobita, Chelicerata, Crustacea, and Uniramia. Each subphylum is subdivided into several classes, orders and families. When we peruse the structures and organs of these creatures, it becomes clear that majority of these are distinct creatures forcefully grouped as if they were all related by evolution. Animals as widely different as the horseshoe crab, scorpion, spider, tick, mite, daddy longlegs (harvestmen), sea spider, water flea, shrimp, barnacle, crayfish, lobster, crab, millipede, centipede, grasshopper, termite, damselfly, aphid, water strider, chafer beetle, hawk moth, paper wasp, and many other creatures are included in this phylum. Zoologists themselves agree that the four subphyla of arthropods are fundamentally distinct from each other. In reality, when we view the distinctness and uniqueness of most of the creatures within the classes and orders of arthropods, we can see that most of the creatures beyond the family subgroupings are distinct and must have been independently born in the primordial pond.

Tardigrades, commonly called water bears (phylum Tardigrada) are also unique organisms with unique body structures and organs. The uniqueness of these creatures can be seen by Mitchell's writing:[26]

> From what little is known, tardigrade development appears to be unique. ... Tardigrades pose many questions in phylogeny. A molted cuticle, legs held under the body, a metameric nervous system, and a hemocoel all indicate that tardigrades are related to arthropods. Yet, their enterocoelous development and resemblance to pseudocoelomates are puzzling. In common with arthropods and nematodes, tardigrades lack functional body cilia. They also have other features that may indicate an affinity with pseudocoelomates. Some zoologists consider the tardigrade body cavity a pseudocoel rather than a hemocoel. Also, tardigrade pharyngeal muscles are radially arranged, like those of nematodes, and the muscle bands of tardigrades are remarkably like those of rotifers. Do these features indicate a phylogenetic relationship among tardigrades, arthropods, and pseudocoelomates? To some zoologists, tardigrades represent a serious challenge to the usual grouping of phyla as pseudocoelomates, protostomes, and deuterostomes.

Tongue worms are worm-like parasites of flesh-eating vertebrates and are grouped into the phylum Pentastomida. Again, these are grouped into a separate phylum because of their unique structures.

Walking worms or velvet worms are caterpillar-like organisms, with many pairs of walking legs and with some segmented internal systems. These creatures are grouped in a separate phylum Onychophora. There are about 80 species in this phylum, whose ancestors date back over 500 million years to the Cambrian period. The structures and organs of the creatures in this phylum show that these are unique. It comes out from Mitchell's writing (italics mine):[27]

> Onychophorans are a *curious mix* of annelid-like, arthropod-like *and unique* features. Their appendages resemble annelid parapodia in being unjointed, but are arthropod-like in holding the body up off the substrate. Their excretory organs are metamerically arranged

metanephridia, similar to those of annelids. ... Onychophorans have many arthropod-like features. ... The onychophoran nervous system resembles the annelid-arthropod type, but is also reminiscent of the ladderlike system of certain flatworms. ... Sensory structures include touch receptors and chemoreceptors concentrated on the antennae, a pair of eyespots at the base of the antennae, and special moisture detectors, called hygroreceptors, on the body surface and antennae.

Although zoologists feel that some of the characteristics are annelid-like, some others arthropod-like and yet others flatworm-like, indeed the onychophorans are unique creatures with unique structures and organs. Note that Mitchell says that in addition to having mixed features, they also have unique features. When we look at these creatures with the view of the new theory of the independent birth of organisms, their distinctiveness will become acceptable. Furthermore, the fact that the phylogeny of the onychophoran phylum is highly speculative becomes clear from the following writing by Mitchell and associates:[28]

> Because they are so much like arthropods and yet have some annelidan features, onychophorans have a special significance in animal phylogeny. They have sometimes been considered a link between the annelids and the arthropods. More likely, however, onychophorans are descendants of a metameric group that was ancestral to arthropods. They may represent one of the earliest arthropod-like phyla that diverged from the stock of segmented protostomes that gave rise to annelids and arthropods.

Three small phyla (Bryozoa, Phoronida, and Brachiopoda) are collectively called lophophorates because of their crown of tentacles called a lophophore. The lophophore, essentially a projection of the body wall that surrounds the mouth, uses cilia to trap food. According to Mitchell et al.,[29]

> the phylogenetic position of the lophophorates is controversial.

Although the Bryozoa, Phoronida, and Brachiopoda are all called lophophorates, the creatures in them are indeed distinct and unique. Bryozoans lack breathing, circulation, and excretion organs. Instead coelomocyte cells and body fluid (propelled by cilia) move nutrients, gases, and wastes. The funiculus, a long cord, may also transport nutrients. They have a ring of nerves surrounding the mouth and a nerve net covering the body.

About 20 species of worm-like animals are classified into the phylum Phoronida. Phoronids have a closed circulatory system, distinguishing them from other lophophorates. Two large pumping blood vessels circulate red blood cells with a type of hemoglobin. Specialized excretory organs collect wastes and expel them near the anus. It has a complex nervous system that allows muscles to act quickly to avoid predators.

One might think that all "worm-like" creatures are related by evolution. But as we have seen above, there are many creatures that are worm-like yet absolutely distinct from each other in their body structures. Consider the tapeworms, ribbon worms, roundworms, hairworms, beard worms, tongue worms, walking worms, spiny-headed worms, peanut worms, and earthworms. Although they are all generically called worms, each is absolutely distinct and unique and is classified in a totally distinct phylum. The tapeworms are phylum platyhelminthes, ribbon worms are phylum nemertea (or rhynchocoela), roundworms are phylum nematoda, hairworms are phylum nematomorpha, spiny-headed worms are phylum acanthocephala, and earthworms are in the phylum annelida. Yet another "worm" is classified into the phylum gnathostomulida. These are distinct and unique creatures with absolutely unique bodily structures that cannot be connected by evolutionary theory. These differences and uniquenesses have been known for over a century, and it is the zoologists, biologists, and evolutionists who have brought out these distinctions, although they have still not come out of the theory of evolution, for lack of a scientifically viable alternative theory.

Obviously, many distinct creatures are generically considered "worms" based on their superficial "wormy" appearance, which gives a feeling that each worm type could have evolved from another type of worm. However, we clearly see that these are entirely unique creatures which simply cannot be connected by evolution. It shows us that although the overall body appearance of creatures may seem similar, each creature has unique structures and therefore unique DG pathways that are absolutely unrelatable by evolution.

Creatures in the phylum Brachiopoda have shells resembling ancient Roman oil lamps, and hence are called lampshells. Brachiopods illustrate how one organism that superficially resembles another can be erroneously assumed to be related to it. Having two shells, they resemble bivalve molluscs, and were once classified in the phylum Mollusca. In fact, bivalve molluscs and brachiopod lampshells are fundamentally different. As Mitchell puts it,[30]

> Actually, bivalve molluscs and brachiapods are fundamentally different, and their superficial similarity results from convergent evolution. ... The valves of bivalve mollusc shells cover the right and left sides of the animal, whereas lampshells have dorsal and ventral valves. Also, bivalve molluscs filter-feed with their gills, while brachiopods use a lophophore.

There are other unique structures that are brachiopod-specific. The resemblance of brachiopod valves and the bivalves also beautifully illustrates how two absolutely distinct creatures can originate similar overall body structures independently from the primordial pond to the extent that people think that they are evolutionarily related. But when we scrutinize a little deeper, they are seen to be completely unique organisms.

When we consider the creatures grouped under the phylum Echinodermata, they are found to be highly unique. Let us see what zoologists Mitchell et al. have to say about these organisms:[31]

> These deuterostomes have three distinctive features. First, they are bilaterally symmetrical as larvae, but are typically radially or biradially symmetrical as adults. As seen in most sea stars, the adult body tends to be pentamerous, meaning that it has five radiating parts. Second, the name *echinoderm*, meaning "like sea urchin" or "spiny skin," refers to the calcerous spines or spicules embedded beneath the skin. The spines are actually components of a hard endoskeleton formed of uniquely structured calcerous ossicles. Third, echinoderms have a hydraulic locomotor system unlike any other system of movement in the animal kingdom. Usually called the water vascular system (although it is not primarily a circulatory system), this is a set of interconnected internal water tubes with many small external projections called tube feet, or podia. The water vascular system provides the slow, steady movement characteristic of sea stars and many other echinoderms, and it has a variety of other functions as well. Radiating bands or grooves of the echinoderm body housing parts of the water vascular system are called ambulacra.

The phylum Echinodermata contains six classes. Again, as we have reiterated, creatures grouped into different classes are unique; they are grouped together based on only superficial resemblance and similarities. For instance, the single species *Xyloplax medusiformis* of the class Concentricycloidea is truly unique except for superficial resemblance to other echinoderms. Sea cucumbers (class Holothuroidea) are also unique, except for the superficial characteristics by which the echinoderms are defined. Let us remember how the Molluscan bivalves (phylum Mollusca) and the Brachiopod lampshell bivalves (phylum Brachiapoda) appear to be extremely similar but are actually completely distinct. Mitchell and associates say about sea cucumbers,[32]

> Sea cucumbers are unique in having two clusters of branched tubes, called respiratory trees, that are the main gas exchange/excretory organs.

Sea urchins and sand dollars (class Echinoidea) have a complex feeding apparatus, called Aristotle's lantern, surrounding the mouth.[33] The lantern has moveable teeth that scrape and bite. Sea urchins and other echinoids have tube feet specialized for breathing. Sea cucumbers are unique in a pair of tubular respiratory trees used in breathing and excreting. Furthermore, when we analyze the anatomy of the creatures classified into various classes of the phylum Echinodermata, we shall find that each of these classes are unique. The phylum as a whole is no doubt unique. As Mitchell and associates themselves put it:[34]

> Despite the wealth of fossils and a voluminous literature on the subject, the phylogeny and affinities of echinoderms remain very speculative and controversial.

Another aim in this chapter is to demonstrate that all these creatures originated in the primordial pond by showing that they all appeared in the Cambrian or Precambrian era. According to Mitchell et al.,[35]

> The echinoderm phylum is ancient, and the hard endoskeletons of these animals have left a rich fossil record. About 20,000 fossil species have been described in about 16 extinct classes. Echinoderms were diverse in the Cambrian (about 600 million years ago), and undoubtedly there were many species during Precambrian times.

As every phylum so far seen comes to support the new theory of the independent birth of organisms, the next phylum we shall see is no different. Our tests — uniqueness, unconnectability, and antiquity — all are proved in the phylum Chaetognatha. We can simply read on what Mitchell et al. has to say about the creatures:[36]

> Chaetognaths, commonly called arrowworms because of their dartlike appearance, are all marine. ... Charles Darwin commented that chaetognaths are "remarkable for obscurity of their [phylogenetic] affinities." Darwin has been studying reproduction in arrowworms of the genus *Sagitta*, and when he wrote his report in 1844, chaetognaths had already been classified in Linnaeu's class Vermes (worms) for 75 years, ever since their discovery in 1769. Today, there is general agreement that arrowworms are deuterostomes, but questions remain about their relationships with other animals. Fossil chaetognaths dating from the Cambrian, nearly 600 million years ago, show that the phylum is ancient. Fossils closely resemble modern species, indicating that the general body form has been successful and has not greatly changed for hundreds of millions of years.

Creatures of the phylum Hemichordata are also unique and support the new theory. There are two subgroups within this phylum: Enteropneusts (acorn worms) and Pterobranchs. Their anatomies show that they are very unique. For want of space we shall not go into details.

Finally, the phylum Chordata, which includes the subphylum vertebrata, also supports the new theory. There are over 47,000 living species in the phylum, including about 2,000 that are marine filter-feeders. The others, having a bony or cartilaginous backbone, are the vertebrates. Four features define the chordates. However, we can see that entirely distinct creatures have been again lumped together under this phylum based on superficial features. For instance, the creatures in the subphyla Cephalochordata, Urochordata, and Vertebrata are distinctly different. Invertebrate chordates are divided into two subphyla, the Cephalochordata (lancelets) and the Urochordata (tunicates). They lack vertebrae and have small or nonexistent coeloms. Their distinctive pharynx is used for filter feeding.

Although the lancelets (subphylum Cephalochordata) have the four diagnostic chordate features, they are unique. They have unique protonephridia, organs for excretion.[37] They lack a heart and instead use the aorta and small contractile bulbs near the gills to pump blood. Instead of having a brain or any sense organ, they have sensory cells located throughout the body.

The creatures defined as subphylum Urochordates are also unique. These creatures have no coelom and have a distinct developmental pathway. One class within this subphylum has a highly unusual open circulatory system, with a heart that pumps in two directions. The blood cells may contain unusually high concentrations of certain rare elements. The pharynx aids in excretion. Creatures of the other classes within Urochordata (Larvacea and Thaliacea) are also likewise unique.

Uniqueness of creatures and their unconnectability to other creatures among invertebrate organisms are what we have demonstrated so far. Another important concept that we illustrate is that these creatures originated at around the time of the origin of life itself. We can see that indeed fossils of these creatures are found in the Cambrian era, and possibly in the precambrian, about 600 million years ago, when multicellular life first appeared. Furthermore, most of these creatures have retained their overall structures with little change. In the following passages, we shall illustrate similar concepts among the vertebrate animals.

Uniqueness among vertebrates

Vertebrates as a group are distinct and unique compared to the invertebrates. As we have seen in Chapter 9, molecular evidence clearly demonstrates that they could not have evolved from any invertebrate creature. The status of the thinking of zoologists on the evolution of vertebrates very well supports and corroborates the findings and concepts in the new theory. Although zoologists, in line with evolutionary thinking, want to show that the vertebrate lineage evolved from some invertebrate ancestor, they are not able to do so. Only conjecture and speculation remains to the extent of confusion. Let us read what zoologists Mitchell et al. have to say about vertebrate evolution (italics mine):[38]

> Zoologists generally agree that the earliest chordates were invertebrates, perhaps gill filter feeders somewhat like urochordate larvae or cephalochordates. Unfortunately, few invertebrate chordates have been fossilized, and there is no direct evidence of the main events in early chordate evolution.
> ... Central questions concerning chordate evolution are: What were ancestral chordates like? What group of nonchordate invertebrates gave rise to the chordate line? If the first chordates were invertebrates, what invertebrate chordates were ancestral to the first vertebrates? A hypothesis developed in the 1920s by British biologist Walter Garstand is generally accepted as the best interpretation of early chordate evolution. .... Assuming that urochordate-like visceral animals were ancestral chordates, what animals might have given rise to them? It would be logical to look for urochordate ancestors among other sessile filter feeders in the deuterostome line. Many zoologists consider the pterobranchs, sessile filter feeders of the phylum Hemichordata, good models of urochordate ancestors. ... Perhaps all these animal groups are distantly related and shared an ancestor with early chordates, but *linking them is purely conjectural*. It is also possible that *they evolved their similarities independently* through adaptation to sessile filter feeding. ... For these reasons, zoologists generally agree that lancelets do not represent ancestral vertebrates.
> ... Another bit of evidence linking invertebrate chordates and vertebrates is seen in the life cycle of lampreys, a group of living agnathans. Larval lampreys, called ammocoetes, have all the fundamental features of chordates and are remarkably similar to urochordate larvae and to lancelets. Much like lancelets, ammocoetes live in sand and filter-feed with gill structures. They also have vertebrate kidneys, a liver, and a pancreas, and their muscles are segmented.

We can see that much pure conjecture is involved in trying to connect vertebrates to invertebrate ancestors. But as we have demonstrated in Chapter 9, none of the invertebrates have the more than 600 proteins in the blood plasma of the vertebrates. In addition, many systems in vertebrates (such as the blood coagulation and immune systems) are totally absent in the invertebrates. Also, most invertebrates have their own unique systems for functions such as blood coagulation and defense mechanisms against infections (see Chapter 9). Furthermore, the agnathans, which are supposed to be the ancestor of the vertebrates, are very ancient and are found in fossils in the Cambrian period. This fact also corroborates the new theory that these creatures must have originated in the primordial pond along with all the other creatures on earth.

Another fact we have to consider here is that the lampreys have a blood plasma similar to that of vertebrates,[39] while all other invertebrates lack all of the vertebrate blood plasma proteins. Thus, if the lamprey is considered to be at the bottom of the vertebrate line, then it cannot be connected to any other invertebrate by evolution (it is another matter that the lampreys themselves are unique animals, unconnectable to other vertebrates).

Having established that vertebrates as a group are unique creatures and did not evolve from invertebrates, we shall now see that subgroups of vertebrates are also unique and must have originated independently of each other.

Diverse creatures grouped as fishes

There are two living classes of vertebrate fishes. Sharks, rays, and other cartilaginous forms make up the class Chondrichthyes, and bony fishes define the class Osteichthyes. Other vertebrates are collectively called tetrapods because they have four limbs. The class Amphibia contains frogs, salamanders, and the worm-like caecilians. Most amphibians can move about on land, yet must reproduce in water. The Reptilia class includes turtles, crocodiles, snakes, and lizards. Unlike amphibians, reptiles are adept at retaining water, with a nearly impermeable skin. Birds (class Aves) are warm-blooded (maintain a constant body temperature) and have feathers. Mammals have distinct characteristics and organs, and constitute the class Mammalia. Reptiles, birds, and mammals are all termed amniotes because they all have special membranes that surround and protect the developing embryo. Agnathans, fishes, and amphibians are anamniotes because they lack such a membrane.

We will not go deeply into the details of the vertebrates, for we have discussed them sufficiently throughout this book so far. We will touch upon some major details that would assist us in forging and verifying our concepts of uniqueness, unconnectability, and ancientness of these creatures.

Besides the two classes of living fishes, Chondrychthyes and Osteichthyes, there are two classes of extinct fishes, the Acanthodii and the Placodermi. Acanthodii, also called spiny sharks, are found in fossils at least 450 million years old. They had a sharp spine at the leading edge of their fins. They also had a bony skeleton, large eyes, bony scales, bony gill covers, paired pectoral fins, and a heterocercal tail, which has a large dorsal lobe that contains the tail end of the spinal column.

Placodermi means "plate skin," referring to the bony armor that covers these fishes. These are found in fossils at least 400 million years old.

Chondrychthyes fishes have skeletons made of cartilage, with some calcium deposits providing extra strength. Other "diagnostic" features are a heterocercal tail, lack of a swim bladder, an intestinal coil called a spiral valve, plate-like scales, and bony teeth. Also, unlike most other vertebrates, chondrichthyans maintain an internal osmotic pressure that is higher than that of sea water. Fertilization is internal. Sharks, skates, and rays form the subclass Elasmobranchii, and the quite different ratfishes make up the subclass Holocephali. These subclasses are quite distinct and have unique anatomic structures for unique functions, but we shall not go into details.

Bony fishes (Osteichthyes) have a complex bony skeleton, with unique structures. Unlike chondrichthyans, most have a swim bladder that traps air for buoyancy, and most species have a homocercal tail (symmetrical, top to bottom). Bony fishes have several kinds of bony scales, all distinct from the plate-like scales of chondrichthyans. There are three subclasses of osteichthyans: Dipnoi (lung fishes), Crossopterygii (lobe-finned fishes), and Actinopterygii (ray-finned fishes), each with unique structures.

To appreciate the uniqueness of organisms classified under fishes, we can recall our discussion concerning the worms. We saw that many distinct, totally unconnected invertebrate creatures (even classified under completely distinct phyla) are called generically "worms" due to their superficial "worm-like" appearance. Similarly, we can see that many distinct organisms are generically called "fishes" because they share "fish-like" characteristics, *i.e.*, they swim in the water. Many unique creatures, which were apparently born independently, are classified under the general name fishes. The difficulty of zoologists to trace the phylogeny of fishes comes to our support:[40]

> Despite a fairly rich fossil record, the phylogeny of fishes is difficult to determine. Clearly, the evolution of jaws from gill arches was a key event in early fish evolution, but we do not know what types of fishes first had jaws. Representatives of major classes appear in the fossil record dating back about 400 million years, but these groups had undoubtedly diversified earlier. It is likely that jaws and the earliest fishes evolved at least 500 million years ago. .... Acanthodians seem more closely related to chondrichthyans than to bony fishes, although this is controversial. The affinities of the placoderms are more difficult to sort out.

It appears that almost all marine vertebrates have been placed under the large group of fishes, except a few obvious marine reptiles and mammals. We saw that the cartilaginous fishes such as sharks, the bony fishes, the lung fishes, and the ray fishes are quite distinct. Many other examples illustrate the uniqueness and distinctness of the creatures classified under fishes. For instance, the deep-sea fish, the striped anglerfish, has a long projection that serves to lure prey.[41] Many deep-sea fishes are also bioluminescent, having light-emitting cells or carrying sacs of bioluminescent bacteria. Luminescence is achieved with a specialized enzyme called luciferase.[42]

It appears that many different independently born organisms, with certain underlying characteristics required for swimming in water, are all seen as related creatures and are named fishes. Suppose numerous creatures are independently born, and that they all had bones to construct their body structures. If they are to live only by swimming in water, then at least most of them can have similar body structures and appearances. When this is so, if portions of the genome of the first-born "fish" can be used as common reagents for building several more viable new genomes, then it should lead to even more similarities among these creatures, although their genomes were otherwise independently constructed in the primordial pond and the creatures were independently born.

Amphibians

Living amphibians are classified into three orders: Anurans (frogs), Caecilians (worm-like amphibians), and Urodeles (salamanders). When we peruse the details of the structures and functions of amphibians, we can see that there are many unique and unconnectable creatures among them that illustrate the new principles of the independent birth of organisms in the primordial pond.

Most living amphibians have a mixture of aquatic and terrestrial characteristics. Many are aquatic as larva, with gills that become replaced with lungs as adults. However, almost an equal number of amphibians do not have an aquatic larval stage. Also, many salamanders never develop lungs. Amphibian lungs are inefficient, so in many species the skin can absorb oxygen directly. Amphibians have a multichambered heart. Many amphibians (especially frogs) have four limbs, while other species have two limbs or none at all. All amphibians must lay their eggs in water or in very damp places.

Although such major common factors have been used by people to define amphibians, it is clear that many unrelated amphibian creatures have been grouped as though they are related by evolution. This becomes even more clear when one carefully looks at their structures and functions (and also their genomes that are widely different in size). When we look at the basic structures of the three orders — frogs, caecilians, and salamanders, they are totally distinct. While frogs have limbs, some caecilians look like worms, without legs, having folds in the skin that makes them resemble segmented earthworms. Some have tentacles as sensory organs. Salamanders are also distinct from frogs and caecilians.

Reptiles

Reptiles are unique creatures, with unique features such as the water-resistant cleidoic (self-contained) egg and the highly efficient amniote kidney. There are four orders of living reptiles: snakes and lizards, turtles and tortoises, the crocodilians (crocodiles, alligators, and their allies), and the tuatara of New Zeland (the sole member of the fourth order).

Turtles (order Chelonia) are unique among reptiles because of their distinctive domed shell, which is fused to their skeleton. Turtles also have a long, flexible neck that aids in feeding. The head can be pulled safely inside the shell. The limbs extend from openings in the shell.

When the tuatara (order Rhynchocephalia) was discovered in 1831, it was classified as a lizard. Later it was found to be a rhynchocephalian, a group that was thought extinct. The fossil record shows that the tuatara has changed little in the past 200 million years.

Lizards and snakes belong to the largest reptilian order, the Squamata. A marked distinction between lizards and snakes is that snakes do not have legs. Additionally, the snake eyes stay permanently open behind a clear cover, whereas lizard eyes open and close; unlike snakes, lizards typically have an external ear; and lizard scales are similar all over the body, whereas snakes have special scales on their undersides.

Chameleons can change their skin color from gray to green, brown, or yellow for camouflage. Geckos have adhesive pads on their toes, enabling them to climb walls. Some geckos, unlike all other lizards and most other reptiles, have a voice. Skinks have flat, overlapping scales making them shiny and smooth. Sometimes a skink will shed its tail, leaving it wiggling in front of a predator, enabling the skink to escape.

The worm lizards resemble earthworms. They are blind, lack ear openings, hind legs, and usually front legs, and live underground.

Snakes are classified into a few families: boas (Boidae), pythons (Pythonidae), cobras, coral snakes and kraits (Elapidae), sea snakes (Hydrophiidae), and vipers (Viperidae). Elapids use fangs to inject deadly venom into their prey. Hydrophiids live in tropical seas, have a modified tail for swimming, and have a method to remove excess salt from their body fluids. Vipers also have a venom injection system.[43] Members of one subfamily, the pit vipers, have a sensory pit on the head that detects the heat of warm-blooded prey.

Crocodilians are semiaquatic. Although the mouth does not seal, flaps at the throat seal off the windpipe and esophagus while under water. Ear flaps and nasal valves provide a similar function. They have an extremely tough skin, reinforced with bony scales, and a muscular tail used for swimming and as a weapon.

There are many groups of extinct reptiles. The cotylosaurs and pelycosaurs (land reptiles), ichthyosaurs and plesiosaurs (marine reptiles), therapsids (mammal-like reptiles), thecodonts (bipedal reptiles), dinosaurs (ruling reptiles), pterosaurs (flying reptiles), ornithischians (bird-hipped dinosaurs) and saurischians (lizard-hipped dinosaurs) are some of these. Many have unique characteristics, but again we shall not go into details.

As with other groups of animals we have seen, reptiles exhibit many unique body structures and characteristics. Turtles lack teeth, but have a horny beak. Most other reptiles have well-developed teeth. The most specialized reptilian teeth occur in snakes. Venomous snakes have grooved or hollow fangs that conduct poison from the poison gland into the prey. The venom has a variety of toxic, often deadly effects.

To conserve water reptiles can secrete a nearly dry, pasty urine. Sea snakes and the Galapagos marine iguana have glands devoted to excreting excess salt. Marine iguanas can actually expel vaporous clouds of concentrated salt solution.

Snakes and lizards have a well-developed sensory structure called Jacobson's organ. It has cavities lined with nerve endings similar to those used for smelling. Snakes flick their tongues in and out to gather molecules for the Jacobson's organ to analyze. This organ is used to follow the chemical trails of prey, and it also plays a part during courtship in recognizing females by males.

The tuatara and many lizards have an unusual eye: it is complete with a lens and retina, but is covered over by scales. While having no visual function, it is thought that lizards use it to sense the presence or absence of sunlight.

Reptiles, unlike most amphibians, have a copulatory organ for transferring sperm. Only the tuatara lacks copulatory organs. Crocodilians and turtles have a single penis, but snakes and lizards have a pair of structures called hemipenes, which can be used individually.

Unlike amphibians and fish, most reptiles are oviparous, laying self-contained eggs. Reptiles lack a larval stage; the hatchling is small, but fully-developed. However, some lizards and snakes have juvenile offspring. Most of these species are ovoviviparous, retaining the eggs internally until they hatch. A few reptiles are viviparous, such as the European skink, which feeds developing embryos through a type of placenta.[44]

The phylogeny of the reptiles is very enigmatic to zoologists. Let us read what Mitchell and associates have to say:[45]

> According to the classical view, the reptiles that appeared on this world stage evolved from labyrinthodont amphibians. However,

recent comparative studies on the development of the middle ear suggest that amphibians and reptiles may have evolved as more or less independent lineages from crossopterygian (lobefinned fish) ancestors. Resolution of the problem of reptilian origins still eludes zoologists, and a fully satisfactory answer may never be found.

BIRDS

Birds appear to share a mixture of characteristics with reptiles and some unique characteristics of their own. Their general body form, scales on their legs, and the nature of their beak and skin make them similar to reptiles. However, feathers, warm-bloodedness, and a uropygeal gland (for secreting oil) are unique to birds.

There are two subclasses of birds. Archaeornithes is extinct, and the only known species is *Archaeopteryx*, about which we have discussed previously. Neornithes includes all living and many extinct birds. The birds are supposed to have originated 200-250 million years ago, but as we shall see in Chapter 11, even a slight error in dating methods can push this backward considerably.

All 8600 species of living birds, classified in about 28 orders, lack teeth. There are two large subgroups. *Ratites* are flightless birds such as ostriches, rheas, emus, cassowaries, and kiwis. Their breastbones cannot accommodate flight muscles. Carinate birds, quite unlike the ratites, have a sternal keel in their skeletons for supporting large flight muscles.

There are many unique features in birds. Digestion begins chemically in a *proventriculus* organ and continues mechanically in the gizzard. Birds keep gravel in their gizzards to help grind food. Many birds also have specialized enzymes to digest chitinous insects and other invertebrates. Some birds have an extra transparent eyelid that covers the eye during swimming and provides enhanced under-water vision. Birds have a unique organ projecting from the retina called the *pecten*. Pigeons and doves secrete a solution of sloughed cells, called pigeon's milk, that they feed to their young. As in mammals, its secretion is directed by the hormone *prolactin*.

We discussed the uropygeal gland and the development of bird feathers in Chapter 3, so we shall not elaborate here. These characteristics are not possible to be evolved in birds from a reptilian ancestor, for reasons we have discussed in Chapter 3.

There are two possibilities for the origin of birds in the new theory:
1. One original bird was independently born in the primordial pond by means of genome mixing; all other birds arose by organismal change from this bird.

2. From the one original bird genome that emerged by means of genome-mixing in the primordial pond, several new genomes with differing characteristics were produced in the pond (through mutation of the genes and sequences in seed cells) over a period of geologic time leading to many different birds directly from them.

It is difficult to say which of these two is correct. In fact, both of these may be possible. A few birds could have originated in the primordial pond by means of genome mixing, and from each of these could have originated many more birds by means of organismal change and descent with modification. Analysis of the genomes of numerous birds could tell the difference. Furthermore, the verification of the new theory by establishing the uniqueness and unconnectability of numerous invertebrates, the fact that the vertebrates arose independently of the invertebrates in the primordial pond, and the demonstration that among vertebrates numerous creatures did originate independently, will all show the possibility that, although all birds seem to be quite similar in terms of their overall body structures, many of them could have arisen directly from the primordial pond by means of changed genomes in seed cells.

Mammals

Mammals distinguish themselves with many unique features. Except for one group of egg-laying creatures, the monotremes, all mammals are viviparous and have internal fertilization. Viviparous mammals are either *marsupials* or *eutherians*. Marsupials give birth to embryonic offspring that must complete their development while attached to the mother's nipple. In half the living marsupials, the young develop in a pouch, the *marsupium*, as in kangaroos. Eutherians are the placental mammals, where embryos remain in the uterus and receive nutrients through the placenta during fetal development. The placenta grows from embryonic cells and makes a complex connection with the mother's blood stream. All mammals, including monotremes, have some amount of internal fetal development and nurse their young.

Marsupials also have a placenta, but with the exception of one group, the bandicoots, it is distinct from that of eutherians. In most marsupials, the yolk sac of the embryo attaches to the uterine wall via a simple *choriovitelline* placenta.[46] This is completely different from the eutherian placenta, and allows little exchange of nutrients and waste. Instead the embryo feeds itself by absorbing a uterine "milk" secreted from the uterine wall. Many marsupian animals would appear superficially similar to the placentals. The unique koala is called the native "bear." The native "cat" and the Tasmanian "wolf" are other examples.

The three living prototherians constitute a single order, the Monotremes. The term monotreme refers to digestive, urinary, and reproductive products being released from a common chamber, the *cloaca*. This feature is also found in birds and reptiles. Monotremes lay eggs similar to those of birds and must incubate them. Additionally, the adults have a toothless beak-like organ. The duck-billed platypus uses its sensitive, flexible bill to probe for prey on lake and stream bottoms. Female platypuses do not have nipples; they secrete milk from pores in their skin. Spiny anteaters (echidnas) grow a temporary pouch to incubate their eggs. As in marsupials, the newborn echidnas remain for a time in the pouch, suckling milk from nipple-like skin projections.

Eutherian mammals remain in the uterus for the entire fetal period. Their *chorioallantoic* placenta provides a firm connection to the uterine wall, and provides exchange of materials between fetal and maternal blood. The hormone *relaxin*, produced by the ovary, softens the birth canal tissues in preparation for delivery. The hormone *oxytocin*, produced by the pituitary gland, stimulates uterine contractions during birth. Specialized mammary glands produce milk under the influence of the hormone *prolactin*. Additional action of oxytocin helps push milk towards the nipple.

From the above details, one can see that at least the major groups of creatures classified as Mammals cannot be connected to each other or a common ancestor by evolution. Zoologists agree that the monotremes, marsupials, and eutherians are not connected to each other.[47,48] They could not have originated from nonmammals, for they contain unique organs and unique proteins. If we happen to find these proteins commonly among these creatures, then one could argue that all these creatures are related. Even under those circumstances, one could see that mammals could not have evolved from nonmammals, for they have unique organs such as the placenta and the mammary gland.

Unless mammals contain proteins and genes that are common to all mammals, it is not possible that all mammals are related to each other, and it is not possible that all mammals originated from one original mammal. However, it is possible that many distinct mammals were independently born, and from each a set of similar species was derived by organismal change. Currently, many animals that resemble each other are termed different species. These could be derived from a distinct creature, independently born from the primordial pond. The typical mammal has many unique organs such as the placenta and the mammary gland and many unique proteins such as the placenta-specific proteins. Thus, we should be able to sift through the data on the mammals and know which are the distinct creatures born independently from the primordial pond and which are the creatures that were derived from them. Where do we draw the line? As we discussed in Chapter 8, we can follow the new definition of an independently born creature. If a creature contains a unique body part or unique gene, then it is an independently-born creature.

With this discussion, let us not go into much detail of the various orders of mammals. We shall only touch upon some details to corroborate our concepts. Order Insectivora includes shrews, hedgehogs, and moles. Each of their four feet bears five digits with claws. The short-tailed shrew and several other species have venomous saliva. Bats are included in the order Chiroptera. Bats have eyes, but many species use echolocation to navigate and locate prey. Order Primates include the lemurs, lorises, galagos, tarsiers, monkeys, marmosets, great apes, and humans. Order Carnivora includes the terrestrial carnivorans (dogs, wolves, bears, raccoons, skunks, mink, otters, weasels, mongooses, hyenas, and cats; about 240 species) and a smaller group (about 35 species) of aquatic carnivorans (seals, sea lions, and the walrus).

Ungulates (hoofed mammals) include the orders Perissodactyla (horses, rhinoceroses, and tapirs) and the much more diverse Artiodactyla (camels, pigs, deer, antelope, hippopotamuses, and cattle). The artiodactyl ruminants (cattle, sheep, goats, camels, giraffes, deer, and antelope) have a digestive system specially designed to allow them to digest cellulose.

Other orders include Proboscidea (elephants), Tubulidentata (aardvark), and Sirenia (sea cows, manatees, and dugongs). Sirenians do not have limbs; resembling whales, they have a horizontally flattened fluke used for swimming. Order Xenarthra includes anteaters, tree sloths, and armadillos. Anteaters have no teeth; instead they capture insects with a long, sticky tongue. Armadillos have distinctive body armor made of bone and horny plates. Rodents comprise the largest order of mammals, with 1760 species. Superficially resembling rodents are the rabbits and pikas in the order Lagomorpha, and the hyraxes (order Hyracoidea).

Cetaceans include the totally aquatic mammals — porpoises, dolphins, and whales. Like fishes (and unlike seals and walruses), cetaceans are efficient swimmers with powerful muscles and tails for propulsion. Unlike fishes, cetaceans undulate up and down rather than from side to side. The cetacean blowholes are a unique feature. Instead of teeth, baleen whales have hardened epidermal plates that filter prey from the sea water. Toothed whales and dolphins use echolocation (sonar) for tracking their prey.

In addition to the unique characteristics we have identified in the various vertebrates, we can add a few more to the list here.[49] In fishes, blood cells are formed by the kidney. In amphibians and other vertebrates, blood cells are produced in bone marrow. Fishes do not have bone marrow. The amniote kidney is distinct in several features, including a pair of ureters, distinct from the urinary ducts of other vertebrates. Similarly, amniote sperm ducts are different from those in most bony fishes. Reptiles, birds, and mammals produce cleidoic (self-contained) eggs, enclosing the embryo, food (yolk and albumin), and water in a protective shell. The shell is also porous, allowing gases to diffuse in and out. Placental mammals, marsupials, and many snakes and lizards do not have a cleidoic egg; embryos depend on the uterus for nourishment and protection.

WHEN THE SUPERFICIAL BUT UNREAL SIMILARITIES AMONG SOME CREATURES AND GENUINE SIMILARITIES AMONG SIMILAR SPECIES OF A DISTINCT CREATURE ARE TAKEN INTO ACCOUNT, IT SHOWS THAT NUMEROUS CREATURES ARE TRULY DISTINCT

Considering that all creatures are based on similar tissues, cells, and biochemistry, it should be realized that numerous creatures can be independently born in the primordial pond by the same mechanisms of genome building. Based on fundamentally the same biochemistry and molecular biology, numerous genomes can be independently built in the primordial pond, leading to numerous creatures fundamentally based on similar biochemistry, molecular biology, tissues, and cells. Thus we should remember that these basic similarities in distinct organisms do not imply an evolutionary connection among them.

We should not forget that one of the main reasons that people are misled that all organisms are related is the genuine relationship among the similar species of a distinct organism. Despite the fact that most creatures are clearly unique and distinct, this has been the main reason for the persistence of evolutionary theory.

Another thing we want to remember is how one can be misled by the superficially similar appearances of creatures that even completely distinct and unconnectable creatures are related by evolution. What we saw above concerning the generic appearance of many distinct creatures that look like worms is an excellent example. Another good example is the superficial similarity between the Brachiopod lampshell bivalves (phylum Brachiapoda) and the totally unrelated Molluscan bivalves (phylum Mollusca) — in fact for a long time they were grouped within the same phylum based on their similarities. We can also find many examples in fishes, for instance, the ray fishes are distinct from the angler fish and the shark.

As an analogy, although we have been building automobiles for several decades, all are based on moving bodies with wheels. Starting from the unicycle, bicycle, tricycle, and automobiles and motorcycles, all have the same overall mechanisms, parts, and appearance (no matter who builds it). Why? All are constrained by the same things: gravity and friction. Similar unifying constraints are imposed on living things, and the result is inevitable similarities.

Consider the various kinds of airplanes. All use the same principle of flying in the air. Imagine airplanes without wheels. Perhaps they would have robotic limbs similar to legs. If we do not have jets or propellers, the airplanes would probably flap their wings. They would then look like birds. Perhaps boats would have flippers and automobiles would have legs. In other words, under certain constraints, all the automobiles, planes, and boats might resemble different animals — those that walk, swim, and fly. And like animals, these machines would have unifying similarities, making many of them appear to be the product of the same design. When we look deeply into the structures of distinct living creatures, they are found to be truly distinct machines, built independently to work in a small constrained environment of the earth.

The most important points considered in previous chapters and verified in this chapter are:

1. A multicellular creature could be born directly from the primordial pond by the direct assembly of its genome in the pond.
2. Numerous invertebrate creatures were born independently in the primordial pond.
3. Vertebrates were born independently of any invertebrates in the primordial pond.
4. Numerous vertebrate creatures originated in the primordial pond independently of each other.
5. Each independently-originating genome could have been changed in its seed cell, as the seed cell divided and produced multitudes of its own copies in the primordial pond. The genome in each copy could be changed and restructured randomly, and some of the changed genomes will lead to new living creatures (Figure 10.5). This could lead to creatures changed slightly or considerably with similarities to the creature that first originated from the independently-assembled prototypical genome.
6. Each independently-originating creature (whether invertebrate or vertebrate) also gave rise to many related similar species by many mechanisms such as natural selection and mutation, that is, by means of change through organismal descent with modification.
7. The constraining physical forces — gravity, land, water, and air — unified all creatures to have overall functional similarities. The independent origin of numerous creatures in the same primordial pond enabled the use of the same biochemical and molecular biological materials, and

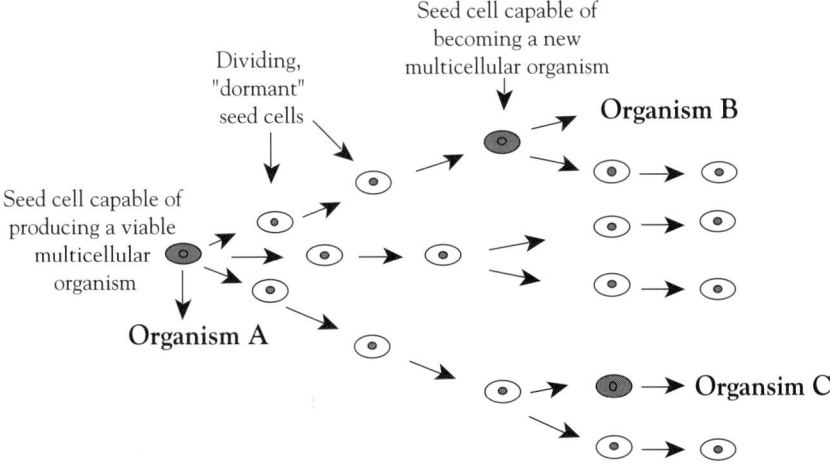

FIGURE 10.5. MODIFICATION OF AN ORGANISM'S GENOME IN ITS FREE-LIVING SEED CELL IN THE PRIMORDIAL POND AND PRODUCTION OF A NEW CREATURE. A seed cell capable of giving rise to a multicellular creature could reproduce for a long geological time and lose its special ability of growing into an organism while undergoing major modifications and repatterning of the genome. Rarely, a descendent seed cell in the primordial pond could give rise to a changed new organism, but with similarity to the organism produced from the starting seed cell.

led to the presence of these substances in the independently-born creatures.

8. If we take representatives of all the distinct animals without taxonomically classifying them, each is an element of a random collection of organisms with random organs and appendages.

CONCLUSION

Why have scientists encountered so many contradictions and inconsistencies in their attempts to explain the origin of organisms on earth in terms of evolution? The origin of creatures classified into higher taxonomic groups has been enigmatic, unexplainable by any evolutionary theory since Darwin first proposed his theory of natural selection 130 years ago. Natural selection explains the narrow-range diversity of very similar species, but the extrapolation of this mechanism to all life on earth does not make even common sense after we set aside the premise that all organisms evolved from a common ancestor. Creatures classified into higher taxonomic categories, usually above the level of the "family," cannot be connected by evolution.

The reality is that numerous creatures were independently born in a common primordial pond, and hence there simply is no common ancestral organism. The genome of each separately born organism was assembled directly in the primordial pond from a common gene pool. Many genomes were assembled independently, and then many more were assembled by combinations of genes from the pool plus fragments of genomes assembled earlier. Our theory thus concedes a "relatedness" of sorts between some distinct organisms, but only at the level of the primordial pond. There are no other organismal relationships — through any organismal mutations or organismal changes — except incidental variations that have yielded varieties of the same fundamental organism. Millions of creatures could have been born separately in the primordial pond. Each then gave rise to many of its own similar species. The predictions of our new theory are precisely what we find to be true in our careful observations of the natural world.

In Chapter 9 we explored our new theory of the independent birth of organisms by examining the available evidence at the molecular level, particularly genes and gene systems, and we emerged from the chapter with our theory stronger. In this chapter we have corroborated our theory by examining the uniqueness of creatures' anatomical structures, and again the theory is strengthened. In Chapter 11 we will further reinforce our theory by a careful review of the fossil record.

11

A New Look at the Fossil Record

Any theory that purports to explain the origin of life on earth should be able to verify its predictions in contemporary evidence, by observations of all the details of the life around us. If a theory explains only a few details, but leaves others unexplained, then it is not a strong theory. The theory of evolution has carried that problem on its shoulders since Darwin, and we should have been more suspicious all along. Until now, however, we have had no realistic, scientifically derived alternative.

Our new theory of the independent birth of creatures can explain almost all of the details of life on earth. In the preceding chapters we have verified the new theory by observations of the creatures living on earth today, and by examining their molecular details and recognizing their uniqueness and unrelatedness. In this chapter, we shall add the fossil record to our already considerable mass of evidence that weighs in on the side of independent birth of organisms. We can learn much from an examination of the remains of creatures that have lived in the past, and whose structures have been entombed, hardened and preserved.

Abrupt Appearance of Numerous Multi-Cellular Life Forms at the Base of the Animal Fossil Record

The fossil record and the Cambrian explosion: A total blow to the evolutionary theories, but an absolute support to the new theory of the independent birth of organisms

A sudden, big burst of unique creatures appeared at the very start of multi-cellular life on earth according to the fossil record. It is totally against Darwin's theory or any other theory of evolution.

The evolutionary theme is that life started at some time during the earth's history in one primitive creature. From the original one creature, many new creatures have evolved over a period of a few hundred million years. One creature could give rise to one or more new creatures by evolution, thus producing numerous related creatures in a branching tree-like pathway. Diversity and multiplicity thus ensued over geological time producing all creatures related to one another.

From the start of life, the remains of an individual animal would be preserved as a fossil if it were buried in soil under certain conditions. The fossils of the first-evolved creatures were deposited first, and newer layers of later-evolved creatures would be deposited above successively over time. Thus, according to the evolutionary theory, the successive strata of fossils should contain organisms related to each other. Because the history of life is supposed to have been written this way, the successive fossil strata is supposed to be a record of life's history on earth — and hence the name "fossil record." The whole geologic "column," however, does not occur in one place in a single piece. It is constructed from many pieces of strata occurring in various places. According to evolutionary theory, the time at which a particular stratum was laid down represents the time at which the particular creature found in it had evolved from another creature that might appear to be similar to it in a stratum below. Thus the succession of different creatures in the fossil record represents the sequence of evolutionary steps that life on earth has taken. Different relative strata are variously named such as the Cambrian, Silurian, Jurassic, etc., based on the geologic times they represent. The fossils could be dated using modern radioactive dating methods, and any errors in these methods would change the age of the fossils.

If the evolutionary theory is correct, it should be reflected in the fossil record, and should thus be verifiable. From the bottom of the fossil record where the first signs of multicellular life are displayed, an evolutionary sequence should be discernible as predicted by the theory. Because, according to the evolutionary theory, it takes millions of years for one organism to change into another creature, in the beginning of multicellular life — i.e., when the original creature gave rise to the first few creatures — it should have taken several tens of millions of years for even a few new creatures to appear in the record. Then, in the course of time, the frequency of newly appearing organisms should increase rapidly.

Do we observe this kind of scenario in the fossil record? Certainly not! On the contrary, numerous distinct and unique creatures are displayed abruptly in the fossil record at its very bottom when multicellular life first appeared. In the 1830s Roderick Murchison discovered that the appearance of the first living beings did not occur gradually with successive addition of more complex forms of life. About 600 million years ago, at the beginning of the Cambrian period, most of the distinct invertebrate creatures simultaneously and suddenly appeared in the fossil record within a short span of a few million years. The Cambrian fossils were numerous, abundant, and distinct — including trilobites, brachiopods such as lampshells, gastropods, bivalves, cephalopods, many poriferans, nemerteans, beard worms, cnidarians, annelids, jellyfish, sponges, sea urchins, sea cucumbers, sea lilies, crustaceans, and many others. Since multitudes of creatures abruptly appeared in a big burst in the record of Cambrian times within about 10–20 million years, it is termed the "Cambrian explosion." In fact, very recently it has been shown that this Cambrian explosion is limited to as few as five million years, beginning from about 540 million years ago.[1] We should also note that the numerous creatures that appeared abruptly in the explosion are fully-formed unique living forms unrelatable to each other.

This Cambrian explosion is a blow to the evolutionary theory. It was a great disappointment to Darwin that his theory's predictions were not verified in the fossil record. When Darwin first proposed his theory he was fully aware of this problem, but he thought that the predicted evolutionary pattern would be verified by creatures that were yet undiscovered beneath the Cambrian strata. He thought that during Precambrian times the world swarmed with living creatures, which were the more primitive ancestors of the Cambrian fauna. But they were not found during his lifetime. It disturbed him so much that he wrote in his last edition of *Origin of Species*:

> The case at present must remain inexplicable; and may be truly urged as a valid argument against the views here entertained.

Has the evolutionary theory been verified by the discovery of its predicted pattern in the fossil record after Darwin's time? No! The fossils of such precursor creatures were never found. Since Darwin's time, rich fossil records of Precambrian life stretching back more than three billion years have been found. However, the fossils of Precambrian periods do not explain the problem of the Cambrian explosion. They include only the simple bacteria and blue-green algae, and some higher plants such as green algae. The sudden appearance of numerous complex multicellular life forms in the Cambrian seems as sudden as ever, and the problem of explaining the mystery of this Cambrian explosion still remains as fresh as when Darwin first wrestled with it.

What is even more destructive to the evolutionary theories is that almost all of the distinct types of creatures that we find living today, and all that have ever lived, appeared right at the beginning of multicellular life according to the fossil record. The fossil record is sort of upside down to what is predicted by the evolutionary theory. The representatives of almost all the creatures that are classified into the higher taxa — all the phyla, classes, orders and families — had originated suddenly and almost simultaneously at the start of multicellular life. This picture is clearly not what is predicted by the evolutionary theories. Evolution says that from one original creature, one or a few new creatures evolve, which are very much related to the first one — a step that would take several million years. From each of these new creatures, one or a few more new organisms should arise — another step taking many millions of years. Thus for the first many tens of millions of years, only a relatively small number of creatures, all very similar and relatable to each other, would be produced. So perhaps after 50 million years, all life on earth would be classifiable into at most a few families within a single order. But what is observed is absolutely opposite to this. In an extremely short period (about five million years) that could be equated to a geological instant, numerous creatures, all unrelatable and classifiable into almost all the known phyla, classes, orders, and families appear on the scene for the first time. This is in total contradiction to the long geological time, and the similarity and relatedness of organisms, that the evolutionary theory predicts.

The recent findings of geochronologist Samuel Bowring and colleagues, published in 1993 in *Science*, shows that the Cambrian explosion is even more abrupt and sudden than previously thought.[2] Let us read a commentary that appeared in *Science* on the original research article:[3]

> ... geochronologist Samuel Bowring of the Massachusetts Institutes of Technology and his colleagues report that two lumps of volcanic rock from Siberia have yielded a new, more recent date for the beginning of the explosion, shrinking its duration to a mere 5 million to 10 million years — less than a third as long as paleontologists had traditionally assumed. "This Big Bang in animal evolution hap-

pened faster than we imagined", says Sepkoski. ... Until now, almost every wall-chart of the geological time scale put the beginning of the Cambrian explosion at 570 million years ago and its end 20 million to 40 million years or more later. ... But Bowring and colleagues have now used a different version of uranium-lead dating to determine a dramatically younger age for the beginning of the Cambrian explosion — 533 million years or less. That result squeezes the explosion down to a few million years.

Thus all the multitudes of creatures that appeared in that geological instant are all distinct and unrelatable to each other. Would such a scenario that is totally opposed to the theory of evolution be such a blow to make people abandon this theory? Yes it should be. But because there has been no scientific theory that can explain all the details of life on earth — molecules, organisms, and fossils — without involving organismal evolution, even such a totally undermining scenario has been put aside by scientists.

The Cambrian explosion fully demonstrates the new theory of the independent birth of organisms

The new theory of the independent birth of organisms clearly explains the fossil record. In fact, all of what are predicted by the new theory are precisely observed in the fossil record. It explains the sudden and simultaneous appearance of numerous creatures at the start of multicellular life. It explains the uniqueness and unrelatedness of all these multitudes of creatures that simultaneously appeared. And it explains the appearance of further unrelatable creatures for a considerable time after the start of life.

In fact, the new theory predicts the explosion of numerous unique creatures and the fossil scenario in general at the start of multicellular life from any primordial pond. In whichever primordial pond there developed conditions for the origin of life, life would have begun in an explosion, or else no life would have started. We shall see that this is true when we discuss the Burgess Shale and Ediacaran fauna.

According to the theory of the independent birth of creatures, an organism was the result of the expression of the genome assembled directly from the primordial pond. A genome was randomly assembled into a single seed cell that could grow into a multicellular creature (Chapters 7 and 8). Therefore, there was no need for transition between the unicellular eukaryote and multicellular organism. Because the genomes of different multicellular organisms were organized independently in the primordial pond, there was no need for transition between the different multicellular organisms either. Each one was an independent entity. There was no evolutionary connection through descent with modification. The gaps among distinct creatures stemmed from the manner in which the genomes of different creatures were independently assembled in the primordial pond.

The biochemical processes of the primordial pond reached a very rich, critical stage that was suddenly capable of producing multitudes of distinct genomes each of which led to a distinct creature. Thus, numerous distinct, unrelatable creatures would be produced suddenly and simultaneously from a primordial pond. This is precisely what must have happened in the Cambrian explosion of numerous unique creatures. Indeed the Cambrian explosion illustrates this phenomenon very well in all respects — the suddenness, uniqueness, and unrelatedness of numerous creatures. In addition, there is no reason for the rich biochemical activities in the primordial pond to stop abruptly after the first big burst of creatures. Although it may slow down after the first peak activity, the primordial pond activities leading to distinct creatures would still continue over a considerable geological time before finally stopping. Figure 11.1 explains this phenomenon.

In the new theory, the formation of life from non-life is not a rare accident. It is a biochemical inevitability of a rich primordial pond. Given the conditions of a biochemically-rich primordial pond, and the vast amounts of genetic sequences available, multitudes of life forms must have been inevitably formed simultaneously. The mechanisms behind the relatively sudden production of numerous and diverse viable organisms from the primordial pond have been described in earlier chapters. Based on those concepts, the scenario of the Cambrian explosion is what one would expect to have happened if this theory is correct. It is clear, based on this theory, why there should be a simultaneous appearance or a big burst of creatures when life first arose on earth. Five to 10 million years was enough time for a large number of possible assortments of genes to be assembled from the universal gene pool of the primordial pond.[4]

The first appearance of living beings in the fossil record would therefore have simultaneously included multitudes of simple and complex organisms, as reflected in the Cambrian explosion of the fossil record. The continuing birth of creatures in the primordial pond over a long geological period led to additional entirely new creatures. This would result in the appearance of entirely new creatures for a further considerable geological time in the fossil record although at a much reduced rate (see Figure 11.1). Every creature born was immutable, confined to a framework of slight variations as long as it lived. These predictions of the new theory fulfill the notable fact observed in the fossil record — that when a new creature appears in the record, it usually does so abruptly and then apparently remains stable for as long as the record of that creature lasts. This is true with the first big burst of organisms as well as the later-born organisms.

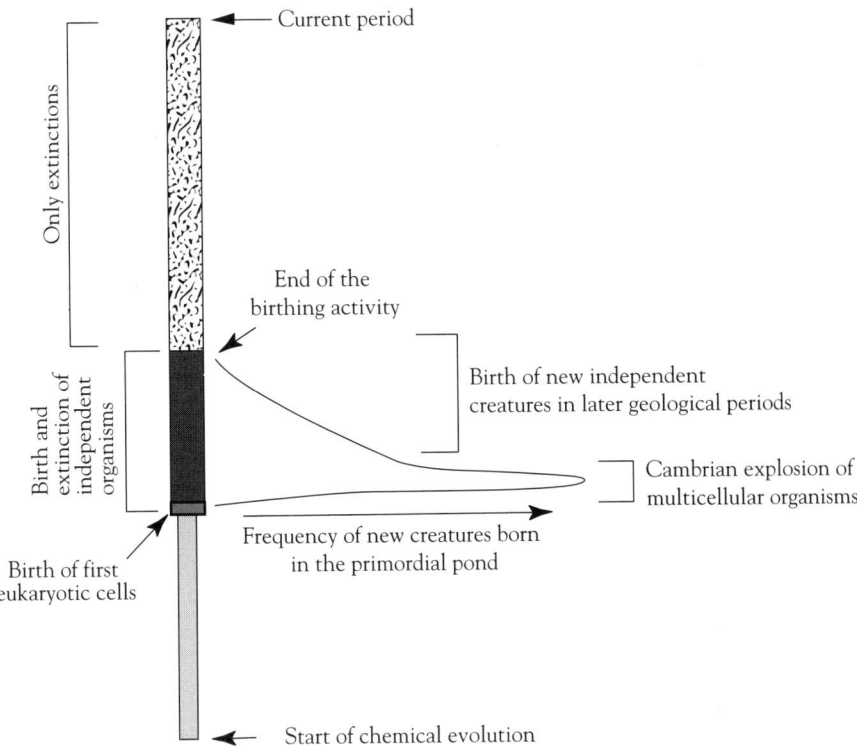

FIGURE 11.1. THE CHRONOLOGY AND TIME TABLE OF THE INDEPENDENT BIRTH OF ORGANISMS. Chemical evolution started when the earth started to cool, approximately 4 billion years ago. When the conditions of the primordial ponds were right (~600 million years ago), great quantities of DNA and cellular machineries were available and the first eukaryotic cells were formed. Soon after this, the seed cells of numerous multicellular organisms were formed. This gave rise to the sudden explosion of multitudes of independent creatures (Cambrian explosion) in a short geological period (~5 million years). New independent creatures continued to be born in later geological periods. But it slowly reduced and ended, a long geological time ago, as the primordial pond became depleted of its rich biochemicals. Extinctions have occurred since the beginning. Thus, there are now only extinctions and no more births of new creatures. The figure is not drawn to scale and depicts only the concept.

The process of independent births would result in a random assortment of creatures, simple as well as complex, at the very beginning of the history of life on earth. The genomes of creatures from the first big burst would then mix in the open primordial pond. Numerous such mixings would lead to creatures with mixed organismal characteristics. Not all such genome-mixings would have become viable organisms, but even if one in a large number of mixings succeeded, it would still lead to many viable mixed creatures. This phenomenon is also supported by the fossil record.

The sudden appearance of multitudes of unrelatable creatures in the Burgess Shale: An enigma to all evolutionary theories, but one of the strongest supports for the new theory

A small quarry in British Columbia, the Burgess Shale, contains the fossil remains of many unique invertebrate organisms, classifiable into about 20-30 separate phyla. They are unlike anything now living, and the creatures contained in it cannot be placed in any modern group.[5] The Burgess Shale represents a period about 30-40 million years after the Cambrian explosion. The uniqueness of creatures in the Burgess Shale is astounding for a small quarry only a city block wide.[6] The anatomical range of the animals from this quarry probably exceeds the entire spectrum of invertebrate life in today's oceans. The relationships among these creatures are far from obvious, and possible ancestral forms to any of these creatures are unknown. Steven J. Gould has noted that each creature in the Burgess Shale fauna appears to be formed as though a builder takes various body structures randomly from a sack of many structures and builds each creature at random:[7]

> What order could possibly be found among the Burgess arthropods? Each one seemed to be built from a grabbag of characters — as though the Burgess architect owned a sack of all possible arthropod structures, and reached in at random to pick one variation upon each necessary part whenever he wanted to build a new creature.

The Burgess Shale represents a geologically very short time. When one tries to find answers to the Burgess mystery based on evolutionary theories, one can only reach a dead end. Just as we discussed for the Cambrian explosion, such a great number of unique creatures could not evolve from an original creature in a geological instant. By Darwin's theory, lineages of one or a few creatures cannot diverge far enough to be classified into many distinct phyla in such a short time. The Burgess Shale remains an enigma for paleontologists and evolutionists. An explanation given by paleontologists is that all the biological niches were empty, so that the very first creatures could diversify rapidly and occupy all these empty niches. However, they do not seem to appreciate that the question is not about the niches, but about the fundamental change of the genomes: What mechanism exists that could change the genome so rapidly and radically?

Since Darwin's gradual evolution cannot give an answer to the Burgess Shale mystery, paleontologists Stephen Gould and Niles Eldredge have tried to explain it by a process called punctuated equilibrium. We have seen in Chapter 4 that this is really not an answer to the problem of either the Cambrian explosion or the Burgess Shale mystery. Gould and Eldredge only restate the problem in another way — that unique creatures suddenly evolve in the periphery of a large population and then stay unchanged for a long geological period. They do not offer any genetic mechanism as to how a unique creature with new body parts can be evolved even over a long stretch of geological time, let alone in a geological instant. They do not have a mechanism as to how new genes or unique DG pathways can evolve within the genomes of organisms. While Gould clearly understands the problems and elegantly and beautifully illustrates them,[8] we can confidently say that the explanations provided by him for the Burgess mystery are not valid.

As we demonstrated in Chapters 3 and 4, there is no mechanism that can change the genome of one creature into that of another distinct creature, let alone changing one genome into many genomes of unique creatures with new and unique body structures through "lineages" by the mechanism of natural selection or any other mechanisms of evolution. Therefore it is out of question for the scenario of the Burgess Shale to be generated by any evolutionary mechanisms.

The theory of the independent birth of creatures explains the Burgess scenario in a similar manner to the way it explains the Cambrian explosion. The sudden appearance of numerous unique creatures in the Cambrian explosion was due to their independent and simultaneous births in a primordial pond. The Burgess Shale seems to represent a period about 30–40 million years after the Cambrian explosion (the "middle Cambrian"). In my opinion, the Burgess Shale could represent creatures that originated in an entirely different pond separate from the "Cambrian pond." Given two different biochemically rich ponds with the same or similar basic biochemical and genetic machineries, life forms would be inevitably born in both of them. The randomness and distribution of life forms in each of them may vary depending upon the size of the DNA sequence pool and environmental conditions in each of the ponds. Similarly, the structures and body parts of the creatures from two distinct ponds could be entirely distinct and absolutely unrelated. In fact, one cannot expect the same types of organisms to be produced if life originated in two distinct ponds. Remember that the distinct creatures born within a pond are unique and unrelated among themselves. Remember that within one pond, the later-born organisms could have been built by the mixing of the genome pieces of the earlier-born creatures, leading to similarities. But between the creatures born in two distinct ponds, there would be no such similarity. However, there could be an overall *functional* similarity among the creatures that were born in two different ponds, because they were born in the same overall physical and chemical environments of the earth (gravity, water, land, and air). Thus, the overall description of some creatures arising from one pond may appear to fit those from another pond in a functional sense — for instance, some of the creatures from both ponds might have eyes, antennae, a mouth and other structures on a head at the front side of the body, some gill structures on the top or bottom, some swimming appendages at the back and sides, and some walking or crawling structures at the bottom. Because many creatures that swim in water would fit these descriptions, many of those distinct creatures from two different ponds would appear to be similar in their basic overall appearance.[9]

A great number of body plans that were possible could be brought about in the beginning of life in any given rich primordial pond by the random assortment of the millions of genes found in it into genomes of viable organisms. The Burgess pond must have been such a rich one, independent of the Cambrian. The new theory predicts that many different body plans can be built with the random assortments of genes for a variety of organs and appendages. This is what seems to have happened in the Burgess Shale creatures. According to the new theory, they would have had common genes and unrelated unique genes among them. Thus they would exhibit some common body plans and some very different and unique body structures and appendages. But overall, the randomness in kind and distribution of the unique anatomical structures and appendages among the Burgess organisms indicate that they are just the organisms that precisely fit the predictions of the theory of the independent birth of creatures.

It is possible that the Burgess Shale creatures originated during a geological period or in a geological region distinct from those of the Cambrian fauna, and that the Burgess creatures entirely died off due to some drastic environmental changes. We should note here, as we discussed in Chapter 8, that because every creature could change only slightly to follow environmental changes and would become extinct if the environment changed drastically, a whole set of creatures that were born and established in a particular set of environments could become extinct if the environment changed drastically.

WHY DO WE NOT FIND ANCESTORS OF THE CAMBRIAN CREATURES IN THE SOFT-BODIED EDIACARAN FAUNA? THE PUZZLE OF THE EDIACARAN FAUNA IS EXPLAINED BY THE THEORY OF THE INDEPENDENT BIRTH OF CREATURES.

Ediacaran fauna — evidence that it is a separate set of living forms that originated independently in the Ediacaran pond

A new group of fauna was discovered in the Ediacara locality in Australia.[10] Its creatures are entirely soft-bodied. It appears at a geological time just before the Cambrian (about 640 million years ago). A number of soft-bodied animals, some of which have been interpreted as polychaete annelid worms, coelenterates, and soft-bodied arthropods, have been found in these fossils. Because this fauna appears at a geological time just before the Cambrian explosion, some evolutionists have tried to evolutionarily connect them to the creatures in the Cambrian explosion, and have placed them in several different phyla of modern creatures. However, Dolf Seilacher, professor of paleontology at Tubingen, Germany, found that the Ediacaran creatures cannot be connected to the Cambrian fauna, despite some superficial similarity of outward form.[11,12] Because the Ediacaran creatures differ considerably from the Cambrian fauna, Seilacher has questioned their identification with them, and feels that no Ediacaran creatures survived into the Cambrian. He concluded that this fauna represents an early animal radiation that was largely extinguished and that the Ediacaran creatures represent an entirely separate experiment in multicellular life — one that ultimately failed in a Precambrian extinction. We should note here that Seilacher's thoughts are under the larger premise of evolutionary theory. That is, to him all the creatures of the Ediacaran fauna evolved from a single original creature in that period.[13]

It also appears that there were numerous distinct invertebrate creatures in the fauna represented in the Ediacaran strata. The possibility that numerous creatures could have existed right at the start of the Ediacaran fossil fauna is illustrated by what Gould says:

> By studying the varied and abundant trace fossils (tracks, trails, and burrows) of the same strata, he [Dolf Seilacher] is convinced that metazoan animals of modern design — probably genuine worms in one form or another — shared the earth with the Ediacara fauna.

Thus, as with the Burgess, several different anatomical possibilities were present right at the beginning.

While these observations are enigmatic and mysterious to the evolutionary theories, they are not only easily explained by the new theory but also are truly strong corroborations to it. In the new theory, there is absolutely no need for simpler organisms to precede complex organisms. Depending upon the biochemical richness and complexity in a primordial pond and its vastness of the universal DNA sequence pool, only a few unique creatures could be born from one pond, and more numerous unique creatures could be born from another pond.

Thus there is no need to try to find the ancestors of the Cambrian fauna in the Ediacaran fauna. The multitudes of complex creatures in the Cambrian explosion did not have more primitive ancestors anywhere. The animals of the Cambrian were born in the Cambrian pond, each of them independently, with their complex bodies and functions. We can see according to the new theory that the Ediacaran fossils originated in another separate primordial pond, the Ediacaran pond, in which different life forms were born independently but all died out for some environmental reasons.

A main theme here is that we need not look for the first one or a few organisms, supposed to be the ancestor of all the creatures in the fossil record, because there was no such ancestor nor descendants. This is because numerous distinct unrelated multicellular creatures were born at the very start of life itself — in every primordial pond where life originated. What matters to us — according to the fossil record — is that multitudes of unique creatures had erupted over a short geological time in each pond. Furthermore, most or all of the creatures could become extinct due to the highly changeable environment at that time of earth's history, and because the creatures were fixed and could not adapt to the changing environment. This is probably what we see in the Ediacaran pond fauna as well as the Burgess Shale fauna — two distinct ponds, separate from the Cambrian pond.[14] The Ediacaran and Burgess pond fauna became totally extinct, while the Cambrian pond fauna survived.

Thus life seems to have originated in many ponds on the primitive earth, when the conditions were right — possibly at distinct locations and at different geological times. In each pond, life originated independently of life in other ponds. And in every pond, numerous unique creatures were born, each independently of others in the pond. The Cambrian, Ediacaran, Burgess, and Tommotian[15] (another possible set of separate life in another pond) all seem to be separate ponds that gave rise to such independent creatures.

Appearance of Entirely New Creatures in the Fossil Record After the Cambrian Period

We have so far seen that the scenario of the fossil record at the start of multicellular life is well explained by, and is a clear support for, the new theory. This theory also predicts that after the big burst of organisms, conditions for the independent birth of creatures should have continued for a long geological time. When we look at the whole scenario with the details of the new theory, they convince us that this is what is reflected in the fossil record after the Cambrian explosion. It is well known in the fossil record that entirely new forms of organisms would abruptly appear, while the "old" ones continued to live essentially unchanged even up to the present day, unless they became extinct. This scenario is an enigma to Darwin's theory. As we saw before, other evolutionary theories such as the punctuated equilibrium theory of Gould and Eldredge does not offer a genetic mechanism as to how new creatures with new organs and body parts could evolve abruptly. Thus we can see that none of the evolutionary theories can really explain the abrupt appearance of new creatures in later geological periods. This picture, however, is clearly predicted and explained by the theory of the independent birth of creatures.

A very long geologic time was required to produce the conditions conducive for the formation of multicellular organisms in a primordial pond. These conditions continued for many millions of years, producing new creatures.

As we discussed in Chapter 8, the primordial broth was boiling and brewing possibly for hundreds of millions of years, increasing in its chemical and molecular complexities. As the primordial broth was becoming increasingly rich, the conditions reached a stage conducive for the birth of creatures. The plethora of genes and the variety of biochemicals and macromolecules in the primordial pond led first to the big burst of organisms. These conditions continued, and more creatures were later produced by exactly the same mechanisms — except that pieces of the genomes of the first-born creatures were used in the construction of the genomes of later-born creatures.

Extinctions and the changing scenario of the fossil record is clearly explained by the theory of the independent birth of creatures. The phenomenon of extinction under the new theory is explained in Chapter 8. Over geological time, especially at the beginning of the birth of creatures, the environmental conditions could have been changing rapidly, leading to extinctions, while new creatures were constantly being born in the primordial pond. The creatures born in the changed environment could be completely different from the previous ones.[16] New, unrelated creatures could be continuously generated in the primordial pond while many creatures were becoming extinct. Thus, a given creature need have neither its "precursor" in the older fauna nor its descendants in the later fauna. Depending upon the extent and rapidity with which environmental conditions changed, extinctions could happen sporadically or in intermittent spurts.

This process is reflected in the fossil record, wherein many new immutable creatures appeared in later geologic time, exhibiting a sequence of appearance of new creatures and disappearance of old creatures.

Earth's age is considered to be ~ 4.6 billion years. It is supposed to have taken ~4 billion years for the very first eukaryotic single cells to appear. More importantly, it took 4 billion years (89% of earth's age) for the multicellular, sexually reproducing organisms to appear in a big burst. There is no reason for us not to accept that the same conditions that took a long time to develop and led to the birth of multicellular organisms in a primordial pond in the first place continued to exist for at least an additional few tens of millions of years.

During this long period, new creatures were born. In fact, all the fossils could have been deposited layer over layer during this considerable geological time. The first layer could have been deposited by the remains of the first-born group of creatures. While many of these immutable creatures continued to live and some became extinct, further new immutable creatures could have been born independently of the first. This phenomenon could continue till the end of the fertile period for the birth of creatures in that primordial pond. This process could lead to a succession of creatures in the fossil record. While "old" creatures could remain in the fossil sequence without any change, entirely new creatures could abruptly appear. Organisms that are different, but that appear to be related or similar to older creatures to different extents also could appear (two similar creatures could be related at the level of the genome by common genes without any evolutionary connection between them, as we discussed in previous chapters). Gradually, the conditions for the birth of new creatures in the pond must have changed and vanished, after which time, no new creatures appeared on earth.

In essence, we see two parts to the scenario of the history of life on earth which is reflected in the fossil record. The first is the sudden burst of organisms independently and simultaneously born in the Cambrian primordial pond. The second is the periodic appearance of entirely new creatures, likewise independently born, starting from the initial explosion of organisms, until the conditions for the birth of creatures ended.

The earth's life history can be likened to a woman with childhood, puberty, fertility, and menopause. A human female, for example, goes through a series of developmental steps in her life. The series of steps that she goes through from her birth seems to fit well with the series of aging processes that the earth follows. The first three billion years seems to be analogous to the childhood of the woman. The next billion years would be the puberty age. The next 50-100 million years or so seems to be the earth's fertile period, when she gave birth to millions of creatures. And now the earth is no longer fertile.

The conditions for the birth of creatures extended for many millions of years from the start of the fertile period. However, these conditions have vanished long geological time ago, and no more new creatures will ever be formed, although extinctions will continue to occur. However, we should remember that many similar species of every distinct creature will be formed always in the future.

The age of fertile primordial ponds is over. The modern oceans and ponds contain only the life that was born millions of years ago. This concept implies that the last organism was born a long geological time ago. This means that the age of birds and mammals may be even older than we believe now from the fossil record. Paleontologists are always finding the same fossils including those of primates in earlier and earlier rocks.[17]

The abrupt appearance of new creatures in the fossil record is well documented in paleontological literature. We shall describe below for the sake of the general reader a brief account of this literature. It will show that the new theory's predictions are well borne out. We will not go into the comparative anatomy of the fossil animals, but will refer to and quote from authorities on paleontology who are in fact strongly committed to evolutionary theory.

Seed cells (eukaryotic single cells) in the primordial pond directly gave birth to multicellular organisms: Support from fossil record

The very first multicellular organisms in the fossil record appear as complex sexually-reproducing organisms. A very large number of creatures appear suddenly in a burst exactly as predicted by the theory. There exist no transitional creatures among them. Furthermore, each creature seems to have lived unchanged as long as it is seen in the fossil record. These findings indicate that the creatures were immutable. Logically then, each must have originated from a single cell (the seed cell), analogous to the zygote. This is precisely what is proposed in the theory of the independent birth of creatures.

We have established in Chapter 7 that it was the eukaryotic gene that originated directly in the primordial pond's genetic sequences, and that it was the eukaryotic cell that first originated in the primordial pond. We also showed in Chapters 7 and 8 that the process of genome assembly by the random assortment of genes in a rich primordial pond would lead to numerous genomes, not only for unicellular but also for multicellular creatures.

All this would require that the multicellular creatures originated in the primordial pond at around the same time the unicellular eukaryotes did. Indeed, this is precisely what we see from the fossil record. Multicellular creatures appeared in the Cambrian explosion at around 600 million years ago. The fossil record shows that the single celled eukaryotes also appeared at around that time. Recently, J.W. Schopf from UCLA reported that he may have found them in 1400-million-year-old rocks, but this is disputed by other paleontologists who feel that this finding is erroneous.

While discussing the history of prokaryotic and eukaryotic cells, Stephen Gould has commented the following:[18]

> UCLA paleobotonist J. W. Schopf believes that he has evidence for eukaryotic algae in Australian rocks about a billion years old. Others contend that Schopf's organelles are really the postmortem degradation products of prokaryotic cells. If these critics are right, then we have no evidence for eukaryotes until the very latest Precambrian, just before the great Cambrian "explosion" of 600 million years ago.

This puts things in clear perspective. Small discrepancies and errors in the timing of these events would thus make it possible that the single-celled eukaryotes originated right at the Cambrian explosion. It is quite clear from the fossil record then that both multicellular creatures and unicellular creatures originated on earth at around the same time. This is absolutely corroborative of the new theory that the genomes of the unicellular eukaryotes and the multicellular creatures were assembled in the primordial pond and that these creatures were born at around the same geological time.

Evolutionary biologists' misunderstandings about the origin of eukaryotic cells

Following the evolutionary theory, scientists have long believed that life started with prokaryotes. Eukaryotic cells evolved by the association of various prokaryotic cells. Eukaryotic cells then led to one or a few multicellular life forms. From this one original, primitive, multicellular creature evolved all other organisms by natural selection or by other means of organismal evolution. Prokaryotes are supposed to have appeared 3.6 billion years ago. Eukaryotic cells are generally assumed to have evolved from them about 600 million years ago. From then on it is assumed to have taken several million years for the first original multicellular creature to appear. For all the enormous time it has taken for these initial processes, there is absolutely no evidence of intermediates in the fossil record.

Evolutionary textbooks and articles clearly show the current state of thinking in this field. Eli C. Minkoff states concerning the origin of the eukaryotic cell,[19]

> Perhaps the greatest unexplained transition in evolutionary history, subsequent to the origin of life itself, lies in the origin of eukaryotic cells. ... The presence of a nuclear envelope between nucleus and cytoplasm has traditionally been considered the hallmark of the prokaryote/eukaryote distinction.

Douglas J. Futuyma writes in his book,[20]

> The earliest fossil indication of life is in South African rocks dated at 3.4–3.1 billion years old, which contain forms that resemble bacteria, including Cyanobacteria (the blue green bacteria or "algae"), and stromatolites — mound-like structures that are still formed in parts of Australia by Cyanobacteria. The earliest known organisms, then, were prokaryotes, apparently capable of photosynthesis.
>
> ... The Cyanobacteria and other prokaryotes appear to have held sway for almost two billion years. The earliest known eukaryotes, probably green algae, are in 0.9 billion-year-old rocks, although there is some evidence that they go back to about 1.5 billion years. The origin of eukaryotes is a major event in the history of life, for it marks the evolution of chromosomes, meiosis, and organized sexual reproduction.
>
> ... How eukaryotic chromosomes and meiosis evolved, however, is entirely mysterious. Almost certainly the several kingdoms into which authors classify eukaryotes — fungi, several kingdoms of protozoans and of algae, plants, and animals — became differentiated during the Precambrian, but their fossil record is far too fragmentary to document their origins.

In general, biologists consider that the origin of the nucleus in the eukaryotic cell to be one of the most enigmatic and important problems in biology. Some evolutionary biologists have revived an old concept that a prokaryote engulfed another prokaryote to evolve a nucleus and become a eukaryotic cell. We have seen in Chapter 7 that such a theory simply does not work. We can see that the evolutionary pathway through which the first life is assumed to have come through is purely an assumption without any evidence.

My theory on the origin of eukaryotic genes and eukaryotic cells that we discussed elaborately in Chapter 7 puts things in clear perspective. Eukaryotic cells originated directly in the primordial pond from their genomes assembled from the vast number of genes existing in the pond. They originated with split genes (exons split by introns) in their genomes and a nucleus that housed the genome within the cell — directly from the primordial pond. By solving the problem of the origin of the eukaryotic cell with its nucleus, we are able to solve many important problems. For instance, we could show that the multicellular creatures were also born directly from their seed cells and genomes assembled in the primordial pond. Except for small discrepancies that can be explained from small observational errors, the fossil record is supportive of what is predicted in the new theory — that unicellular eukaryotes and multicellular creatures appeared at around the same time.

Our aim here in showing that numerous creatures appeared abruptly and simultaneously on earth at the start of multicellular life is to illustrate the new theory that numerous creatures were born independently in a primordial pond. We could demonstrate that these invertebrate creatures were unrelated to each other by looking at the details of the creatures in the Cambrian explosion or in the Burgess fauna. In the Cambrian fauna, for instance, numerous fossil samples for each distinct creature have been found. They all show that each creature appeared in the fossil record, for the first time, fully formed. In other words, the snail, the sea cucumber, the sea lily, the lamp shell, the bivalve, and the jelly fish each appeared as it was. There are absolutely no intermediate or transitional forms between any two creatures, nor to an assumed ancestor. There is really nothing more needed to demonstrate that these numerous creatures must have been born independently and simultaneously.

INDEPENDENT BIRTHS OF VERTEBRATES

The prediction of the new theory that creatures were born directly from their seed cells in the primordial pond must be true for any distinct organism, at any level of complexity. Thus it predicts that all the distinct vertebrate creatures were born independently of the invertebrates and also independently of each other. We can see that this is what must have happened from the fossil record. Darwin's theory requires that transitional forms existed between organisms that appeared in the fossil record in a sequential manner. When Darwin proposed his theory, he claimed that the fossil record at that time was incomplete and that transitional forms would eventually be found. But the fossil record is more than complete now and the systematic gaps in the fossil record and the absence of transitional forms between organisms at successive steps on the assumed evolutionary ladder have become even more contrary to Darwin's expectations.[21] In contrast, this scenario is what is exactly predicted by the new theory. The following fossil evidence shows that each distinct vertebrate organism must have been born independently of the others.

Lack of fossil ancestors for fishes

The first vertebrate organisms that are assumed to have evolved from the invertebrates are the fishes. The fossil record shows no evidence of ancestral forms to the fishes. According to Alfred Romer, an authority on vertebrate paleontology,[22]

> At one time or another the ancestry of the vertebrates has been sought in almost every invertebrate group. ... A theory of descent of the vertebrates from the annelid worms has been advocated. The arthropods, most highly organized of invertebrates, have also been strongly advocated as vertebrate ancestors, especially the arachnids, a group including not only the spiders but the scorpions and a number of such water-living types as the horseshoe crab and the extinct water scorpions - the eurypterids. ... The echinoderms — starfishes, sea urchins, sea lilies, and the like — seem the most unpromising of all as potential ancestors of the vertebrates.

Romer's reasons are based on comparative anatomy and embryology that indicate how none of these creatures could have been the ancestors of the fishes. Thus, the assumed ancestors of fishes in the fossil record is based only on speculation trying to fit the fossil record to the evolutionary theory. We can clearly see that there are no ancestors in the fossil record for the fishes or for any other vertebrate organisms. We should remember here our molecular evidence that several hundred genes for the proteins of vertebrate plasma are fully absent in any of the invertebrates (Chapters 4 and 9). Structurally too, none of the fishes could be connected to any of the invertebrates (Chapter 10). Thus, there should be absolutely no doubt that vertebrates were born independently of any invertebrates in a primordial pond — although, vertebrates were all born in the same pond as the invertebrates.

Many distinct types of fishes were born independently and directly from the primordial pond

We said in Chapter 10 that there could be many distinct fishes roaming in our seas, rivers, and ponds that are unique and unrelated to each other. We then said that it was possible that these had originated independently in the primordial pond. Now, here we can see the evidence for our predictions and conclusions in the fossil record.

Paleontological analyses of the vertebrates have shown that many major groups of fish are distinct from each other with no transitional forms or evolutionary connections. Here we shall select a few examples wherein organisms suddenly appear already fully developed without any traceable transitions or ancestors. According to Romer, when we first see the fishes, they had already acquired bone and are divided into several distinct groups. Romer states,[23]

> In sediments of late Silurian and early Devonian age, numerous fishlike vertebrates of varied types are present, and it is obvious that a long evolutionary history had taken place before that time. But of that history we are still mainly ignorant. .... These tantalizing fragments tell us that vertebrates had appeared on the evolutionary scene at this early stage of the Paleozoic and (a matter of particular interest) had already acquired bone, but tell us little more. Obviously, major evolutionary events were occurring in vertebrate history during the Ordovician and Silurian, but we are still in almost complete ignorance regarding them. ... When we first see these ostracoderms, they already have a long history behind them and are divided into several distinct groups.

It is important for us to note here that vertebrates have appeared in the fossil record at a very early stage. From what Romer says, they could have appeared soon after, or probably at around the same time as the Cambrian explosion. And this would indicate the near simultaneity of the birth of various creatures — including the vertebrates.

Let us read what Romer says about gnathostomes in the Devonian:[24]

> A considerable majority of the fish population of that period, however, belonged to groups now long extinct and peculiar in structure: the arthrodires, with heavy armor in articulated head and thoracic segments; the antiarchs, grotesque little creatures which look like a cross between a turtle and a crustacean; a series of odd forms which are armored caricatures of modern skates and rays; the acanthodians, sharklike in superficial appearance but most un-sharklike in various anatomical features. Where to place these curious creatures has been a vexing problem. One or the other of these types has at times been thought allied to the ostracoderms, to the sharks, to the lungfish, to the "ganoids;" but in each case the supposed likenesses have been more than outweighed by the obvious differences. There are few common features uniting these groups other than the fact they are, without exception, peculiar.

What Romer writes about the evolutionary position of the placoderms fits beautifully with the new theory:[25]

They appear at a time — at about the Silurian-Devonian boundary — when we would expect the appearance of proper ancestors for the sharks and higher bony fish groups. We would expect "generalized" forms that would fit neatly into our preconceived evolutionary picture. Do we get them in the placoderms? Not at all. Instead, we find a series of wildly impossible types which do not fit into any proper pattern; which do not, at first sight, seem to come from any possible source, or to be appropriate ancestors to any later or more advanced types. In fact, one tends to feel that the presence of these placoderms, making up such an important part of the Devonian fish story, is an incongruous episode; it would have simplified the situation if they had never existed! But they did exist; and we must attempt to fit them into the vertebrate evolutionary story.

We can very well see how the various orders of the fishes are quite distinct from each other. Further, as the new theory predicts, there have been quite grotesque creatures unrelated to any other organisms. As much as Alfred Romer was an authority in paleontology and showed that many creatures were evolutionarily unconnectable based on fossil record, he remained an evolutionist. It is then clear that even authorities of paleontology such as Romer, who showed that there are crucial and numerous problems of evolutionary unconnectability in the fossil record, have tried to fit the fossil record into the preconceived evolutionary story. This is revealed in his following writing:[26]

> The varied grotesque types which evolved during the Devonian from such a hypothetical ancestor present, in themselves, a remarkable story. But are any of these odd creatures antecedent to the fishes of later time? At first one would be tempted to a vigorous denial of the possibility. But, as we have seen, we must consider seriously the possibility that at least the sharks and chimeras of later days may have descended from such impossible ancestors.
> How and where do the acanthodians fit into the general picture of early fish evolution? They were surely descended from some jawless type, but show no special connections with any of the known ostracoderm orders; their early development presumably took place in Ordovician-Silurian fresh waters from which no sedimentary deposits have persisted.

What Romer writes of the bony fishes is also corroborative of their sudden and independent appearance on earth:[27]

> The appearance of the typical bony fishes in the geologic record is a dramatically sudden one. ... In the Middle Devonian, however, all the major types — ray-finned forms, crossopterygians and lungfishes — appear full fledged and diversified, and at once dominate the scene.

... The common ancestor of the bony fish groups is unknown. There are various features, many of them noted above, in which the two typical subclasses of bony fish are already widely divergent when we first see them — features such as fin structure, scale structure, and so on. So marked are these differences that it has been suggested that the Osteichthyes are an artificial assemblage and that ray-finned fishes and sarcopterygians represent two or three progressive lines which have arisen separately from an archaic gnathostome stock.

THE INDEPENDENT BIRTH OF AMPHIBIANS

Evolutionary theory says that amphibians evolved from fish ancestors, developing limbs for use on land from the fins of fish. A rhipidistian crossopterygian fish is assumed to have given rise to the first amphibian.[28] However, no real intermediate between the fin of the fish and the limb of the amphibian has ever been found.

Living amphibians include three groups: the frogs and toads, the salamanders and newts, and some limbless, worm-like creatures. Romer writes regarding salamanders:[29]

> The oldest known salamander is a late Jurassic genus. It is disappointing that even the older fossil salamanders show no primitive characteristics. The modern structural pattern of the urodeles was, it would seem, established by Jurassic time; there has since been little important evolutionary advance.

From this it is clear that creatures such as urodeles originated abruptly, with all the structures of today's living urodeles and have never changed after that. Our predictions of the independent birth of unique creatures and their constancy thereafter are well illustrated here.

THE REPTILES

When the bones of two fossil animals are different it is clear that the structures of the animals would also have been different. But if the bones are similar between two different animals, it does not mean that the soft parts were also similar. There can be quite a variation in the soft parts between two animals which have similar bony structures. This is well exemplified when we analyze the reptiles and amphibians.

Evolutionists say there is ample evidence for the evolution of reptiles from amphibians in the fossil record. However, the major characteristic that distinguishes reptiles from amphibians is that the reptilian egg is an amniote egg that can be laid on land.[30] The amphibian egg cannot be laid on land and it can fertilize and develop only in water. The mechanisms of development of the amphibian egg and the reptilian egg are totally different. The differences could not have been brought about by descent with modification through random genomic changes. The differences in the soft parts of the bodies of the two classes living today are also sharp.

As we saw in Chapter 10, based on comparative studies zoologists feel that reptiles could have evolved directly from fishes. This is also supportive of the fact that the reptiles are not descendants of amphibians.

Although the bony structures of reptiles are similar to those of amphibians, the differences that exist in the soft body parts among these two living groups indicate that they are distinct. These two groups, according to the new theory, could be related at the level of their genomes in the primordial pond. But by the several reasons we have discussed, the reptiles could not have evolved through organismal descent with modification from the amphibians or from the fishes.

THERE ARE NO INTERMEDIATES BETWEEN REPTILES AND BIRDS: STRONG EVIDENCE THAT BIRDS ORIGINATED AS BIRDS IN THE PRIMORDIAL POND FROM THEIR SEED CELLS

According to evolutionary theory, wings have evolved in four groups of animals — insects, reptiles (extinct flying reptiles), birds, and mammals (bats). It should be noted that no intermediates for wings exist in the fossil record for any of these groups of creatures. Evolutionists claim that birds evolved from reptiles and that the fossil animal *archaeopteryx* is an intermediate between the two. But the fact is that this fossil animal is a true bird and is not a transitional form from the reptile.

The presence of claws at the front edge of each wing in *archaeopteryx* is claimed to be a reptilian characteristic. Otherwise, *archaeopteryx* is a fully-developed bird.[31] The claws need not indicate that *archaeopteryx* is an intermediate between the reptiles and birds, because there are living birds that have this characteristic. Ostrich has claws on its wings. The young of two living birds — the hoatzin and the touraco — possess two claws on the front edge of their wings. These birds are also poor flyers, a characteristic cited of the *archaeopteryx*.

Fossil *archaeopteryx* has teeth and has an extension of the vertebrae along the tail. These features are claimed to be reptilian. It should be however noted that teeth is a random characteristic in many kinds of animals: fishes, amphibians, reptiles, birds, and mammals. Thus from this feature one cannot say that *archaeopteryx* is reptile-like. As described in the new theory, these characteristics are expected in a random manner among various creatures. In fact the discovery of the fossil of a true bird in rocks of the same geological period as *archaeopteryx* indicates that birds had already appeared by that time.[32]

Studying the geometry of the *archaeopteryx* claw, Alan Feduccia of the University of North Carolina has shown recently that *archaeopteryx* is fully a bird. Let us read from his 1993 research paper in *Science:*[33]

> Archaeopteryx probably cannot tell us much about the early origins of feathers and flight in true protobirds because Archaeopteryx was, in the modern sense, a bird.

Let us read what W.E. Swinton, an expert on birds, says about the origin of birds:[34]

> The origin of birds is largely a matter of deduction. There is no fossil evidence of the stages through which the remarkable change from reptile to bird was achieved.

It can be thus seen that the fossil record does not show the assumed transition from reptile to bird, but indicates that birds appeared abruptly. In fact the *archaeopteryx* is an animal with mixed characteristics, just as the duck-billed platypus and others (see Chapter 8), which is supportive of the theory of the independent birth of creatures.

There are many differences between the bony skeleton of reptiles and that of birds. Furthermore, the most distinguishing characteristic between the bird and the reptile is the feather. Feathers are unique and extremely complex structures. The development of the feathers (present only in birds) and the scales (present in the reptiles) follow totally different developmental genetic pathways (their development may even require some unique genes, see Chapter 3). Also we should remember the unique uropigeal glands in the birds. Furthermore, almost all the reptiles have a penis for copulation, lacking in birds.[35] As we saw in Chapters 3 and 4, it is not possible to evolve these bird characteristics from the genome of a reptile. The characteristics of the two distinct animals could only have originated in their genomes when they were independently born from the primordial pond.

Flying structures in insects, flying reptiles (now extinct), and flying mammals (the bats) also have no evolutionary intermediates in the fossil record

The assumed transitional intermediates between the reptiles and extinct flying reptiles have also not been found in the fossil record. The flying reptiles appear in fossils full-fledged — without any intermediate forms. It is assumed that the bat is evolved from an insectivore such as the moles, shrews, and hedgehogs. However, no transitional forms that would represent an intermediate is found in the fossil record. When the bat appears in the fossil record for the first time, only a full-fledged bat is found.

Evolutionist Olson says about insects:[36]

> There is almost nothing to give any information about the history of the origin of flight in insects.

Absence of fossil evidence for mammalian evolution

We have quoted from noted paleontologists who are followers of Darwin's evolutionary theory and they clearly recognize the major problems posed by the fossil record for the theory of evolution. They indicate that there are clear gaps in the fossil record and that for almost all the groups of creatures there have been neither ancestors nor descendants in the fossil record. George Gaylord Simpson says about each order of mammals:[37]

> The earliest and most primitive known members of every order already have the basic ordinal characters, and in no case is an approximately continuous sequence from one order to another known. In most cases the break is so sharp and the gap so large that the origin of the order is speculative and much disputed. ... This regular absence of transitional forms is not confined to mammals, but is an almost universal phenomenon, as has long been noted by paleontologists. It is true of almost all orders of all classes of animals, both vertebrate and invertebrate.

These gaps support the new theory that among mammals distinct organisms could have directly originated in the primordial pond. It should be however remembered here that the genomes of these creatures could be related at the level of the primordial pond.

Fossils of the rodents also show that they are distinct groups of mammals. Rodents appear in the fossil record as rodents. There are neither ancestors nor transitional forms found in the fossil record for the rodents. Let us read what Romer says about the origin of rodents:[38]

> The origin of the rodents is obscure. When they first appear, in the late Paleocene, in the genus *Paramys*, we are already dealing with a typical, if rather primitive, true rodent, with the definitive ordinal characters well developed. Presumably, of course, they had arisen from some basal, insectivorous, placental stock; but no transitional forms are known. To perfect the dental and other features of the order, a considerable period of time — perhaps the whole extent of the Paleocene — seems necessary. But in what region or environment this occurred, we do not know.

It should be noted that transitional forms among the different groups of rodents are also absent.

Hares and rabbits were considered originally to be a suborder of the rodents. Now they are grouped in a separate order Lagomorpha. According to Romer,

> The lagomorphs show no close approach to other placental groups, and the ordinal characters are well developed in even the oldest known forms.

ALL DISTINCT CREATURES APPEARED AT THE BEGINNING OF MULTICELLULAR LIFE ON EARTH. ONLY SLIGHTLY CHANGED SIMILAR SPECIES OF EVERY DISTINCT CREATURE HAVE BEEN PRODUCED SINCE THEN, AS EVIDENCED BY THE FOSSIL RECORD.

Our prediction that numerous distinct creatures originated independently in the primordial pond, and that each unique creature then gave rise to its own set of similar species is well borne out in the fossil record. The fossil record shows that similar species of each distinct creature increase in number over geological time, but truly distinct organisms have not been produced. If at all, the distinct creatures have reduced in numbers by extinction. This fact is naturally puzzling to paleontologists and evolutionists. Let us read a quote from Stephen Gould in his book *Wonderful Life* (pages 47 and 64) that demonstrates the fossil evidence for our predictions:

> The current earth may hold more species than ever before, but most are iterations upon a few basic anatomical designs. (Taxonomists have described more than a half million species of beetles, but nearly all are minimally altered Xeroxes of a single ground plan.) In fact, the probable increase in number of species through time merely underscores the puzzle and paradox. Compared with the Burgess seas, today's oceans contain many more species based upon many fewer anatomical plans.
>
> ... In a geological moment near the beginning of the Cambrian, nearly all modern phyla made their first appearance, along with an even greater array of anatomical experiments that did not survive very long thereafter. The 500 million subsequent years have produced no new phyla, only twists and turns upon established designs.

WHY IS THE FOSSIL RECORD THOUGHT TO REFLECT THE ERRONEOUS CONCEPT OF SPECIATION AND EVOLUTION?

Why has Darwin's theory been accepted for the past 130 years? Because, as Darwin's theory seemed to fit all other aspects of the scenario of life on earth, it also seemed to fit the fossil record. In other words, just the sequence of various creatures in the fossil strata ("the geological succession of organic beings," as Darwin called it) is sufficient to induce a feeling of acceptance of the evolutionary idea of one creature giving rise to another in geological time. But in reality what we see in the fossil remains is the superficial and deceptive appearance that old creatures give rise to new ones. Such superficial appearances make one believe that Darwin's theory explains the fossil record and that the fossil record supports Darwin's theory — even ignoring the mountain of problems posed by the fossil record such as the Cambrian explosion.

Biologists have always been taught to believe that evolution takes a path from simple to complex systems. Because of the frame of mind and the world view about evolution and the fossil record, scientists were forced to take the simple ⟶ complex evolutionary route to explain how complex life evolved starting from the "primitive" prokaryotes. Life on earth took the following route: prokaryote ⟶ eukaryotic single cells ⟶ one multicellular animal ⟶ many further descendent creatures (with the evolution of sex, new organs, and further complexity). Furthermore, the "breathing of life into the first creature" (the evolution of the first multicellular, sexually reproducing creature that enabled Darwin's mechanisms to begin operating) was an evolutionary accident, which could not be repeated easily. In fact to evolutionists, the origin of the first simplest microbial life form itself was possible only by a freak accident.[39]

If we come out of this world view and the misunderstandings of the fossil record scenario, we can, without being biased, take a broader look at the truth. We can open our minds to the extremely high potentials of the primordial broth: very large quantities of DNA sequence, availability of complex molecules first chemically produced and then by the DNA-coded machineries, the availability of a plethora of genes, and the extremely high probability for the millions of different genes to combine into myriads of genomes almost simultaneously giving "birth" to numerous diverse creatures — all in one primordial pond. We can then begin to appreciate the need for, and the validity of, such a radically different approach.

As we have seen, it is only the independent and simultaneous assortment of genomes capable of being born into different organisms that led to the sudden explosion of a variety of creatures at the very beginning of multicellular life on earth. The conditions for such a birth must have been brewing and developing for several hundred million years. When these conditions reached a conducive stage, then it started to give birth to an assortment of creatures almost simultaneously in the primordial pond. If these conditions took a very long geological time to develop and could give rise to varied forms of life, there is no reason why these conditions in the primordial broth could not continue further in time. What took possibly hundreds of millions of years to develop into a rich broth need not vanish quickly after giving rise to the first one or a few creatures. The pond could give rise to more new creatures until the conducive conditions for this process were lost.

There are two levels at which we must view the scenario of life on earth: 1) the independent birth of immutable creatures in a primordial pond; 2) the production of artificial breeds, natural varieties and similar species of an immutable creature after it was born, within the framework of limited variations permitted by its genome — i.e., similar creatures which do not basically vary from the fundamental prototype creature. Thus, the presence of similar creatures in the fossil record only indicates the production of similar species of a distinct creature, born independently in a primordial pond.

In the new theory, the simultaneity of the independent birth of numerous creatures is meant in the geological sense. In this scenario, there will be an initial burst of diverse creatures distributed over a few million years and then the frequency of distinct organisms being newly born will reduce (see Figure 11.1). The creatures born from the start of multicellular life will continue to live without any significant change, unless they become extinct. This process will generate a predictable distribution of creatures in the fossil record. This is clearly what we find in reality in the fossil record.

It is crucial to realize that life could not have arisen if genes had to evolve from shorter sequences in prebiotic times. Nor could genes evolve from shorter sequences within the first cell. First of all, the first cell could not even come into existence without a minimal set of complete genes. Thus, there is no question of the origin of a primitive creature on earth — as an improbable freak accident as evolutionists believe — that served as the original ancestor for Darwin's mechanisms or any other evolutionary mechanisms to operate on it. Only because numerous complete genes could directly occur in the vast amount of random DNA sequences in the primordial pond was life made possible at all on earth. Only by such a process were the assembly of numerous genomes and the independent births of multitudes of creatures possible. And thus, only by the independent birth of creatures could all the numerous living things on earth have originated.

Conclusion

So the fossil record, too, verifies the concepts and predictions of our new theory. The sudden appearance of numerous creatures in the fossil record without any common ancestors is a strong indication that all of these creatures originated simultaneously. The fact that the creatures in the Cambrian explosion were unique and evolutionarily unconnectable, and the appearance of entirely new creatures later in the fossil record, also support the new theory, while repudiating all evolution theories. And finally, many creatures that originated eons ago have remained virtually unchanged through the millennia — a contradiction of evolution, but a firm corroboration for the new theory.

Although many details in the fossil record discredit evolution, evolutionists have always tried to force these details into the domain of evolutionary theory — because so far there has been no scientific infrastructure to offer any other type of explanation. These attempts have been largely unsuccessful, but that should come as no surprise when we remember that the effort has constituted a complete inversion of the scientific method. To try to fit evidence to a foregone conclusion is antithetical to all that we know about sound, logical scientific inquiry. Nevertheless the practice persists, but as we have seen in this chapter, many evolutionary biologists and paleontologists themselves clearly recognize the problems presented by the fossil record.

Our new theory of the independent birth of creatures comprehensively explains the evidence in the fossil record, while Darwin's theory does not. Our new theory also cohesively and consistently explains genomic and other molecular details of all living creatures, and explains the uniqueness and unrelatedness of living creatures, while evolution theory does not. The only reasonable conclusion we can draw from these exhaustive investigations is that our new theory of the independent birth of ogranisms is true, while evolution theory is not.

12

Conclusion

The beauty, complexity and diversity of earth's life forms have fascinated the human mind for centuries. Yet the origin and diversity of life on earth have remained persistent enigmas. Although many, beginning with Charles Darwin, have believed this problem to be solved by the theory of evolution, we now know that evolution explains only a small portion of what it purports to explain. While evolution can account for only some aspects of life, many other extremely crucial and fundamental questions are left unanswered. The theory of evolution only superficially appears to explain the origin of organisms on earth, but in fact it is fundamentally flawed.

We have here formulated a new theory that declares that multitudes of creatures were born independently in the primordial pond. The predictions of this theory fit very well with our observations of life on earth in all respects: genes, biomolecules, organisms, and fossils. Whereas evolutionary theories face many major problems in trying to explain the scenario of life, the new theory does not seem to face any substantial problems. The principles of our theory are completely and fundamentally distinct from those of any of the evolutionary theories. While "The Independent Birth of Organisms" may appear radical at first, we have seen that its principles are absolutely reasonable and valid, and the theory itself becomes more convincing the more we understand its principles.

The new theory says that a vast number of unique genes occurred in the random primordial DNA sequences in at least some of the primordial ponds on earth. The rich primordial ponds contained enough DNA material to permit the random formation of numerous genes, which in turn formed multitudes of independent genomes. If the primordial pond's genetic sequences could, probabilistically, contain even one gene with a specific nucleotide sequence that could code for a particular protein, which in turn specified a particular biochemical function, then the pond would have been rich enough in DNA sequences for multitudes of such genes to have formed independently. We have provided ample evidence to corroborate this in Chapters 6 and 7.

The genomes of numerous distinct creatures, both multicellular and unicellular, could have been assembled directly from the gene pool in a DNA-rich primordial pond. We know that there is not much difference in complexity between the genomes of unicellular eukaryotes and those of multicellular eukaryotes, nor is there much difference in complexity among the genomes of various multicellular creatures — whether anatomically and functionally simple, like the worm, or complex, like the human. Thus, if the genome of one unicellular eukaryote could be assembled by random gene combinations, then the genomes for many multicellular creatures could also form independently with the same probability. The genomes for structurally and functionally complex organisms are no more complex than the genomes for relatively simple single cells.

According to the new theory, every independent genome assembled in a distinct seed cell analogous to a zygote, or fertilized egg. We see similar seed cells, as zygotes, in all living creatures today — convincing us that the direct assembly of genomes into distinct seed cells and the direct birth of each creature separately from such seed cells must have been a reality. All these principles are brought forth convincingly in Chapter 8.

The new theory also states that either the independent births of numerous multicellular creatures happened in a rich primordial pond, or no life — even of a simple unicellular creature — could be formed at all. In other words, if a primordial pond is minimally rich enough in genes and prebiotic biochemical processes to give rise to any life at all, even the simplest unicellular creature, then it should be able to give rise to multitudes of creatures, unicellular and multicellular. It is an all-or-none law. Such a phenomenon would lead to the simultaneous birth of numerous unique multicellular creatures from a given primordial pond. If the creatures survived, it would present a fossil picture of a sudden appearance of multitudes of complex distinct creatures at the start of multicellular life on earth. We showed in Chapter 11 that this is what is precisely observed in the fossil record — in the Cambrian explosion and in the Burgess fauna.

The process of independent genome assembly from a common pool of genes would lead to a particular genomic and molecular scenario in independently-born creatures. While many similar genes would be found in various independently-born creatures, there should also be many unique genes. We demonstrated this in Chapter 9. The presence of unique genes and proteins in numerous creatures is truly astonishing, fully validating the new theory. We showed that no theory could explain this scenario by any mechanism of evolutionary change.

Finally, if the new theory is correct, then multitudes of highly distinct creatures should be found in the living world, which is shown to be the case in Chapter 10. And this should be reflected in the fossil scenario, as demonstrated in Chapter 11.

Thus we have shown that the entire scenario of life on earth perfectly fits the new theory of the independent birth of organisms.

In fact, if we destroy the notion that similar genes can be explained only by the evolutionary theory, then there is nothing left to support evolutionary theory. As we have seen, the new theory can explain the presence of both unique genes as well as similar genes in distinct organisms, whereas the evolutionary theory can explain only the presence of similar genes. We must now shed our beliefs in evolutionary theory, which we have imbibed for over a century. There are many facts that easily mislead us to believing in evolution — even if creatures were in fact independently born from a common primordial pond:

- All organisms are made up of genes, proteins, cells, and tissues.
- Several genes in many organisms are functionally the same and nearly identical in sequence.
- Many genes in distinct organisms have similarity in some regions and not in other regions, giving the impression that these have changed through organismal evolution.
- There is considerable similarity in the metabolisms and biochemistry among distinct creatures.

As we have seen, even if numerous creatures were independently born from the common pool of genes in the primordial pond, they will all share the fundamentals of genes and proteins, biochemistries and metabolisms, and cells and tissues. Indeed when we probe deeper into the scenario of life, we see that all evidence is against evolutionary theory and for the theory of the independent birth of organisms:

- Complex eukaryotic genes could directly originate from random primordial sequences, and complex eukaryotic cells could directly arise from primordial ponds.
- The Cambrian explosion illustrates the simultaneous births of numerous distinct creatures in the primordial pond when multicellular creatures first appeared.
- The origin or "rooting" of all the genes in today's living organisms — even by molecular evolutionists' computations — at about 600 million years to one billion years ago indicates a common primordial origin for all genes.
- Unique genes are found in distinct genomes, and they are absolutely unrelatable by any means of evolution.
- Unique protein and cell systems are found throughout the living world.
- Our computer simulation studies show that similar genes, which seem to be evolutionarily related, can appear independently in random genetic sequences.
- The genome sizes in various organisms are randomly distributed — the C-value paradox.
- The presence of absolutely unique organisms with unique structures and body parts forces their classification into multitudes of higher-level taxonomic categories, which are unrelatable by evolution.
- The constancy of organisms since their origins as shown by the fossil record demonstrates that the DG pathways of all creatures are constant.

Consider the reactions that took place in the primordial pond as analogous to chemical reactions occurring in a self-contained pot. Imagine a large pot in which existed very fertile conditions for the organization of genomes — numerous genes coding for enzymes and structural proteins (which in turn can synthesize membranes and small molecules such as sugars, amino acids, and vitamins). Imagine that first many independent genomes are assembled randomly from the large set of genes available in the pot, which are capable of giving rise to living creatures. These assembled genomes are also available openly in the pot while genome assembly continues. Now, pieces of the first successful genomes and genes from the gene pool react and recombine among themselves randomly, and out of these recombinations emerge further new successful genomes. These newly-emerging genomes contain a mixture of existing genomes, and the creatures that arise from the new genomes will share some characteristics with the creatures that came out of the older genomes. But these genomes are unique in their own right, because they are derived by the random mixing of the first available genomes and not derived through organismal evolution.

It is interesting to note that the probability of forming genomes by recombining pieces of the first available genomes is far greater than the probability of the first genomes to have emerged by *de novo* processes. In fact, this process of producing new genomes by mixing existing genomes would rapidly accelerate with time, once this mixing reaction starts. This is because the number of available genomes increases over time, all of which can take part in the continuing mixing process. This partly explains the explosion of new creatures in a given primordial pond, as we have seen in the Cambrian explosion.

We can see that the rich conditions needed for genome assembly could have certainly existed in many of the primordial ponds — from the scientific explorations and research that have been undertaken in the past several decades for proving that chemical evolution did take place. We can also note that life could have originated in more than one primordial pond. There is evidence that it indeed happened, such as the unique Ediacaran and Burgess Shale organisms. The Cambrian fauna seem to have originated in yet another distinct pond. The fauna from the Ediacaran and Burgess ponds have all become extinct, while the Cambrian fauna survived as the creatures we see on earth today.

Thus the new theory puts the origin of multicellular organisms in the same context as the origin of life itself. If we consider that all the DNA conglomerations and genome formations took place within a fairly closed vessel, it is easy to discern that mixing of the genes and parts of various genomes can easily happen. This would lead to the commonality of parts of genomes in distinctly-organized genomes as well as the presence of both the same and unique genes. In this context, one can visualize that similar domains with similar sequences can easily occur in various proteins both by means of their independent origins from distinct sequences and by means of their derivation from common sequences by random recombinations within the closed vessel.

Some of the questions that could be asked of the new theory are: How could genes be simply available in the primordial pond? Even if genes were available in the primordial pond, how could they assemble to form the genome of a living organism? How could all the things needed to form a complex organism be available in the primordial pond and come together, all starting from inanimate matter? We have described answers to these questions in Chapters 6, 7, and 8. Let us remember, even under evolutionary theory, the very same questions apply for the origin of the very first, single living organism, whether it is the simplest multicellular creature, or the simplest unicellular microorganism. We have shown that the genomes of *all* organisms — from amoeba to elephant — are highly but equally complex.

When we look at the complexity of the genome of even the simplest unicellular organism, it becomes clear that no freak accident could create it, as claimed by some evolutionists. They say that genes were improbable in the primordial soup, and therefore life was improbable. The probability for the origin of the genome of the simplest creature, on the contrary, was inevitable. We have sufficiently discussed the details of the genome that indicate that the genome of the simplest unicellular creature cannot simply occur by any kind of freak accident. It has to be assembled from numerous distinct genes, all of which must have occurred probabilistically in a primordial pond. And if conditions for such a probability did exist, then we have shown that the assembly of genomes for numerous multicellular creatures is subject to the same inevitability.

Evolutionists believe that the first creature had a root stock of genes that gave rise to all the genes of all the future organisms — including the multitudes of new and unique genes. While they believe that even one gene cannot occur probabilistically, they expect that many genes needed to construct the simplest original microbe must have appeared on earth accidentally. This expectation has absolutely no scientific basis. Whether it is a prokaryote, unicellular eukaryote, or a multicellular eukaryote, it should contain the necessary number and kind of genes for the most primitive life. It should contain the genes for all the proteins for the minimum biochemical functions that any living creature requires. Even if we conservatively estimate this requirement to be only 100 distinct genes, each as complex as the genes in today's living creatures, at least 100 genes must have existed in the primordial pond. And the probability for 100 genes existing is the same as that for thousands or even millions of genes, so the primordial pond must have been far richer than evolutionists believe.

Remember that at least one simple organism — at least a unicellular creature — must originate somehow for the evolutionary theory to operate upon it. When we logically try to explain the origin of the simple microorganism, we can see that the same explanation will be sufficient to explain the origin of many multicellular creatures simultaneously and independently of each other in a primordial pond. Thus, if the simplest microorganism can and must originate in the primordial pond under the theory of evolution, then there is absolutely no need for an evolutionary theory to explain the origin of the complex multicellular organisms! In other words, if any one single simple creature arose in a primordial pond, then it is inevitable that numerous creatures would similarly arise, simultaneously and independently.

Interestingly, Darwin never addressed the origin of life, except to state that it was irrelevant to his theory.

We can see thus that the new theory is radically distinct from any of the evolutionary theories. There is no need for the origin of a simple unicellular organism first and then the evolution of a multicellular creature, and, from this original multicellular creature, the evolution of many multicellular creatures. If there were genes available to construct the simplest possible organism, then the same process would inevitably lead to the genomes of all the complex organisms on earth.

Thus the theory of independent birth of creatures shatters the widely held concept that it is improbable for even a small gene to occur in random sequences purely by chance. Consequently it shatters the deeply-rooted feeling that life on earth had to be a freak accident. It demonstrates that life was not an accident but was an inevitable consequence of the rich molecular reactions in the primordial ponds.

The ultimate crux of the evolutionary theories is that genetic mutations bring about entirely new genes and unique body parts. However, we have demonstrated that this is virtually impossible. When we probe deeper, we find it is improbable to bring about even one entirely new gene within an organism even over geological time, let alone the multitudes of unique genes present in numerous organisms. Likewise, unique body structures and organs cannot be brought about by means of evolutionary change based on genetic mutation. We demonstrated this in Chapters 3 and 4. Only slight changes and variations can be brought about by mutations and changes in gene sequences. Mutations and sequence changes can occur only in a constant set of genes that exist in the genome of a distinct independently born creature. Mutations can only lead to slightly changed similar species of a distinct creature by various means such as natural selection, geographical isolation, and genetic drift. The scenario of creatures living today attest to these principles. There are numerous sets of similar species, but each set is distinct from the others, unconnectable by evolution. It is the similarity within each set of similar species that has misled Darwin to extend this phenomenon across all creatures, and to say that all creatures on earth had evolved from one or a few ancestral creatures.

According to the new theory, each independently originating unique creature could give rise to many similar creatures by means of organismal change. For instance, an independently-born snail could give rise to many kinds of snails. Only here evolutionary mechanisms such as natural selection and adaptation come into play. Only in such instances do evolutionary theories hold good. Thus the major scenario of life on earth — the presence of multitudes of distinct creatures — is explained by the new theory, and the minor scenario of sets of very similar species is explained by evolutionary theory. The primary means by which all the numerous distinct creatures originated on earth is explained by the new theory. The secondary means by which a set of similar species originated from each distinct, independently born creature is explained by evolutionary theory.

We should note that we are not against the concept of molecular evolution in the primordial pond. We do not say that evolution did not occur at the level of prebiotic organization. Instead, we say that all the molecular evolution happened prebiotically and not within organisms. We are opposed to the claim of Darwin and the evolutionists that all creatures originated from one or a few original creatures by organismal descent with modification. We are against the concept of organismal evolution of unique creatures, by whatever mechanism this is claimed to have happened.

There are many other difficulties for the evolutionary biologists in explaining the scenario of life on earth. For instance, the origin of higher taxa is a great mystery for the theory of evolution. These problems that have bothered evolutionists for decades are very well solved by the new theory. Similarly, there are a number of major problems in the fossil record, such as the Cambrian explosion and the unconnectability of creatures appearing suddenly in later geological periods. Again, these scenarios are very well explained by the new theory. These are not minor details in the scenario of life that may be ignored. These important observations must be addressed by any theory that tries to explain the origin of creatures. While these problems are detrimental to any theory of organismal evolution, the very same observations are well explained by the new theory and indeed strongly support it.

Why do we say that all the later modifications to Darwin's basic evolutionary theory are also incorrect? Because these theories do not show how a new gene or a new body structure could be brought about by evolutionary mechanisms. These modified theories can only explain how mutations might change one snail into another similar snail, but not how to bring about a new gene or a new, truly distinct organism.

One might wonder: If all creatures are based on tissues and cells, and on the same biochemical and biological principles of DNA and proteins, would this picture not logically mean that one creature first originated, and then changed into all the different organisms that we find today? To answer this question, we should go back to the period when life itself originated in the primordial pond. When we probe this question deeper and deeper, it becomes clear that every distinct, independently-born creature would originate by the same processes under similar constraints in the same primordial pond, so that all would share similar tissues, cells, and fundamentally the same biological processes.

The fact that vertebrates in their plasma contain more than 600 proteins that are totally absent in the invertebrates says an important thing: vertebrates did not evolve from invertebrates. Just this fact alone tells us that if vertebrates — which most of us feel are the most complex among all creatures — could originate independently of the invertebrates, then it should be equally probable that numerous invertebrates could originate separately of each other, and that numerous vertebrates also could originate independently of each other. The fact that numerous invertebrates have many unique proteins in their plasma and many other unique cellular systems corroborates this and says that they are distinct, unrelated creatures that have nothing to do with vertebrates. We are making a reasonable extrapolation here. If just one or a few proteins are different between the vertebrates and invertebrates, then we can think about the possibility of some other mechanisms. But the whole set of plasma proteins and whole cellular systems are distinct (Chapter 9). We demonstrated in Chapter 4 that it is highly improbable for any genetic mechanism to create such vast changes in the genome.

When we look at the whole molecular scenario with this clear background in mind, the facts that essentially identical genes (such as the globin genes) exist in many invertebrate and vertebrate organisms is not a problem. If we look at only the scenario of the same genes across some creatures, then naturally we would think that these creatures must be related by organismal evolution. But this is only one piece of the puzzle. The presence of unique genes in various creatures is another important piece, which can exist only through independent origins. When we look at the whole picture, the only way to resolve this puzzle is to agree that identical copies of genes as well as unique genes were included in genomes that were independently assembled in the primordial pond. When we bring together all the details of life on earth — molecules, organisms, and fossils — it is only the theory of the independent birth of organisms that can explain the whole scenario of life (see Figure 12.1).

Considering the fossil record, we need to sit back and look at established "facts" with a new view. Until recently it was thought that prokaryotes were the first living creatures, which originated 3.6 billion years ago, and that single-celled eukaryotes appeared on earth about a billion years ago. Even many textbooks and research articles continue to say this.[1] However, molecular biologists now essentially agree that single-celled eukaryotes were the first to originate, and that the prokaryote must have been derived from them. That is, even considering what the fossil record seems to say about the origin of prokaryotes and single-celled eukaryotes, molecular biologists are able to accept something that is fully contrary to

Which theory is valid?
The New Theory of the Independent Birth of Organisms or the Theory of Evolution?

Scenario of life on earth	Is the scenario explained by...	
	Darwin's theory or any other theory of evolution?	The new theory of the independent birth of organisms?
Molecular scenario:		
Unique genes in distinct organisms	No	Yes
Similar but unrelated genes in distinct organisms	No	Yes
Identical and relatable similar genes in distinct organisms	Yes	Yes
C-value paradox	No	Yes
Uniqueness of genomes	No	Yes
Organismal scenario:		
Origin of unique structures and body parts	No	Yes
Origin of higher taxa	No	Yes
Distinct, unrelatable sets of creatures	No	Yes
Origin of highly complex organs	No	Yes
Origin of instincts and intelligence	No	Yes
Artificial breeds and natural varieties	Yes	Yes
Similar species of a distinct creature	Yes	Yes
Fossil scenario:		
Cambrian explosion — sudden, simultaneous explosion of creatures at the start of multicellular life	No	Yes
Constancy of organisms in fossil record	No	Yes
Sudden appearance of new creatures at later geological periods	No	Yes
Burgess shale fauna	No	Yes

FIGURE 12.1.

it. In a similar manner, when we take a fresh look at the fossil record based on the new theory of the independent birth of creatures, we see that it precisely fits the new theory. We have merely reinterpreted the data, showing that it presents none of the problems that evolutionary theories have faced. Indeed, if any data needs modification, it is only that concerning the total time represented by the fossil record because of the possibility of some errors concerning the ages of fossils.

The Cambrian explosion, an enigma under Darwin's theory — is clearly explained by the new theory. All the major taxa and body plans appear on earth in an extremely short geological period of 5–10 million years, at the start of multicellular life. This clearly illustrates that all the multitudes of unique creatures directly, independently, and simultaneously originated from the primordial pond — especially combined with the fact that such a sudden burst of creatures can never be explained by evolution.

The new theory states that after the first big burst of creatures, the primordial pond continued to be active in the birth of new creatures, except that in the later periods of such activity, more and more genome mixing or genome restructuring occurred, giving rise to newer creatures that were related in their genomes to earlier creatures (note that these are not related by descent with modification). The Cambrian explosion of multitudes of creatures occurred within about five million years. Let us accept that the births of all creatures in the primordial pond actually occurred within only about 50 million years. The fossil record clearly shows that all the new and unique creatures have appeared within the first 30–50 million years from the start of the Cambrian explosion — although some creatures such as birds and mammals seem to appear later. It is quite possible that in the future we may find reasons for this discrepancy. Either small errors in dating method may be found (if they were even 50 percent erroneous the discrepancy would be enormous), or the dates of the later fossils may be pushed back by new fossil finds, which happens all the time. Thus, the new theory is conceptually acceptable and correct.

Each organism termed a species is not necessarily distinct. We do not say, for example, that many different snail species are distinct and that they were all born independently of each other in the primordial pond. There are organisms that are fundamentally similar and organisms that are fundamentally distinct. Each set of fundamentally similar organisms form a distinct group. The prototype of each distinct group (the snails, for example) is unique and distinct from the prototypes of other distinct groups (such as the crabs). There is no doubt that vertebrates are distinct from invertebrates, and that there are numerous distinct groups within each. This is consistent with the fact that there are many invertebrate phyla, and that all the vertebrates form only one half of one phylum. We only need to know which are the distinct groups. At a gross level we can guess, but to be precise, it perhaps needs research.

We see that usually creatures defined within a family could be similar creatures "evolutionarily" derived from an independently-born creature. However, we are open to the possibility that sometimes creatures classified within even a larger taxonomic category such as an order could be derived from one organism. However, if we accept that the vertebrates are not at all related to any invertebrate by means of evolution, then we accept that at least a simple vertebrate, however primitive, could be born independently in the primordial pond, even if it had to use parts of invertebrate genomes. If we accept this possibility, then why should we not accept that many vertebrates arose by the same mechanism? Thus, we should keep open the possibility that creatures within an order originated independently by this means of genome restructuring in the primordial pond. In other words, dogs and bears could be either related organismally, or they are not related organismally but are related through their genomes in the primordial pond. As long as we stick to the definition of the independently-born creature — unique developmental genetic pathways, unique body structures, and a constant set of genes — we can correctly determine which organisms were independently born in the primordial pond and which were then derived from the independently born creatures by means of modification and change.

Perhaps the problem with Darwin's evolutionary theory lies in the fact that he searched for the origin of "species." The true species as we see in light of the new theory is a variety of an independently born creature. It is the resemblance among these varieties (*i.e.*, similar species) of a single organism that misled Darwin. It is the extension of this organic relationship among the varieties of a truly independent creature that has led to the misinterpretation of the whole scenario of life on earth. Thus what we have done here in this book is to trace the origin of the numerous distinct independent creatures and not that of similar species.

We may ask if we can prove the new theory by conducting some simulated primordial pond experiments and producing some independent organisms. Unfortunately, it is not possible to conduct such experiments because the amount of random DNA and other materials that are needed for even the simplest life form to come about is far too large. However, we should realize that the DNA sequence information of living creatures is the ultimate information that we need to prove the theory. Until the early 1980s, one could not have formulated the new theory, because the minimal amount of DNA and protein sequence information from living organisms required for proving such a theory was not available. But now we have the minimum amount of such information with which we can prove the theory, and this is all we will ever need to show what happened in the primordial pond eons ago.

The new theory is largely based on genetic sequences of living organisms, which are far more reliable than the fossilized bones of extinct organisms layered in the earth, a process about which we know very little. Even if we wait for a million years, the best information we can get for analyzing the origin of creatures is their DNA sequence information, with which we will have to derive the history of life. The complete genomic sequence information of several organisms, which should be available in the future, should verify the validity of the theory even further.

Out of a billion species that have so far appeared on earth (including those that have become extinct), several million creatures may have originated independently in the primordial pond, and from each distinct creature came many varieties and similar species by a number of mechanisms — natural selection, genetic drift, mutations in trivial genes such as those that affect the coat thickness, color, or body size. It is possible that some "new" creatures can originate from the independently-born creatures by losing body parts or functions that are not crucial to the life of the organism. Slight modifications of the body parts or functions should also have been possible. Thus, evolution has played only a minor role in the origin of organisms on earth, and it is the independent birth of organisms that has played the major and most important role.

Each creature that originated in the primordial pond has lived forever as fundamentally the same creature, diversifying only slightly to produce incidental variations of essentially similar species. Perhaps the greatest practical significance of this new theory lies in our recognition that new creatures will not and cannot evolve to replace organisms that become extinct. Except for similar variations of existing organisms, no new creatures will ever arise again! The fossil record confirms that no fundamentally new creatures have appeared on earth in a very long geological time, and this is a sobering reminder that we must do everything humanly possible to preserve these beautiful creatures for future generations to cherish and enjoy.

The new theory is important because it breaks down the conventional barrier to biological thought: that all organisms on earth evolved from one original ancestral creature. This new notion is likely to have significant ramifications because it brings out an entirely new way of thinking. Our new theory accommodates most of the evidence that cannot be reconciled with evolutionary theories, and paves the way for more meaningful biological and biomedical research. It seems likely to carry tremendous implications in research, education and environmental protection, and in the general philosophy and culture of people, for it will significantly change the conventional Darwinian way of thinking in every walk of life. In the past several decades virtually all biological processes have been explained in terms of evolution, and several times in this century the concept has even been extended to sociology and economics, where the doctrine of "survival of the fittest" has been cited to justify a number of political agendas. Our theory of the independent birth of organisms stands all of this conventional wisdom on its ear.

Our new theory shows that the biological connections among living things date back to the prebiotic processes in the common primordial pond. It shows that the unifying principles of biology are derived not from evolutionary connections among creatures descended from a single common ancestor, but rather from their origins in a common primordial pond eons ago.

Our explorations in this book have shown us that the endless beauty of life on earth appeared all at once. Incredible as it may seem, this immense splendor of the biosphere came into being in a geological instant. And it happened inevitably from the chemical reactions of a single primordial pond. Many ponds may have produced life during that fertile period of the earth eons ago, but the life from only one pond survived until today. All living creatures suddenly erupted from that pond, and simply walked, swam, flew or flowered away to fill the earth with the awesome power and beauty of organic Nature.

At the dawn of life, Nature set the stage. Millions of diverse organisms arrived at the same time to play scene after scene in the complex ecological drama that unfolded. Even today, the original players remain virtually unchanged, although over the millennia many have permanently left the stage. This has been the truth of life on earth from the beginning, and will remain the truth of life forever. We can only hope that these fresh insights into the origins of life on earth will better equip us to preserve life in all its diverse glory, and will motivate us to the task — and sensitize us to its urgency. Let us not perturb Nature, lest more and more participants leave the stage. We must rededicate ourselves to forestalling the final curtain for as long as we can, so that our future generations may enjoy what we cherish today.

Appendix: Genetics Primer

The new theory of the independent birth of organisms is corroborated by what we know about genes and genomes, and about how they function to develop an organism. This book deals with the origin and diversity of organisms at the most fundamental level of life — the origin of genes and genomes. A general knowledge of molecular genetics would greatly aid readers in following and thoroughly understanding the principles discussed in this book.

Please be aware that much of what follows is gross simplification. A complete inventory of our current knowledge of genetics, genes and genomes could easily fill thousands of pages — in very small print — but complete technical details are unnecessary to a full understanding of the principles of our new theory. We therefore provide only a general overview of these topics here.

THE ORGANISM, ORGANS, TISSUES AND CELLS

A MULTICELLULAR ORGANISM IS MADE UP OF ORGANS, TISSUES, AND CELLS

Any organism, whether a butterfly or a snail, an elephant or a blue whale, is made up of organs and tissues. Each organ carries out a specialized function. The function of an organ could be purely biochemical such as the liver regulating blood chemistry, or more physical, such as the heart pumping blood. Organs also include structures like the skin, the eye, or the leg. An organ is composed of tissues, which are groups of cells working together to carry out a function. The heart, for example, is mostly muscle tissue. The cell is the fundamental unit of all body parts, whether it be the brain, bone, skin, eye, muscle, or tongue. A living creature is made up of various types of tissues and cells. For instance, the human body is made up of approximately 300 distinct tissues and body parts, consisting of about ten trillion cells.

AN INDIVIDUAL MULTICELLULAR CREATURE STARTS TO GROW FROM A SINGLE CELL

Usually, an organism starts its growth from a single cell. Take for example a chicken. When an egg is laid, it contains only one single cell, from which the chicken will grow. The rest of the egg is food for the growing embryo. The single cell absorbs the food, grows, and divides into two cells, then into four cells, then into eight, sixteen, thirty-two, and so on, and grows in a programmed manner into a fully-developed chicken before it hatches out of the egg (see Figure 1). The first single cell contains the programmed instructions for it to convert the food in the egg into the cells of the chicken, and to develop the particular body parts of the chicken.

THE PRECISE INSTRUCTIONS FOR THE GROWTH AND DEVELOPMENT OF AN ORGANISM ARE CONTAINED IN ITS GENES AND THE GENOME, AND THE TOOLS FOR GROWTH AND DEVELOPMENT ARE THE PROTEINS

The chicken consists of many body parts: legs, wings, head, eyes, beak, heart, lung, stomach, and so on. When the chicken grows from the single cell, all these body parts must be developed at the right places in the chicken. No mistakes can be made either in the shape and size of these body parts or in their positions in the body of the chicken. How does the first cell achieve this, and where do these instructions come from?

It is the genome that contains all these instructions. Each instruction is contained in units of what are called genes. Genes, however, do not directly carry out any function. They simply carry these instructions, like an instruc-

GENETICS PRIMER 537

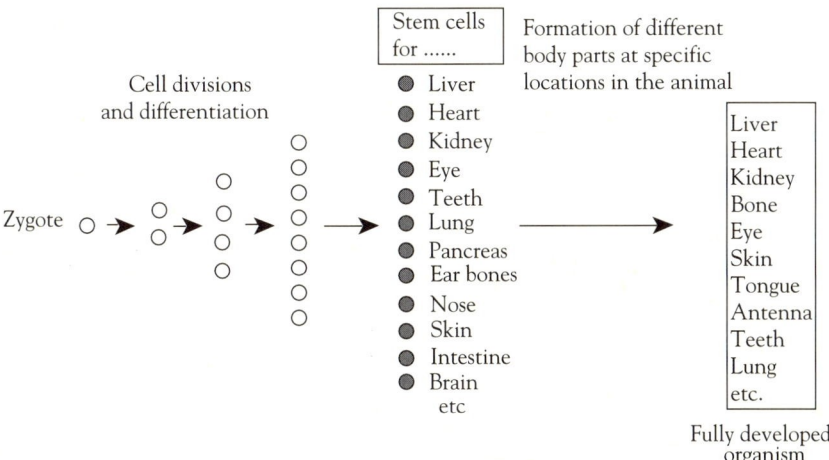

FIGURE 1. THE DIFFERENTIATION OF THE ZYGOTE INTO MANY DIFFERENT TYPES OF CELLS FORMING VARIOUS TISSUES AND ORGANS. The zygote is a single cell, containing a full complement of the genome of an organism. Only a small subset of the genes in the genome is expressed in it. As the zygote multiplies into many cells, at some very early stage the commitment of different cells to become particular body parts occurs. This happens by the differential activation of genes present in the genome that most probably starts at the first few divisions of the zygote.

tion manual for assembling a bicycle, only billions of times more complicated. Every living cell in an individual contains the same complete set of instructions.

Each instruction is usually biochemical in nature, and carried out by molecules called proteins. For instance, the first cell in the chicken egg has to take up the sugar in the food around it and break it down to smaller sugars to derive energy. For this purpose it needs some specific proteins called enzymes. In the case of breaking down a sugar like sucrose, it would require an enzyme called sucrase. This is a particular protein with a particular structure. It can bind the sugar sucrose and break it down to its component sugars glucose and fructose. There are thousands of such specific biochemical reactions that occur in the first cell, each requiring a specific enzyme protein. The message for constructing each of these protein molecules is contained in specific genes. All these genes that are required to construct every cell, tissue, and organ of the chicken make up the genome.

The Protein

What is a protein?

At the grossest level, the tissues, organs and body parts carry out the function of the organism through their external and internal coordination. The chicken walks, sees, eats, and copulates, which are external organismal functions. The chicken's stomach digests the food, the heart pumps its blood, the lung helps in breathing, and the ovary secretes the ovum. These are the internal organismal functions.

At a finer level, however, it is the unit of life, the cells, that make possible the functions of all these tissues and organs — for instance, eye cells, muscle cells, liver cells, and brain cells. Ultimately, it is the proteins in the cells that carry out the functions of all the cells — liver proteins, eye proteins, and brain proteins. Although often small molecules carry out some biochemical functions in the cells, such as the vitamins and steroids, it is the proteins that synthesize these small molecules, and when necessary, break them down. Thus, it is the proteins that are the true workhorses of the cells, tissues, and organs, and in fact, the whole organism.

Amino acids and peptides

A protein is a molecule made up of a linear chain of smaller units called *amino acids*. There are 20 distinct amino acids that make up the proteins. All 20 amino acids have the same basic structure with slight distinguishing variations.

One amino acid can be coupled to another by a chemical reaction forming a *peptide bond*, giving rise to a *dipeptide*. The dipeptide can be coupled with another amino acid molecule forming a tripeptide. This reaction can be extended to form a linear *polypeptide* chain containing thousands of amino acids (see Figure 2). In fact, polypeptides found in nature vary from a length of approximately 30 amino acids to about 4000 amino acids. Any amino acid can be coupled with any other amino acid in a growing polypeptide chain, so that any amino acid sequence is possible. The terms protein and polypeptide are generally interchangeable. The term protein usually indicates the polypeptide chain in its specific three-dimensional shape. In some cases two or more separate polypeptides can link loosely together to form one large protein.

A unique amino acid sequence of a polypeptide leads to a unique three-dimensional structure of a protein

The physicochemical properties of the 20 different amino acids are different. Some can be positively charged, and some negative, while some are

Ala + Gly ⟶ Ala-Gly   Dipeptide

Ala-Gly + Tyr ⟶ Ala-Gly-Tyr   Tripeptide

Ala-Gly-Tyr + Arg ⟶ Ala-Gly-Tyr-Arg   Tetrapeptide

Ala-Gly-Tyr-Arg-Ser-Thr-Val-Leu-Glu-Arg-Ile   A polypeptide

FIGURE 2. THE SYNTHESIS OF A POLYPEPTIDE FROM AMINO ACIDS. An amino acid can be coupled to another by forming a peptide bond resulting in a dipeptide. The dipeptide can be coupled with another amino acid by a similar reaction forming a tripeptide. This reaction can be extended to form a polypeptide chain of thousands of amino acids.

hydrophobic (water-hating) and others are hydrophilic (water-loving). A hydrophilic group is attracted to water whereas a hydrophobic group repels water. In addition, the three-dimensional structure, shape, and size of each of the amino acids also varies.

The amino acids in a polypeptide chain can interact among one another in many different ways. In living organisms, a highly specific set of interactions are found to occur among the different amino acids, resulting in unique three-dimensional (3-D) structures of the polypeptides. This "folding" of a polypeptide chain into a 3-D protein structure is a complex, poorly understood phenomenon. When we see the structure of a polypeptide, it reveals a "back bone" of peptide bonds. The back bone seems to contribute to the specific folding pattern of the protein, at least by holding the different amino acids in place and by applying certain restrictions in its bending. In addition, the interaction of the amino acids with water molecules greatly contributes to the folding phenomenon. The hydrophobic groups come together and coalesce, so that they form an interior core protected from water, while the hydrophilic amino acids fall outside of the protein structure, so that they face the surrounding water molecules (see Figure 3). The folding dynamics of a given protein molecule seem to be such that a unique structure is formed from a given amino acid sequence under a given set of chemical conditions.

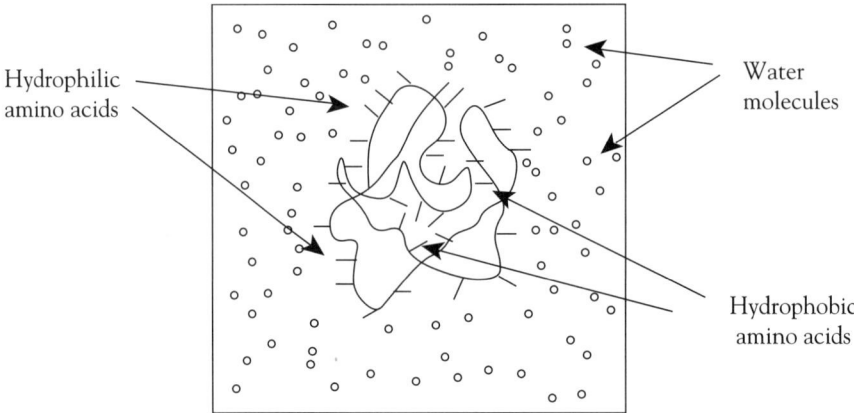

Figure 3. The main forces in the folding of a polypeptide into the three dimensional structure of a protein are the hydrophobic and hydrophilic interactions of the different amino acids. Although other interactions among the amino acids play roles in folding a polypeptide, the main forces are the hydrophobic and hydrophilic interactions. A polypeptide tends to fold so that most of the hydrophobic amino acids are located in the interior of the protein, and most of the hydrophilic amino acids are outside facing the water molecules.

Although the individual amino acids have no catalytic or enzymatic activity themselves, when a chain of amino acids is organized into the protein macromolecule in a specific shape and size, and with a particular arrangement of amino acids in the three-dimensional structure, the protein can have catalytic activity or binding activity with another macromolecule or small molecule. The primary amino acid sequence determines the specific arrangement of a polypeptide chain in a specific shape and structure, which in turn determines the protein's function. The sequence of different amino acids in the protein chain therefore confers specific functions.

A note about catalytic or enzymatic reaction is needed here. If we leave a solution of the sugar sucrose at room temperature by itself in a closed sterile vessel, it will remain sucrose for many years. However, we say that a particular protein called sucrase has the ability to break the sucrose down into glucose and fructose. How is this achieved? This protein has some specific amino acids in its 3-D structure placed in such a manner that allows it to grab a sucrose molecule and break it apart into the two smaller units. It is simply like a particular machine that carries out a particular function.

Likewise, another protein can grab two specific small molecules when they come near it and can create a chemical bond between them, thereby uniting the two small molecules. In this case, this enzyme is said to have a synthetic activity.

The amino acid sequence of a given protein can vary greatly without affecting its structure or function: the degeneracy of amino acids in a protein

In the above we said that a unique amino acid sequence leads to a unique three-dimensional structure of a protein. However, amino acids at many places in a linear sequence can be replaced with some other similar amino acids without changing the 3-D structure of the protein. At the same time, changes at some other positions in the linear sequence have drastic consequences. When we change such an amino acid, it will dramatically alter the 3-D structure and thus the protein's biological function. These are called invariant amino acid locations. The positions where certain amino acid changes do not alter either the structure or the function of the protein significantly are called the variable amino acid locations. At such locations the variability of the amino acids is called amino acid degeneracy (see Figure 7.13).

In reality we see that most positions in a protein are variable. Only very few are invariant. In other words, amino acids in many positions in a given protein are degenerate. This is an important phenomenon in understanding the origin of life and the origin of diverse organisms.

As an example, some of the acceptable variations in the protein λ repressor are shown in a short region of 17 amino acids (see Figure 7.13). It can be seen that a large number of variations at most of the positions in the amino acid sequence does not affect the three-dimensional structure of the protein, and therefore its function. This figure is shown only to illustrate the high level of degeneracy typical of a protein.

The many different kinds of protein functions

Proteins function in many different ways in living organisms. One of the main functions of proteins is to perform specific biochemical reactions. As we saw above, proteins with such specific catalytic functions are called *enzymes*. For instance, the enzyme glutamine synthetase converts the amino acid *glutamate* to *glutamine*.[1] The enzyme *urease* catalyzes the hydrolysis of urea, and *arginase* catalyses the hydrolysis of the amino acid arginine.[2] So far, several thousand different enzymes, each capable of catalyzing a different kind of chemical reaction, have been discovered in living creatures. Other kinds of proteins include transport proteins in blood that bind and carry specific molecules or ions from one organ to another. For instance, the *hemoglobin* protein of red blood cells picks up oxygen from the lungs and carries it to the tissues where the oxygen is released, and carries back carbon dioxide to the lungs to be exhaled. Proteins also serve as nutrient storage devices. *Ovalbumin*, the major protein of egg white, and *casein*, the major protein of milk, are examples of nutrient proteins. The *ferritin* in animal tissues stores iron. *Actin* and *myosin* are filamentous proteins functioning in the contractile system of muscle.

Many proteins serve as the structural building components of cells and body parts. *Collagen* is the major structural protein of tendons and cartilage. *Elastin* can stretch for use in ligaments. *Keratin* is a tough protein material that makes up hair, fingernails, and feathers. *Fibroin* is used to make silk fibers and spider webs. *Immunoglobulins* (antibodies) defend against infection in vertebrates. Hormones, such as *insulin, growth hormone*, and *thyroxin*, regulate sugar metabolism and other pathways. Other regulatory proteins, such as *repressors*, regulate the operation of genes and comprise many structures associated with genes (see below). In summary, proteins carry out a variety of molecular operations in cells, organs, and the whole organism, enabling the existence of living systems.

Proteins are the most abundant molecules in cells and constitute over half the dry weight of most organisms.[3] As will be discussed later, while the proteins carry out the actual work of the cells and the organisms, it is the genes which are the passive carriers of information for the proteins. We still do not know how many kinds of proteins are present in even one organism, such as the chicken or the human. However, we are able to roughly estimate that there may be up to 30,000 different proteins in a typical creature.

Some proteins contain chemical groups other than amino acids[4]

Many proteins contain only amino acids and no other chemical groups; these are called simple proteins. However, some proteins contain one or more chemical components in addition to amino acids; these are called conjugated proteins. The non-amino-acid part of a conjugated protein, usually a small molecule, is called a *prosthetic group*. Conjugated proteins are classified on the basis of the chemical nature of their prosthetic groups. Lipoproteins contain lipids, glycoproteins contain sugar groups, and metalloproteins contain a specific metal, such as iron, copper, or zinc.

Some enzymes require a separate nonprotein cofactor for their function; this could be an organic molecule called a coenzyme or some inorganic component such as a metal ion. Vitamins are precursors of various coenzymes. For example, riboflavin (vitamin B_2) is a component of the coenzymes called flavin mononucleotide (FMN) and flavin adenine dinucleotide (FAD). Enzymes that require a metal ion for activity are called metalloenzymes. For instance, there are many iron-requiring enzymes.

PROTEINS SYNTHESIZE ALMOST ALL THE COMPONENTS OF LIVING ORGANISMS

The structures of an organism are constructed mainly with a few different types of large molecules. They are proteins, lipids, and carbohydrates. In addition, hard parts of the body such as bones are built with minerals such as calcium. The mechanism by which the animal produces these structures is by the action of proteins on precursor molecules found in food. For instance, a set of proteins called the bone morphogenetic proteins (BMPs) are responsible for the synthesis of bone from ingested calcium. This is the way that proteins function as the workhorses of the cell, synthesizing the molecules required to build and maintain the body. They synthesize the lipids and carbohydrates and assemble the cells. In fact all the work of an organism at the molecular, cellular, and organismal levels is ultimately carried out by the proteins.

The genes: The fundamental units of a genome

How are these proteins produced in the organism? They are contained as coded messages in the genes. One gene, for example, may contain the information needed to construct one protein. Genes are made up of DNA, another type of large molecule, which resembles a long twisted ladder (a double helix) containing a specific sequence of different "rungs," called *nucleotides*. Information or instructions as to when the different proteins should be precisely produced, is also ingrained in the DNA nucleotide sequences in the form of genetic circuits or pathways. Because the expression of a gene can be switched on or off by the action of another gene, a set of genes can be connected through a genetic network. Thus a set of genetic switches can be integrated in a specific genetic network. DNA functions as the memory-storing molecules for the proteins, and the proteins do the actual work of the cells and the organism.

THE GENOME

The genome contains the complete complement of an organism's DNA. When we speak about a genome, sometimes it refers to that of a species and sometimes to that of the individual of that species, in appropriate contexts. The genome of an organism contains all the genes that are needed to build the organism, but also contains considerable nongenic DNA. Genes constitute only a small proportion, typically less than five percent of the whole genome. These are only rough estimates. The rest is considered "junk DNA," whose origin has been so far unclear.

The number of genes in a genome

Although at present it is difficult to determine the total number of genes used, or expressed, in a genome, there is an overall consensus about this.[5] The human genome is believed to express approximately 50,000 genes, out of which 10,000 genes may be required to build and maintain a basic cell — we call them here the "basic-cell genes,"[6] all of which are used in every cell in an animal and are required for the existence of the simplest living cell. Therefore, if we divide the remaining 40,000 genes evenly among the approximately 300 different tissue types, about 130 tissue-specific genes are needed to build and maintain each tissue (an organ is built with many tissues and therefore needs correspondingly more genes).

The total number of different species living on earth is approximately in the tens of millions. A large number of basically the same proteins are present in many of these organisms — whether they are bacteria, plants, or animals. But there are also a large number of unique proteins in many organisms that are not found in other organisms. It is thought that the total number of genes found in the living world may be one to ten million.[7] What is important to remember is that different organisms express unique sets of genes in addition to expressing a common set of genes.

Most of the DNA in a genome is junk: The genes exist only as small islands in large oceans of meaningless DNA

A genome contains not only genes but also sequences that are not genes and have no function. These are called nongenic or intergenic sequences. Almost all the genes are dispersed in a genome with long intergenic sequences separating consecutive genes. The length of the intergenic sequence between consecutive genes is not defined; it can be many times longer than genes. The origin of the intergenic, junk DNA sequences is so far not clearly understood. In fact, the origin of the more important parts of a genome, the genes, itself has not been fully understood until recently. (See Chapter 7 for a new explanation on the origin of genes and Chapters 8 and 9 on the origin of junk DNA.)

The genes are organized successively one after the other on the linear DNA molecule, with junk DNA separating them. However, the genes in a genome do not seem to be organized in any functionally recognizable order, except in rare occasions. Tissue-specific genes do not seem to be grouped together. For example, the eye-specific genes are not together and separate from liver-specific genes and so on. Some other types of genes such as globin genes or immunoglobulin genes are, however, clustered together.

The genome in a multicellular creature is usually separated into many chromosomes

The genome of a multicellular creature is usually made up of several separate DNA molecules. Each DNA molecule is complexed with proteins and is called a *chromosome*. A chromosome contains a very long single DNA molecule, usually running a few millions to a few hundred millions of nucleotides. A chromosome of a multicellular creature is almost always associated with proteins that function to maintain its structure, or to regulate the genes in it. *Histones* are a set of proteins that "package" the DNA and lead to an ordered chromosomal structure called *chromatin*; the DNA in the chromatin is very tightly associated with the histones. Non-histone chromosomal proteins are unique proteins that are thought to function in activating and deactivating the expression of different genes by binding to specific DNA sequences.

DNA is the genetic material. RNA is usually used for copying purposes and temporary uses of passing on messages.

In 1953 James Watson and Francis Crick postulated the double helical, twisted ladder structure of the deoxyribonucleic acid (DNA). Their hypothesis, in addition to solving the structure of the DNA molecule, showed how it could be duplicated exactly (see Figure 4). The sequence information in a given DNA molecule can be exactly duplicated because of the fundamental structure of the DNA. Each of the two sides, or strands, of a DNA ladder contain a nucleotide sequence that is precisely complementary to the other. Therefore, a complete double-stranded DNA sequence can be derived from either strand. Watson's and Crick's discovery soon led to the *central dogma* of molecular genetics.[8] This dogma defines three major steps in the processing of genetic information. The first is *replication*, the process of copying the parent DNA into two daughter DNA molecules with nucleotide sequences identical to those of the parent DNA. The second step is *transcription*, the process in which the genetic message in DNA is copied into ribonucleic acid (RNA) molecules. The third step is *translation*, in which the protein sequence message contained in the RNA code is translated into protein.

Each strand of the DNA molecule is a linear chain consisting of repeating units of one of four possible nucleotides, which are distinguished by their *bases*. The bases adenine, guanine, cytosine, and thymine are symbolized as A, G, C, and T. The two strands of a double-stranded DNA are complementary to each other and are paired between the nucleotide bases on each strand through hydrogen bonding. The underlying rule is that A on one

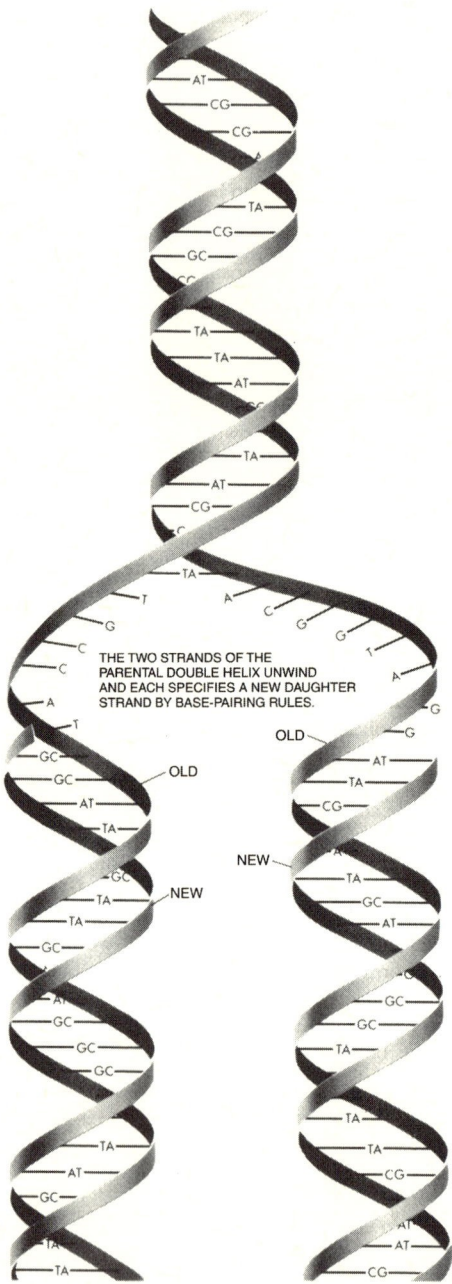

FIGURE 4. THE DNA DOUBLE HELIX AND ITS REPLICATION. The DNA is a double helical molecule with complementary base-pairing between the bases A and T, and C and G, producing the A•T and G•C base pairs. When it replicates, identical daughter double helices are generated from the parent DNA molecule. [From *RECOMBINANT DNA: A SHORT COURSE* by Watson, Tooze, and Kurtz. Copyright © 1983 by James D. Watson, John Tooze and David T. Kurtz. Adapted with permission of W. H. Freeman and Company.]

strand always pairs with T on the other strand and, likewise, C pairs with G. In its linear sequence of nucleotides, DNA stores the information for the linear sequence of amino acids in proteins. The stretch of DNA containing the message for a protein, and the sequence associated with its control of expression, is called a gene. A long DNA can contain many genes.

RNA (ribonucleic acid) is similar to DNA. However, it is almost always single-stranded. The sugar portion in RNA is ribose, whereas in DNA it is deoxyribose. While DNA functions as the memory keeper for proteins, RNA has many active functions in the cell. An RNA copy of the gene, the *messenger RNA* (mRNA), is used for the translation of the gene's coding sequence into its corresponding protein, after which the mRNA is destroyed. It is similar to making a photocopy of a master document, using and then destroying the copy, leaving the master document unaffected. *Ribosomal RNA* (rRNA), complexed with many specific proteins, functions in the ribosomes, the machinery that translates the mRNA coding sequences into proteins. *Transfer RNAs* (tRNAs) are the true "decoder" molecules which provide the physical link between the mRNA and the amino acids they code for. *Spliceosomal RNA*, in combination with many proteins, functions in the spliceosomes, the machinery that cuts out unwanted sequences from the primary mRNA. In this manner, RNA can function much like a protein.

DNA SEQUENCES: VARIOUS KINDS

A *gene* is a stretch of DNA sequence that can code for the sequence of a protein molecule. The natures of the protein and the DNA molecules are such that the information for the linear chain of amino acids in a protein can be coded in the four-letter linear DNA sequence. There are 64 three-letter codes possible from the four nucleotides in a DNA sequence, which are called *codons*. Each consecutive codon in a linear DNA sequence determines which amino acid of the 20 that make up proteins should be added onto a linear protein chain, or if the growing protein chain should be terminated. Three out of the 64 codons are the chain terminators or *stop codons*. The rest code for the twenty amino acids — in some cases one codon codes for one amino acid, and in other cases more than one codon codes for the same amino acid. Table 1 is called a Codon Table, showing the different codons and the specific amino acids they code.

A gene thus contains a DNA sequence that codes for a protein chain, which is usually called a *coding sequence*. It also contains a stretch of sequence near the coding sequence, the *promoter*, that indicates where the gene's coding sequence starts. It is at the promoter sequence that the enzyme called *RNA polymerase* binds and copies the gene into a mRNA molecule. Near

		SECOND BASE			
		U	C	A	G
FIRST BASE	U	UUU } Phe UUC UUA } Leu UUG	UCU } Ser UCC UCA UCG	UAU } Tyr UAC UAA } STOP UAG	UGU } Cys UGC UGA STOP UGG Trp
	C	CUU } Leu CUC CUA CUG	CCU } Pro CCC CCA CCG	CAU } His CAC CAA } Gln CAG	CGU } Arg CGC CGA CGG
	A	AUU } Ile AUC AUA AUG Met	ACU } Thr ACC ACA ACG	AAU } Asn AAC AAA } Lys AAG	AGU } Ser AGC AGA } Arg AGG
	G	GUU } Val GUC GUA GUG	GCU } Ala GCC GCA GCG	GAU } Asp GAC GAA } Glu GAG	GGU } Gly GGC GGA GGG

TABLE 1. The Codon Table. A = adenine, G = guanine, C = cytosine, and U = uracil. U in RNA replaces T in DNA. Codons are usually represented as RNA codes.

the promoter, there is another sequence that regulates when it should be switched on, i.e., when the gene's coding sequence should be expressed into a protein. This is called a *regulatory sequence* or *operator*. It specifies the "switch" for a gene, which can be switched on or off by the binding of one or more *regulatory proteins* to it.

The typical gene of a multicellular creature is split into coding and non-coding sequences

In all multicellular organisms, the coding sequence of genes is interrupted by non-coding sequences (see Figure 5). The pieces of coding sequence are

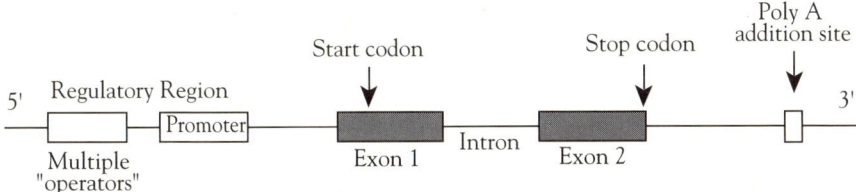

FIGURE 5. THE STRUCTURE OF A EUKARYOTIC GENE. The coding sequence is discontinuous — it is split into coding exons and noncoding intervening introns. The exons are usually short, with a maximum of about 600 nucleotides, whereas the introns are usually long, up to many thousands of nucleotides. The longest known gene contains about one million nucleotides. Usually more than one protein binds to the regulatory region in a eukaryote, especially in multicellular organisms. The start codon for translation of the coding sequence is usually located somewhere within the first exon, and the stop codon is usually the last codon of the last exon. Typically a gene contains many exons (so far up to about 100 exons are known in one single gene). Processing of the primary RNA requires a poly A tail which is added at a site after the last exon. The figure shows the main features of a eukaryotic gene, but their lengths are not drawn proportionally.

called *exons*, and the intervening sequences, which apparently have no function, are called *introns*. Introns are similar to intergenic junk DNA, except introns occur within individual genes rather than between genes. The average length of an intron is approximately ten times longer than the average length of an exon.

The sequences around the junctions of exons and introns are highly conserved, meaning that almost the same sequence is found at the junctions of every exon and intron in all genes. A specific sequence of nine nucleotides is found at the junction of every exon and intron (at the end of an exon and at the start of an intron). This is called the *5' splice site*. Similarly, a short sequence of four nucleotides is present at the junction of the intron and exon (at the end of an intron and start of the next exon). This is called the *3' splice site*. These sequences are thought to be the signals for the splicing machinery indicating where exactly to splice the exons together, editing out the introns.

Figure 6 shows a stretch of human DNA sequence containing a gene. This illustrates that although the sequence appears to be simply a random stretch of the four nucleotide characters, different sequence regions have distinct biological meaning, and code for a specific protein with a particular biochemical function.

```
AAGCTTTTTGATGTGCTGCTGGATTCGGTTTGCCAGTATTTTATTGAGGATTTTTGCATC
AATGTTCATCAAGGATATTGGTCTAAAATTCTCTTTTTTGGTTGTGTCTCTGCCCGGCTT
TGGTATCAGGATGATGCTGGCCTCATAAAATGAGTTAGGGAGGATTCCCTCTCTTTCTAT
TGATTGGAATAGTTTCAGAAGGAATGGTACCAGTTCCTCCTTGTACGTCTGGTATAATTC
GGCTGTGAATCCATCTGGTCATGGACTCTTTTTGGTTGGTAATCTATTGATTATTGCCAC
AATTTCAGATCCTGTTATTGGTCTATTCAGAGATTCAACTTCTTACTGGTTTAGTCTTGG
GAGAGTGTATGTGTCGAGGAATTTATCCATTTCTTCTAGATTTTCTAGTTTATTTGCGTA
GAGGTGTTTGTAGTATTCTCTGATGGTAGTTTGTATTTCTGTGGGATCGGTGGTGATATC
CCCTTTATCATTTTTTATTGCATCTATTTGATTCTTCTCTCTTTTTTTCTTTATTAGTCT
TGCTAGCGGTCTATCAATTTTGTTGATCCTTTCAAAAAACCAGCTCCTGGATTCATTAAT
TTTTTGAAGGGTTTTTTGTGTCTCTATTTCCTTCAGTTCTGCACTGATTTTAGTTATTTC
TTGCCTTCTGCTAGTTTTGAATGTGTTTGCTCTTGCTTTTCTAGTTCTTTTAATTGTGAT
GTTAGGGTGTCAGTTTTGGATCTTTCCTGCTTTCTCTTGTGGGCATTTAGTGCTATAAAT
TTCCCTCTACACACTGCTTTGAATGTGTTCCAGAGATTCTGGTATGCTGTGTCTTTGTTC
TCGTTGGTTTCAAGAACATCTTTATTTCTGCCTTCATTTTGTTACGTACCCAGTAGTCAT
TCAGGAGCAGGTTGCTCAGTTTCCATGTAATTGAGCGGTTTTGAGTGAGTTTCTTAATCC
TGAGTTCTAGTTTGATTGCACTAAAATTTTTAAAAAGTAAAAAAAATACATGTGGTTTAA
TACAATTCATGCCAACTCATTCCCTCGTTTTTTGCTATAAACCTTGCAAGGAGATGAATA
ATCCAAGGCTCTTGGATAAGATAAGGGCCCCATCCATCTTGCTCCTCTCAGCCCTGGAGG
AGGAGGGAGAGTCCTTTTCCCCTGTCTACGCTCATGCACCCCCAATGAGTCCCTGCCTCC
AGCCCTGACCTCTGCCCTCGGTCTCTCAGGCAGATCCAGGGCCAGTTCTCCCATGACGTG
ATCCCTCCCGAAGGCAAGGCACCAGGCAAGATAAAAGGATTGCAGCTGAACAGGGTGGAG
GGAGCATTGGAATGGCACTCAGGGCAAAGGCAGAGGTGTGCATGGCAGTGCCCTGGCTGT
CCCTGCAAAGGGCACAGGCACTGGGCACGAGAGCCGCCCGGG
```

FIGURE 6. A STRETCH OF HUMAN DNA SEQUENCE. A portion of the human gene for a cytochrome protein.

How does the genome relate to the organism?

Let us recapitulate the interplay of the genomic information to build an organism and maintain it — the relationship of the whole animal to that of its organs, cells, genomes, and genes. An organism has many organs and tissues, each having a specific physical or biochemical function and organized in a particular shape and size. A coherent coordination of the different organs, tissues, and appendages makes the functioning of an animal possible. At a lower level, the cells, which are the fundamental living unit of organisms, make the functions of the different organs and tissues possible. At the lowest level, it is the proteins which make the function of cells, and therefore the organs and organisms, possible. But the information for all the proteins, and when and where in the body during development and maintenance of the animal they should be synthesized, is stored in the genomic DNA. Thus, the genome is the master of the cell and the organism. The genome contains all the genes required to construct and maintain the organism. Each cell in an organism contains its own copy of the genome, but the sets of genes expressed in different types of cells, such as a liver cell or a muscle cell, are different.

THE EUKARYOTE AND THE PROKARYOTE REPRESENT TWO DISTINCT KINDS OF CELLS IN THE LIVING WORLD.

The living world is made up of two distinct kinds of cells, *prokaryotes* and *eukaryotes*. Prokaryotes (bacteria) are always single-celled organisms, which are comparatively small and contain their DNA, proteins and all other molecules within a single sack of cell wall. The DNA usually makes up a single chromosome.[9] Eukaryotes, however, can be both single-cellular or multicellular organisms. The typical cell is large and contains the DNA, usually in multiple chromosomes, within a specialized sack called the *nucleus*. The distinct presence of nucleus in a cell is considered to be the hallmark of the eukaryotes; its origin has been the subject of debate (see Chapter 7 for a new explanation on the origin of the nucleus). The eukaryotic cell also contains organelles, such as the mitochondria, which function as powerhouses producing the energy needed by the cell. Figure 7 illustrates the typical structures of prokaryotic and eukaryotic cells.

Eukaryotic genes are split into exons and introns whereas prokaryotic genes are not split

The genes of all eukaryotic organisms, whether multicellular or unicellular, are split into introns and exons (see Figure 5).

In the prokaryote, the gene is not split — the coding sequence is continuous, containing no introns (see Figure 8). Consequently, the prokaryotic gene is about 10 times shorter than the eukaryotic gene. However, the average amount of usable sequence in a gene in prokaryotes and eukaryotes is approximately the same. As a result, the average length of the proteins in the prokaryote and the eukaryote is also the same. The regulatory sequences in the prokaryote are short and compact, whereas those in the eukaryote can be very long. Furthermore, it appears that there is no junk DNA in the prokaryote; the consecutive genes are organized tightly in its genome without any intergenic DNA.

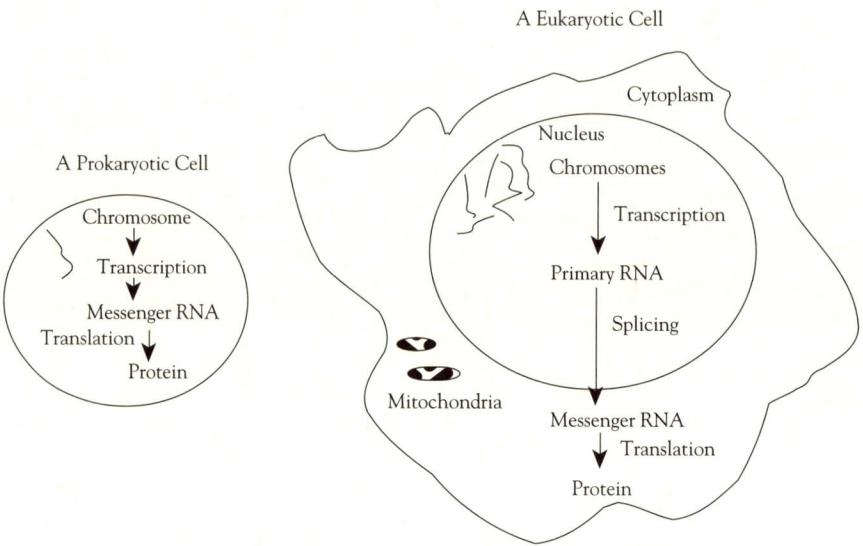

FIGURE 7. THE PRIMARY DIFFERENCES BETWEEN A PROKARYOTIC AND A EUKARYOTIC CELL. The eukaryotic cell contains a nucleus which is absent in the prokaryotic cell. The eukaryotic gene contains introns, which are absent in prokaryotic genes. The prokaryotic cell is far smaller than the typical eukaryotic cell. The prokaryote usually contains a single chromosome (about 1–5 million nucleotides). The eukaryotic genome is far larger (~50 million to 300 billion nucleotides) and is contained in several chromosomes. The prokaryotic genes are short, with lengths of a few hundred to a few thousand nucleotides. The eukaryotic genes are longer, from a few thousand up to a million nucleotides. Transcription and translation of the gene are separated in the eukaryotic cell by the nuclear membrane, whereas in the prokaryotic cell both happen in the same environment.

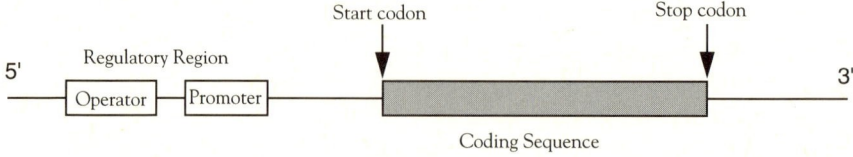

FIGURE 8. THE STRUCTURE OF A PROKARYOTIC GENE. Unlike eukaryotic genes (see Figure 5), the coding sequence is contained in a single piece. The longest known prokaryotic gene contains about 10,000 nucleotides.

The cellular genetic machineries

All cells express their proteins by means of some cellular machineries that are synthesized and used within the cell. These function in copying a gene sequence into an RNA molecule (*transcription*), editing out the introns thereby splicing the exons together (*RNA splicing*) and, after transporting this edited mRNA from the nucleus into the cytoplasm of the cell, decoding the message into an amino acid sequence (*translation*). When a cell divides into two, machineries are also needed for DNA replication, whereby genomic DNA is completely duplicated.

The copying of the gene's DNA sequence into the RNA sequence: transcription

The gene is kept as a "read-only memory" in the genome — that is, it can only be read, not changed, except by rare mutations. When the gene is expressed in a bacterial cell, an RNA copy of it is made, used in decoding the message, and then degraded (see Figure 9). The process of copying the gene into the mRNA is called transcription. An enzyme called *RNA polymerase* carries out this function. Transcription in prokaryotes and eukaryotes is similar, but that of the latter is far more complex.

How is the beginning and end of the gene, i.e., the "transcription unit," specifically recognized? A specific sequence at the beginning of a gene indicates the start of transcription, and likewise, another sequence at the end of the gene is indicative of the end of the transcription. At the transcription *initiation site*, the RNA polymerase enzyme binds to the DNA and starts the transcription process as it moves along the DNA, "reading" the sequence; at the transcription *termination site*, it stops. In almost all the genes, the promoters have a short similar sequence, TATAAT (called a TATA box), which is recognized by the RNA polymerase enzyme.[10]

The important thing about gene expression in a multicellular organism is that all the genes are not expressed at all times in all cells. Specific genes are expressed at specific times in particular cells developing specific tissues during embryonic development. Similar tissue-specific expression of genes is maintained after development. When there are tens of thousands of genes in a genome, how is the transcription of specific genes achieved? As discussed before, this is achieved by particular sequences present at the front end of each gene, which are bound by specific proteins elaborated differently in the different cells in the developing embryo. The differential switching on and off of genes in various cells in a developing embryo is a complex phenomenon, and poorly understood. Some genes are expressed in all developing tissues, some are expressed only in a few tissues and others

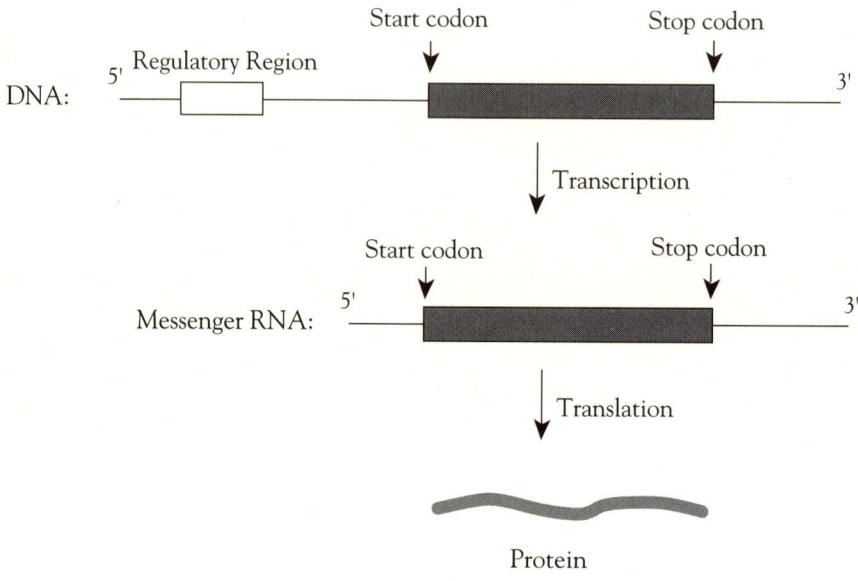

FIGURE 9. GENE TRANSCRIPTION AND TRANSLATION IN THE PROKARYOTE. The gene is first copied into RNA by the RNA polymerase enzyme (transcription). The coded message in this RNA is decoded by the ribosomes into the corresponding protein (translation).

are expressed only in one tissue. In fact, the specific expression of a gene in one or a few specific cell types is achieved by many sequences at the front end of the gene, each binding one regulatory protein. It appears that the timing and location of the expression of a particular gene in an embryo is accomplished by the binding of multiple proteins — not a single protein — to its regulatory sequence.

It is at the level of transcription that the regulation of gene expression is primarily accomplished. Most genes are regulated by sometimes allowing, and at other times preventing the binding of RNA polymerase. Under most situations, once the RNA copy of the entire gene is available in the nucleus, the spliceosomes will recognize it and start to splice the exons together. The mRNA will move to the cytoplasm, and the ribosomes will start to act on it and translate it into its corresponding protein. Although some control of gene expression operates after the transcription of a gene into its RNA copy, it is rare compared to the regulation at the level of transcription.

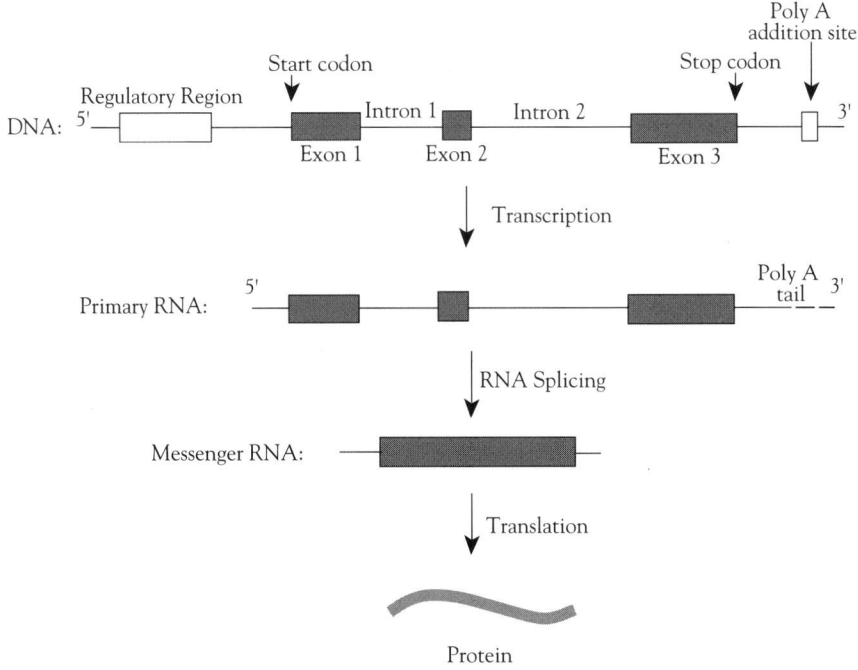

FIGURE 10. GENE TRANSCRIPTION, RNA SPLICING, AND TRANSLATION IN THE EUKARYOTE. In the eukaryote, the process of decoding the message contained in the gene has an extra step compared to that in the prokaryote. The gene is first copied into RNA (the primary RNA), which contains all the exons and introns. This happens in the nucleus, where the chromosomes are housed. From this copy, the introns are edited out and the exons are spliced together consecutively. This forms the messenger RNA, which is then transported to the cytoplasm, where the ribosomes translate it to the corresponding protein.

Putting the exons together: RNA splicing

In coding for a protein from the DNA message, first a complete copy of the gene with all of its exons and introns is transcribed into an RNA, which is called the *primary RNA* transcript of the gene. The introns are precisely edited out and the consecutive exons are spliced together from this primary RNA, leading to the messenger RNA (mRNA) containing the complete coding sequence in one piece. This activity is called *RNA splicing* (see Figure 10). The spliceosomal RNAs along with some proteins constitute the *spliceosome* machinery. RNA splicing occurs only in eukaryotic cells and not in prokaryotic cells because prokaryotic genes lack introns.

Decoding the amino acid sequence information of a protein from the mRNA codes: translation

The spliced exon sequence, the mRNA, is transported from the cell's nucleus to its cytoplasm, where the code is read by a machinery called the *ribosome*, which manufactures proteins. This activity is termed translation (see Figure 10).

The message for the consecutive amino acids in a polypeptide sequence is contained in the mRNA as consecutive three-base codons. The consecutive codons are read by the ribosome, and each codon is aligned with a transfer RNA (tRNA) containing a complementary *anticodon* along with the appropriate amino acid. As the consecutive codons are read, and different tRNAs are brought in, the corresponding amino acids attached to the tRNA molecules are consecutively linked by peptide bonds, thus forming a growing polypeptide chain. When all the mRNA is read completely, the polypeptide is completely "synthesized." This polypeptide then folds into its unique three-dimensional shape, ready to carry out its protein function.

Regulation of Gene Expression

For many reasons, all the genes in the genome are not expressed in every cell of an individual. In the chicken zygote (the fertilized egg cell), only a subset of the genes in its genome are expressed. The complete set of instructions in the chicken genome for the development of the chicken embryo into the full-grown chicken can be compared to a big computer program. A computer program is nothing more than a list of instructions, written in a language that the computer understands. Starting from the first instruction in the list, each instruction is taken and executed exactly as it is stated. In a more complex computer program, each instruction in a master list of instructions can point to sub-lists of instructions to be executed. A very complex list can consist of thousands or millions of instructions, many lists nested and networked in a complex manner. It is such a complex computer program to which we can compare the genetic instructions in a genome. We call the execution of the genome program *genome expression*.

Out of the approximately 50,000 genes in the chicken genome, only a small subset, perhaps 5000, are switched on in the zygote. In a computer program, one instruction can point to a sub-list of instructions by what is called a "GO TO" statement. Such GO TO instructions are already present in the first cell in the form of proteins that control the switching on or off of particular genes. These proteins are brought into the zygote by the sperm and ovum during fertilization. So, the first list of instructions is poised to launch a big genome program expressing itself into the embryo and later into the full-grown chicken. We can already see that the genome program is an extremely complex one, with numerous instructions and GO TO statements nested and networked. Various genes, proteins, genetic regulatory switches, and small molecules participate in executing this complex program.

We shall see below, how one instruction, such as SWITCH ON one particular gene, is executed in a genome. We shall illustrate this first in a simple bacterial cell, and then go on to more complex cells.

Control of gene expression in the prokaryote

When we speak about the expression of a gene, it is the coding sequence of the gene that is expressed. Therefore regulation of a gene means the regulated switching on or off of the synthesis of the protein encoded by the gene. It was Francois Jacob and Jacques Monod who first suggested that the expression of a gene is regulated by a protein encoded by another gene. On the basis of experiments with the bacterium *Escherichia coli*, they suggested a hypothesis known as the *operon* concept (see Figure 11). It was known that when *E. coli* was growing in a medium containing sugars such as glucose, but not the milk sugar lactose, no enzyme that could utilize lactose was synthesized. When the bacterium was exposed to a medium containing only lactose, it quickly started to produce lactose-digesting enzymes. Jacob and Monod suggested that a different protein, encoded by another gene, switched on the previously dormant genes in response to lactose.[11]

Their concept was that there exists a specific sequence at the front end of the coding sequence of a gene that specifically binds the RNA polymerase, the enzyme that transcribes the gene's coding sequence. This sequence is called the promoter sequence. Close to this sequence exists another sequence, the operator, that binds another protein, the repressor. The repressor protein, when bound to the DNA operator, physically blocks the binding of the RNA polymerase to the promoter, and thereby shuts off the transcription of the coding sequence. When an inducer, such as lactose, binds to the repressor protein, the shape of the protein is changed such that it no longer can bind to the DNA operator. Once the operator becomes freed of the repressor, the RNA polymerase has no hindrance to bind to the promoter, and to transcribe the messenger RNA of the coding sequence.

The operon concept of Jacob and Monod proved to be correct with many examples of gene regulation systems in *E. coli*. In all these cases, the regulatory proteins were found to be DNA-binding proteins, which bound to specific DNA sequences near the genes, and controlled their expression

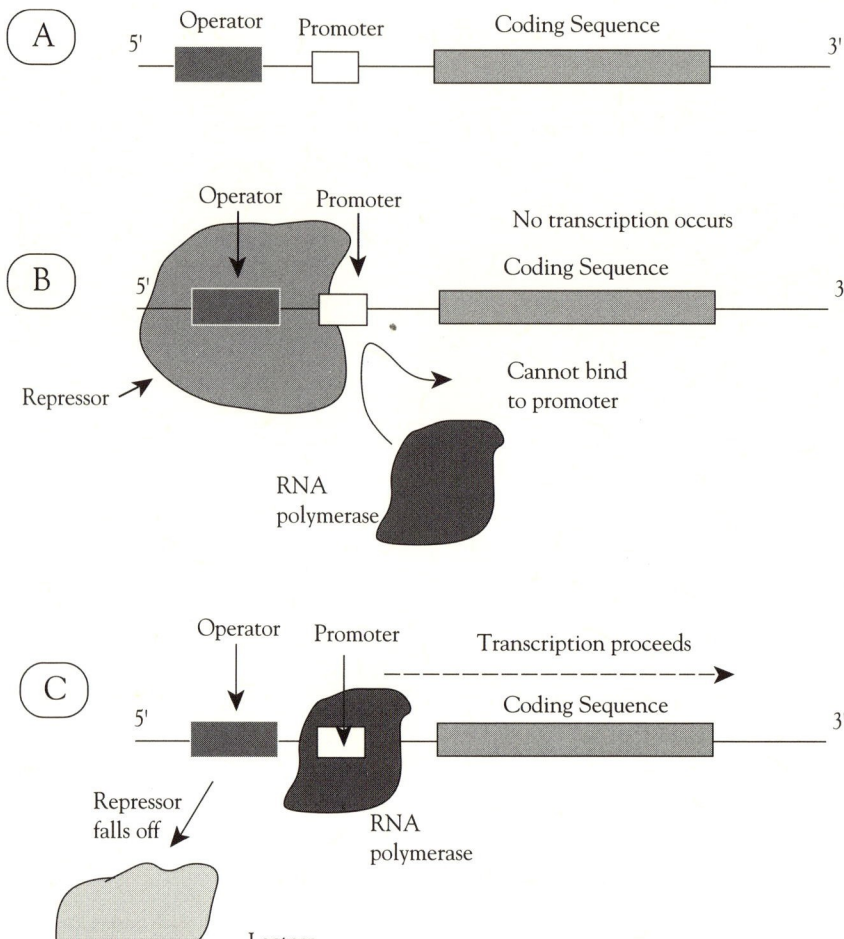

FIGURE 11. REGULATION OF GENE TRANSCRIPTION IN THE PROKARYOTE. (A) Regulation is achieved by the binding of a protein to the regulatory sequence called the operator. (B) Under normal circumstances, a repressor protein is bound to the operator, which physically hinders the binding of the enzyme RNA polymerase. Therefore, the coding sequence is not transcribed and the gene is "off." (C) When a biochemical that can induce the transcription of the gene binds to the repressor, the repressor's structure is altered such that it can no longer bind to the operator. Thus, the promoter is freed to bind with the RNA polymerase, which starts to transcribe the gene's coding sequence. The gene is now "on."

either by repressing (switching off) or promoting (switching on) the transcription of the nearby gene. In the genome containing many genes there exists a coordinated program of gene control. Not all the genes contained in a bacterium's genome are expressed at all the times during its short life span of about 20–40 minutes. The bacterium makes the whole program of its growth far more efficient, by switching different genes on and off rapidly in response to changing conditions and at different times during its life cycle. For example, only when the sugar lactose is available in its surroundings, the bacterium will switch on the set of genes required for utilizing this particular sugar. This may be a way of streamlining the available resources for the synthesis of the specific proteins required at specific times only. Such a program of selective use of proteins at specific times may also be necessary to avoid interference among proteins.

Proteins such as the repressor thus behave like switches that can specifically switch on or off particular genes necessary only at specific times during the life cycle of the bacterium. A similar but far more complex control mechanism seems to operate in the cells of multicellular organisms. The organism starts with a single cell and grows into many different types of cells which make up the many different tissues, organs, and appendages. This is achieved by a complex regulation of the expression of specific genes present in the genome of a multicellular organism — by differential expression of different series of genes at different times in different embryonic cells during development.

REGULATION OF GENE EXPRESSION AND DEVELOPMENTAL GENETICS IN MULTICELLULAR ORGANISMS

All multicellular organisms grow and develop by a specific genomic program. There are two things that we must note in the life of an organism. The first and most important is the development of the embryo from the zygote to the fully-formed "baby," which contains all the organs and appendages of the adult organism. The second is the growth and aging of this fully-formed organism.

Let us assume that there are 50,000 genes in the genome of a given animal, say the chicken. Although all the genes are present in the genome of the zygote, only a subset, perhaps 5,000 genes, is expressed in it. Nonetheless this is a large set of genes expressed when the zygote starts to divide, because a very large number of genes are needed to construct even a single, basic, eukaryotic cell. As the early embryonic cell divisions are occurring, different cells are committed to form the different tissues, organs and body parts. We must recognize that the cells constituting the different organs in the chicken, such as liver, eye, brain, and kidney, are different because they contain different sets of proteins (albeit somewhat an overlapping set) and function differently. Such tissue- or organ-specific proteins must be synthesized specifically in different cells committed to form different organs during development. Thus, in addition to expressing the basic 5,000 genes in all the cells in the developing embryo, different subsets of genes will be expressed in different cell lines leading to the various body parts. This process is termed *developmental genetics*.

Although the commitment as to which cell will eventually form which organ or body part occurs early in the development of the embryo, when and how exactly it occurs in the first few cell divisions is still unclear. However, recent investigations reveal a number of details as to how this might happen.

We saw earlier the mechanism as to how a gene is specifically expressed in a bacterium due to the induction by the sugar lactose. Let us now consider that the inducer is a protein, instead of lactose. This protein is synthesized in a particular cell that is located in a specific three-dimensional coordinate in the developing embryo. Then, only in that cell this protein will specifically induce the gene in question. The switching-on of this gene can lead to a cascade of further new genes switched on and other genes switched off in that cell and cells subsequently derived from it. This type of genetic cascade can rapidly expand one cell into many different types of cells in a few cell generations (see Figure 1).

Early embryogenesis and morphogenetic fields

Before discussing the complex mechanism of gene expression during development, let us first look at some interesting details of developmental biology when the zygote divides and the embryo develops. The development of the South African clawed frog, *Xenopus laevis*, is a good example of how a vertebrate develops. A fertilized *Xenopus* egg cell divides rapidly into two cells after about 90 minutes. The cells then divide synchronously into four, eight and so on, every 30 minutes. When there are 4,000 cells, the embryo is called a *midblastula* and resembles a hollow sphere. Although the cells all look superficially identical, certain cells are already committed to become a layer called the *mesoderm*.[12] During the process called *gastrulation*, two other layers with distinct developmental potentials are defined: the *endoderm* and the *ectoderm*. Most of the body, including the vertebral column, the muscles, and the bones, originate from the mesoderm. The endoderm gives rise to the digestive tract as well as various other organs, such as the lungs, liver, and pancreas. The ectoderm produces the skin and nervous system.

Some protein growth factors released by the surrounding yolk cells are responsible for the formation of the mesoderm.[13] Some maternal factors (factors from the mother) or maternal genes can in some cases influence development from outside the embryo.

Within each layer, there are specific regions destined to express a specific body structure. Such a region is called a *morphogenetic field*. The mesoderm of the *Xenopus* at its "neurula stage" is subdivided into morphogenetic fields for many organs: gills, ears, limbs, the tail, and so on.

Within each morphogenetic field, the potential for forming an organ varies gradually. That is, if one removes the center of a field, the corresponding organ or body part will still be formed; if instead the entire region is completely removed, the organ or body part will not be formed. It is therefore proposed that each morphogenetic field contains a gradient of information for specifying an organ. These gradients correspond closely with patterns of gene expression.

Multiple genes control one gene through a "transcription complex" of many proteins

In reality, instead of one protein from a given gene switching on (or off) another gene, two or more proteins (encoded by different genes) are needed to ensure the switching on of another gene. It is as though many repressor binding sites are present at the front end of a gene, and only when all repressors are released from the bound state will the gene be switched on. This is like a combination lock, which will open only when all the numbers are correctly entered. This mechanism seems to be needed because thousands of genes are communicating with one another and cross talk must be avoided to ensure that the right genes are switched on or off. It is however not yet completely understood as to how the specific "regulating" proteins themselves are differentially synthesized in the first few cells derived from the zygote. As we shall see below, some clues have been gained. One mechanism is proposed in Chapter 3 (see Figure 3.3).

To build a complex multicellular organism, the cells in a developing embryo have to "know" where they are and which genes to turn on. DNA-binding proteins are most probably responsible for such complex gene control and use the same fundamental mechanism as in bacterial operons, except that in eukaryotes, not one but many proteins control the expression of one gene. Also, some eukaryotic regulatory proteins can bind as far as 40,000 nucleotides from the target gene and still control that gene. There may be as many as 20 DNA-binding regions for one gene.[14] The fundamental mechanism in a eukaryotic organism, therefore, is that a gene is controlled through multiple binding sites and multiple regulatory proteins (see Figure 12). The precision of gene regulation can be increased when many proteins

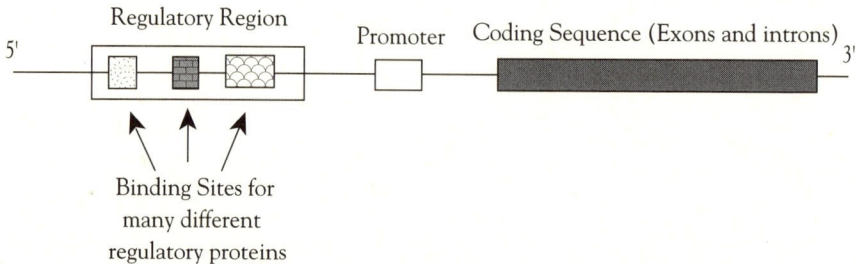

FIGURE 12. REGULATION OF GENE TRANSCRIPTION IN THE EUKARYOTE. Instead of the usual one regulatory binding site in the prokaryote, the regulatory region in the eukaryote usually contains many specific sites for the binding of many different regulatory proteins. It is similar to the repressor situation in the prokaryote, but many proteins are bound instead of one, and all of them must be released to open the gene for transcription. Another possibility is that the RNA polymerase by itself cannot bind to the promoter, and only when the whole complex of the different regulatory proteins and RNA polymerase are bound to the DNA, can the transcription start.

have to come together and bind among themselves to form a large complex that has the required specific shape and function in controlling the gene.

If every gene needed one specific protein for its control, it would have to carry a very large number of different genes to produce these proteins. But if multiple proteins can control one gene, then only a small number of different controlling proteins are needed to regulate many genes. Consider, for example, a transcription complex formed by a complex of two different proteins. With only four proteins A, B, C, and D, then, as many as 10 different complexes can be formed: AA, AB, AC, AD, BB, BC, BD, CC, CD, and DD. Thus four proteins can regulate 10 different genes. With only 20 different regulatory proteins, and with five proteins on average for each transcription complex, more than a million genes (20^5) can be controlled differently.

The Homeobox and Homeotic Genes

Some genes, when disrupted by a mutation, causes a body part to be misplaced or duplicated. Such changes are called *homeotic* mutations. For example, *bithorax* mutant flies have two pairs of wings instead of one; *Antennapedia* mutants have extra legs growing where their antennae should be. Single mutations may lead to such homeotic transformations even though hundreds of active genes develop the abnormally placed body parts. This is because homeotic mutations disrupt *master* genes that control many *subordinate* genes. A homeotic gene, therefore, is a gene that controls a set of genes that develop a particular body structure.

All multicellular organisms — invertebrates, vertebrates, and plants — contain a similar DNA region in their homeotic genes. This similar region is called the *homeobox*, which corresponds to a 60-amino-acid portion of homeotic proteins, called the *homeodomain*. The homeodomain binds to specific DNA sequences in genes regulated by the homeotic genes. A homeodomain protein seems to specifically bind to its subordinate genes, activating or repressing their expression in a cell.

In general, homeotic genes are vastly different except for their common homeodomains. Therefore, current opinion of developmental geneticists is that the protein regions that are different are rather more important than the similar homeodomains[15] — the homeodomain simply recognizes DNA, while the rest of the protein determines how and which genes to regulate. In other words, the DNA binding and transcription is a general phenomenon operated by the homeodomain, and the rest of the protein is responsible for the specific phenomenon of activating and repressing distinctly different genes.

An interesting characteristic of homeobox genes is that they are arranged together in a precise order on the linear DNA molecule that makes up a chromosome: those located at one end of a complex are expressed in rearward parts of the body and those at the other end are expressed closer to the head. Additionally, similar homeotic gene clusters occur in different organisms.

The development of the frog *Xenopus* illustrates how homeobox genes work. The entire forelimb is derived from the area of mesoderm that expresses the homeotic gene called *XlHbox1*. A small forelimb bud appears on *Xenopus* within three weeks after fertilization. Although at this stage the mesoderm bud appears to be uniform, it contains a gradient of the *XlHbox1* protein. The protein is most abundant in cells along the front side of the limb bud — the side that gives rise to the thumb — and least abundant in cells on the rear side, which gives rise to the smallest digit. As the bud extends and takes shape, the concentration of *XlHbox1* protein stays highest near the shoulder. In contrast, protein from another gene establishes a gradient that is highest along the rear side and distal end of the limb — a pattern precisely the reverse of that for *XlHbox1*. Note that other homeobox genes are also involved in the forelimb development.

Genetic mutations

A DNA sequence in a living creature can be changed by a few different mechanisms. One or more nucleotides in a DNA sequence can be deleted from it, or one or more nucleotides can be added to it at a specific location. If this happens within a gene, it may affect the function of the gene, either positively, negatively, or not at all. Usually such a change in a gene is termed a gene mutation.

The gene mutations are categorized into *addition* mutations, *deletion* mutations, *frame-shift* mutations, *translocations*, and *inversions*. When the reading frame of a coding sequence is altered by the addition or deletion of one or more nucleotides in it, it is called a frame-shift mutation (see below). If a subsequence is deleted in one place of a sequence and introduced into another, it is called a translocation. If a subsequence is inverted within a long DNA sequence, it is called an inversion.

A mutation that results in the substitution of one amino acid for another is called a *missense* mutation. If a codon is replaced by another that codes for the same amino acid (e.g., a change from CUA to CUG, both of which code for leucine), it results in *samesense* mutations (also called "silent" mutations). If a codon that codes for an amino acid is replaced by one that codes for chain termination, resulting in the premature termination of the synthesis of the protein chain, it is called a *nonsense* mutation.

Even more catastrophic effects result from *frame-shift* mutations. Normally, the cellular machinery simply starts at one end of a coding sequence and defines a codon *reading frame* simply by counting off three nucleotides at a time. A frame-shift mutation results from an insertion or deletion of a nucleotide in a coding sequence. This can shift the reading frame by one nucleotide, resulting in different codons and a drastically changed amino acid sequence in the protein product. For example, consider the following DNA sequence: ...TACGTCAAGTCG... Its original reading frame might define these four codons: .../TAC/GTC/AAG/TCG/... If an insertion mutation occurs in the second codon, adding an extra T, for example, the reading frame will be shifted at the point of insertion, changing every codon that follows: .../TAC/GT**T**/CAA/GTC/G...

Mutations may be caused by chemical agents, called mutagens, or by physical agents, such as ultraviolet radiation and x-rays. Errors in the normal DNA replication process can also generate mutations. Mutations may lead to congenital abnormalities, genetic diseases and cancer (see Figure 4.11 and Figure 13).

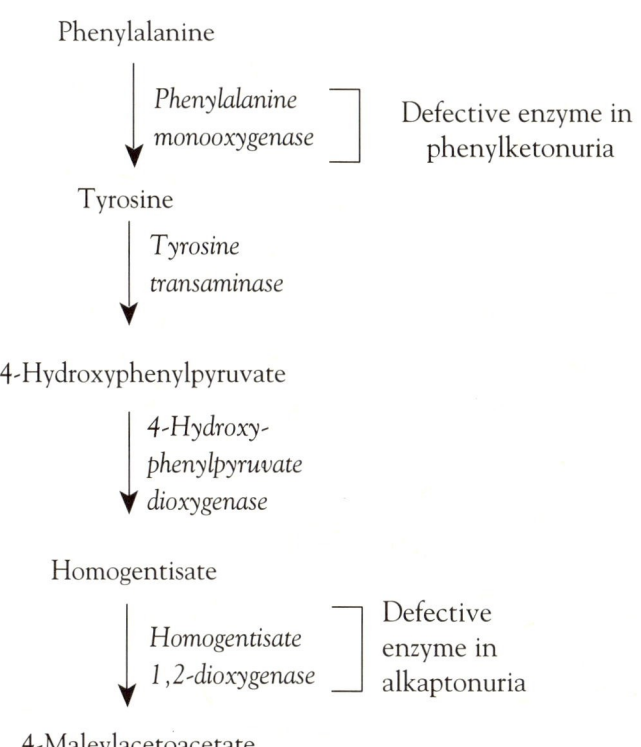

FIGURE 13. MUTATIONS IN GENES CAN LEAD TO GENETIC DISEASES. DNA sequence mutations can change the amino acid sequences of proteins and result in defective enzymes which are part of a metabolic cycle. Lack of an enzyme can result in the accumulation of the reactant for that enzyme which can cause disorders including mental retardation. For instance, in the normal pathway for the metabolism of phenylalanine, a defect in the first enzyme leads to phenylketonuria, and in the last enzyme leads to alkaptonuria.

APPENDIX

SOME ANECDOTES

THE AMOUNT OF DNA MATERIAL IN ONE INDIVIDUAL ORGANISM IS ENORMOUS

Single eukaryotic cells contain far more DNA than a prokaryote. A single cell of a slime mold has over 10 times the DNA of an *E. coli* cell (a prokaryote). Cells of the fruit fly *Drosophila* have about 25 times as much DNA as *E. coli* cells. The cells of human beings have about 600 times as much DNA as *E. coli*. And the cells of some amphibians such as the salamander contain nearly 50 times the amount of DNA in human cells.

The total physical length of all the DNA in a single human cell is about two meters. Compare this with 1.4 millimeters for the *E. coli* DNA. The adult human body consists of about 10^{13} cells. Thus, the total length of all the DNA in a human body is about 2×10^{13} meters or 10^{10} kilometers. The distance between the earth and the Sun is about 10^8 kilometers. Thus if the DNA in one human individual is laid end to end, it will cover a distance about 100 times longer than that from the earth to the Sun. And the DNA in one salamander individual, laid end to end, will cover a distance about 5,000 times longer than that from the earth to the Sun.

The genome of a human being contains 3×10^9 nucleotides. Since each cell contains a pair of chromosomes, and each chromosome contains two strands of DNA, a single human cell contains 12×10^9 nucleotides. Thus, the total amount of DNA material contained in the ten trillion cells of the typical human body is $12 \times 10^9 \times 10^{13} = 12 \times 10^{22}$ nucleotides. However, chemistry tells us that one *gram molecular weight* of any molecule (for a nucleotide it is ~300 grams, taking the average molecular weight of all four nucleotides to be ~300) would contain 6.02×10^{23} nucleotides (this is known as the *Avogadro number*). One kilogram of DNA would therefore contain approximately 2.5×10^{24} nucleotides, and the total mass of DNA material in a typical human being is about 60 grams. We can compute, by crude estimation from the quantity of all the available organic material on earth, that if all the earth's organic material is converted into DNA, there would be over 10^{46} nucleotides and the weight of this DNA would be approximately 10^{22} kilograms.

NOTES AND REFERENCES

CHAPTER 1

1. This question truly addresses not only the origin of the diverse organisms on earth but also the origin of life itself from inanimate matter.

2. Darwin, C., 1859, *On the Origin of Species*, John Murray, London.

3. Darwin essentially believed that there should have been one original ancestor from which all the organisms on earth had evolved by descent with modification. He stated in *Origin of Species* (Darwin, C., 1859, *Origin of Species*, 1979 Edition, Avenel Books, New York, page 454):

   > I believe that animals have descended from at most only four or five progenitors, and plants from an equal or lesser number. Analogy would lead me one step further, namely, to the belief that all animals and plants have descended from some one prototype. But analogy may be a deceitful guide. Nevertheless all living things have much in common, in

their chemical composition, their germinal vesicles, their cellular structure, and their laws of growth and reproduction. We see this even in so trifling a circumstance as that the same poison often similarly affects plants and animals; or that the poison secreted by the gall-fly produces monstrous growths on the wild rose or oak-tree. Therefore I should infer from analogy that probably all the organic beings which have ever lived on this earth have descended from some one primordial form, into which life was first breathed.

Even the modern belief is that all life on earth should have begun from one single primordial cell.

4   Darwin wrote:

> There is grandeur in this view of life, with its several powers, having been originally breathed into a few forms or into one; and that, whilst this planet has gone on according to the fixed law of gravity, from so simple a beginning endless forms most beautiful and most wonderful have been, and are being, evolved.

Reference: Darwin, C., 1859, *Origin of Species*, 1979 Edition, Avenel Books, New York, page 459.

Chapter 2

1   All humans are similar in overall physical characteristics with only minor variations, so they are grouped into one species, the human species. Chimpanzees and gibbons are similarly grouped, as separate species. But these ape species are similar to each other, so they are grouped within a classification category called the family. Many such families that are similar are grouped as primates, a still higher classification category called the order. The orders primate and bovine, although quite different from each other, have some other similar structures such as the placenta and breasts. They are grouped together in a class called mammals. The mammals are grouped in the subphylum called vertebrates, along with the classes reptiles, birds, amphibians and fishes — based on their common feature of having a backbone. All the organisms on earth are classified in this way.

2   The French scientist (1744-1829) who proposed the theory of inheritance of acquired characteristics. The classic example is how the giraffe got its long neck. A short-necked giraffe managed to develop a longer

neck during its lifetime by repeatedly stretching its neck to get at the tender leaves at the tops of acacia trees. The longer neck of the animal was passed on to its descendants. Thus, whatever useful changes in the length of its neck the giraffe managed to acquire during its own life would show up in a slightly longer-necked offspring. This process repeated over a number of generations would result in a long-necked giraffe.

3   Darwin, C., 1859, *Origin of Species*, (John Murray, London), 1979 Edition, Avenel Books, New York.

4   Darwin, C., 1859, ref above, page 130.

5   ibid, page 132.

6   Although Darwin felt that there could have been four or five original organisms, he preferred the possibility of only one organism. See note 3, Chapter 1.

7   Numerous unique creatures classifiable into almost all the major taxonomic groupings appear within 5–10 million years in the fossil record at the very start of multicellular life.

8   See for example, Gould, S. J., 1989, *Wonderful Life*, W. W. Norton & Company, New York.

9   Darwin, C., 1859, ref above, page 217.

10 Lewontin, R. C., 1982, Adaptation, in *The Fossil Record and Evolution*, Scientific American Books, with introduction by Laporte, L. F., page 16.

11 Futuyma, D. J., 1986, *Evolutionary Biology*, Second Edition, Sinauer Associates, Inc., Sunderland, Massachusetts, page 9.

12 Lamarck's theory states that physical characteristics of an organism that are acquired by the organism through its actions are heritable. Thus, a giraffe acquired its long neck by continuously trying to reach tall trees for food over numerous generations. See note 2, Chapter 2.

   Also see: Lamarck, J. B., 1809, *Philosophie Zoologique, ou Exposition des Considerations Relatives a l'Histoire Naturelle des Animaux, etc.*, Paris, Translation with an introduction by H. Elliot, 1963, New York: Hafner.

13 See for instance, Minkoff, E. C., 1983, *Evolutionary Biology*, Addison-Wesley Publishing Company, Reading, Massachusetts; and Futuyma, D. J., ref above.

14 Minkoff, E. C., 1983, ref above, page 96; Futuyma, D. J., ref above, page 10.

15 We need not be concerned here as to what these processes mean. They will become clear when we read Chapters 3 and 4.

16 Minkoff, E. C., ref above, page 244.

17 Innovations here mean the origin of new features such as a new physiological characteristic or a new organ. To cite a striking example, the appearance of even the most primitive eye in an eye-less animal such as a worm or the appearance of a primitive horn in an animal without a horn is considered an evolutionary innovation.

18 Minkoff, E. C., ref above, page 244.

19 Quoted in Minkoff, E. C., ref above, page 245.

20 Minkoff, E. C., ref above, page 245.

21 Bush, G. L.,1982, What Do We Really Know About Speciation? in *Perspectives on Evolution*, Sinauer Associates, Inc., Sunderland, Massachusetts, page 125.

22 Gould, S. J., 1982, The Meaning of Punctuated Equilibrium and Its Role in Validating a Hierarchical Approach to Macroevolution, in *Perspectives on Evolution*, Sinauer Associates, Inc., Sunderland, Massachusetts, page 88.

23 Eldredge, N., and Gould, S. J., 1972, Punctuated Equilibria: An Alternative to Phyletic Gradualism, in T. J. M. Schopf, ed., *Models in Paleobiology*, page 82-115, Freeman, Cooper and Company, San Francisco.

24 Gould, S. J., 1982, ref above, page 84.

25 ibid, page 88.

26 Minkoff, E. C., ref above, page 349; Futuyma, D. J., ref above, page 403.

27 Gingerich, P. D., 1976, Paleontology and Phylogeny: Patterns of Evolution at the Species Level in Early Tertiary mammals, *American Journal of Science*, 276:1-28.

Gingerich, P. D., 1977, Patterns of Evolution in the Mammalian Fossil Record, In A. Hallam, ed. *Patterns of Evolution as Illustrated by the Fossil Record*, Amsterdam: Elsevier, pages 469-500.

Gingerich, P. D. and Schoeninger, M., 1977, The Fossil Record and Primate Phylogeny, *Journal of Human Evolution*, 6:483-505.

[28] Futuyma, D. J., ref above, page 143.

Kimura, M., 1983, The Neutral Theory of Molecular Evolution, in *Evolution of Genes and Proteins*, Nei, M. and Koehn, R. K., eds., page 208.

[29] It is customary to believe under evolutionary theory that every genetic change that appears is the outcome of a natural selection process. But Kimura feels that only a fraction of these changes are actually selected through the process of natural selection. Other changes — selectively neutral mutations — are not subjected to natural selection and they actually aid in the evolution of organisms more than natural selection itself. We shall discuss more on this subject in Chapter 4.

[30] Kimura, M., 1983, ref above, page 209.

[31] Bush, G. L., 1982, ref above.

[32] Goldschmidt, R., 1940, *The Material Basis of Evolution*, Yale University Press, New Haven, Connecticut.

Chapter 3

[1] Futuyma, D., 1986, *Evolutionary Biology*, Sinauer Associates, Inc., Sunderland, Massachusetts, page 76.

[2] Genes involved in many metabolic pathways are functionally common to organisms as widely different as plants, bacteria, invertebrates, and vertebrates. For instance, many of the enzymes involved in glucose metabolism could be the same in the rose, the bacterium *E. Coli*, the dragonfly, and the human.

Lehninger, A. L., 1982, *Principles of Biochemistry*, Worth Publishers, Inc., New York, page 124.

[3] Lehninger, A. L., 1982, ref above, page 124.

[4] Minkoff, E. C., 1983, *Evolutionary Biology*, Addison-Wesley Publishing Company, page 175–76.

[5] This is computed as follows. Let the probability of an event happening be p. Then the probability of the event not happening is $1-p$. Thus, the probability the event not happening in n number of tries is $(1-p)^n$. This means that the probability of the event occurring at least once during the n tries is $1-(1-p)^n$.

[6] Futuyma, D., 1986, ref above, page 72.

Alberts, B., Bray, D., Lewis, J., Raff, M., Roberts, K., and Watson, J. D., *Molecular Biology of the Cell*, 1983, Garland Publishers, New York, page 214.

[7] If mutations can never lead to new genes, what do they do? A majority of mutations have no effect because they do not change the fundamental structure and function of the protein — changing the gene only to its normal variants. The only thing that mutations can otherwise do is cause defective genes, which might result in disease or developmental defects. These defects have absolutely nothing to do with evolutionary change.

[8] Mitchell, L. G., Mutchmor, J. A. and Dolphin, W. D., 1988, *Zoology*, The Benjamin/Cummings Publishing Company, Inc., Menlo Park, California, pages 583-584.

[9] Once we understand that an organism with a new and unique body part cannot evolve from an organism lacking in it, then we can clearly see that even the supposedly similar organisms, which do not contain any anatomically new body part — such as the rat and human — could not have evolved from a common ancestor. These organisms have distinct gap between them, and evolution cannot be accepted as capable of explaining the origin and diversity of such organisms.

[10] The embryonic cell lineage map is a genetic road map of where each cell starting from the single-cell zygote will be placed in the developed body of an organism, and how they develop successively and chronologically in a timed sequence. This can indeed be equated, under our discussions, to the developmental genetic pathway, because these space-timed sequences for various body parts are genetically determined.

[11] Davidson, E. H., 1990, How Embryos Work: A Comparative View of Diverse Modes of Cell Fate Specification, *Development*, 108: 365-389.

[12] Even developmental gene mutations (mutations in genes which have direct function in the embryonic development of body parts in an organism) can only produce developmental defects and can have no contribution to the evolution of an entirely new body part. Such genes are called "homeotic genes," and may contain "homeobox" sequences. See Genetics Primer.

[13] This is in direct opposition to Darwin's theory, because the organisms are supposed to be constantly changing in the changing physical and ecological environments.

[14] Futuyma, ref above, page 398; Minkoff, ref above, page 328.

[15] Another interesting property of individual variations is that among the same sized individuals in a given species, a trait or an appendage can be quite varying in proportion; for instance, in sheep, in one individual the horn can be quite short and in another comparatively long. However, even such differences fall within a particular range. Although environmental factors such as food or temperature can affect these characteristics, the propensity to express them is genetic and inherited.

[16] Darwin, C., 1859, *Origin of Species*, 1979 Edition, Avenel Publishers, page 82.

[17] Hyman's *Comparative Vertebrate Anatomy*, Univ. of Chicago Press, ed. M. H. Wake, 1979.

[18] As far as is known, there has been no variety or breed produced with a new outgrowth, even a useless one. Extra limbs or body parts should not be considered to be new outgrowths as discussed later in the text.

[19] It is true that individual variations in a population can be steered in any direction by artificial selection, thereby quickly producing breeds which would look quite different from one another. But this cannot be extended to mean that they will become different organisms after a very long time. For example, in dogs, man has produced many different looking breeds, but no breed has or will ever become a new distinct organism. Not one of them has ever developed or will develop, for instance, horns. The breeds will never become anything other than dogs. Artificial selection works only within the framework of the given species.

[20] Lehninger, A. L., 1982, *Principles of Biochemistry*, Worth Publishers, Inc., New York, page 404.

[21] De Vries, H., quoted in Minkoff, E. C., 1983, ref above, page 100.

[22] Minkoff, E. C., 1983, ref above, page 125.

[23] Bush, G. L., 1982, in *Perspectives in Evolution*, Milkman, R., ed., Sinauer Associates, Inc., Sunderland, Massachusetts, page 125.

Goldschmidt, R. B., 1940, *The Material Basis of Evolution*, Yale University Press, New Haven, Connecticut.

I have critically questioned many surgeons and specialists concerning many different organs if in their experience they have come across any extraneous outgrowths. The ophthalmologists (eye specialists), gynecologists (female reproductive system), nephrologists (kidney), and gasteroenterologists (digestive tract) vouch that except for cancers and other known well-defined abnormalities, they have only found normal variations throughout their years of practice, and have never seen any extraneous outgrowths. Note that developmental aberrations of existing body parts are not counted as extraneous outgrowths. This is also the same experience of veterinarians who work with many different animals.

[24] Smith., D. W., 1982, *Recognizable Patterns of Human Malformation*, W. B. Saunders Company.

[25] Darwin, C., 1859, ref above, page 90.

[26] ibid, page 73.

[27] ibid, page 91.

[28] ibid, page 78.

[29] ibid, page 101.

[30] If the mutation does not concern a developmental gene, then it may concern metabolic genes and basic cell genes. The appearance of a flower with a different color or the change of a white moth into a black moth is caused mostly by a single gene mutation. Such genes are part of the metabolic pathway leading to color pigments. A great number of genetic diseases are also produced by single gene mutations. For example, thalasemias, galactosemia, and phenylketonuria are all diseases caused by mutations occurring in a single gene.

[31] See Futuyma, D. J., ref above, and Minkoff, E. C., ref above.

[32] Doolittle, R. F., 1989, Similar amino acid sequences revisited, *Trends in Biochemical Sciences*, pages 244-45.

[33] Mitchell, L. G., et al., 1988, ref above, pages 778-779.

[34] Bischof, P. and Klopper, A., 1983, Placental Proteins, in *Progress in Obstetrics and Gynecology*, Edinborough, pages 57-72.

Bischof, P., 1984, Placental Proteins, in *Contribution to Gynecology and Obstetrics*, Keller, P. J., ed., Karger, Basel, pages 1-96.

Chegini, N., Lei, Z. M., Rao, C. V. and Bischof, P., 1991, The Presence of Pregnancy-Associated Plasma Protein-A in Human Corpora Lutea: Cellular and Subcellular Distribution and Dependence on Reproductive State, *Biology of Reproduction*, 44:201-206.

[35] Farin, C. E., Imakawa, K., Hansen, T. R., McDonnell, J. J., Murphy, C. N., Farin, P. W., and Roberts, R. M., 1990, Expression of Trophoblastic Interferon Genes in Sheep and Cattle, *Biology of Reproduction*, 43:219.

Roberts, R. M., Farin, C. E., and Imakawa, K., 1989, Embryonic Mediators of Maternal Recognition of Pregnancy, in *Blastocyst Implantation*, Yoshinaga, K., ed., Adams Publishing Group, Ltd., Boston.

Flemming, S., O'Neill, C., Collier, M., Spinks, N. R., Ryan, J. P. and Ammit, A. J., 1989, The Role of Embryonic Signals in the *Control of Blastocyst Implantation*, Yoshinaga, K., ed., ref above.

[36] Roberts, R. M. and Bazer, F. W., 1988, The Functions of Uterine Secretions, *Journal of Reproductive Fertility*, 82:875-892.

Bazer, F. W., Simmen, R. C. M. and Simmen, F. K., 1991, Comparative Aspects of Conceptus Signals for Maternal Recognition of Pregnancy, in *The Primate Endometrium*, Bulletti, C. and Gurpide, E., eds., Annals of the New York Academy of Sciences, New York, pages 202-211.

Hansen, P. J., Ing, N. H., Moffatt, R. J., Baumbach, G. A., Saunders, P. T. K., Bazer, F. W. and Roberts, R. M., 1987, Biochemical Characterization and Biosynthesis of Uterine Milk Proteins of the Pregnant Sheep Uterus, *Biology of Reproduction*, 36:405-418.

[37] Tullner, W. W., 1974, Comparative Aspects of Primate Chorionic Gonadotropins, in *Contributions to Primatology*, Karger, Basel, 3: 235-257.

Hearn, J. P., 1986, The Embryo-Maternal Dialogue During Early Pregnancy in Primates, *Journal of Reproduction and Fertility*, 76: 809-819.

Murphy, B. D. and Martinuk, S. D., 1991, Equine Chorionic Gonadotrophin, *Endocrine Reviews*, 12: 27-44.

[38] In my theory of the Independent Birth of Organisms described in Chapter 8, the random distribution of different eyes in the animal world, which does not follow any evolutionary sequence, is in fact predicted.

[39] Darwin, C., 1859, ref above, page 217.

[40] Mitchell, L. G., et al., 1988, ref above, page 279.

[41] ibid, page 275.

[42] We shall see in Chapters 8 and 10 that the very same observations strongly corroborate the new theory of the Independent Birth of Organisms.

[43] Salvini-Plawen, L. v. and Mayr, E., 1977, On the Evolution of Photoreceptors and Eyes, *Evolutionary Biology*, 10: 207-263.

[44] The thesis that Plawen and Mayr set out to prove was that the various gradations of eyes already known to exist in the invertebrate world

could be connected by organismal evolution as predicted by Darwin. But, on the contrary, they ended up with confusion instead of concretion. Although there was a repertoire of eyes from simple to complex, they were present in a haphazard and random manner not following an evolutionary sequence. Because they believed in evolution, they finally managed to say that eyes must have originated in more than 100 different independent lines of evolution, each starting from an animal that did not have even photoreceptors. However, each of those lines is totally imaginary, based on extremely superficial assessments, and with the assumption that one structure of an eye can be easily replaced or restructured. On top of this, many unsolved problems remain.

[45] Wessells, N. K. and Hopson, J. L., 1988, *Biology*, Random House, page 318.

[46] Bush, G. L., 1982, ref above, pages 119-128.

[47] Futuyma, ref above, page 425.

[48] Darwin, C., 1859, ref above, page 217.

[49] Evolutionists find a gene that is common to most organisms and study its sequence variations in different organisms assuming that such variations are what is responsible for evolution. Although variations of the same gene exist in different individuals of the same species, which is called genetic polymorphism, the evolutionists assume that the variations in the same gene in different creatures are the primary cause of the change of one creature into another. As we have seen, this is absolutely erroneous. A given gene carries out a particular biochemical reaction or function whether it is present in the different individuals of the same species or in distinctly different organisms, whether it is a wheat stalk or a rat. We will scrutinize each of the mutational mechanisms in the next chapter.

[50] Minkoff, ref above, page 177; Futuyma, ref above, page 158.

[51] This is the same phenomenon with many different color variations in different body parts such as the eye or the skin of individuals, breeds, or varieties of a species. Furthermore, the evolution of drug resistance is shown to be proof of evolution. But this is incorrect. For instance, in Futuyma's book, page 158, he says:

> Populations of hundreds of species of insects have evolved resistance to various insecticides ...

He even shows that a change of just one to three nucleotides in the gene is responsible for this. It is very important to understand that these changes do not lead to a new gene, but to a gene whose product will have altered binding (reduced or no binding) to the herbicide or insecticide. We shall deal with this in more detail in the next chapter.

52 Evolutionists conduct ecological experiments, in which a species changes slightly in one or more of their characteristics under new ecological environments. They immediately jump at the results and claim that this is due to the evolutionary principles. What they do not see is that this is bound to happen due to the hidden potential that a species has, but which potential is confined within the closed constant framework of the individual variations and the closed framework of the genome. For example, under a new ecological environment in which a competitor for food resources is removed, a mouse species expectedly becomes slightly larger. Where has the species changed in this example? Evolutionists have all along mistaken the changes occurring within the population of a species of an immutable organism for evolutionary changes leading to a new distinct organism.

53 New body parts are present not only in such higher taxonomic categories. Within almost all the taxonomic categories other than species and family, we can find this. We shall deal more elaborately on this point in chapter 10.

54 See for instance, Futuyma, D. J., ref above, page 419.

Chapter 4

1 Cohen, S. A. and Shapiro, J. A., Transposable Genetic Elements, *Scientific American*, February 1980, pages 40-49.

2 Fedoroff, N. V. Transposable Genetic Elements in Maize, *Scientific American*, June 1984, pages 84-98.

3 In McClintock's studies, the gene responsible for the synthesis of the purple pigment is termed C locus. The transposable element in maize that caused the mutation is *Ac* or *Ds*. *Ac* is an autonomous element, meaning that it contains all the necessary enzymes as well as the terminal repeats to transpose itself. *Ds* is a nonautonomous element, which means that it has the terminal repeat which can insert and excise into and out of a sequence, but does not itself have the enzymes required for this process. It operates when the *Ac* is present in the same cell and provides the enzymes to *Ds*.

4   Watson, J. D., Tooze, J. and Kurtz, D. T., 1983, *Recombinant DNA, A Short Course*, Scientific American Books, pages 140-41.

5   Lewin, B., 1990, *Genes IV*, Oxford University Press, page 649.

6   This has been studied well in the bacteria. One can expect that the activities of the transposons (not the transposons themselves) in animals and plants can be similar to those in the bacteria.

7   Wessler, S. R., 1988, *Science*, 242:399-405.

8   Wessler, S. R., 1988, ref above.

9   Transposon footprints of 6 or 9 nucleotides in revertants of the gene *wx-m1(Ds)* have been directly related to the reduced enzymatic activity of the *waxy* protein, UDP-glucosyl transferase, and provides a molecular explanation for the displayed intermediate characteristics.

Wessler, S. R., Baran, G., Varagona, M. and Dellaporta, S. L., 1986, *EMBO Journal*, 5:2427.

10 The reduced activity of a revertant of the gene (allele) *adh-2F11(Ds)* results from the addition of two amino acids, one of which is a cysteine, to a cluster of Cys codons that are believed to be involved in zinc binding.

Chen, C. H. and Freeling, M., 1986, *Maydica* 31:93.

11 It appears that transposons in eukaryotes only move the sequences belonging to themselves, and it is unlikely that they move any other passenger sequence. Further, they do not leave a copy at their original location. Thus, the question of moving different sequences and combining them into a new gene does not even exist. For argument sake, even if they leave a copy at their original location and move a passenger sequence, the location from and to which a sequence is transposed should be random, because transposition cannot be a directed process. Among such random movements should be the regulatory switches. Hence the probability that even one switch could be moved to a precise position in the DG pathway of a new organ is negligibly small. The probability that a new developmental genetic network, consisting of only a few hundred genes, could be generated by transposition mechanisms is far too low to be meaningful.

12 There are indications that in some cases the transpositions occur at nonrandom locations in the genome, because they are dependent upon certain target sequences. It will, however, only reduce the effect of transposition in evolutionary change even further.

¹³ In fact, if such a thing is happening in living organisms, almost always the development of an embryo would be defective. The fact, fortunately, that most often the embryos grow into normal full-grown babies unequivocally show that such a thing is not happening in the genomes of organisms.

¹⁴ Darnell, J., Lodish, H. and Baltimore, D., 1986, Molecular Cell Biology, Scientific American Books, page 445.

¹⁵ A locus is defined as a DNA region containing one or a few genes on the chromosome. Generally, scientists do not know whether the region contains only one gene or a few genes which are involved in the production of the phenotypic trait, and, therefore, they denote it into one region called the locus.

¹⁶ Usually in the metabolic pathway for the production of a colored pigment, the pathway goes through various pigment molecules that are colored differently. As a consequence, mutations in genes at different positions in the pathway may result in the termination of that pathway at that gene, resulting in the accumulation of a different colored pigment in the cells. In the case of the fruit fly the normal brick color of the eye is produced by the presence of two pigments (red and brown). A number of enzymes are involved in the synthesis of these two pigments. Different mutations in the genes for these enzymes cause the variation in the eye color (such as the white eye [no pigment], deep reddish yellow eye, pinkish eye, lemon yellow eye, pale yellow eye, cream colored eye, bright red eye, orange eye, and so on).

See Phillips, J. P. and Forrest, H. S., 1980, Ommochromes and Pteridines, in The Genetics and Biology of Drosophila, Ashburner, M. and Wright, T. R. F., eds. Volume 2d. Academic Press, New York, pages 541–623.

¹⁷ For example, if mutation due to a transposon occurs in the master gene for a DG pathway leading to a body part making it defective, the DG pathway will be switched off as a whole, and the development of that body part will be abolished.

¹⁸ Watson, J. D., et al., 1983, ref above, page 140.

¹⁹ So far, there has been no clear cut way of proving that the transposons can or cannot contribute to organismal evolution because such methods have to have mathematical precision. This is only possible with our approach here in this book, because we can dissect out all the possible genetic effects of transposons by involving the essential requirement for the evolution of a new gene and a new DG pathway for a new body part. Only

by this requirement we can show with mathematical precision that transposons, or for that matter any mechanism of genetic mutation, can never bring about the evolution of all the organisms living on earth from an original one or a few ancestors as Darwin had proposed.

[20] Darnell, J. E., et al., ref above.

[21] Cohen, S. A. and Shapiro, J. A., ref above.

[22] Fedoroff, N. V., ref above.

[23] Suzuki, D. T., Griffiths, A. J. F., Miller, J. H. and Lewontin, R. C., 1989, *An Introduction to Genetic Analysis*, Fourth Edition, W. H. Freeman, page 541.

[24] Doolittle, W. F. and Sapienza, C., 1980, Selfish Genes, the Phenotype Paradigm and Genome Evolution, *Nature*, 284:601-603.

[25] As we saw before, any mutational change that does not happen in the germ cells is not useful in the assumed evolutionary change of genes or organisms.

[26] Watson, J. D., et al., 1983, ref above, page 117.

[27] Lewin, B., 1990, *Genes IV*, Oxford University Press, page 708.

[28] Darnell, J. E., et al., 1986, ref above; Watson, J. D., et al., 1983, ref above.

[29] Lewin, B., 1990, ref above, page 730; Watson, J. D., et al., 1983, ref above, page 146.

[30] Alberts, B., Bray, D., Lewis, J., Raff, M., Roberts, K. and Watson, J. D., 1983, *Molecular Biology of the Cell*, Garland Publishing, Inc., New York & London, pages 771-773.

[31] Such duplications are explained under the new theory to have happened in the primordial pond while assembling the independent genomes of various organisms (Chapter 8).

[32] Alberts, B., et. al., 1983, ref above, pages 119, 470.

[33] In these cases, the evolutionists' conclusion that the genes with similar sequences have resulted by the duplication from a common ancestral gene is purely an inference based on sequence similarity of these genes and these proteins as well as the functional similarity of the proteins in present day organisms. Other than these inferences, there is absolutely no proof whatsover that these genes have evolved by duplication from an ancestral gene.

NOTES AND REFERENCES   581

[34] Note that these computations of match include several gaps, that is, insertions and deletions.

[35] Amino acid sequence similarity at the 40-50% level can be shown to exist in totally independent proteins in the primordial pond (see Chapters 7 and 9) — showing that we do not need to connect such proteins by evolution.

[36] Anderson, N. L., Tracy, R. P. and Anderson, N. G., 1984, High Resolution Two-dimensional Electrophoretic Mapping of Plasma Proteins, in *The Plasma Proteins*, Second Edition, Volume IV, Putnam, S. W., ed., Academic Press, New York, pages 221-270.

[37] As shall be discussed later, blood coagulation in many invertebrates is carried out in an entirely different manner by some gelling substances which are also distinct from the blood coagulation proteins.

[38] Doolittle, R. F., 1984, in *The Plasma Proteins*, Second Edition, Volume IV, Putnam, S. W., ed., Academic Press, New York, pages 317-360.

[39] See Doolittle, R. F., 1984, ref above, for additional sources.

[40] Doolittle, R. F., 1984, ref above.

[41] For all Doolittle's misunderstandings arising from his following Darwin's theory of evolution, he has clearly understood the problems existing at the molecular level. In fact, his research has provided many molecular details from living organisms supporting and corroborating our concepts we have described here.

[42] Doolittle, R. F., 1984, ref above.

[43] Ibid.

[44] Ibid.

[45] Ibid.

[46] Ibid.

[47] Ibid.

[48] However, all these findings support my theory (independent birth of organisms, Chapter 8), by showing that the different invertebrates must have been born independent of each other in the primordial pond; and that vertebrates did not evolve from invertebrates, but both were born independently.

[49] Doolittle, R. F., 1979, Protein Evolution, in *The Proteins*, Neurath, H. and Hill, R. L., eds., Academic Press, New York, Volume IV, page 99.

[50] It does not seem possible that the protein has become smaller by deletion. Rather it is possible that they are distinct independent proteins although functionally similar.

[51] Alberts, B., et al., 1983, ref above, page 470.

[52] Lewin, B., 1990, ref above, page 499.

[53] Ibid., page 506.

[54] If the evolutionary theory is correct, then these patterns should be correlatable by tracing them through evolutionary change at least to some extent. But the fact that the various genes in the different clusters are truly randomly distributed argues against the theory. Furthermore, the pseudogenes cannot be precisely excised from one genome, so that if a pseudogene is present in a lower organism, say the mouse, then it should be present in all the higher organisms assumed to be descended from the mouse. The rabbit, mouse, goat, and human are all supposed to be closely related to one another within the class of mammals, which is within the subphylum of vertebrates, which is within the phylum of chordates, which is placed among 34 distinct phyla of organisms. Even evolutionists would certainly agree that it is not possible to produce or delete pseudogenes precisely leading to the observed scenario by means of organismal evolution, or to produce the various β-like globins in the different organisms with variable numbers and sequences.

[55] Lewin, B., 1990, ref above.

[56] Futuyma, D. J., 1986, *Evolutionary Biology*, Second Edition, Sinauer Associates, Sunderland, Massachusetts, pages 405-406.

[57] Doolittle, R. F., Similar Amino Acid Sequences Revisited, *Trends in Biochemical Sciences*, 14 July, 1989, pages 244-45.

[58] It is very important to realize that the general characteristics of the genes and genomes of the supposedly ancestral, most primitive organisms is no different from the characteristics of the genes and genomes of the supposedly most highly evolved human organism. We shall determine and elaborate this important general principle in Chapter 8, which will show that there is no need to invoke the evolution of organisms from a common ancestor in order to account for the genomic scenario or the complexity of the supposed higher organisms from the lowest original organism.

[59] Henikoff, S., Haughn, G. W., Calvo, J. M. and Wallace, J. C., 1988, *Proceedings of the National Academy of Sciences*, USA, 85:6601-6606.

Kofoid, E. C. and Parkinson, J. S., 1988, *Proceedings of the National Academy of Sciences*, USA, 85:4981-4985.

60 Futuyma, D. J., ref above, page 67.

61 Note that the genome is 90-99.5% junk DNA. Thus random duplication of sequences should duplicate mostly junk DNA. Furthermore, saying that a new gene can evolve from a duplicate copy is equivalent to saying that a new gene can evolve from a duplicated junk DNA sequence. The reason is that mutations cannot distinguish genes from junk. Even mutations in a sequence of a fully functional gene does not distinguish whether it is a coding sequence (exon), noncoding sequence (intron), splice junctions, and regulatory sequences (promoters, poly A sites etc.). Most of the mutations (>90%) should occur within introns, since introns constitute 90% of a gene. Greater than 90% of mutations in exons do not affect the function of proteins, because these changes only lead either to synonymous codons or to silent amino acid changes (see Genetics Primer and Chapter 7). The remaining 10% of the mutations in the invariant amino acids of the protein will change the protein to a defective protein — not to a new protein.

62 It is also important to understand that the rare gene which is duplicated — if it is not operative for the same function just as another copy — can only become a pseudogene by mutations and not to a new gene. Thus pseudogenes can be generated within the genome of an organism, even if the organism is immutable. Therefore, the presence of pseudogenes within the genomes of organisms is not evidence of evolutionary activity leading to new genes.

63 Lewin, B., 1990, ref above, page 509.

64 Ibid, page 510.

65 This scenario can be explained under the theory of the independent birth of organisms. The histone genes were independently selected in the genomes of various organisms at the time of the independent assembly of their genomes in the primordial pond. The spacers were also selected independently in each genome, except that the spacers in each genome were different. For some biological reason, the spacers within a genome had to be constant during the life of each organism. They thus remain constant in each genome, while they are distinct in different genomes.

66 See for instance, Darnell, J. E., et al., ref above, page 420.

67 Gilbert, W., 1978, *Nature*, London, 271:501-502.

[68] Blake, C. C. F., 1978, *Nature*, London, 273:267; Blake, C. C. F., 1979, *Nature*, London, 277:598.

[69] Doolittle, R. F., 1986, *Cold Spring Harbor Symposia on Quantitative Biology*, Volume L1, pages 447-455.

[70] Almost all the proteins so far claimed to be showing any shuffling of exons in their genes are vertebrate proteins. All these appear to be related in their function, participating in animal defense system such as blood clotting and tissue repair — e.g. epidermal growth factor precursor, factor IX, factor X, tissue plasminogen activator, plasminogen, urokinase, and prothrombin. The similar protein domains present in some of these proteins may have similar functions.

Doolittle, R. F., 1985, *Trends in Biochemical Sciences*, 10:233-237.

Holland, S. K. and Blake, C. C. F., 1990, Proteins, Exons, and Molecular Evolution, in *Inervening Sequences in Evolution and Development*, Stone, E. M. and Schwartz, R. J., eds., Oxford University Press, New York, pages 10-42.

[71] Dorit, R. L., Schoenbach, L. and Gilbert, W., 1990, How Big Is the Universe of Exons? *Science*, 250:1377-1382.

[72] Anderson, N. L., Tracy, R. P. and Anderson, N. G., 1984, High Resolution Two-dimensional Electrophoretic Mapping of Plasma Proteins, in *The Plasma Proteins*, Second Edition, Volume IV, Putnam, F. W., ed., Academic Press, New York, pages 221-270.

[73] Senapathy, P., 1986, Origin of Eukaryotic Introns: A Hypothesis, Based on Codon Distribution Statistics in Genes, and Its Implications, *Proceedings of the National Academy of Sciences*, USA, 83:2133-2137.

Senapathy, P., 1988, Possible Evolution of Splice-junction Signals in Eukaryotic Genes from Stop Codons, *Proceedings of the National Academy of the Sciences*, USA, 85:1129-1133.

Senapathy, P., 1988, Distribution and Repetition of Sequence Elements in Eukaryotic DNA: New Insights by Computer Aided Statistical Analysis, *Molecular Genetics (Life Sciences Advances)*, 7:53-65.

Senapathy, P., Shapiro, M. B. and Harris, N., 1990, Splice Junctions, Branch Point Sites, and Exons: Sequence Statistics, Identification, and Applications to the Genome Project, in *Methods in Enzymology, Computer Analysis of Protein and Nucleic Acid Sequences*, Doolittle, R. F., ed., 183:252-278.

Harris, N. and Senapathy, P., 1990, Distribution and Consensus of Branch Point Signals in Eukaryotic Genes: A Computerized Statistical Analysis, *Nucleic Acids Research*, 18:3015-3019.

[74] Holland, S. K. and Blake, C. C. F., 1990, Proteins, Exons, and Molecular Evolution, in *Inervening Sequences in Evolution and Development*, Stone, E. M. and Schwartz, R. J., eds., Oxford University Press, New York, page 32.

[75] Although Colin Blake thinks that my theory explains the origin of introns and split genes in the primordial ancestor (i.e. the first original organism), it really explains their origins in genes which originated in the primordial DNA sequences even before the first living cells were ever formed (see Chaper 7).

[76] Doolittle, R. F., 1986, *Cold Spring Harbor Symposia on Quantitative Biology*, Volume L1, page 447-455.

[77] Blake, C. C. F., 1985, "Exons and the Evolution of Proteins," *International Review of Cytology* 93:149.

[78] Dorit, R. L., Schoenbach, L. and Gilbert, W., 1990, How Big is the Universe of Exons? *Science*, 250:1377-1382.

[79] Gibbons, A., 1990, Calculating the Original Family of Exons, *Science*, 250:1342.

[80] Dorit, R. L., Schoenbach, L. and Gilbert, W, 1990, ref above.

[81] Lehninger, A. H., 1982, *Principles of Biochemistry*, Worth Publishers, Inc., page 196.

[82] Lehninger, A. H., 1982, ref above, pages 196, 241 and 541.

[83] Some of the other human genetic diseases in which specific enzymes are defective are: galactosemia, Tay-Sachs disease, alkaptonuria, and albinism.

[84] Busslinger, M., Moschonas, N., and Flavell, R. A., 1981, *Cell*, 27:289-298.

[85] Reddy, E. P., Reynolds, R. K., Santos, E., and Barbacid, M., 1982, *Nature*, London, 300:149.

Santos, E., Martin-Zanca, D. Reddy, E. P., Pierotti, M. A., Porta, G. D. and Barbacid, M., 1984, *Science*, 223:661-664.

[86] In the genome, genes are distributed like small islands in an ocean. That is, there exist very long DNA regions between consecutive genes that

do not have any function, and therefore are considered to be nonsense or junk. Mutations in these regions do not have any effect.

87 A spontaneous error frequency of one base-pair change in roughly 10^9 base-pair replications is estimated by experiments. In mammals this figure is equivalent to about three changes per haploid genome for each cell generation. Consequently, a gene encoding an average size protein containing about 10^3 coding base pairs would require about 10^6 cell generations to accumulate a mutation. This indicates a rate estimate of one mutation in an average gene every 200,000 years, if we assume about five cell generations per year in an average germ line (from parental egg to daughter egg).

(see Alberts, B. et al, 1983, ref above, page 214.)

88 Alberts, B. et al, 1983, ref above, page 214.

89 The probability for the occurrence of a mutation at a particular site in a gene of 1,000 characters is 1/1,000. The probability of a particular nucleotide change occurring at a given site is 1/3. Thus, the probability of the given sequence of 100 characters (it does not matter that it is spread at different locations within the gene) is: $(1/3)^{100} \times (1/1,000)^{100} = \sim 10^{-350}$.

90 The probability for the occurrence of a mutation at a particular site in a gene of 10,000 characters is 1/10,000. The probability of a particular nucleotide change occurring at a given site is 1/3. Thus, the probability of the given sequence of 100 characters (it does not matter that it is spread at different locations within the gene) is: $(1/3)^{100} \times (1/10,000)^{100} = \sim 10^{-450}$.

91 If we actually take into account the split nature of the gene, greater than 99% of mutations in a gene will not affect the structure or the function of the protein, since only 5-10% of the gene's sequence is exons, of which only ~10% specify critical, invariant amino acid sequences. See Chapter 7 and Primer.

92 During the action of an enzyme, particular amino acids at crucial positions in the three-dimensional structure of the protein carry out the specific biochemical reaction. These amino acids are called active site amino acids. If these amino acids are changed, the protein's biochemical function is fully destroyed. Some amino acids at a few other positions, which are involved in the binding of cofactors or other accessory molecules required in the biochemical reaction, are

also as important as the active site amino acids. These also should be placed in precise positions in the three dimensional structure of the protein.

[93] If the regulatory protein binds to the regulatory DNA sequence in the normal situation, when the regulatory sequence becomes defective the protein cannot bind to it.

[94] At most, the amount of protein synthesis can vary slightly depending upon the strength of binding of the regulatory sequence with the regulatory protein. This does not alter the DG pathway of the organism in any way. This contributes, in my opinion, only to the normal individual variations of the organism.

[95] Dawkins, R., 1986, *Blind Watchmaker*, W. W. Norton, New York.

[96] Kuppers, B., 1989, *Information and the Origin of Life*, The MIT Press, Cambridge, Massachusetts.

[97] Quoted in Kuppers, B., 1989, ref above, page 84.

Eigen, M., 1976, Wie entsteht Information? Prizipien der Selbstorganisation in the Biologie, *Berichte der Bunsengesellschaft fur Physikalische Chemie*, 80:1059.

[98] Kuppers, B., 1989, ref above, page 87.

[99] Hall, B. G., 1982, Evolution of new metabolic functions in laboratory organisms, in *Evolution of Genes and proteins*, Nei, M. and Koehn, R. K., eds, Sinauer Associates, Sunderland, Massachusetts, page 241

[100] See Hall, B. G., 1982, ref above.

[101] A set of genes which are under the control of one regulatory protein and which are responsible for the degradation and utilization of the sugar lactose when the bacterium encounters this sugar.

Lewin, B., 1990, ref above, page 242.

[102] Molecular geneticists think that these translocations and inversions could bring about an effect called position effect. That is, rarely a given gene may have a slightly varying phenotypic effect depending on what other genes are located next to it. But the effects are so minor that they could not be useful in the evolution of unique body structures.

[103] Futuyma, D. J., 1986, ref above, page 62.

[104] ibid.

[105] Minkoff, E. C., 1983, *Evolutionary Biology*, Addison-Wesley Publishing Company, Reading, Massachusetts, page 125.

[106] Minkoff, E. C., 1983, ref above, pages 100, 245.

[107] This ensures that the sperm or the egg contains a mixture of the set of gene variants from the father and the mother. Thus when the sperm of a male and the egg of a female unites to produce an offspring, the offspring will now contain variants from the two grandfathers and the two grandmothers. The proportion of the variants from each of the four grandparents differ in different grandchildren.

[108] Only because of this process, the sons and daughters of a father and mother vary among themselves.

[109] Chromosomes of the same sort are called homologous chromosomes. Almost in all multicellular organisms, each chromosome is present in two copies, one from the father and one from the mother. The same chromosome from the father and the mother are called homologous. For instance, there are 22 pairs of chromosomes and a female (X) and a male (Y) chromosome in the human. Chromosome 17 from the father and chromosome 17 from the mother are called homologous chromosomes. In contrast, different chromosomes, say chromosome 3 from the father and chromosome 12 from the mother, are called heterologous chromosomes.

[110] Futuyma, D. J., 1986, ref above, page 51.

[111] ibid, page 67.

[112] Senapathy, P., Tratschin, J-D. and Carter, B. J., 1984, Replication of Adeno-associated virus DNA: Complementation of naturally occurring *rep* mutants by wild-type genomes or an *ori* mutant and correction of terminal palindrome deletions, *Journal of Molecular Biology*, 179:1-20.

Senapathy, P. and Carter, B. J., 1984, Molecular cloning of Adeno-associated variant genomes and generation of infectious virus by recombination in mammalian cells, *Journal of Biological Chemistry*, 259:4661-4668.

[113] Minkoff, E. C., 1983, ref above, page 130.

[114] Still rarely, both copies of a gene on the two chromosomes may be defective, when it is called a recessive mutation. In such a case, the defective gene may result in the defective protein it encodes, and may lead to a genetic disease such as the phenylketonuria.

[115] Futuyma, D. J., 1986, ref above, page 67.

[116] Pleiotropic mutations cause a finite, defined, set of aberrations or errors from mutations in existing genes or genetic pathways. This is because the number of pleitropic genes in a genome is constant. What we need for Darwin's proposed mechanisms to work are bizarre aberrations caused by the expression of new genes and new genetic pathways which are indeed nonexistent. Darwin's mechanisms can work only if such truly bizarre individual differences exist in a population. The real bizarre aberrations that we are talking about here have never been found in nature, or even under experimental conditions. (One should not confuse the aberrations of existing limbs and organs with the extraneous monstrous outgrowths we are discussing here.)

[117] Remember, whenever we refer to immutable genome, we mean that the genome of one organism is not changeable to the genome of another organism with a new gene or a new body structure. The genes and the sequences in the genome can mutate, but the genome itself, speaking functionally, does not mutate.

[118] Because such aberrations can be produced in the fruit fly, Darwin's theory is very convincing on the face of it. Because what Darwin has predicted seemingly comes true in these studies, it has been so far difficult for many to deny the theory. This is how Darwin in the first place, and later the geneticists, have been misled because the results are highly misleading.

[119] The mutations and their morphological effects of the fruit fly are definable and cataloguable. If Darwin's theory is correct, they should not be so. They should be totally random. That is the only way variety and diversity could result from random morphological changes.

[120] Futuyma, D. J., 1986, ref above, page 60-61.

[121] This suggestion is based on indirect work. (Futuyma, D. J., 1986, ref above, page 456).

[122] Futuyma, D. J., 1986, ref above, page 60-61.

[123] ibid, page 78.

[124] Kimura, M., 1983, The neutral theory of molecular evolution, in *Evolution of genes and proteins*, Sinauer Associates, Sunderland, Massachusetts, pages 208-233.

[125] Kimura, M., 1983, ref above.

This data was originally obtained by Motoo Kimura with the aim of showing that mutations were selectively neutral. But, whether mutations are selective or neutral, this data does not in any way show that the human or the shark has evolved from an earlier ancestor. We are using the same data to illustrate our point here.

[126] Darnell, J. E., et al, 1986, ref above, page 1157.

[127] Futuyma, D. J., 1986, ref above, page 78-79.

[128] We must understand that the situation of single-celled organisms is different. In bacteria, genes for similar biochemical function may be transferred between different bacteria by any possible agent. But again, where is the production of a new gene here? How can such a mechanism bring about a unique new DG pathway in a multi-cellular organism?

[129] Dobzhansky, T. (1937) *Genetics and the origin of species*, The 1982 edition from Columbia University Press, New York, page 192.

[130] Futuyma, D. J., 1986, ref above, page 143.

Kimura, M., 1983, ref above.

Nei, M. 1983. Genetic polymorphism and the role of mutation in evolution, in *Evolution of genes and proteins*, Nei M. and Koehn, R. K., eds., Sinauer Associates, Sunderland, Massachusetts, pages 208-233.

[131] As we shall see in Chapter 7, a protein can tolerate a great number of amino acid changes without absolutely affecting its structure or its specific biochemical function. It is this sort of amino acid changes that Kimura misunderstands to be his neutral mutations. Certainly, as we have determined, these variations do contribute to individual variations within a species — and are thus responsible for the production of varieties and similar species within an immutable organism. Thus, Kimura has misunderstood the ability of these mutations capable of leading to the varieties and similar species as also extendable to the formation of entirely new organisms with new genes and with new organs and body parts. Thus, Kimura is making the very same mistake that other evolutionists are making — just in another form.

[132] Kimura, M., 1983, ref above.

[133] Note that the biochemical function of a particular protein is qualitatively unaffected by change of its sequence in most of its sequence positions. As we shall see in Chapter 7, most sequence changes at greater than 90% of the sequence positions in a given protein do not affect the

structure or function of the protein qualitatively. It is equally important to note that when changes occur in the remaining 10% of the positions, which are crucial to both the structure and function of the protein, they only lead to a defective protein, which loses its capability to carry out the particular biochemical function altogether. They can never alter the protein's biochemical function into another biochemical function.

[134] Eldredge, N., 1971, The allopatric model and phylogeny in Paleozoic invertebrates, *Evolution*, 25:156-167.

Eldredge, N. and Gould, S. J., 1972, Punctuated equilibria: an alternative to phyletic gradualism, in Schopf, T. J. M. ed., *Models in paleobiology*, Freeman, Cooper and Company, San Francisco, pages 82-115.

Gould, S. J. and N. Eldredge, 1977, Punctuated equilibria: the tempo and mode of evolution reconsidered, *Paleobiology*, 3:115-151.

Stanley, S. M., 1979, Macroevolution: pattern and process, Freeman, San Francisco.

[135] Gould, S. J., 1991, Opus 200, in *Natural History*, August issue, pages 12-18.

[136] We can see that Gould does not seem to recognize that new genes can never evolve in any geological time — let alone in ten thousand years. If they do not originate in the first place, how then can they spread in a population?

[137] Gould, S. J., 1982, The meaning of punctuated equilibrium and its role in validating a hierarchical approach to macroevolution, in *Perspectives on Evolution*, Sinauer Associates, Sunderland, Massachusetts, pages 83-104 (page 84).

[138] Gould, S. J. 1982, ref above, page 88.

[139] Goldschmidt, R., 1940, *The material basis of evolution*, Yale University Press, New Haven, Connecticut. (1982 publication with an introduction by Stephen Jay Gould.)

Futuyma, D. J., 1986, ref above, page 419-421.

[140] Minkoff, E. C., 1983, ref above, page 244-45.

Futuyma, D. J., 1986, ref above, page 420.

[141] Gould, S. J., 1982, ref above, pages 84, 88.

[142] Bush, G. L., 1982, What do we really know about speciation? in *Perspectives on Evolution*, Sinauer Associates, Sunderland, Massachusetts, pages 119-128.

[143] Again remember that neutral mutations are really neutral in terms of the specificity of the biochemical function of a given protein. They are not only selectively neutral as Kimura thinks (meaning that they do not serve as substrates for natural selection), but, absolutely, they are also evolutionarily neutral (meaning that they have nothing whatsover to do with evolution of new organisms at all). In fact, we can categorize all mutations into two groups: defective mutations and evolutionarily neutral mutations.

[144] Cohen, S. and Shapiro, J. A., 1980, ref above.

[145] Note that we are forced to use the requirement for a new gene or a new DG pathway whenever we speak about a new organism. We have extensively discussed the reasons for this in Chapter 3 that otherwise one might misunderstand two varieties of the same species or two similar species of the same organism to mean two distinct organisms, and get confused of our arguments.

[146] Alberts, B., et al, 1983, ref above, page 216.

[147] When they say "it is not very surprising, therefore, that the proteins of mammals as different as whales and humans are very similar," they imply that the slight variations in these genes are responsible for the whale and the human to have changed from their assumed common ancestor. They do not seem to realize that these variations are only sequence variants of exactly the same proteins in terms of their function — which can therefore never effect anything related to evolution. Furthermore, in these arguments, they do not at all speak about the absolutely unique genes occurring in the multitudes of organisms, the origin of which is certainly not explainable by means of organismal evolution.

Chapter 5

[1] Darwin was plain about his inability to address this question. He noted this in more than one instance in his book *Origin of Species*. For instance, when he spoke of the difficulty in evolving highly complex organs, he says:

> How a nerve comes to be sensitive to light, hardly concerns us more than how life itself first originated.

(Darwin, C., 1859, *Origin of Species*, 1979 Edition, Avenel Books, New York, page 217).

In trying to explain instincts, he notes,

> I must premise, that I have nothing to do with the origin of the primary mental powers, any more than I have with that of life itself.

(Darwin, C., 1859, ref above, page 234).

However, one has to acknowledge that in the complete absence of the fundamental knowledge about genes, it is simply impossible for anyone to understand anything concerning the origin of life itself.

[2] Senapathy, P., 1986, Origin of Eukaryotic Introns: A hypothesis, based on codon distribution statistics in genes, and its implications, *Proceedings of the National Academy of Sciences USA*, 83:2133-2137.

Senapathy, P., 1988, Possible evolution of splice-junction signals in eukaryotic genes from stop codons, *Proceedings of the National Academy of the Sciences USA*, 85:1129-1133.

Senapathy, P., 1988, Distribution and repetition of sequence elements in eukaryotic DNA: New insights by computer aided statistical analysis, *Molecular Genetics (Life Sciences Advances)*, 7:53-65.

[3] We can compute the amount of DNA material that could have existed in a primordial pond on the primitive earth based on certain reasonable assumptions, as we do in Chapter 6.

[4] Senapathy, P., references above.

[5] See also Senapathy, P., Shapiro, M. B. and Harris, N., 1990, Splice junctions, branch point sites, and exons: Sequence statistics, Identification, and Applications to the Genome Project, in *Methods in Enzymology, Computer Analysis of Protein and Nucleic Acid Sequences*, Doolittle, R. F., ed., 183:252-278.

Harris, N. and Senapathy, P., 1990, Distribution and consensus of branch point signals in eukaryotic genes: A computerized statistical analysis, *Nucleic Acids Research*, 18:3015-3019.

[6] It is possible that we may be able to design and construct the genomes of entirely new creatures by genetic engineering techniques in the future, and thus create entirely new organisms that never before existed.

Chapter 6

[1] Many aspects of these chemical evolution mechanisms were proposed long ago by Sydney Fox, Cyril Ponnamperuma, A. I. Oparin, and others, and strong experimental support has been obtained. However, all

these scientists have carried out their research subscribing to the evolutionary theory of Darwin. They were working to find proof for the chemical evolution to have led to the first cell, on which Darwin's mechanisms are believed to have operated and to have brought about the rest of the biota on earth.

2. The term biotic pertains to biochemical processes occurring in living cells after living cells were first formed in the primordial pond. Thus, prebiotic means the chemical syntheses that took place before any living cell was formed.

3. The following are the three main cellular processes of decoding biological information contained in the DNA molecule in the cell. The DNA replication is carried out by an enzyme called DNA polymerase, which duplicates the DNA molecule. The protein information in the DNA is copied into an RNA molecule by an enzyme called RNA polymerase. The sequence that codes for a protein in the gene is decoded from the RNA message by a large machinery called the ribosome. The ribosome is very large compared to an RNA or protein molecule, and is formed by the combination of many RNA and protein molecules.

4. Quoted in Lewin, R., 1982, The thread of life: The Smithsonian Looks at Evolution, *The Smithsonian Books*, Washington, D.C.

    Minkoff, E. C., 1983, *Evolutionary Biology*, Addison-Wesley Publishing, Reading, Massachusetts, pages 407-415.

5. Minkoff, E. C., 1983, ref above, page 407-415.

    Stansfield, W. D., 1977, *The science of evolution*, Macmillan publishing company, New York, pages 50-65.

6. See Minkoff, E. C., 1983, ref above; Stansfield, W. D., 1977, ref above.

7. Personal communication.

8. Jiang, H., Kumar, S., Honda, Y. and Ponnamperuma, C., 1990, Search for biological activity in prebiotic synthesis, Presented at the *American Chemical Society* meeting, April 1990, Boston, Massachusetts; Also, personal communication with Prof. Ponnamperuma.

9. Personal Communication.

10. Lewin, B., 1990, *Genes IV*, Oxford University Press, New York.

11. Lewin, B., 1990, ref above.

Mapping and sequencing the human genome, 1988, published by the National research council, National Academy Press, Washington, D. C.

[12] The exact number of species that live on earth is unknown. An approximate estimation is 10 to 50 million species. See for instance: McNeely, J. A., Miller, K. R., Reid, W. V., Mittermeire, R. A. and Werner, T. B., 1990, *Conserving the world's biological diversity*, Published by the International Union for Conservation of Nature and Natural Resources, page 18.

[13] Although it was possible that these ponds remained isolated for long periods of time, it was also possible that their contents were mixed due to constant raining and flooding, especially considering the turbulent environment of the fierce primordial earth. But that does not affect the increase in chemical and biochemical complexity of the prebiotic ponds. In fact such a mixing of the primordial ponds would homogenize the types of the biochemical reactions and the resulting macromolecules in the various primordial ponds, and perhaps enrich them, and at the same time allow different independent prebiotic processes to take place for considerable amounts of time in each of the ponds.

[14] In fact, these methods have been automated in recent times. One can synthesize a DNA with a specific sequence longer than 1000 nucleotides in a pure form. If specific sequence is not a concern, these methods can synthesize DNA several thousands of nucleotides long.

[15] Fox, S., 1988, *The emergence of life: Darwinian evolution from the inside*, Basic Books, New York.

Stansfield, W. D., 1977, ref above; Minkoff, E. C., 1983, ref above;

[16] Darnell, J. E., Lodish, H. and Baltimore, D., 1986, *Molecular Cell Biology*, Scientific American Books, New York, page 1141.

[17] See, for instance, Waldrop, M. M., 1989, Did life really start out in an RNA world? *Science*, 246:1248-49.

[18] See, for instance, Darnell, J. E., et al, 1986, ref above, page 1158.

[19] Alberts, B. et al, 1983, *Molecular Biology of the Cell*, Garland Publishers, New York, page 10.

Lehninger, A. L., 1982, *Principles of Biochemistry*, Worth Publishers, New York, page 317.

[20] Fox, S., 1988, ref above.

Also see Fox, S. W., 1974, Coacervate droplets and proteinoid microspheres, in *The Origin of Life and Evolutionary Biochemistry*, Dose, K., Fox, S. W., Deborin, G. A. and Pavlovskaya, T. E., eds., Plenum Press, New York, page 123.

Chapter 7

[1] Even the simplest cell needs DNA, amino acids, nucleotides, proteins and other molecules to construct the structure of the cell such as cell membranes, and cellular machineries such as ribosomes. The cell needs enzymes to synthesize these molecules and metabolize nutrients. Without such absolutely basic things, no living cell can ever come into existence. When we compute the absolute minimum number of genes required for these fundamental functions for the simplest living cell, it runs into the thousands.

[2] Genes can be tested for their functional usefulness only in living systems. Therefore, even if we accept Darwin's first ancestral organism, it is important to realize that the first living cell must have the full set of complete genes minimally required for the simplest living entity; and such a full set of complete genes must have occurred in the primordial pond's genetic sequences before this cell could come into existence. Even under evolutionary theory, genes cannot evolve unless and until there was a living organism, only in which the functionality of a gene could be tested. One who says that genes had somehow evolved from shorter coding sequences to form the first living cell is therefore certainly incorrect.

[3] Evolutionists say that new species arise by random mutations in the genomes of organisms. They say that genes could not have occurred purely by chance in the primordial pond. Therefore, to them, even the first life on earth was not a probabilistic outcome, i.e., it is the result of an improbable, freak accident.

[4] How can small parts of a gene have a meaning before the gene itself is fully formed? Unless a sequence codes for a protein with specific biochemical function, even as primitive or feeble as possible, it cannot be called a gene. Consequently, modifications and evolution can happen only on a sequence which already specifies a gene. A gene cannot evolve from shorter coding sequences which are purported to be small-

er functional parts of a gene, because the smaller functional parts of a gene have no meaning until they are parts of the complete gene.

5. Monod argued that only the origin of the first living system, say a single cell, was a freak accident, that is, one that did not have an *a priori* probability but occurred as an odd event. But he was a strong believer of Darwin: he believed that Darwin's mechanisms started from that first life that was formed somehow by such an accident. See for instance the discussion by Kuppers, B., 1989, *Information and the Origin of Life*, The MIT Press, Cambridge, Massachusetts, pages 11 and 60-61.

   Monod, J., 1970, *Chance and Necessity*, 1972 edition, Vintage Books, New York.

6. Dawkins, R., 1986, *Blind Watchmaker*, W. W. Norton, New York, page 46.

   Kuppers, B., 1989, *Information and the Origin of Life*, The MIT Press, Cambridge, Massachusetts, page 59-62.

7. It is sufficient if each gene occurs as a complete entity somewhere in the USP. Random recombinations between sequences (by biochemical processes) could assemble many different random combinations of sets of genes. However, unquestionably, most of these combinations would be meaningless. But there would be rarely some combinations that have the right genes in the right developmental genetic pathways which would express them to develop a viable organism. Thus, it is the random processes which can identify the viable genomes among the myriads of meaningless genomes. In fact, as shown in the next chapter, such successful combinations are inevitable.

8. But remember that even a computer which carries out a trillion operations per second will take 140 million years to generate such a sequence. However, there are modern computers with many processors working in parallel, using which we can carry out parallel searches from different positions of the long random sequence to find sentences in them.

9. The expected mean length of the random sequence for the three-letter word "not" in this sentence to occur is 26^3 (17,576) random characters, running 5 printed pages. However, when we run this experiment several times on different random English sequences, frequently the sentence is found in shorter lengths of the random sequence. One such run is shown here in this figure.

10. The expected length of the random sequence for a two-letter word like GO to occur is 26^2 (676 characters). It means that on average, one

GO will occur in 676 random characters. However, in reality, if we walk on different random sequence streams, sometimes it will occur within 50 characters, and at other times may not occur even in 4000 characters. However, we can compute, based on certain mathematical equations we have worked out, that in a random sequence 6 times the length of the expected mean length for a specific sequence (here 6 × 676 = 4056 English characters), that specific sequence will occur 99.9999% of the time.

See Senapathy, P., 1988, Distribution and repetition of sequence elements in eukaryotic DNA: New insights by computer aided statistical analysis, *Molecular Genetics (Life Sciences Advances)*, 7:53-65.

[11] A sentence may occur more than once within the random sequence of our search.

[12] At the maximum, if the sentence contains more than one word with six characters, the expected mean length of the random sequence may increase as many times as the number of the longest words in the sentence; say three times if there are 3 such longest words. Yet this increase is negligible compared to the difference between the expected mean lengths for the sentence as a whole and the sentence in word pieces.

[13] This was theorized by some scientists two decades ago, notably by Lynn Margulis.

See Margulis, L., 1970, *The origin of eukaryotic cells*, Yale University Press, New Haven, Connecticut.

[14] Senapathy, P., 1986, Origin of Eukaryotic Introns: A hypothesis, based on codon distribution statistics in genes, and its implications, *Proceedings of the National Academy of Sciences, USA*, 83:2133-2137.

Shapiro, M. B. and Senapathy, P., 1987, RNA splice junctions of different classes of eukaryotes: Sequence statistics and functional implications in gene-expression, 1987, *Nucleic Acids Research*, 15:7155-7175.

Senapathy, P., 1988, Possible evolution of splice-junction signals in eukaryotic genes from stop codons, *Proceedings of the National Academy of the Sciences, USA*, 85:1129-1133.

Senapathy, P., 1988, Molecular Genetics, ref above.

Senapathy, P., Shapiro, M. B. and Harris, N., 1990, Splice junctions, branch point sites, and exons: Sequence statistics, Identification, and Applications to the Genome Project, in *Methods in Enzymology, Com-*

puter Analysis of Protein and Nucleic Acid Sequences, Doolittle, R. F., ed., 183:252-278.

Harris, N. and Senapathy, P., 1990, Distribution and consensus of branch point signals in eukaryotic genes: A computerized statistical analysis, *Nucleic Acids Research*, 18:3015-3019.

[15] While greater than 90% of the eukaryotic genome is unused sequence, the bacterial DNA organization is very tight. In addition, bacterial cells can divide more quickly. Based on these characteristics, Ford Doolittle speculated that the single-celled eukaryotes came first in evolution and that yeasts and prokaryotes evolved from them by losing introns (Doolittle, W. F., 1978, *Nature* 272:581). However, he later suggested that the first cells were progenote — cells that did not contain a nucleus. According to him the prokaryotes evolved from the progenotes by losing introns. Eukaryotes later evolved from the progenotes by gaining a nucleus (Darnell, J. E., Jr., and Doolittle, W. F., 1986, Proceedings of the National Academy of Sciences, USA, 83:1271-1275). One can see from our analyses in this chapter that Ford Doolittle's latter proposals are incorrect.

[16] A reading frame signifies the contiguous portion of a DNA sequence that can code for a protein without interruption by a stop codon. Out of the 64 possible codons, 61 code for one of the 20 different amino acids. The codons TAG, TAA and TGA do not code for any amino acids. Consequently, wherever one of these occurs in a DNA (or RNA) sequence, no amino acid is coded, and growth of the protein chain is terminated. Thus, the sequence that exists between two successively occurring stop codons, which are separated by a sequence that is a multiple of 3, is called a reading frame (RF).

[17] GenBank is the databank containing information on all the nucleic acid sequences known so far from all the living organisms. This databank is currently compiled by the Center for Biotechnology Informatics, National Institutes of Health, Bethesda, Maryland.

[18] The splicing, as far as we know, occurs only in RNA. However, the spliced RNA can be converted back into DNA by an enzyme called reverse transcriptase. It is reasonable that such an enzyme was present in the primordial pond.

[19] Let us not be concerned about whether the primordial soup was dilute or concentrated, because primordial ponds came in a wide range of concentrations. And for our purposes, certainly there must have existed

many primordial ponds with the right conditions for such interactions. Remember that what forms the primordial pond is simply the atoms and molecules of the earth itself.

[20] Senapathy, P., et al, 1990, ref above.

[21] I thank Dr. Irving Miller of the National Institutes of Health for pointing to this fact.

[22] I thank Marvin Shapiro of the NIH for writing the computer programs for these analyses while I was at the NIH.

[23] Even most of the other codons beginning at this position start with T.A or T.G, the first 2 nucleotides of the three stop codons.

[24] Nomi Harris and I began this study when she was a summer student at the NIH Division of Computer Research and Technology.

[25] Holland, S. K. and Blake, C. C. F., 1990. Proteins, exons, and molecular evolution, in *Intervening sequences in evolution and development*, Stone, E. M. and Schwartz, R. J., ed., Oxford University Press, New York, page 32.

[26] Colin Blake and Walter Gilbert are proponents of the exon shuffling hypothesis, which explains a rare function of the introns subsequent to their origin in genes. Colin Blake states in his article (see ref 25, page 26):

> It is important to distinguish between the role and origin of introns, noting that the gene-shuffling hypothesis relates only to possibly an incidental intron function, in response to evolutionary pressures, and not to the origin of the split gene; otherwise the evolutionary potential inherent in the theory would imply non-Darwinian, anticipatory evolution.

[27] Blake fully appreciates my theory that the split genes originated in the primordial pond as the first genes occurring randomly in the primordial genetic sequences. Based on Blake's comments, however, we can see that he thinks that the split genes thus originated were the cause for the primordial ancestor, meaning the original primitive organism proposed in Darwin's theory as the progenitor of all organisms. But, as we have been discussing so far, in my new theory of the independent birth of organisms, they were the cause for not just one original organism, but for all the multitudes of independently-born organisms.

[28] Recent experiments have begun to show that there may be many proteins whose chains are longer than 3000 amino acids.

[29] Senapathy, P., et al, 1990, ref above.

[30] If an amino acid is coded by more than just one out of the 64 possible codons, then in an amino acid sequence coded by a random DNA sequence, that amino acid will occur as frequently as these codons. An analysis of the proportion of amino acids in real proteins show that they are actually coded for from an overall random DNA sequence. This greatly increases the probability for the occurrence of a real protein sequence in a random DNA sequence.

[31] The probability of one codon is 1/64. If the codon is degenerate such that on an average there can be three codons at each amino acid position, the probability of a codon at one position = $3 \times (1/64) = 0.0468$. The probability for two consecutive codons with this probability = $(3/64)^2 = (0.0468)^2 = 0.0022$. Similarly, the probability for the 200 consecutive codons without degeneracy is approximately 10^{-370}, and the same with degeneracy is 10^{-270}, the difference being 10^{100}. Keep in mind that when the codon degeneracy increases from one to three, the probability of one codon is only tripled. But when we compute the probability of a sequence of codons, the individual probabilities are multiplied, which greatly magnifies the difference between the probability of one codon and three codons at one given position.

[32] King, J. L. and Jukes, T. H., 1969, *Science*, 164:788.

[33] Doolittle, R. F., 1981, *Science*, 214:149-159.

[34] Bowie, J. U., Reidhaar-Olson, J. F., Lim, W. A. and Sauer, R. T., 1990, Deciphering the message in protein sequences: Tolerance to amino acid substitutions, *Science*, 247:1306–1310.

[35] The function of the *lac* repressor is to bind a specific DNA sequence, as well as the sugar lactose, reversibly.

[36] Miller, J. H. et al., 1979, *Journal of Molecular Biology*, 131:191.

[37] Some variations are not permitted in some specific sequences. However, this does not happen to the extent that would significantly affect our computations later in the chapter.

[38] We simply multiply the number of variable amino acids at every position in order to obtain this number.

[39] In a simulation experiment where we search for such a 24-nucleotide-DNA sequence coding for the given 8 amino acids with allowed variations, the probability that it would occur in a 40,000-nucleotide-long random DNA sequence is 99.9999%.

[40] Even if it takes 40,000 nucleotides (nts) to find it once, there will be more than a billion (10^9) possible 24-nt sequences, each capable of encoding that portion of the λ-repressor, in that length of 10^{14} nts; keep in mind that we needed the whole length to find it just once when the 24-nt sequence was invariant.

[41] Senapathy, P., 1986, 1988, references above.

[42] The frequency of waiting intervals longer than 6 times the mean is close to zero. It means that within 6 times the expected mean length for a given subsequence, the subsequence will almost certainly occur. For instance, the expected mean length for the occurrence of ATG is 192 nucleotides, so on average in a random sequence, one ATG will occur once in 192 nucleotides. About 70% of the times a given subsequence would occur within the expected mean length (here 192 nts) and the rest 30% of the times it requires longer than the expected mean length. However, we can compute that in a random sequence 6 times the length of the expected mean length for a specific sequence (here 6 x 192 = 1152 nucleotides), that specific sequence will occur with a likelihood of 99.9999%. See Senapathy, P., 1988, Molecular Genetics, ref above.

[43] I thank Dr. Ganesan Ramalingam for assisting in these studies. He was a graduate student in computer science at the University of Wisconsin-Madison when he wrote the computer programs for these studies. He is currently a computer scientist at the IBM Watson Research Center in New York.

[44] I would like to thank Dr. Sandy Orlow of the National Institutes of Health and Mr. Rob Farber of Los Alamos National Laboratory for writing computer programs in the initial phase of this study.

[45] Although the vast sequence pool of the USP occurs in innumerable pieces, there will be many DNA pieces several millions of nucleotides long, and genes will occur in these pieces. Our approach essentially finds the shortest occurrence of a given gene in these long DNA sequences.

[46] This is computed by adding the expected mean lengths of the 2 exons.

[47] The probability of each exon is determined by first computing the probability of each amino acid location (taking into account of the codon and amino acid degeneracy) and then multiplying the probability of each amino acid position in the complete exon. The expected mean length is the reciprocal of this probability, and the expected

mean lengths of each exon is added up to give the expected mean length of the complete gene. The expected mean length of the gene varies significantly according to how, and in how many positions, the gene is split.

[48] The length of the gene found is not dependent upon the length of the random sequence in which it occurs. The gene may occur soon after the search is started, but its length in this case may be very long (i.e. the intron(s) may be very long). In contrast, the gene may not occur for a considerable length in the random sequence, and when it occurs it may be a very short gene.

[49] Such experiments are in progress.

[50] Senapathy, P., et al, 1990, ref above.

[51] The frequency of a degenerate codon at a given codon position is proportional to the degeneracy of the codon. Only one codon can occur at an amino acid position that is coded by only one codon (i.e. the codon is nondegenerate). For an amino acid coded by two codons, any of the two codons can occur; and so on. From the Codon Table, compute the number of degenerate codons for each amino acid. Square this number for each amino acid and total all of them. The result is $2(1 \times 1) + 9(2 \times 2) + 1(3 \times 3) + 5(4 \times 4) + 3(6 \times 6) = 235$ (there are 9 amino acids with 2 degenerate codons, one amino acid with 3 degenerate codons, five amino acids with 4 degenerate codons and three with 6 degenerate codons and two with no degenerate codons). The average codon degeneracy per codon position is $235/61 = 3.85$.

[52] Note that the expected mean length for a 20-amino-acid invariant sequence increases only to about 10^{24} nucleotides.

[53] The expected mean length for any given 400-nucleotide exon is the same as that for $16 \times 2/3 = 10.67$ amino acids, which is $\sim 10^{13}$ nucleotides. Thus, to find this length of exon in a 1000-nucleotide random sequence, we need to iterate approximately 10^6 to 10^9 times. This is because, from our studies we find that for every order of magnitude increase in the number of iterations, the length in which we find the consecutive exons reduces about 1-1.5 orders of magnitude.

[54] If each pond can have 10^{32} nucleotides, and if unique sequences are repeated a thousand times on average, then there will be approx. 10^{29} nucleotides in unique sequences in each pond.

Chapter 8

1. This material process has no power to pick and choose. The chemical and molecular interactions simply obey the laws of chemistry and physics, and among myriads of such interactions among genes, one may be successful in forming something meaningful, as though it resulted by design.

2. The increase in the probability of a gene in a random sequence due to the above principles compared to an invariant gene is not just a few orders of magnitude, but several hundreds to several thousands of orders of magnitude, making the final probability clearly realistic.

3. There are many articles written by authorities on this subject, many of which agree on this. See for instance, the recent book *Intervening sequences in evolution and development*, 1990, Stone, E. M. and Schwartz, R. J., eds., Oxford University Press, New York.

4. The fossil record seems to indicate the contrary, but we have the best possible information concerning the origin of life in the genetic sequences of organisms living today, far more reliable and far better analyzable than that of the fossil record.

5. Lewin, B., 1990, *Genes IV*, Oxford University Press, New York, pages 467-68.

6. These diverse multicellular life forms are simpler in their body organization and do not have tissues or organs, but their different cell types have a division of labor.

    See Mitchell, L. G., Mutchmor, J. A. and Dolphin, W. D., 1988, *Zoology*, The Benjamin/Cummings Publishing Company, Menlo Park, California, pages 470 and 583-584.

7. Whatever we discuss regarding animals here are applicable to plants. However, the autotrophs — cells and organisms (mainly plants) that are capable of directly converting the earth's chemicals to food (by photosynthesis) without eating other organisms — must have originated in the primordial pond at least at around the same time as, if not before, the animals. Once autotrophs originated in the primordial pond by the same principles of the independent birth of animals we have alluded to, then the heterotrophs, animals that eat other living forms for food, can become viable.

8. Lewin, B., 1990, ref above.

[9] This principle has so far not been directly analyzed or stated in the literature. However, it can be derived by analyzing the data and information even from textbooks on modern molecular biology.

[10] Lewin, B. 1990, ref above, page 466.

[11] Futuyma, D. J., 1986, *Evolutionary Biology*, Sinauer Associates, Sunderland, Massachusetts, page 48.

Lewin, B., ref above, pages 467-468.

[12] Futuyma, D. J., ref above, page 48.

[13] The exact number of genes present in different organisms are not yet known, but it is so far clear that the different organisms have distinct numbers of unique genes and quite different amounts of DNA in their genomes.

[14] Viruses are not independent living forms. They contain far fewer genes, and can only reproduce as a parasite within a living cell. Outside the cell, they are akin to any chemical. They do not, therefore, represent a minimum living entity.

[15] Although mycoplasmas are the simplest living cells, they are not free-living — they normally lead a parasitic existence with animal and plant cells. They are supposed to contain ~750 genes (see Alberts, B., et al, 1983, *Molecular Biology of the cell*, Garland Publishing, New York, page 10). Even if this is the minimum number of genes for a living cell, it is quite large — even if 100 specific genes could probabilistically occur in the primordial pond, it should inevitably contain innumerable genes.

[16] This does not happen inside the organisms.

[17] Schopf, J. W. and Oehler, D. Z., 1976, *Science*, 193:47-48.

[18] See for instance, Gould, S. J., 1977, *Ever Since Darwin*, W. W. Norton & Company, New York, page 115.

[19] Most laid eggs are encapsulated, but some fertilized eggs (zygotes) are not encapsulated but are surrounded with some nutrient deposition.

[20] Only extremely rarely do organisms not start their development as a single cell. There are simple invertebrate organisms that reproduce by the splitting of one individual into two or more pieces (e.g. flatworms such as *Dugesia* [Mitchell, L. G., et al, ref above, page 522], sea anemones, and sea stars [page 328]). Furthermore, many of the simplest invertebrate living forms and some vertebrates reproduce both sexually and

asexually; even in these cases, the development of the individual starts from a single cell.

[21] Mitchell, L. G., et al, ref above, page 794.

[22] ibid, page 768.

[23] Goodeaux, L. L., Moreau, J. D., Anzalone, C. A., Thibodeaux, J. K., Cranfield, M. R., and Rousset, J. D., 1991, Nonsurgical uterine flushing technique in seven species of nonhuman primates, *Assisted Reproductive Technology and Andrology*, Volume II, pages 287-292.

Pope, V. Z., Pope, C. E., and Beck, L. R., 1983, A 4-year summary of the nonsurgical recovery of baboon embryos: A report on 498 eggs, *American Journal of Primatology*, 5:357-364.

[24] Boatman, D. E., 1987, In vitro growth of non-human primate pre- and peri-implantation embryos, *The mammalian preimplantation embryo*, Bavister, B. D., ed., Plenum Publishing Corporation, New York, pages 273-308.

Wolf, D. P., Thomson, J. A., Zelinski-Wooten, M. B, and Stouffer, R. L., 1990. In vitro fertilization-embryo transfer in nonhuman primates: The technique and its applications, *Molecular Reproduction and Development*, Wiley-Liss, Inc., New York.

[25] *Mammalian Development, A Practical Approach*, 1987, Monk, M., ed., IRL Press, Washington D.C.

McLaren, A., 1987, Reproductive options, present and future, in *Embryonic and Fetal Development*, Austin, C. R. and Short, R. V. eds., Cambridge University Press, New York, pages 176-192.

[26] Hearn, J. P., 1983, The common marmoset, in *Reproduction in New World Primates*, Hearn, J. P., ed., MTP Press, Ltd., Lancaster, pages 181-215.

[27] Dukelow, W. R., Pierce, D. L., Rodebush, W. E., Jarosz, S. J. and Sengoku, K., 1990, In vitro fertilization in nonhuman primates, *Journal of Medical Primatology*, 19:627-639.

Wolf, D. P., et al., ref above.

Iritani, A., 1988, Current status of biotechnological studies in mammalian reproduction, *Fertility and Sterility*, 50:543-551.

[28] McLaren, A., 1987, ref. above.

29 McLaren A., 1987, The embryo, in *Embryonic and Fetal Development*, Austin, C. R. and Short, R. V. eds., Cambridge University Press, New York, pages 1-25.

30 McLaren, A., 1987, Reproductive options, present and future, in *Embryonic and Fetal Development*, Austin, C. R. and Short, R. V. eds., Cambridge University Press, New York, pages 176-192.

31 Millions of genomes can be freely available from 1) copying of a genome by free enzymes in the primordial pond, and 2) genomes from the many copies of a seed cell and from the trillions of cells in every individual that died in the primordial pond.

32 Even if a creature is started by only one pair of male and female individuals, the coordinates of its framework of individual variations is intrinsically defined.

33 Natural selection can produce many similar species looking quite different from each other within the allowed limits of a distinct creature. Wide variations in skin color, coat thickness, and body size may be among them, misleading the observer to believe that even distinct creatures can arise from one creature through natural selection.

34 Entirely distinct organisms, like the rose, lobster, or rat, all may have several identical and similar proteins (such as the cytochromes). But obviously they also have many unique genes. This proves that many genes of different creatures could be similar because of functional constraints and because their genomes were assembled from the same pool of genes — not because of evolution.

35 Evolutionists try to connect the plants and animals through one single cell — believed to be the progenitor of all organisms on earth — because of the presence of many almost identical and similar genes in both animals and plants, without realizing that it is simply improbable to bring about the unique genes present in these multicellular organisms through evolution from the supposed single-celled ancestral organism.

36 This can happen both in the seed cell copies and in the individual organisms. When the seed cell multiplies, gene mutational changes can occur, some of which may not affect its ability to develop into the organism. Thereby it may lead to some variations of the individuals of an organism born directly from the seed cells.

37 When one genome becomes successful in producing a viable organism, its set of genes, physically linked and arranged in a complex genetic net-

work, becomes preserved in multiple copies. As cells die, they may release their genomes into the primordial pond. The chromosomes break down into smaller pieces, many of which will remain intact for a long time. They can be used along with the conglomeration of genes randomly from the UGP in the construction of other genomes.

38 Genes that construct body parts do not occur together, rather they are distributed throughout the different chromosomes. They are only functionally linked as a genetic circuit.

39 Life could have originated in many separate primordial ponds. Because of geological upheavals, many of these ponds could have been sometimes mixed and at other times isolated. It is possible that life that originated in different ponds could have survived for different lengths of geological times. It also appears from what we know so far that the creatures living on earth now were all formed in one pond. The life in different ponds must have been formed during a short span of geological time, owing to the requirement for a similar set of earth's overall physical and chemical environments for the life-building primordial soups.

40 In fact one has to only peruse a zoology book (such as that of Mitchell, L. G., et al, ref above), with the view of the theory of independent birth of organisms, to discern the uniqueness in the wide variety of these organisms — forgetting about the large artificial groupings of phyla, classes, and orders.

41 Such a great repertoire of simple unrelated invertebrates clearly attest to their independent births. Given that these organisms (whose genomes were not much less complex compared to all other organisms on earth) originated independently — that the primordial pond's UGP was sufficient to lead to a large number of them — then, automatically, by our probabilistic arguments, it can be seen that all the complex organisms, from the worm to the human, must have originated in the primordial pond independently.

42 Wessels, N. K. and Hopson, J. L., 1988, *Biology*, Random House, New York, page 590.

43 Willmer, P., 1990, *Invertebrate Relationships, Patterns in Animal Evolution*, Cambridge University Press, Cambridge, page 329.

44 Mitchell, L. G., et al, ref above, page 275.

45 Attenborough, D., 1979, *Life on Earth*, Little, Brown and Company, Boston, page 207.

⁴⁶ The independent organisms born at the start of multicellular life were invertebrate organisms of the kind we see in the Cambrian explosion or the Burgess Shale fauna. Although the independence of the birth of organisms in the primordial pond is true in the case of later-born organisms, such as the cow or rabbit, their genomes were possibly built by using pieces of genomes of already-born organisms in addition to using the UGP. Although their genomes were related at the level of the primordial pond, organismally they were born independently in the primordial pond. This is corroborated by the fact, as we established in Chapter 4, that vertebrates did not evolve from invertebrates. Among vertebrates, likewise, organisms belonging to different classes, orders and families were also born separately.

⁴⁷ Gould, S. J., 1989, *Wonderful Life*, W. W. Norton, New York, page 160.

⁴⁸ ibid, page 184.

⁴⁹ ibid, page 196.

⁵⁰ ibid, page 206.

⁵¹ ibid, page 216.

⁵² These are animals belonging to the phylum chordata, comprising the true vertebrates (animals with a backbone) and some invertebrates having a notochord. The notochord is a group of cells that enables the animal to stiffen its back.

⁵³ Not all animals under a phylum follow a strict body structure; animals that follow some of the basic and superficially defined features (not all of the features) are lumped together into a phylum.

⁵⁴ Wessels, N. K. and Hopson, J. L., ref above, page 597.

Gould, S. J., 1989, ref above, page 102.

⁵⁵ A seed cell dividing and living for a geological time freely in the primordial pond may undergo drastic modifications and restructuring of its genome leading to a changed organism, which may be similar to the previous organism.

⁵⁶ In this process, DNA pieces that contain linked gene sets for a whole structure or function would have a better selection value. Thus, it is possible that in the later born organisms, genes that develop particular organismal structures or functions might be physically linked. The availability of a set of genes that build a particular body part in one

DNA piece or even in a few DNA pieces, if found to occur in living systems, could have only happened in the primordial pond by random processes.

[57] If something is improbable but it occurs, it is a freak accident; but when something is extremely probable and it occurs it is inevitable.

[58] The organs and appendages of each organism appear to be best suited to the particular environment in which they live, but considering the earth's environment on the whole, it does not seem to be true. For instance, humans, dogs, cats, and horses would be better off with wings — both to escape from predation as well as to find food. According to the evolutionary theory, these would someday in the future evolve wings. But, according to my theory, organisms were simply born with their sets of body parts. If they were suitable for life in one or more environments, they survived and lived in those environments forever. They cannot change and use other environments. Thus, we shall never fly even in trillions of years.

[59] There exists a wide spectrum of organisms that clearly exemplifies this concept. A whole repertoire of invertebrate organisms attest to this. If we view these organisms with the new theory in mind, we can see that these are indeed multicellular masses, truly bizarre and grotesque — but have a minimum number of structures for the minimally required functions for a viable organism. Many invertebrates indeed look grotesque in this sense, and contain a mouth and an anus and very few internal organs and external body parts. The following examples can be found in zoology books (page numbers from Mitchell, L. G., et al, ref above, are given in parentheses): fan worms or feather duster worms (14), hydra and jelly fish (390), planktonic urochordates, sea urchins and sponges (470), hydrozoans (491), sea nettles (492), sea anemones (493), anthozoans (494), sea slugs and sea butterflies (560), some tunicates (679), *Xyloplax medusiformis* (658), sea cucumbers (659), and sea urchin (662). Starting from a simple level of complexity in organisms with no specialized tissues or body parts such as in Trichoplax adhaerens (defined under phylum placazoa), or orthonectids and dicyemids (phylum mesozoa), there is a wide spectrum with different complexities of anatomical structures and with intricate functions. When we look at these in light of our discussions, truly these are evolutionarily unconnectable.

[60] Futuyma, D. J., ref above, page 498.

[61] In this connection, a quote from Gordon Rattray Taylor, former chief science advisor to BBC television, in his 1983 book is relevant:

> Darwinism is not so much a theory as a subsection of some theory as yet unformulated.

Taylor, G. R., 1983, *The Great Evolution Mystery*, Harper and Row, New York.

[62] This applies to each of the similar species of a distinct organism, because each is a true representative of the distinct creature. As a creature traverses the various environments, it may spawn many similar species within its defined framework.

[63] For example, people live within a temperature range of 40 to 110 degrees Fahrenheit. If the temperature of the whole world stays at 120 degrees for some months, and no one has air conditioning, most of us would die, but some would survive. The survivors, even if they are only a few, would be sufficient to bring back the population of the human beings.

[64] Body parts that do the same function that are obviously different in structure, such as millipede legs and those of the rat or dragonfly wings and those of the bird are called analogous, meaning that they were convergently evolved to do the same function. Body parts that appear to be similar in different organisms, and which evolutionists erroneously think are derived from one another or from a common ancestor, such as the forelimbs of horse and man, are called homologous. Let us remember, based on our analyses in Chapters 3 and 4, that no new body parts, whether the legs of a millipede or the foot of a snail can be evolved.

[65] This principle can also be applied when quite different organisms have some structures that are similar in function and in their overall appearance such as the elongated snout of quite disparate anteaters belonging to entirely different classes, or the tube feet with suckers of different invertebrates.

[66] The DNA sequence that binds a protein in living cells is usually fairly short, starting from approximately 10 nucleotides, for example the promoter sequences, and they tolerate a great deal of variability (see Genetics Primer).

[67] Our discussions will be essentially the same even if the gene-controlling molecule is RNA instead of protein, as suggested by Eric H. Davidson.

[68] Often, a protein with a DNA-binding domain has another domain with another biochemical function such as binding to another small molecule; it can be shown that such proteins also have a high probability of occurrence in the UGP.

[69] It is important to remember that the biochemistry of the subtissues of the eye is no more complex than that of any other cell type, tissue or organ in the body.

[70] Darwin, C., 1859, *The Origin of Species*, 1979 Edition, Avenel Books, New York, page 217.

[71] Darwin, F., 1888, The life and letters of Charles Darwin, Volume II, in *The Sources of Science*, Volume 102, J. Murray, London, page 273.

[72] The concept under evolutionary theory that sex in one organism evolved from that in another is erroneous. There are many unique sex organs and associated reproductive systems in the animal world that cannot be connected evolutionarily. For instance, how could the breast evolve in the mammal, which is lacking in the reptiles? How could the placenta, as we already discussed in Chapter 3, evolve from a reptile?

[73] At the same time, it is possible for sex to have originated in many different organisms born first in the primordial pond, whose genomes were assembled totally independently.

[74] Mitchell, L. G., et al, ref above, pages 600-601.

[75] Attenborough, D. ref above, pages 64-65.

[76] Mitchell L. G., et al, ref above, pages 578-79.

[77] ibid, pages 346-47.

[78] ibid, page 729.

[79] ibid, page 768.

[80] ibid, page 742-43.

[81] The methods of reproduction in the animal world are broadly defined into only a few categories such as oviparity, viviparity, and ovoviviparity. However, in fact, all imaginable methods of reproduction are found in the living world. All these originated in the primordial pond in the repertoire of independently born organisms because all were possible from the vast UGP simply through the formation of almost equally probable DG pathways.

[82] Futuyma, D. J., ref above, page 427.

Mitchell, L. G., et al, ref above, page 742.

[83] Futuyma, D. J., ref above, page 427.

[84] Pointing to the abolition of some structures and functions in different organisms, Douglas Futuyma says,

> In this as in many other instances, it is possible to explain an evolutionary event that results from the loss of a developmental mechanism (i.e. failure to produce or respond to an inducing influence); it is more difficult to understand how the ancestral developmental system evolved in the first place.

Futuyma, D. J., ref above, page 427.

[85] Doolittle, W. F. and Sapienza, C., 1980, Selfish genes, the phenotype paradigm and genome evolution, *Nature*, 284:601-603.

[86] The table shows only some representative organisms. But this phenomenon is present throughout the living world.

[87] Lewin, B., ref above, page 467.

[88] Bowring, S. A., Grotzinger, J. P., Isachsen, C. E., Knoll, A. H., Pelechaty, S. M., and Kolosov, P., 1993, Calibrating rates of early Cambrian evolution, *Science*, 261:1293-98.

Kerr, R. A., 1993, Evolution's big bang gets even more explosive, *Science*, 261:1273-74.

Chapter 9

[1] Davidson, E. H., 1986, *Gene Activity in Early Development*, Academic Press, New York.

[2] Davidson, E. H., 1990, How embryos work: a comparative view of diverse modes of cell fate specification, *Development*, 108:365-389.

[3] ibid, page 365.

[4] ibid, page 366.

[5] ibid, page 381.

[6] ibid, page 384.

[7] These common strategies can originate in the different genomes organized independently in the primordial pond by taking the genes from

other successful genomes. Alternatively, the fundamental processes for early embryonic development may be narrowly defined, so that independently assembled genomes could use somewhat similar developmental strategies.

[8] Here Davidson is incorrect. Such a reassembly can be achieved only in the open primordial pond from pieces of naked genomes of priorly successful organisms, and can never be accomplished within the genomes of organisms.

[9] Tullner, W. W., 1974, Comparative Aspects of Primate Chorionic Gonadotropins, in *Contributions in Primatology*, 3:235-257, Krager, Basel.

Hearn, J. P., 1986, The embryo-maternal dialogue during early pregnancy in primates, *Journal of Reproduction and Fertility*, 76:809-819.

Murphy, B. D. and Martinuk. S. D., 1991, Equine chorionic gonadotrophin, *Endocrine Reviews*, 12:27-44.

[10] Xu, X. and Doolittle, R. F., 1990, Presence of a vertebrate fibrinogen-like sequence in an echinoderm, *Proceedings of the National Academy of the Science, USA*, 87:2097-2101.

[11] Brehelin, M., ed., 1986, *Immunity in Invertebrates*, Springer-verlag, New York.

[12] Valembois, P., Roch, P. and Lassegues, M., 1986, Antibacterial molecules in annelids, in *Immunity in Invertebrates*, Brehelin, M., ed., Springer-Verlag, New York.

[13] Olafsen, J. A., 1986, Invertebrate Lectins: Biochemical heterogeneity as a possible key to their biological function, in *Immunity in Invertebrates*, Brehelin, M., ed., Springer-Verlag, New York, page 94.

[14] Valembois, P., et al, 1986, ref above.

[15] Travis, J., 1993, Tracing the immune system's evolutionary history, *Science*, 261:164-165.

[16] Evolutionists might say that such unique organisms have evolved these unique systems to cope with the unique environments. But when we scrutinize the genes and proteins responsible for these unique systems and discover that they require absolutely unique and distinct genes and proteins to build these systems, it becomes obvious that these organisms should have originated independently in the primordial pond.

[17] Willmer, P., 1990, *Invertebrate Relationships: Patterns in Animal Evolution*, Cambridge University Press, Cambridge, pages 83-85.

[18] Xu, X., and Doolittle, R. F., 1990, ref above.

Miyata, T., Hiranaga, M., Ulmezy, M. and Iwanaga, S., 1983, *Annals of the New York Academy of Sciences*, 408, 651-654.

[19] Wistow, G. J. and Piatigorsky, J., 1988, Lens crystallins: The evolution and expression of proteins for a highly specialized tissue, *Annual Review of Biochemistry*, 57:479-504.

[20] Piatigorsky, J., Horwitz, J., Kuwabara, T. and Cutress, C. E., 1989, The cellular eye lens and crystallins of cubomedusan jellyfish, *Journal of Comparative Physiology A*, 164:577-587.

[21] Wistow, G. J. and Piatigorsky, J., 1988, ref above.

[22] It is possible that a given biochemical reaction may require fairly stringent requirements for the active site of the protein, i.e. the three-dimensional positioning of certain amino acids that take part in the biochemical reaction. The rest of the protein body aids in bringing these active-site amino acids in the required three-dimensional space. Many distinct protein structures, with variations in length and amino acid sequence, may be able to create the same active site. Thus, entirely distinct, evolutionarily unrelated proteins could carry out exactly the same biochemical function. However, because of the constraints in bringing together the active site in 3-D space, there could be considerable amino acid sequence similarity among such distinct, independent proteins.

[23] Cornish-Bowden, A., 1982, Related genes can have unrelated introns, *Nature*, 297:625-626.

[24] Tolstoshev, P., and Solomon, E., 1982, Collagen genes, *Nature*, 300:581-582.

Vuorio, E., and de Crombrugghe, B., 1990, The family of collagen genes, *Annual Review of Biochemistry*, 59:837-72.

[25] These parameters include various lengths and amino acid sequences of proteins that would specify the same active sites. Except for the specific features we observe in particular proteins, we cannot theoretically predict the protein-parameters for a given biochemical function.

[26] What we need to keep in mind here is that we can demonstrate the occurrence of a given protein's gene only as a split gene. Although the

protein subtilisin is from bacteria whose gene does not contain introns, it will exist in a random DNA sequence with a split structure.

[27] Cornish-Bowden, A., 1982, ref above.

[28] ibid.

[29] Bork, P., Sander, C. and Valencia, A., 1992, An ATPase domain common to prokaryotic cell cycle proteins, sugar kinases, actin, and hsp70 heat shock proteins, *Proceedings of the National Academy of Science, USA*, 89:7290-7294.

[30] A kinase is an enzyme that catalyzes the addition of a phosphate group to a given molecule such as a sugar.

[31] Julius, D., MacDermott, A. B., Axel, R. and Jessell, T. M., 1988, Molecular characterization of a functional cDNA encoding the serotonin 1c receptor, *Science*, 241:558-564.

[32] We do not know the gene structures (i.e. the exon-intron positions) of all these proteins. When we know them we might see that they are distinct from each other, corroborating our concepts. On the other hand, even if the gene structures are somewhat similar, we can still show how such genes could have originated independently in the primordial pond. During the formation of the USP, unique sequences were duplicated multitudes of times. These sequences were recombined among themselves and with the unique sequences. In such a process, there would have been enormous DNA sequences with similarity over great portions. From millions of such sequences, very similar genes could have originated. We should also remember the use of genome pieces from already successful creatures in later born organisms from the open primordial pond.

[33] Krueger, N. X. and Saito, H., 1992, A human transmembrane protein-tyrosine-phosphatase, PTPζ, is expressed in brain and has an N-terminal receptor domain homologous to carbonic anhydrases, *Proceedings of the National Academy of Science, USA*, 89:7417-7421.

[34] Schubert, F. R., Nieselt-Struwe, K., and Gruss, P., 1993, The Antennapedia-type homeobox genes have evolved from three precursors separated early in metazoan evolution, *Proceedings of the National Academy of Science, USA*, 90:143-147.

[35] Doolittle, R. F., 1986, Protein sequence data banks: The continuing search for related structures, in *Protein Engineering, Applications in Science, Medicine, and Industry*, Academic Press, New York, pages 15-27.

36 Although we do not show the results of our simulation here, our conclusions are derived as follows. If we obtain many different coding DNA sequences for the same protein sequence (with all its amino acid variability), the best matching sequences among these can have very high similarity.

37 Lewin, B., 1990, *Genes IV*, Oxford University Press, Oxford, page 494.

38 ibid.

39 However, the presence of discrepancies in the organization of the fetal and adult globin genes and their pseudogenes in various vertebrates show that this scenario could not have arisen through organismal evolution, but only through their independent births in the primordial pond.

40 An example of such a reaction is the interaction of a gene's promoter sequence with the enzyme RNA polymerase. Another is the interaction of the ribosomes with the messenger RNA while translating it into protein. Yet another is the proteins of the ribosomes interacting among themselves to build the ribosomes.

41 This is especially true when a biochemical function puts constraints in the structure of a protein, which in turn constrains the amino acid sequences of the protein, at least in some regions.

42 Doolittle, R. F., 1981, Similar amino acid sequences: Chance or common ancestry? *Science*, 214:149-159.

Doolittle, R. F., 1989, Similar amino acid sequences revisited, *Trends in Biochemical Sciences* , (14 July 1989), pages 244-45.

Dayhoff, M. O., et al., *Atlas of Protein Sequence and Structure*, National Biomedical Research Foundation, Washington, D.C.

43 Lehninger, A. H., 1982, *Principles of Biochemistry*, Worth Publishers, page 419.

44 ibid, page 531.

45 The structures and sequences of all these genes have not been worked out yet. Our purpose in discussing these examples is to illustrate the concepts.

46 Doolittle, R. F., 1989, ref above.

47 In the new theory, the different enzymes that carry out similar biochemical reactions need not be related in sequence. They can indepen-

dently occur from the universal sequence pool without sequence similarity but with similar structure and function.

[48] Doolittle, R. F., 1989, ref above.

[49] ibid.

[50] Let us consider the total number of words in all the books in a particular library. They may contain many billions of words. However, the total number of distinct words used in all these books can be only in the hundred thousands or millions. In the context of living things, the number of proteins used may be in the millions. However, the basic subfunctions which are used in biochemical reactions, such as for cofactor-binding, coenzyme-binding, and DNA-binding may be very small.

[51] Doolittle, R. F., 1981, ref above.

[52] Numerous proteins that are minimally required for any living cell to come into existence are functionally similar.

[53] Doolittle writes:

> In this regard, a systematic consensus search of a data base resulted in the uncovering of a large family of bacterial activator proteins, and a similar study gathered in an unsuspected clutch of bacterial signaling proteins.

Doolittle, R. F., 1989, ref above.

[54] e.g. Doolittle, R. F., 1981, 1989, references above.

Dayhoff, M. O., et al, ref above.

[55] Doolittle, R. F., 1989, ref above.

[56] Doolittle, R. F., 1979, Protein evolution, in *The Proteins*, Neurath, H. and Hill, R. L., eds., Academic Press, New York, Volume 4, pages 1-118 (page 81).

[57] Doolittle, R. F., 1981, ref above.

[58] The problems that arise in database searching, which can mislead one into taking a false similarity to be genuine, can be categorized as follows. 1) sequence gaps, 2) assumed convergence and divergence, 3) assumptions of shuffling segments, 4) randomly occurring redundancies, and 5) false similarity in large database searches. See Doolittle, R. F., 1981, ref above.

[59] Because codons are degenerate in the third base position, a mutation at the third base position is often silent in terms of amino acid change.

Therefore, these mutations are more commonly accepted than others. One can witness this effect by counting the number of the third base, second base, and first base changes between two related genes within an organism (alpha and ß hemoglobin genes within human, for instance) or between two different organisms (histone genes between different species).

See for instance, Li, W-H. and Graur, D., 1991, *Fundamentals of Molecular Evolution*, Sinauer Associates, Sunderland, Massachusetts.

60 It is easy to obtain coding similarities in two independent genes in which intron locations and sequences can be entirely different — if the sequence of a gene is split into exons and introns. Almost all the similarities of coding sequences other than the exactly identical genes in two different genomes can be explained in this manner.

61 Doolittle, R. F., 1984, Evolution of the vertebrate plasma proteins, in *The Plasma Proteins*, Putnam, S. W. ed., 2nd edition, Volume 4, pages 317-360.

62 How order in the very first original primitive organism originated is not at all explained even by modern molecular evolutionary theory. The level of order in the genome of even the most primitive organism is no less than that of so-called highly complex organisms. When such a high level of extreme order as found even in the genome of a worm can arise from the chaos of random genetic sequences, it is not comparatively more difficult to create the order in the genomes of organisms such as the complex human. Under the new theory, such a high level of order found in the genomes of organisms (albeit there is much random, junk DNA still existing in them) can only arise from the chaos of the random primordial genetic sequences.

63 The same defective genes such as the globin pseudogenes can be present in some organisms because their genomes could be related at the seed cell level — not due to organismic evolution.

64 The cells of eukaryotic organisms usually contain two of each gene, making the genome diploid; a haploid genome contains only one copy of each gene.

65 Some scientists have suggested that the C-value paradox may be partly due to repetitive DNA sequences, which are repeated in different genomes to different extents. But evolutionists themselves feel that it cannot account for the extremely wide variations in the C values in var-

ious organisms. We can indeed see that the wide C value difference between genomes is largely due to the content of the unique DNA sequences of the different genomes, whether genic or nongenic.

See Li, W-H. and Graur, D., 1991, *Fundamentals of Molecular Evolution*, Sinauer Associates, Sunderland, Massachusetts, page 210.

[66] Syvanen, M., 1984, The evolutionary implications of mobile genetic elements, *Annual Review of Genetics*, 18:271-93.

[67] Gierl, A. and Saedler, H., 1989, Maize transposable elements, *Annual Review of Genetics*, 23:71-85.

[68] Senapathy, P., Tratschin, J-D. and Carter, B. J., 1984, Replication of Adeno-associated virus DNA: Complementation of naturally occurring *rep* mutants by wild-type genomes or an *ori* mutant and correction of terminal palindrome deletions, *Journal of Molecular Biology*, 179:1-20.

Senapathy, P. and Carter, B. J., 1984, Molecular cloning of Adeno-associated variant genomes and generation of infectious virus by recombination in mammalian cells, *Journal of Biological Chemistry*, 259:4661-4668.

[69] The fact that transposons are found in the simplest single-celled organisms support this notion. Many transposons are found in bacteria, yeast, and other unicellular organisms.

[70] Li, W-H. and Graur, D., 1991, *Fundamentals of Molecular Evolution*, Sinauer Associates, Sunderland, Massachusetts.

Jelinek, W. R. and Schmid, C. W., 1982, Repetitive sequences in eukaryotic DNA and their expression, *Annual Review of Biochemistry*, 51:813-44.

[71] The few details it seems to explain do not constitute a proof that organismal evolution has taken place — although this is precisely what evolutionists have taken them to be.

Chapter 10

[1] Attenborough, D., 1979, *Life on Earth*, Little, Brown and Company, Boston, pages 62-64.

[2] Evolutionists explain the presence of functionally-similar structures such as the feet of the snail and rat to be what they call convergent evo-

lution of these structures. The idea is that all organisms are evolved from a common ancestor, but in different lineages, different creatures on distant branches of the evolutionary tree evolve functionally-similar structures for the same function due to the same environmental constraints.

[3] Mitchell, L. G., Mutchmore, J. A. and Dolphin, W. D., 1988, *Zoology*, The Benjamin/Cummings Publishing Company, Menlo Park.

[4] Attenborough, D., 1979, *Life on Earth*, Little, Brown and Company, Boston.

[5] Mitchell, L. G., et al, ref above, page 633.

Minkoff, E. C., 1983, *Evolutionary Biology*, Addison-Wesley Publishing Company, Reading, Massachusetts, page 475.

[6] Willmer, P., *Invertebrate Relationships: Patterns in Animal Evolution*, 1990, Cambridge University Press, Cambridge.

[7] Mitchell, L. G., et al, ref above, pages 462-63.

[8] ibid, pages 488-506.

[9] ibid, page 506.

[10] ibid, page 528.

[11] ibid, pages 533-34.

[12] ibid, page 552.

[13] ibid, page 538.

[14] ibid, page 542.

[15] ibid, page 545

[16] ibid, page 548.

[17] ibid, page 551.

[18] This is an example of mixed characteristics, although the resemblance noted here may actually be superficial or purely functional.

[19] Mitchell et al., ref above, page 552.

[20] The authors also note the following (brackets are mine):

> They [creatures in the phylum entoprocts] also bear some resemblance to sedentary rotifers [another phylum] and have protonephridia similar to those of flatworms [yet another phylum] and rotifers. In general

appearance, however, entoprocts most resemble a phylum of coelomates called the Bryozoa (unique creatures which are classified to still another phylum). Accordingly, some zoologists consider the phyla Entoprocta and Bryozoa to be closely related. In placing the Entoprocta here, we are subscribing to the alternative view that the similarities between the two phyla probably resulted from convergent evolution.

When the creatures are so highly unique and cannot be connected by evolution, they certainly subscribe to our concepts of the independent birth of organisms. Furthermore, the shared characteristics could have originated by the mixing of genomes in the open primordial pond. However, the entoprocts may be totally unique, and their mixed similarities to distinct creatures in many other phyla may be only superficial, and could have originated independently from the primordial pond due to functional constraints imposed by the same physical environment.

[21] Scientists sometimes call these the oarlike extensions of the foot (see for instance, Mitchell, L. G., et al, ref above); it shows how scientists tend to modify organs to fit them to the evolutionary theory.

[22] Cephalopod eyes superficially resemble those of vertebrates but they are actually distinct. Each has a transparent cornea, an iris diaphragm, a lens, and a retina. Unlike vertebrates however, cephalopods have a direct retina, meaning that the photoreceptive cells of the retina face the front of the eye and receive light directly. The optic nerve forms from the back side of the photosensory cells. This means there is no blind spot where the optic nerve passes in front of the retinal cells, as in the vertebrate eye.

(See Mitchell, L. G., et al, page 577 for the details of cephalopod eyes).

[23] The scallop, a bivalve, also has complex eyes, but they appear to be different from those of the cephalopods. Scallop eyes have a cornea, lens and retina (see Mitchell, L. G., et al, page 577).

[24] Mitchell, L. G., et al, page 595.

[25] In birds also there is an organ called a gizzard, which grinds and mixes food with the help of ingested gravel. This clearly shows how in entirely distinct creatures similar organs could arise independently.

[26] Mitchell, L. G., et al, ref above, page 642.

[27] ibid, page 643.

[28] ibid, page 644.

[29] ibid, page 651.

[30] ibid, page 654.

[31] ibid, page 656.

[32] ibid, page 665.

[33] ibid, page 664.

[34] ibid, page 669.

[35] ibid.

[36] ibid.

[37] ibid, page 677.

[38] ibid, page 681.

[39] Doolittle, R. F., 1984, Evolution of the vertebrate plasma proteins, in *The Plasma Proteins*, Putnam, S. W., ed., 2nd edition, Volume 4, Academic Press, New York, pages 317-360.

[40] Mitchell, L. G., et al, ref above, page 722.

[41] ibid, page 713.

[42] ibid, page 721.

[43] The highly specialized system of a venom gland and the injection apparatus cannot simply evolve through organismal evolution from a creature that completely lacks such a system.

[44] Mitchell, L. G., et al, ref above, page 768.

[45] ibid.

[46] ibid, page 814.

[47] ibid, ref above, page 817.

[48] There are marsupial animals that superficially resemble the mammalian cats, wolves, and bears. However, marsupials as a group are fundamentally distinct in not having a true placenta, having a marsupium, and other characteristics. Zoologists themselves say that the placentals are not derived from the marsupials or vice versa. It appears that the marsupials are called by the generic names of the mammals such as the cat and wolf, but they are truly not related.

[49] Mitchell, L. G., et al, page 700.

Chapter 11

1. Bowring, S. A., Grotzinger, J. P., Isachsen, C. E., Knoll, A. H., Pelechaty, S. M., and Kolosov, P., 1993, Calibrating rates of early Cambrian evolution, *Science*, 261:1293-98.

   Kerr, R. A., 1993, Evolution's big bang gets even more explosive, *Science*, 261:1273-74.

2. Bowring, S. A., et al, 1993, ref above.

3. Kerr, R. A., 1993, ref above.

4. When I formulated my theory in 1981, I had no knowledge of the Cambrian explosion or the Burgess Shale mystery. Based on molecular sequence research I had come to feel that missing links were not necessary if creatures could have been born separately from the primordial soup from a large amount of genetic sequence. I had a basic knowledge of evolutionary theory, but had no formal training in it and did not know all the details of the fossil and organismal scenarios. I systematically learned evolutionary biology since that time, and I was astonished to find all the fossil evidence that fits the new theory and its explanations perfectly. Thus, these findings indeed came as a surprising corroboration to the new theory, which strengthened my confidence in it.

5. Some paleontologists might say that these Burgess creatures are represented in modern animals. However, the similarity is only superficial, which could be explained in the new theory.

6. Gould, S. J., 1989, *Wonderful Life*, W. W. Norton, New York.

7. Gould, S. J., 1989, ref above, page 160.

8. Gould, S. J., 1989, ref above.

9. Some of them may be similar to the extent that they can even fit such broad descriptions as annelids or arthropods.

10. Cloud, P. and Glaessner, M. F., 1982, The Ediacaran period and system: Metazoa inherit the earth, *Science*, 217:783-792.

    Also see Gould, 1989, ref above, page 311.

11. Seilacher, A., 1984, Late Precambrian and early Cambrian Metazoa: preservational or real extinctions? in *Patterns of Change in Earth Evolution*, Holland, H. D. and Trendall, A. F., eds., Springer-Verlag, Berlin, pages 159-168.

[12] This observation is supportive of the new theory that creatures born in two entirely different ponds could have superficial similarity — because in whichever pond creatures were born, they all were born in the same earthly environment. It would thus support our thesis that some creatures from the Burgess pond could superficially look like some creatures from the Cambrian pond, and each of these could look like some creatures in the Ediacaran pond.

[13] Evolutionists believe that life has tried to start a few times on earth, but all have failed except the one that started in the Cambrian period (see also Raup, D. M. and Valentine, J. W., 1983, Multiple origins of life, *Proceedings of the National Academy of Science USA*, 80:2981-2984). It should be noted, when they say this, they mean that however and whenever life started, it started with one simple original creature and evolved into many new creatures based on evolutionary mechanisms. According to them, after such evolution over a geological time, each set of fauna became totally extinct, such as in a mass extinction — except the one in the Cambrian.

[14] Looking at the geological times of these pond fauna, they all appear to have originated at varying times, however, within about one hundred million years (Ediacaran — 640 million years ago, Cambrian — 540 million years ago and Burgess — 530 million years ago). Thus, this period (~640-530 million years ago) seems to be the period when the earth was ripe to form biochemically rich primordial ponds.

[15] Gould, S. J., 1989, ref above, pages 59 and 314.

[16] This is due to the random perfection of organisms in different environments. See Chapter 8.

[17] Martin, R. D., 1991, New fossils and primate origins, *Nature*, 349:19-20.

Beard, K. C., Krishtalka, L. and Stucky, R. K., 1991, *Nature*, 349:64-67.

[18] Gould, S. J., 1977, *Ever Since Darwin, Reflections in Natural History*, W. W. Norton, New York, page 115.

[19] Minkoff, E. C., 1983, *Evolutionary Biology*, Addison-Wesley Publishing Company, Reading, Massachusetts, page 427.

[20] Futuyma, D. J., 1986, *Evolutionary Biology*, Sinauer Associates, Sunderland, Massachusetts, page 323.

[21] See for instance, Minkoff, E. C., 1983, ref above.

Futuyma, D. J., 1986, ref above.

Mitchell, L. G., Mutchmore, J. A. and Dolphin, W. D., 1988, *Zoology*, The Benjamin/ Cummings Publishing Company, Menlo Park, California.

Romer, A. S., 1966, *Vertebrate Paleontology*, The University of Chicago Press, Chicago, Third Edition.

[22] Romer, A. S., 1966, ref above, pages 12-13.

[23] ibid, pages 15-16.

[24] ibid, page 24.

[25] ibid, page 33.

[26] ibid, page 34.

[27] ibid, pages 52-53.

[28] ibid, page 86.

[29] ibid, page 101.

[30] ibid, page 102; Minkoff, E. C., 1983, ref above, pages 491-92.

[31] Romer, A. S., 1966, ref above, page 166.

[32] Mark, J. L., 1978, *Science*, 199:284.

[33] Feduccia, A., 1993, Evidence from claw geometry indicating arboreal habits of archaeopteryx, *Science*, 259:790-792.

[34] Swinton, W. E., 1960, in *Biology and Comparative Physiology of Birds*, Volume 1, Marshall, A. J., ed., Academic Press, New York, page 1.

[35] Mitchell, L. G., 1988, ref above, page 767.

[36] Olson, E. C., 1965, *The Evolution of Life*, The New American Library, New York.

[37] Simpson, G. G., 1944, *Tempo and Mode of Evolution*, Columbia University Press, New York.

[38] Romer, A. S., 1966, ref above, page 303.

[39] Kuppers, B-O., 1990, *Information and the Origin of Life*, The MIT Press, Cambridge, Massachusetts, pages 59-61.

Monod, J., 1971, *Chance and Necessity*, Vintage Books, New York.

Chapter 12

[1] e.g. Li, W-H. and Graur, D., 1991, *Fundamentals of Molecular Evolution*, Sinauer Associates, Sunderland, Massachusetts.

Genetics Primer

[1] Lehninger, A. L., *Principles of Biochemistry*, 1982, Worth Publishers, page 546.

[2] ibid, page 210.

[3] ibid, page 121.

[4] ibid, page 125.

[5] Doolittle, R. F., 1981, Similar amino acid sequences: Chance or common ancestry? *Science*, 214:149-159.

Lewin, B., 1990, *Genes IV*, Oxford University Press, Oxford, page 468.

[6] Scientists commonly use the term house-keeping genes to denote those that maintain the function of an already-constructed cell. To denote the minimum set of essential genes needed to build any cell, we can coin a term basic-cell genes. This set, logically, must be expressed during the construction of every cell in every body part during the development of the animal, although most of these genes may later become inactive. After development, only the house-keeping and tissue-specific genes will be expressed under normal conditions.

[7] Doolittle, R. F., 1981, ref above.

[8] Watson, J. D., Tooze, J. and Kurtz, D. T., 1983, *Recombinant DNA: A Short Course*, Scientific American Books, New York.

Lehninger, A., 1982, ref above, pages 791-792.

[9] Lehninger, A., 1982, ref above, page 17.

[10] Lewin, B., 1990, *Genes IV*, Oxford University Press, Oxford, page 225.

[11] Watson, J. D., et al, 1983, ref above, page 46.

[12] In fact in the first few divisions from the zygote, only the nuclei divide in the common cytoplasm of the embryo. For instance, the *Drosophila* embryo remains a common cytoplasm for the first 9 divisions. However,

each nucleus seems to know its relative three-dimensional position in the developing embryo. That is, the differential expression of genes are occurring in the different nuclei although they are in the common cytoplasm during early development. (See Lewin, B., 1990, ref above, page 752).

[13] Davidson, E. H., 1990, How embryos work: a comparative view of diverse modes of cell fate specification, *Development*, 108:365-389.

Davidson, E. H., 1986, *Gene Activity in Early Development*, Academic Press, New York, page 45.

Lewin, B., 1990, *Genes IV*, pages 752-755.

[14] Beardsley, T., August 1991, Trends in Biology: Smart Genes, *Scientific American*, pages 86-95.

[15] Lewin, B., ref above, page 770.

INDEX

A

Absolute limitation, 53–55
Adaptation, 2, 3, 14–29, 60, 337–38
Alberts, B., 195
Albumin, 128–29
Amino acids, 538
    degeneracy in proteins in primordial pond, 254–56
    See also Proteins
Amphibians, 480
    origin of, 512
Artificial selection, 15–16, 53–55

B

Behavior patterns
    evolution of, 16, 20

Biochemical functions
    number of proteins and, 420–21
    unique, 35
    unrelated gene code in different organisms for same, 396–99, 422–30
Birds, 483–84
    unlikely evolution of from reptiles, 74–77, 513–15
Blake, Colin C. F., 145, 147, 148, 151, 247
Blending inheritance, 22
Bowring, Samuel A., 494–95
Briggs, Dereck, 329
Burgess fauna, 324, 328–29, 498–501
Bush, Guy L., 25, 27–28, 92–93, 191–92

C

Cambrian explosion, 3, 5, 10, 18–19, 38, 183, 324, 368–69, 441, 492–502, 531
Cancers, 154–55
Carson, 27, 28
Cell divisions
   embryonic, 42
Cellular genetic machineries, 553–56
Ceruloplasmin, 133
Chemical evolution, 207, 215
Chromosomal rearrangements, 166–67
Classification of organisms, 3
Coadaptation, 27, 28
Coagulation, 390–91
Cohen, Stanley A., 118–19, 194
Computer simulation of primordial pond, 430–31
Consensus sequences, 414
Condons
   degeneracy of, 254–56
   stop, 233, 236–37, 242–47
Cri-du-chat syndrome, 167
Crossing over of chromosomes, 171–74
Crystallin proteins, lens, 392–94
C-value paradox and junk DNA, 443–45

D

Darwin's theory of evolution, 2–8, 13–29
   current status of, 27–28
   false premises of, 432–40
   on origin of life, 95
   recent modifications to, 23–27
   See also Adaptation; Natural selection
Davidson, Eric H., 45, 380–82
Dawkins, Richard, 160–63
Developmental genetic pathway (DG pathway), 6, 7, 39–52
   improbable first organism contained all in future, 55–58, 105–6, 184–85
   lack of understanding of, 93–94
   myriad of permutations and combinations of, 335, 344–48
   rigidity of for each organism, 313–16, 379–83
   and transposition mechanism, 113–15
De Vries, Hugo, 24, 62–63, 167
DNA
   amount in one organism, 566
   definition of, 294–95, 543–47
   illustration of, 546
   junk, 443–45, 448–49, 451
   replication of, 546
   sequences of and origin of life, 4–5, 296–98, 547–50, 553–55
Dobzhansky, Theodosius, 22, 27, 28, 178–79, 185
Doolittle, Russell F., 128–32, 141, 148–49, 149–50, 387–88, 414, 435–36, 438, 450
Doolittle, W. F., 120
Downs syndrome, 166
Duplication of chromosomes, 166
Dysgenesis, hybrid, 116–17

E

Ediacaran fauna, 501–2
Eigen, Manfred, 160, 162
Eldredge, Niles, 187–89
   See also Gould-Eldredge theory; Punctuated equilibrium
Embryogenesis, 43–44, 560–61
Eukaryotic cells, 551–52
   as first cells in primordial pond, 230–50, 299–300, 306–7, 383–84, 505–6, 506–8
Evolution
   as a never-ending process, 91–92

only within a species, 316–18
Evolutionary theory, 2–8
    chart comparing with independent birth of organisms theory, 530
    See Darwin's theory of evolution
Evolutionary tree, 460
Evolutionists
    misconceptions of, 96–98
Exon
    assembling together, 555
    shuffling, 144–51
Extinctions, 10, 339–42, 503–4
Eyes
    improbability for evolution of, 79–87
    origin by genome assembly in primordial pond, 348–353

F

Fedoroff, Nina V., 108, 109, 119
Feduccia, Alan, 514
Fibrinogen, 129–32, 387
Fibronectin, 133
Fishes
    diverse creatures grouped as, 478–80
    lack of fossil ancestors for, 509–12
Fission of chromosomes, 166
Fossil record, 10, 16, 17–19, 38, 50–52, 491–92
    Cambrian Period, 492–502
    post Cambrian, 503–19
    See also Cambrian explosion
Fox, Sydney, 200, 208, 218
Fusion of chromosomes, 166
Futuyma, Douglas J., 93–94, 184, 507

G

Galapagos Islands, 14, 60
Garstand, Walter. 477
Gene expression, regulation of, 556–63
Gene families, 422–30, 449
Genes

abundance of in primordial pond, 221–91, 298–99
complexity of for various organisms, 348–52
and duplication of mutations, 123–44
homeobox and homeotic, 562–63
and intron lengths, 266–88
function loss and retrogression of organism, 364–65
fundamental to all living things, 221
organisms share from primordial pond, 319–20, 321–32, 384–85, 416–20
random assembly of in primordial pond, 312–20, 383–84
rearrangement of, 121–22
redundancy in genetic information, 428–30
sharing of, 329–32
similar genes for functionally similar proteins in primordial pond, 430–31
split in primordial pond, 230–50, 252–53
unique in numerous creatures, 34, 55–58, 385–96
unrelated codes of in different organisms for same biochemical function, 396–99, 422–30
Genetic drift, 22
Genetics Primer, 535–66
Genomes
    blind as to niches and enviroments, 89–91
    definition of, 295, 543–52
    flexibility of, 179–80
    modification of, 489
    of organisms closed to evolutionary changes, 31–102, 105–6
    and origin of life, 4–5, 440–42
    patterns of repetitive sequences in, 447–48
    random mutations in, 71–89

and relationship to the organism,
    550–51
similar complexity in all animals,
    301–6, 348–52
size, 91
transposons in, 445–47
Gignerich, P. D., 26
Gilbert, Walter, 145, 149–50, 151
Globin, 154
Goldschmidt, Richard B., 23–25,
    189–93
Gould-Eldredge theory, 25–26,
    187–89, 193, 499
Gould, Stephen J., 25, 187–89,
    190–91, 328–29, 498, 501–2, 506,
    516
Griffiths, Anthony J. F., 119

H

Haft, Daniel, 258
Haldane, J. B. S., 207, 208
Hall, Barry G., 165
Harris, Nomi, 246–47
Hemoglobin, 137
"Hopeful monsters" hypothesis, 24,
    189–93
Huxley, Julian, 22
Hybrid dysgenesis, 116–17
Hybridization, 177

I

Immune systems, 388–89
Immunoglobulins, 134
Independent births of organism from
    primordial pond, 293–373
    reasons for similarities in organisms
        originating by, 342–44
Industrial melanism, 97–98
Instincts, origins of, 353, 359–61
Intelligence, origins of, 359–61
Intragenic recombination, 168
Introns, 266–88

Invertebrates
    do not evolve into vertebrates,
        508–16
    uniqueness among, 458–76

J

Jacob, 557
Junk DNA, 443–45, 448–49, 451

K

Kimura, Motoo, 26–27, 186–87
Kleinfelter syndrome, 167
Kuppers, Bernd-Olaf, 160–63
Kurtz, David T., 110, 123

L

Lamarck, Jean Baptiste, 13
LeJune's syndrome, 167
Lens crystallin proteins, 392–94
Lewin, Benjamin, 195–96
Lewontin, Richard C., 20–21, 119
Life
    definition of, 304–6
    molecular scenario of, 375–451
    origin of, 1–11, 13, 95, 199–204,
        205–19, 293–373
    See also Seed cell
Living fossils, 181

M

McClintock, Barbara, 107, 119
Mammals, 484–87
Mayr, Ernst, 22, 24, 27, 28, 79, 84, 86,
    87, 190, 326
Mendel, Gregor, 22
Metabolic products, 394–96
Miller, Jeffrey H., 119
Miller, L. G. et al., *Zoology*, 262,
    464–77, 482–83

Miller, Stanley, 200, 208, 209, 215, 218–19
Minkoff, Eli C., 24, 170–71, 177–78, 507
Missing links, 3, 5, 16, 17, 338–39
Modern synthesis of evolution, 22–23, 70–71
Molecular scenario of life, 375–451
Monod, Jacques, 223, 557
Monstrous variation, 60–65
Morgan, Thomas H., 171, 175
Morphogenetics, 560–61
Muller, Herman J., 175
Murchison, Roderick, 493
Mutations, 6, 7–8, 9, 33–52
    chromosomal rearrangements, 166–67
    classes of, 106
    crossing over, 171–74
    evolutionary effect limited to finite framework, 58, 96–97, 103–97, 321, 527
    and exon shuffling, 144–51
    and gene duplication, 123–44
    genetic, 564–65
    genetic mechanism of, 105–6
    homeotic, 562–63
    interrelated mechanisms of, 106
    mass, 24
    pleiotropic, 174–76
    point, 151–65
    polyploidy, 176–79
    random, 71–89
    rates of, 37–38
    recombination, 168–71
    structural-gene, 46–52
    transposon mechanism of, 107–23

N

National Biomedical Research Foundation, 201
Natural selection, 2, 3, 13–29, 32, 59–60, 65–70, 337–38
    contrasted with random perfection, 335–36
Negative exponential distribution of sequence waiting intervals in primordial pond, 266–88
Neo-Darwinism, 14, 19, 22, 70–71
Neutral theory of evolution, 26–27
Niches and evolution, 89–91
Nucleic acids
    and origin of life, 294

O

Olson, E. C., 515
Oparin, A. I., 200, 207, 208, 215
Organisms
    changes in, 462
    complexity of genomes similar in all, 301–6, 348–52
    as distinct and evolutionarily unrelated, 453–90
    molecular scenarios reflect independent origins in primordial pond, 440–42
    modification of, 489
    reason for similarities in, 342–44
    relationship among, 461–62
    and relationship of genome to, 550–51
    retrogression of, 364–65
    share genes from primordial pond, 319–20, 321–32, 384–85, 416–20
    unique genes in numerous creatures, 34, 55–58, 385–96
    unrelated gene codes in different organisms for same biochemical function, 396–99, 422–30
Organs
    complex, 16, 19–20
    mutations will not lead to new, 33–52
    probability of evolution of, 87–89
    and random perfection, 351–53

Origin of life, 1–11, 13, 95, 205–19, 293–373
  molecular scenario of, 375–451
  See also Seed cell

P

Pangenesis, 21
Phylogenetic studies, 97
Piatigorsky, Joram, 394
Placenta
  evolution of, 78–79
Pleiotropic mutations, 174–76
Point mutations, 151–65
Polyploidy, 176–79
Ponnamperuma, Cyril, 200, 208, 209, 210, 215
Port Jackson shark, 181–83, 186
Primordial pond
  abundance of genes in, 221–91, 298–99
  common basis of life in, 310–12
  computer simulation of, 430–31
  eukaryotic cells originate in, 230–50, 299–300, 306–7, 383–84
  independent birth of multitudes of organisms from, 293–373, 440–42
  and origin of life, 205–19
  and origin of vertebrates, 508–16
  random assembly of genes in, 312–20
  organisms share genes from, 319–20, 321–32, 384–85
  See also Seed cells
Prokaryote cells, 551–52
  control of gene expression in, 557–59
Proteins
  and coagulation, 390–91
  and consensus sequences, 414
  definition of, 538–43

  functionally-distinct but with similar subfunctions and amino acid subsequences, 399–416
  lens crystallin, 392–94
  and origin of life, 294
  in primordial pond, 430–31
  and possible number of biochemical functions, 420–21
  sequences and randomness, 296–98
Punctuated equilibrium, 25–26, 187–89, 190, 499

R

Ramalingam, Ganesan, 274
Random fluctuations, 22
Randomness
  in genetic sequences, 33–34, 222–30, 296–98, 312–20, 327–29, 383–84
  in mutations, 71–89
  emergence of order out of, 369–70
Random perfection, 204, 332–37
  of complex organs, 351–53
Reading frames (RFs), 233–39, 244
Recombination, 168–71
Redundancy in genetic information, 428–30
Renner complex, 63, 167
Reptiles, 481–83
  origin of, 512–13
  unlikely evolution of into birds, 74–77, 513–15
Respiratory systems, 389–90
Retrogression of organism, 364–65
RNA
  sequence, 553–55
  splicing, 555
Romer, Alfred S., 186, 509–12

S

Salvini-Plawen, L., 79, 84, 86, 87, 326

Sapienza, C., 120
Schoeninger, M., 26
Schopf, J. W., 506
Schubert, F. R., 411–12
Seed cell
    and theory of origin of life, 5, 307–10, 358–59, 489, 505–6
Seilacher, Dolf, 501
Senapathy, Periannan, 148, 246–47
Sequence waiting intervals in primordial pond, 266–88
Sex and sex organs, origins of, 353–59
Shapiro, James A., 118–19, 194
Sickle cell anemia, 154
Simpson, George Gaylord, 22, 515
Speciation, process of, 23, 27–28
    never documented, 92–93
    sudden, 23–25
Species
    individual variation of, 52–55, 55–58, 316–18
Splice junctions, 242–47
Split genes in primordial pond, 230–50, 252–53
Stebbins, Ledyard, 22
Stop condons, 233, 236–37, 242–47
Sturtevant, Alfred, 171
Suzuki, David T., 119, 194
Swinton, W. E., 514

T

Termites, 361–62
Thalasemias, 154
Tooze, John D., 110, 123
Transcription, 553–55
Transferrin, 132–33
Translocation of chromosomes, 166
Transposons
    in various genomes, 445–47
    and mechanism of mutations, 107–23
Trisomy, 166

Turner's syndrome, 167

U

Uniqueness
    among amphibians, 480
    biochemical functions, 35
    among birds, 483–84
    among invertebrates, 458–76
    among mammals, 484–87
    of most organisms, 453–90
    among reptiles, 481–83
    unique genes in numerous creatures, 34, 55–58, 385–96
    among vertebrates, 476–78
Universal gene pool (UGP), 217, 298–99
Universal sequence pool (USP), 202, 205–19
Urey, Harold, 208

V

Variations
    meaning and source of, 55–58
    within a species, 52–55
Vertebrates
    origins of, 508–16
    uniqueness among, 476–78

W

Watson, James D., 110, 123
Wessler, Susan R., 113
Wright, Sewall, 22

Z

Zygote, 40–41, 57, 94, 309–10